PROGRESS IN BRAIN RESEARCH

VOLUME 109

CHOLINERGIC MECHANISMS:
FROM MOLECULAR BIOLOGY TO CLINICAL SIGNIFICANCE

PROGRESS IN BRAIN RESEARCH

VOLUME 109

# CHOLINERGIC MECHANISMS: FROM MOLECULAR BIOLOGY TO CLINICAL SIGNIFICANCE

EDITED BY

## JOCHEN KLEIN and KONRAD LÖFFELHOLZ

*Department of Pharmacology,
Johannes Gutenberg-University of Mainz, Mainz, Germany*

ELSEVIER
AMSTERDAM – LAUSANNE – NEW YORK – OXFORD – SHANNON – TOKYO
1996

© 1996 Elsevier Science B.V. All rights reserved.

ISBN 0-444-82166-x (volume)
ISBN 0-444-80104-9 (series)

Published by:
Elsevier Science B.V.
P.O. Box 211
1000 AE Amsterdam
The Netherlands

1000880195

Printed in The Netherlands on acid-free paper

# List of Contributors

E.X. Albuquerque, Department of Pharmacology, School of Medicine, University of Maryland, 655 West Baltimore Street, Baltimore, MD 21201-1559, USA

M. Alkondon, Department of Pharmacology, School of Medicine, University of Maryland, 655 West Baltimore Street, Baltimore, MD 21201-1559, USA

R. Anand, Department of Neuroscience, Medical School, University of Pennsylvania, Philadelphia, PA 19104-6074, USA

Ch. Andres, INSERM-Unité 316 "Système nerveux du foetus à l'enfant, développement, circulation, métabolism", 3è étage Bât. Vialle, Faculté de médicine, BP 3223, 2bis, Bd Tonnellé, 37032 Tours Cedex, France

G.C. Andrioli, Divisione di Neurochirurgia, Ospedali Galliera, Via Volta 8, 16128 Genova, Italy

A. Anselmet, Laboratoire de Neurobiologie, Ecole Normale Supérieure, C.N.R.S. URA 1857, 46 Rue d´Ulm, 75005 Paris, France

L. Bartolini, Department of Preclinical and Clinical Pharmacology, University of Florence, Viale Morgagni 65, 50134 Florence, Italy

R. Beeri, Department of Biological Chemistry, Institute of Life Sciences, The Hebrew University of Jerusalem, Jerusalem 91904, Israel

J.L. Benovic, Department of Pharmacology, Jefferson Cancer Institute, Thomas Jefferson University, Philadelphia, PA 19107, USA

S. Berrard, Laboratoire de Génétique Moléculaire de la Neurotransmission et des Processus Neurodégénératifs, C.N.R.S., 91198 Gif sur Yvette, France

W. Betz, Department of Physiology, School of Medicine, University of Colorado, Box C-240, Denver, CO 80262, USA

V. Bigl, Paul-Flechsig-Institute for Brain Research. Medical Faculty, University of Leipzig, Jahnallee 59, 04109 Leipzig, Germany

N.J. Birdsall, National Institute for Medical Research, Mill Hill, London, NW7 1AA, UK

A. Björklund, Department of Medical Cell Research, University of Lund, Biskopsgatan 5, 22362 Lund, Sweden

N. Blin, Laboratory of Bioorganic Chemistry, Natl. Institute of Diabetes and Digestive and Kidney Diseases, Bethesda, MD 20892, USA

S. Bon, Laboratoire de Neurobiologie, Ecole Normale Supérieure, C.N.R.S. URA 1857, 46 Rue d´Ulm, 75005 Paris, France

R. Bonfante-Carbarcas, Department of Pharmacology, School of Medicine, University of Maryland, 655 West Baltimore Street, Baltimore, MD 21201-1559, USA

T.I. Bonner, Laboratory of Cell Biology, NIMH, NIH, Bethesda, MD 20892, USA

F. Casamenti, Department of Preclinical and Clinical Pharmacology, University of Florence, Viale Morgagni 65, 50134 Florence, Italy

M.P. Caufield, Wellcome Laboratory for Molecular Pharmacology, Department of Pharmacology, University College London, London WC1E 6BT, UK

P. Cavazzani, Divisione di Neurochirurgia, Ospedali Galliera, Via Volta 8, 16128 Genova, Italy

R. Cervini, Laboratoire de Génétique Moléculaire de la Neurotransmission et des Processus Neurodégénératifs, C.N.R.S., 91198 Gif sur Yvette, France

J.-H. Cheng, Laboratory of Neurobiology, NINDS, NIH, Bethesda, MD 20892, USA

B. Collier, Department of Pharmacology and Therapeutics, McGill University, 3655 Drummond Street, Montreal, Quebec H3G 1Y6, Canada

P. Correia-de-Sá, Laboratory of Pharmacology, ICBAS, University of Porto, L. Prof. Abel Salazar,2, 4000 Porto, Portugal

F. Coussen, Laboratoire de Neurobiologie, Ecole Normale Supérieure, C.N.R.S. URA 1857, 46 Rue d'Ulm, 75005 Paris, France

A. C. Cuello, Department of Pharmacology and Therapeutics, McGill University, 3655 Drummond Street, Montreal, Quebec H3G 1Y6, Canada

R.A. Cunha, Laboratory of Pharmacology, Gulbenkian Institute of Science, Rua da Quinta Grande 6, 2781 Oeiras Codex, Portugal

S.K. DebBurman, Department of Molecular Pharmacology & Biology Chemistry, Medical School, North Western University, 303 E Chicago Avenue - S215, Chicago, IL 60614, USA

F. Diebler, Laboratoire de Neurobiologie Cellulaire et Moléculaire, C.N.R.S., 91198 Gif sur Yvette, France

L.E. Eiden, Section of Molecular Neuroscience, Laboratory of Cell Biology, NIMH, NIH, Bethesda, MD 20892, USA

J. Erickson, Section on Molecular Neuroscience, Laboratory of Cell Biology, NIMH, NIH, Bethesda, MD 20892, USA

C. Felder, Laboratory of Cell Biology, National Institut of Mental Health, Bethesda, MD 20892, USA

M. Frotscher, Institute of Anatomy, University of Freiburg, P.O. Box 111, 79001 Freiburg i. Br., Germany

M. Geiszt, Institut für Pharmakologie, Universitäts-Klinikum Essen, Hufelandstr. 55, 45122 Essen, Germany

V. Gerzanich, Department of Neuroscience, Medical School, University of Pennsylvania, Philadelphia, PA 19104-6074, USA

E. Giacobini, Department of Geriatrics, University of Geneva, School of Medicine, Route de Mon Idée, CH-1226 Thonex-Geneva, Switzerland

L. Giovannelli, Department of Preclinical and Clinical Pharmacology, University of Florence, Viale Morgagni 65, 50134 Florence, Italy

G. Glas, Department of Pharmacology, Pharmacia AB, 751 82 Uppsala, Sweden

U. Hacksell, Department of Organic Pharmaceutical Chemistry, Uppsala University, 75123 Uppsala, Sweden

A. Hausinger, Zoologisches Institut, AK Neurochemie, Johann Wolfgang-Goethe Universität, Biozentrum Niederursel, Marie-Curie-Str, 9/Geb. N210, 60439 Frankfurt, Germany

B. Heimrich, Institute of Anatomy, University of Freiburg, P.O. Box 111, 79001 Freiburg i. Br., Germany

A.W. Henkel, Department of Physiology, School of Medicine, University of Colorado, Box C-240, Denver, CO 80262, USA

Ch. Herrmann, Zoologisches Institut, AK Neurochemie, Johann Wolfgang-Goethe Universität, Biozentrum Niederursel, Marie-Curie-Str, 9/Geb. N210, 60439 Frankfurt, Germany

L.B. Hersh, Deparment of Biochemistry, University of Kentucky, 800 Rose Street, Lexington, KY 40536, USA

M. Hosey, Department of Molecular Pharmacology & Biology Chemistry, Medical School, North Western University, 303 E Chicago Avenue, S215, Chicago, IL 60614, USA

T. Huberman, Department of Biological Chemistry, Institute of Life Sciences, The Hebrew University of Jerusalem, Jerusalem 91904, Israel

H. Inoue, Deparment of Biochemistry, University of Kentucky, 800 Rose Street, Lexington, KY 40536, USA

M. Israel, Laboratoire de Neurobiologie Cellulaire et Moléculaire, C.N.R.S., 91198 Gif sur Yvette, France

K.H. Jakobs, Institut für Pharmakologie, Universitäts-Klinikum Essen, Hufelandstr. 55, 45122 Essen, Germany

G. Johansson, Department of Organic Pharmaceutical Chemistry, Uppsala University, 751 23 Uppsala, Sweden

A. Karczmar, Department of Pharmacology, Medical Center, Loyola University, Maywood, IL 60153, USA

H. Kilbinger, Pharmakologisches Institut, Johannes Gutenberg-Universität, Obere Zahlbacher Str. 67, 55101 Mainz, Germany

J. Klein, Pharmakologisches Institut, Johannes Gutenberg-Universität, Obere Zahlbacher Str. 67, 55101 Mainz, Germany

E. Krejci, Laboratoire de Neurobiologie, Ecole Normale Supérieure, C.N.R.S. URA 1857, 46 Rue d´Ulm, 75005 Paris, France

S. Lazareno, MRC Collaborative Centre, Mill Hill, London, NW7 1AD, UK

C. Legay, Department of Biochemistry, University of Kentucky, 800 Rose Street, Lexington, KY 40536, USA

J. Lindstrom, Department of Neuroscience, Medical School, University of Pennsylvania, Philadelphia, PA 19104-6074, USA

R. Linke, Institute of Anatomy, University of Freiburg, P.O. Box 111, 79001 Freiburg i. Br., Germany

J. Liu, Laboratory of Bioorganic Chemistry, Natl. Institute of Diabetes and Digestive and Kidney Diseases, Bethesda, MD 20892, USA

K. Löffelholz, Pharmakologisches Institut, Johannes Gutenberg-Universität, Obere Zahlbacher Str. 67, 55101 Mainz, Germany

A. Maelicke, Institut für Physiologische Chemie, Johannes Gutenberg-Universität, 55099 Mainz, Germany

J. Mallet, Laboratoire de Génétique Moléculaire de la Neurotransmission et des Processus Neurodégénératifs, C.N.R.S., 91198 Gif sur Yvette, France

M. Marchi, Istituto di Farmacologia e Farmacognosia., Universitá di Genova, Viale Cembrano 4, 16148 Genua, Italy

S. Marchese, Istituto di Farmacologia e Farmacognosia., Universitá di Genova, Viale Cembrano 4, 16148 Genua, Italy

M. Marchioro, Department of Pharmacology, School of Medicine, University of Maryland, 655 West Baltimore Street, Baltimore, MD 21201-1559, USA

J. Massoulié, Laboratoire de Neurobiologie, Ecole Normale Supérieure, C.N.R.S. URA 1857, 46 Rue d´Ulm, 75005 Paris, France

M. Matsubayashi, Department of Pharmacology, School of Medicine, University of Maryland, 655 West Baltimore Street, Baltimore, MD 21201-1559, USA

H. Matsui, MRC Collaborative Centre, Mill Hill, London, NW7 1AD, UK

U.J. McMahan, Department Neurobiology, School of Medicine, Sherman Fairchild Soc., Stanford University, Stanford, CA 94305-5401, USA

M. Mesulam, The Cognitive Neurology and Alzheimer's Disease Centre, Department of Neurology and Psychiatry, Medical School, Northwestern University, 320 E. Superior Street, Room 11-450, Chicago, IL 60611-3010, USA

F.A. Meunier, Laboratoire de Neurobiologie Cellulaire et Moléculaire, C.N.R.S., 91198 Gif sur Yvette, France

F.-M. Meunier, Laboratoire de Neurobiologie Cellulaire et Moléculaire, C.N.R.S., 91198 Gif sur Yvette, France

G. Milligan, Molecular Pharmacology Group, Division of Biochemistry and Molecular Biology, Insitute of Biomedical and Life Sciences, University of Glasgow, Glasgow, G12 8QQ, UK

J. Molgo, Laboratoire de Neurobiologie Cellulaire et Moléculaire, C.N.R.S., 91198 Gif sur Yvette, France

I. Mullaney, Molecular Pharmacology Group, Division of Biochemistry and Molecular Biology, Insitute of Biomedical and Life Sciences, University of Glasgow, Glasgow, G12 8QQ, UK

E. Mutschler, Pharmakologisches Institut, Biozentrum Niederursel, Johann Wolfgang-Goethe-Universität, Marie-Curie-Str. 9/Geb. N260, 60053 Frankfurt, Germany

N.M. Nathanson, Department of Pharmacology SJ-30, School of Medicine, University of Washington, Box 357750, Seattle, WA 98195-7750, USA

T. Naumann, Institute of Anatomy, University of Freiburg, P.O. Box 111, 79001 Freiburg i. Br., Germany

E. Neale, Laboratory of Developmental Neurobiology, NICHD, NIH, Bethesda, MD 20892, USA

B.M. Nilsson, Department of Organic Pharmaceutical Chemistry, Uppsala University, 75123 Uppsala, Sweden

L. Nilvebrant, Department of Pharmacology, Pharmacia AB, 75182 Uppsala, Sweden

G. Nordvall, Department of Organic Pharmaceutical Chemistry, Uppsala University, 75123 Uppsala, Sweden

M.L. Nguyen, Department of Chemistry, University of California, Santa Barbara, CA 943106, USA

R. Pals-Rylaarsdam, Department of Molecular Pharmacology & Biology Chemistry, Medical School, North Western University, 303 E Chicago Avenue - S215, Chicago, IL 60614, USA

S.M. Parsons, Department of Chemistry, University of California, Santa Barbara, CA 943106, USA

X. Peng, Departement of Neuroscience, Medical School, University of Pennsylvania, Philadelphia, PA 19104-6074, USA

G. Pepeu, Department of Preclinical and Clinical Pharmacology, University of Florence, Viale Morgagni 65, 50134 Florence, Italy

E.G. Peralta, Department of Molecular & Cellular Biology, Harvard University, 7 Divinity Avenue, Cambridge, MA 02138, USA

E.F.R. Pereira, Department of Pharmacology, School of Medicine, University of Maryland, 655 West Baltimore Street, Baltimore, MD 21201-1559, USA

M. Plaschke, Institute of Anatomy, University of Freiburg, P.O. Box 111, 79001 Freiburg i. Br., Germany

M.-M. Poo, Department of Biological Sciences, Sherman Fairchild Center for Life Science, Columbia University, New York, NY 10027, USA

M. Raiteri, Istituto di Farmacologia e Farmacognosia., Universitá di Genova, Viale Cembrano 4, 16148 Genua, Italy

J.A. Ribeiro, Laboratory of Pharmacology, Gulbenkian Institute of Science, Rua da Quinta Grande 6, 2781 Oeiras Codex, Portugal

R.M. Richardson, Department of Molecular Pharmacology & Biology Chemistry, Medical School, North Western University, 303 E Chicago Avenue - S215, Chicago, IL 60614, USA

E. Rojas, Laboratory of Cell Biology and Genetics, National Institut of Diabetes and Digestive and Kidney Diseases, Bethesda, MD 20892, USA

S. Roßner, Paul-Flechsig-Institute for Brain Research. Medical Faculty, University of Leipzig, Jahnallee 59, 04109 Leipzig, Germany

U. Rümenapp, Institut für Pharmakologie, Universitäts-Klinikum Essen, Hufelandstr. 55, 45122 Essen, Germany

C. Scali, Department of Preclinical and Clinical Pharmacology, University of Florence, Viale Morgagni 65, 50134 Florence, Italy

M. K.-M. Schäfer, Department of Anatomy and Cell Biology, Philipps University, Robert-Koch-Str. 3, 35033 Marburg, Germany

R. Schliebs, Paul-Flechsig-Institute for Brain Research. Medical Faculty, University of Leipzig, Jahnallee 59, 04109 Leipzig, Germany

M. Schmidt, Institut für Pharmakologie, Universitäts-Klinikum Essen, Hufelandstr. 55, 45122 Essen, Germany

T. Schöneberg, Laboratory of Bioorganic Chemistry, Natl. Institute of Diabetes and Digestive and Kidney Dieseases, Bethesda, MD 20892, USA

A.M. Sebastião, Laboratory of Pharmacology, Gulbenkian Institute of Science, Rua da Quinta Grande 6, 2781 Oeiras Codex, Portugal

M. Sendtner, Neurologische Universitätsklinik, Universität Würzburg, Josef-Schneider-Str. 11, 97080 Würzburg, Germany

M. Shani, Department of Biological Chemistry, Institute of Life Sciences, The Hebrew University of Jerusalem, Jerusalem 91904, Israel

D. Singer-Lahat, Laboratory of Cell Biology and Genetic, National Institut of Diabetes and Digestive and Kidney Diseases, Bethesda, MD 20892, USA

H. Soreq, Department of Biological Chemistry, Institute of Life Sciences, The Hebrew University of Jerusalem, Jerusalem 91904, Israel

R. Stoop, Department of Biological Sciences, Sherman Fairchild Center for Life Science, Columbia University, New York, NY 10027, USA

S. Sundquist, Department of Pharmacology, Pharmacia AB, 751 82 Uppsala, Sweden

P. Svoboda, Institute of Physiology, Czech Academy of Sciences, AVCR, 14220 Praha 4 KRC, Czech Republic

L. Thal, Department of Neurosciences, School of Medicine, University of California San Diego, 9500 Gilman Drive, San Diego, La Jolla, CA 92093-0624, USA

H. Varoqui, Laboratoire de Neurobiologie Cellulaire et Moléculaire, C.N.R.S., 91198 Gif sur Yvette, France

W. Volknandt, Zoologisches Institut, AK Neurochemie, Johann Wolfgang-Goethe Universität, Biozentrum Niederursel, Marie-Curie-Str, 9/Geb. N210, 60439 Frankfurt,, Germany

G. Wang, Department of Neuroscience, Medical School, University of Pennsylvania, Philadelphia, PA 19104-6074, USA

E. Weihe, Department of Anatomy and Cell Biology, Philipps University, Robert-Koch-Str. 3, 35033 Marburg, Germany

G. Wells, Department of Neuroscience, Medical School, University of Pennsylvania, Philadelphia, PA 19104-6074, USA

J. Wess, Laboratory of Bioorganic Chemistry, Natl. Institute of Diabetes and Digestive and Kidney Diseases, Bethesda, MD 20892, USA

L. Williamson, Laboratory of Developmental Neurobiology, NICHD, NIH, Bethesda, MD 20892, USA

J. Yun, Laboratory of Bioorganic Chemistry, Natl. Institute of Diabetes and Digestive and Kidney Dieseases, Bethesda, MD 20892, USA

H. Zimmermann,. Zoologisches Institut, AK Neurochemie, Johann Wolfgang-Goethe Universität, Biozentrum Niederursel, Marie-Curie-Str, 9/Geb. N210, 60439 Frankfurt, Germany

# Preface

This volume offers a comprehensive update and overview of the field of cholinergic transmission as presented by some thirty distinguished investigators who were recruited for their task from Germany, Great Britain, Canada, USA, Sweden, Israel, France and Italy. Exciting new discoveries, described in this volume, are due to recent methodological breakthroughs. These discoveries throw a new light on many areas of cholinergic mechanisms, whether involving central or peripheral nervous system, as molecular, electrophysiological, biophysical, neuronal, behavioral and cognitive aspects of the cholinergic field are considered in detail.

The volume is based on the proceedings of the Ninth International Cholinergic Symposium (ISCM) which was held in Mainz, Germany, June 7–10, 1995. The first ISCM which took place in Skokloster, Sweden, was organized by Professor Edith Heilbronn, some 25 years ago; it could not be easily foreseen, at that time, how prominent in the area of neurosciences the ISCMs would become! The ISCMs meet every three years; prior to the Mainz Symposium, we had gathered in Sweden, USA, Switzerland, Italy, Great Britain and Canada. The next 10th ISCM which is being organized by Professors Jean Massoulié and Jean-Pierre Changeux, will take place in the Summer of 1998 in France.

The 9th ISCM generated a very positive response from its 200 participants. This response, and the success of the 9th ISCM are undubitively due to the effort of the International Advisory Committee and members of the 'Cholinergic Club', especially B. Collier, A.C. Cuello, E. Giacobini, D.J. Jenden, A. Karczmar, H. Kewitz, G. Pepeu, S. Thesleff, S. Tucek, P. Waser and V.P. Whittaker, whose knowledge of the field and expertise helped the Scientific Committee with shaping the program and rendering it exciting and up-to-date. Many thanks are due the Scientific Committee, composed of V. Bigl, H. Kilbinger, K. Loffelholz, A. Maelicke, E. Mutschler and H. Zimmermann for the care and input they put forth for this project.

The logistics of the 9th ISCM were the responsibility of the Organizing Committee. An ISCM is a complex undertaking, and that is was run smoothly and effectively, and that relatively few feathers were ruffled is entirely due to the talent and effort of the Organizing Committee; we owe thanks particularly to Dr. Ruth Lindmar. We would like also to acknowledge the effective work of the Secretariat (Ms. D. Brinkmann) and of the Graduate Students of the Department of Pharmacology of the Johannes Gutenberg-University of Mainz.

The success of the 9th ISCM depended on many financial contributors and sponsors. The 9th ISCM has been held under the auspices or sponsorship of the Johannes Gutenberg-University of Mainz, the state of Rheinland-Pfalz, the Deutsche Forschungsgemeinschaft (DFG), the Deutsche Gesellschaft für experimentelle und klinische Pharmakologie und Toxikologie (DGPT) and the International Society for Neurochemistry (ISN). *Major industrial sponsors*: Bayer AG (Germany), Boehringer Ingelheim (Germany), MSD (Germany). *Sponsors*: Byk Gulden (Germany), Hoffmann-La Roche (Germany), Labotec (Germany), Madaus (Germany), Marion Merrell (Germany), Merz & Co. (Germany), Pharmacia (Germany), Pharmacia Biotech (Germany), Roland (Germany), Schwabe (Germany), SmithKline Beecham (Germany) and Upjohn (USA). *The Industrial Exhibition* was sponsored by Axel Semrau (Germany), Biotrend Chemikalien (Germany), Research Biochemicals International (USA) and Polaroid (Germany).

As superb as may be the quality of the scientific contributions to the 9th ISCM, the financial success of a volume such as this depends on the editorial help and knowledge, and we are very much obliged to Dr. Nello Spiteri and to the staff of Elsevier Science Publishers for their professional effort with respect to this volume.

Jochen Klein and Konrad Löffelholz
Mainz, Germany, 1996

# Contents

# The Otto Loewi Lecture

OTTO LOEWI (1873–1961) (photo courtesy of Dr. Fred Lembeck, Graz, Austria)

"One of my dearest memories from those school years [in Frankfurt am Main] is the place where until 1890 we regularly spent our summer vacation. It was my father's estate, which consisted of an old manor, a large, enchanting garden, and some vineyards. It was situated on the slope of the Haardt Mountains in the Palatinate" (1).

The Ninth International Symposium on Cholinergic Mechanisms (9th ISCM, 1995) has been held in Loewi's native landscape, in Mainz, which is the capital of Rhineland-Palatinate and next to Frankfurt. Otto Loewi was born in Frankfurt as son of the wine merchant Jakob Loewi and Anna Willstätter, went to the Frankfurt ´Humanistisches Gymnasium´ and became medical doctor at the city hospital. In 1909, Loewi accepted the chair in pharmacology at Graz University (Austria). From there, he opened up the field of cholinergic mechanisms with a series of 14 articles published between 1921 and 1936, when he shared the Nobel prize with Sir Henry Dale. In 1938, forced by the Nazi terror, he left Graz, where he had spent almost 30 years, and went to London, bereft of all his belongings. Finally, in 1941, he settled in the United States. For all of these considerations, it was compelling for us, the Scientific Committee of the 9th ISCM, to honor the great scientist Otto Loewi by a Memorial-Lecture.

The Otto Loewi Lecture was given by Professor Alexander Karczmar who was born in Warsaw, Poland, 9 May 1918. In the 1940s, he met Otto Loewi in New York when both belonged to the New York University academic staff. Karczmar has contributed important discoveries and insights into many areas of the expanding cholinergic field including ontogenesis, the modulation of synaptic transmission, anticholinesterases, the role of cholinergic mechanisms in behaviour and finally the development of new drugs. Karczmar is a witness of the development of the cholinergic field following Loewi's discovery and an open-minded observer of the current metamorphosis of the field due to new basic disciplines and methods.

The Editors

1. Otto Loewi, An Autobiographic Sketch. In: *Perspectives in Biology and Medicine*, Vol. IV, pp.1–25, 1960.

J. Klein and K. Löffelholz (Eds.)
*Progress in Brain Research*, Vol. 109
© 1996 Elsevier Science B.V. All rights reserved.

THE OTTO LOEWI LECTURE

# Loewi's discovery and the XXI century

## Alexander G. Karczmar

*Research Services, Hines VA Hospital, Hines, Illinois, and Department of Pharmacology,*
*Loyola University Medical Center, Maywood, IL 60153, USA*

## Introduction

First, let me extend my heartfelt thanks to Konrad Löffelholz and his distinguished Committee for inviting me to deliver the First Loewi Lecture at the IX International Symposium on Cholinergic Mechanisms (ISCM).

This is a great honor which I do not deserve. Indeed, I can think of only minor reasons for my selection as the Otto Loewi lecturer. I assume that one must be an ancient cholinergiker to give an Otto Loewi Lecture, and I certainly can qualify on this ground, since Giancarlo Pepeu, Konrad Löffelholz, Jean Pierre Changeux or Edith Heilbronn are simply too young for the post in question. Yet, there are obviously ancient "cholinergikers" who merit this honor more than I.

Perhaps I can claim, however, one pertinent qualification: as Jews, both Otto Loewi and I, escaped the Nazis to land in USA, and we were both penniless. As well known, following Germany occupation of Austria the Nazis threw Loewi, at the time Professor of Pharmacology in Graz, in prison and did not allow him to leave Austria till he "wrote an order to the Swedish Bank in Stockholm to transfer the Nobel prize money, deposited with the bank in 1936, to a prescribed Nazi-controlled bank" (Loewi, 1960); so, Loewi's penniless state in 1939 had a more glamorous base than mine. On this particular ground Sir William Feldberg deserves the honor of being a Loewi Lecturer infinitely more than I, however, he is not in a position to fulfilling this task.

## A few personal characteristics of Loewi

I met Loewi briefly when he served in the early forties as Research Professor in the Department of Pharmacology of New York University, as on a few occasions he held conversations with graduate students, including myself. We all became immediately aware of his remarkable personality and wit, even before we became impressed with his knowledge, insight and culture. As to his humor, Loewi belonged to the "park bench" school of wits: when ready for one of these conversations with us, he would take a walk in Washington Square, holding forth on multiple topics, and he seemed to time his stories so as to end each upon approaching a park bench; after a short rest, he would proceed to the next bench, telling another story. There was still another trait: Loewi would speak softer and softer as he approached the punch line, to become almost inaudible as he got to it – but, the stories were good, and it was worthwhile for the audience to strain their ears for the denouement. I noticed later that Loewi shared these two traits with his Middle European contemporary scientists and intellectuals, such as Heinrich Klüver of the Klüver-Bucy Syndrome, Marcel Goldenberg, a cardiologist, and Franz Verzar, an endocrinologist, these two latter being personal friends of Loewi.

In the course of these conversations I noted another aspect of Loewi's personality, namely his scepticism. He could have had the missourian "show me" as his motto, and he reacted with caution and doubt to what appeared to him to be half-grounded hypotheses. I remember in fact that he confronted with rather acerbic wit the notion raised at that time of the transmitter nature of histamine.

And still another aspect. Many of Loewi's biographers stressed Loewi's modesty, democratic behavior, ready accessibility and good humor (Held, 1948; Koelle, 1986); on the basis of our superficial and brief acquaintance I could confirm these observations. But, this does not mean that Loewi was bland or universally benevolent. There was an air of authority about him and, indeed, of latent aggressivity. One would not dream – particularly if one was a graduate student or a junior Faculty member! – to take liberties with Professor Loewi, even if he were not a Nobel Prizeman! So, I was not surprised to learn subsequently that Loewi's household in Graz was "aristokratisch" and that, as a student, he belonged to a fraternity, participated in the customary duels, and received "zwei Schmisse" (Arnsberg, 1973, 1986)!

Let me add that in the early 1950s I contributed, with the late Theodore Koppanyi, a paper for Loewi's Festschrift (Karczmar and Koppanyi, 1953), and Loewi graciously rewarded me with a handwritten, kind note. Parenthetically, many Loewi's friends – such as Hans Kaunitz and William van der Kloot – remarked on the promptitude and kindness with which Loewi attended to his personal correspondence and to the acknowledgments that he felt he owed, and this habit persisted into the last days of his life, when he was infirm and debilitated (Lembeck and Giere, 1968).

*The scope of this presentation*

I wish to touch upon the following topics which all concern Loewi's demonstration of chemical transmission and of acetylcholine (ACh) as transmitter: My first topic deals with certain characteristics of that demonstration and of the mode of thinking that enabled Loewi to generate it. Second, I will emphasize the heuristic significance of this demonstration, as on its basis the cholinergic lore expanded so magnificently, and I will dwell on the present status of the field. Finally, in this light of the current image of the cholinergic area I will present my sense of the problems that will be solved in the XXI Century and I will comment on the pertinence of Loewi's mode of discovery with respect to their solution.

## Loewi's discovery of autonomic cholinergic transmission, its background and creative nature

*Loewi's accomplishments*

In the course of his career Loewi has made a number of important discoveries, besides that of humoral, ACh-driven autonomic transmission. Even in the case of autonomic regulation, Loewi's experiments went beyond those concerning the humoral action of ACh, as some of his studies which were carried out in parallel with those concerning the "Vagusstoff", laid the basis for establishing sympathetic transmission to the heart via the action of "Acceleransstoff" [Loewi, 1924; the contribution of Loewi's studies for establishing the nature of the sympathetic regulation was stressed by Loewi's companion as Nobel Prize winner, Sir Henry Dale (Dale, 1934)].

Now, as to Loewi's contributions outside the autonomic area. Loewi investigated the role of cations in the regulation of heart action and urinary secretion. His late discovery in this area concerns the facilitatory dromo- and ino-tropic action of sodium fluoride on the frog heart (Loewi, 1952). This finding, as minor as it may be in the list of Loewi's accomplishments is close to my heart, as many years later Kyozo Koketsu and I (Koketsu and Karczmar, 1966; Koketsu, 1966) described a facilitatory action of fluoride at the ganglion and at the Renshaw cell, unaware of Loewi's publication on this matter. Loewi (o.c.), most apropos, ascribed this effect of fluoride to its membrane, $Ca^{2+}$-dependent action. It appears today that the effect in question is, indeed, membrane directed and may involve $Ca^{2+}$-fluxes. Loewi (o.c.) referred to this fluoride-induced phenomenon as "sensitization"; again, I was not aware of this "first" of Loewi and I thought that I invented the term in the 1950s (Karczmar, 1957). It should be added that the organizer and perennial member of these Symposia, Edith Heilbronn, demonstrated early that NaF has the additional property of being a reactivator of organophosphorus-inhibited ChE (Heilbronn, 1964); as, on its face value, this effect of fluoride does not appear to be related to its sensitizing action, Edith is justified in not quoting Loewi in her paper!

Another, non-autonomic series of studies of Loewi concerned the topics of metabolism, insulin function and nutrition (Dale, 1962; Lembeck and Giere, 1968; Geison, 1973). Here, Loewi endeavored to clarify the role of nutrition and metabolism in conversion of fats and amino acids into proteins, in calorigenesis and in urinary secretion as well as the mechanism of the action of insulin on glucose. Two aspects of these studies are particularly pertinent with respect to one of the main issues of this presentation, namely the question of the mode of Loewi's creative thinking. First, in the course of these studies Loewi sought to find the common principle or mechanism that may have explained jointly all of the phenomena in question (Loewi, 1938, 1960); second, he managed to find a direct approach to resolve the problem concerning protein synthesis in animals, a problem which at the time baffled the nutritionists (Loewi, 1902, 1960; Geison, 1973). I will return to these two points subsequently.

*Background of Loewi's discovery*

Prior to Loewi's crucial investigations of the early 1920s much work was extended for some 70 years concerning the vascular, neuromyal, salivary and, indeed, central phenomena which were pertinent for Loewi's work. Yet, no common denominator, no experiment that could unify the copious, available data in terms of a common mechanism was forthcoming. How come? Let us briefly review the pertinent story.

This story should be initiated with emphasizing that certain speculations advanced already in the XIX Century were most pertinent for the matter at hand. In the seventies of that Century Emil Du Bois-Reymond entertained the possibility that the motor nerves act via liberating a chemical substance. Similarly pertinent were the concepts of Du Bois Reymond's great contemporary, Claude Bernard (clearly defined in his diagrams; Bernard, 1850, 1858); as he analyzed the action of a curare extract on the motor system he came up with a notion that a receptor unit located between the motor nerve terminal and the skeletal muscle intervenes in the elicitation of contraction, a notion similar to that posited by Langley and Dickinson (1889) some 40 years later. In fact, in 1905 Langley stated that curare interacts at the skeletal muscle with an "exceptionally excitable component...or...receptive substance". This work relates to that of one of the great bacteriologists of the turn of the Century, Paul Ehrlich (Holmstedt and Liljestrand, 1963) concerning chemical nature of receptor-drug binding; Ehrlich's concepts were then exploited by the pharmacologists such as the Stedmans and, particularly, Alfred Clark, who, in 1926, applied the receptor concept to the quantitative aspects of ACh's action on the smooth muscle (Clark, 1937).

Then, there were the investigations of the late XIX and early XX Century; they may have had less theoretical significance than those of Bernard, Langley and Du Bois-Reymond, but they laid a firm empirical basis for what was still to come. Thus, Schmiedeberg with Koppe, Gaskell, Dixon, and Harnack with Meyer found that muscarine and pilocarpine on the one hand and vagal stimulation on the other affected the heart similarly, and that their effects were antagonized by atropine (Dale, 1934; Brazier, 1959; Holmstedt and Liljestrand, 1963; Koelle, 1986).

There was another aspect of the nature of autonomic function that had to be considered for the final solution of this question, and that aspect concerned the actions of physostigmine, or, speaking contemporaneously, the extracts of calabar bean. The effects of calabar bean that resemble in many respects those of autonomic activity were known since the investigations in the eighteen twenties to eighteen sixties of Edinburgh pharmacologists and medical men, such as Robert Christison and Thomas Fraser (Dale, 1934; Holmstedt and Liljestrand, 1963; Karczmar, 1970, 1986; Holmstedt, 1972); their work actually embraced also somatic and central effects of the extract. Their studies were continued in Germany in the last decades of the XIX Century by Hamer, Lenz, Bezold and Goetz, Anderson, Arnstein with Sustchinsky, Winteberg, and Harnack with Witkowski (Karczmar, 1970) as they found that physostigmine facilitated or potentiated the effects of the faradic stimulation of the parasympathetic nerves (as they would be called today) on the iris, salivary gland and the heart, and of the motor nerves on the muscle; similar data were actually obtained subsequently by Loewi (Loewi and Mansfeld, 1910). It is important that these investigators demonstrated that physostigmine could evoke the miotic or cardiac inhibitory action in the absence of faradic stimulation of the pertinent nerves, and that the effects of physostigmine or of the extract, whether facilitatory or caused on its own, could be blocked by atropine.

The early XIX century work of Reid Hunt, Walter Dixon, William Howell, Thomas Elliott and, finally, Sir Henry Dale was even more pertinent for subsequent investigations of Loewi (Dale, 1934; Holmstedt and Liljestrand, 1963; Karczmar, 1970, 1986; Holmstedt, 1972). Hunt studied the vasodepressant actions of the constituents of the suprarenal gland as well as those of choline, and he associated the two phenomena. However, he could not find enough choline in the extract of the suprarenal gland to account for the vasodilation, and this led him to the most heuristic suggestion that an unstable derivative of choline may be present in the extract (Hunt, 1901). This notion led him to examine the pharmacology of 19 esters of choline, including ACh (Hunt and Taveau, 1906). He was impressed by the fact that this ester was one hundred thousand times more active than choline in its depressor and negative chronotropic action as he observed that a dose of 0.01 mg of ACh produced cardiac slowing which bore "a remarkable resemblance to that of a brief faradization of the vagus nerve".

Hunt did not theorize as to the physiological significance of his experiments, in fact, he and Taveau suggested (1906) that their findings may bear on certain pathological states, but their contemporaries, Howell and, particularly, Dixon and

Elliott stated, quite clearly, that "a local hormone" may underlie the autonomic function. Thus, Elliott (1904), after having demonstrated the mimicry between the action of ergotoxin on the effect of the faradic stimulation of the sympathetic nerves on the one hand and its action on the response to adrenaline on the other, "advanced ... the daring idea that sympathetic nerve fibers liberate adrenaline ..., to act as ... immediate agents of their effects" (Dale, 1934). Similarly, Dixon (1906, 1907) argued, on grounds similar to those underlying Elliott's hypothesis, that also the parasympathetic nerves must act by releasing a chemical transmitter, his candidate being muscarine. However, Dale (1934) stated that Dixon informed him that in 1906 he "concentrated, and partially purified an extract" of the dog's heart, the vagus of which was stimulated beforehand, and that the extract inhibited the isolated frog's heart, this effect being blocked by atropine! While Dixon never published this result, he gave Dale a pertinent record reprinted by Dale in his 1934 review. Perhaps Dixon was attempting to vindicate his priority with respect to the discovery of humoral transmission.

And then came Dale. He found in 1913, with the help of A.J. Ewins (Dale, 1934) that ACh is a constituent of samples of ergot, a substance of particular interest to Dale; as ACh appeared to be "a product of nature" (Dale, 1914, 1933b,c, 1934), he decided that it may be important physiologically. As he administered ACh intravenously to cats, he became impressed with its extremely potent actions as a vasodilator and somewhat less potent, but spectacular negative chronotropic actions (Dale, 1914). Potent but evanescent! This was the other side of ACh's action, and Dale was struck no less with this aspect of ACh lore than with its pharmacologic potency.

Altogether, the mimicry or the parallelism between the pharmacology at autonomic receptors of muscarine, pilocarpine, atropine and physostigmine and of ACh itself on the one hand, and the pharmacology of the effects of faradic stimulation of pertinent nerves on the other, as well as speculations of Elliott, Howell, Dale and Dixon seemed to beg for a direct demonstration of humoral transmission; similarly, the evidence concerning the "evanescent" action of the "Vagusstoff" and physostigmine potentiation of the function of the vagus nerve invited the notion of the antiChE action of the latter. "The actors were named, and the parts allotted" (Dale, 1938). Yet, as stressed by Dale (1934), "there … was … a gap" between early XX Century and the crucial 1921 year as to "direct contribution to the theory of chemical transmission." What was needed was "experimental ingenuity" (Karczmar, 1970) and a "simple, elegant and convincing demonstration" (Dale, 1934) that would internet the available evidence and the pertinent speculations.

*Loewi's discovery*

To turn now to Loewi's investigations, cited by the Nobel Prize Committee in 1936 as establishing jointly with Sir Henry Dale's work, the humoral, in this case cholinergic, synaptic transmission (or neurotransmission). Loewi's work in question was presented in the 14, 1921–1926 studies which appeared in the Pflüger's Archiv für die gesamte Physiologie. The apex of this work was the double heart-double cannula experiment (Loewi, 1921; Geison, 1973); as is well known, this was the experiment which was based on a dream (Loewi, 1960; Lembeck and Giere, 1968; Geison, 1973). "The nocturnal design" (Loewi, 1960) was as follows: "The hearts of the two frogs were isolated, the first with its nerves, the second without. Both hearts were attached to Straub cannulas filled with a little Ringer solution. The vagus nerve of the first heart was stimulated for a few minutes. Then the Ringer solution that had been in the first heart during the stimulation of the vagus was transferred to the second heart. It slowed and its beats diminished just as if its vagus had been stimulated". It is important to stress here that the Ringer solution did not contain physostigmine, even though Loewi used physostigmine many years before (Loewi and Mansfeld, 1911) and was aware, both because of his own studies and those of others, of its – or the Calabar bean extract's – effectiveness in mimicking or potentiating the autonomic function; actually, in the case of his experiments on the "Vagusstoff" Loewi did not employ physostigmine

until 1926 (Loewi and Navratil, 1926a,b). This matter will be returned to later.

Loewi stated clearly in his 1921 and 1922 "Mitteilungen" that the phenomena involved depended on the release of a chemical substance, the "Vagusstoff" (Loewi, 1922), thus enunciating the concept of chemical, in this case cholinergic neurotransmission; actually, Loewi referred to humoral rather than transmittive role of the "Vagusstoff", and the term "cholinergic" was provided by Dale (1933). Yet, for establishing a complete proof of this concept, Loewi needed to identify the substance in question, and he achieved it by appropriate pharmacological as well as chemical analysis. The pertinent experiments may be classified as concerned with the mimicry and identification criteria of chemical transmission; that is, Loewi demonstrated that the effects of the perfusate and of the suspected transmitter were identical as to their pharmacological characteristics, and that the extract of the perfusate could be, chemically or via an assay, identified as ACh. This approach to the proof of the function of a given neurotransmitter at an identified synapse constitutes what is now considered as a classical approach to the matter (Volle, 1966).

Another important aspect of this particular series of investigations was that they contributed significantly to Dale's notion of the evanescent action of ACh; in fact, they demonstrated the presence of a hydrolytic activity of the heart or its perfusate vis-à-vis of ACh (Loewi and Navratil, 1926a), and Loewi was impressed that the action of ACh "fades so quickly" (Loewi, 1960). This aspect of Loewi's studies of the humoral transmission led far, jointly with those of Dale (1914; Karczmar, 1970), toward the recognition of cholinesterase (ChE) as the natural hydrolysant of ACh, and of the antiChE action of physostigmine, although Loewi did not use this particular term (Loewi and Navratil, 1926a,b). There is still another feature of Loewi's work, and that is its felicity, an ingredient which is a condition sine qua non of any major discovery. In this context it must be stressed first of all that to accomplish the results in question without physostigmine is no mean trick. In fact, Loewi himself failed occasionally in his attempts to do so (Loewi, 1924; Geison, 1973), as did his contemporaries, and he had to fend himself against their arguments and negative findings; while he presented convincing results as to his essential findings, he had to admit that "der Inhalt aus Vagusreizperioden bald wirksam ist, bald nicht" (Loewi, 1924). Many of my contemporaries including Theodore Koppanyi and myself could not readily duplicate, without the aid of physostigmine, Loewi's experiment. There are two aspects of this difficulty and they both relate to the good fortune that accompanied Loewi in his endeavor.

First, there is the chronobiological aspect. A colleague of mine, the late Alexander Friedman, emphasized (Friedman, 1971) that the success of the experiment in Loewi's hands depended on the seasonal, chronobiological nature of the frog's heart sensitivity to ACh, and other investigators stressed that the ChE content of the heart of certain strains of frogs is low, particularly in the Spring (van der Kloot, 1972), and that this phenomenon underlies Loewi's success. However, Loewi did not need our help in this matter and he was well ahead of the game, as he clearly understood and enunciated explicitly the matter of seasonal variation (Loewi, 1924).

Second, there is the species-related aspect of the good luck attached to Loewi's discovery. As pointed out to me by Konrad Löffelholz, "Loewi was fortunate to use the frog heart which, similar to … the reptilian or … avian but in contrast to the mammalian heart has a dense cholinergic innervation of the ventricle (90% of the cardiac mass) … and … reveals a high sensitivity to the negative chronotropic effect of ACh!" (Löffelholz, personal communication; Feldberg and Krayer, 1933; Antony and Rotmann, 1968; Dieterich et al., 1977, 1978; Protas and Leontieva, 1992). However, as already pointed out, it is not easy to duplicate Loewi's experiment in the frog and in the turtle as well!

Final aspect of Loewi's 1921–1926 work concerns the question of receptors. Loewi (see Loewi and Navratil, 1924) clearly stated that, as atropine blocks the vagal action, this effect "gar nicht am Vagus angreift". He similarly perceived that the "Vagusstoff" acts directly on the "Erfolgsorgan", and that cardiac effects of muscarine, pilocarpine and other agonists reflect a similar site of action, as he emphasized that these substances affect the heart even after the division and the degeneration of the vagus nerve (Loewi and Navratil, 1925); in fact, he formulated the option that these agonists and antagonists affected, via "adsorption", a membrane site; he followed here, knowingly or otherwise, the lead of Claude Bernard and John Langley (see above).

*Creativity-related aspects of Loewi's discovery*

As I have already stated, one of the main objects of my paper is to salient the creative, cognition-related elements of Loewi's discovery. These elements are reflected in the parsimony of Loewi's concept of humoral transmission; many investigators, including Dale (1934) commented on the simplicity and the elegance of the concept, and the question of elegance is characteristic for a certain mode of consciousness and creative thinking. Actually, the inhibitory action of the vagus led Loewi's several contemporaries to embrace complicated schemes for explaining the effects in question, including the postulate of several mechanisms or substances being involved in the various phases of vagal action (Hofmann, 1917; Karczmar, 1970), and Loewi's concept constituted a simple explanation of the phenomena in question, including the mechanism of action of physostigmine, whether in facilitating the function of the vagus or potentiating the action of ACh (Loewi referred to this effect as "Sensibilisierung"; Loewi and Navratil, 1926a,b). This simplicity and directness, and his capacity to achieve parsimony with respect to what appeared to be complex problems characterized generally Loewi's empiricism. This was true, for example, with respect to his work, already referred to, on nutrition. In the last decades of the XIX Century attention was directed at the enigma of protein synthesis by the animals or man. Many investigators attempted, unsuccessfully, to maintain animals on either protein diet or the diet consisting of protein or certain products of protein digestion. Accordingly, very complex schemes were recurred to, to explain certain nutritional paradoxes involving protein synthesis. As Loewi became interested in this question, he had, "overnight" (Loewi, 1960; did he, in this case as well, dream of the pertinent procedure?) an insight that complete protein digestion by means of trypsin will yield elementary products of protein metabolism that could be used by the animals in synthesizing protein which the animals could not accomplish when the investigators used protein or its immediate products in their feed.

Now, the idea in question was inspired, as stated by Loewi (1960), by a paper of Kutscher (1903) concerning a trypsin digestion of protein. Thus, the mode of creation employed by Loewi was to connect a technique which had at the time no direct link with nutrition, with a problem which was investigated and written about vigorously at the time and which was a subject of his interest for many years – a close analogy with what happened in the case of his discovery of autonomic chemical or humoral transmission.

This matter of technique should be especially emphasized. Some time ago, in a letter to me, Konrad Löffelholz stressed the role of methodology in the discovery process. Indeed, at least in empirical sciences, discovery may depend on the association of the available and appropriate technique with the given problem, and such an association certainly occurred in the case of Loewi's demonstrations, both in the case of his nutritional and autonomic studies. Actually, Loewi was very cognizant of this heuristic nature of methodology; at the turn of the Century, he became aware of the "simplicity" (Loewi, 1960) of the methods that he employed at that time and he "decided to remedy this shortcoming" by becoming a student in Ernest Starling's laboratory [Loewi (1960) claimed that, at that time "Germany ... lost its leadership" in methodology to England. Paradoxically, he found Starling's equipment to be "primitive"!]. It should be added that, in another domain of human endeavor, certain thinkers, particularly those with a socialist or a Marxist viewpoint, such as Bernal (1956), pointed out that the societal, industrial and economic progress of the society depends on its ability for technical creativity. In fact, new methods, new "Produktionsmittel" are needed and indeed generated by, not only economic but also scientific requirements of the day.

Arguably the most important of all the components of the creative mode and the discovery process is a subconscious function that terminates with what Arthur Koestler (1964, 1967) referred to as the "Eureka" or "Aha!" moment. In Loewi's case, the creative process leading to the demonstration of humoral transmission was very long; according to Loewi (1960) it originated with his 1903 conversation with Walter Fletcher (Fletcher, 1926) and culminated in his famous dream; in this dream he associated the potential solution of the problem, hidden as it were in the available evidence, with an appropriate method [Loewi thought in 1960 that he "used ... the technique ... in question ... in 1920"; perhaps his memory in 1960 did not serve him right, as it is more likely that this technique was described in his 1918 paper). The subconscious nature of this associative process was stressed by Loewi (1960) himself, as he considered a psychoanalyst, Ernest Kris, to be a catalyst of his demonstration. A similar concatenation of methodology (which in this case involved as well certain biochemical procedures), available evidence and capacity to uncover the meaning hidden within the copious, published results led him to state in 1926 (Loewi and Navratil, 1926a,b) that the in vivo action of physostigmine is due to its "Spaltungshemmung" of the "Vagusstoff".

Actually, Loewi was a cautious and skeptical investigator. His scepticism has already been referred to; he was opposed to the early speculations concerning, for example, the existence of chemical, cholinergic transmission at the sites of a rapid response, namely at the nicotinic sites, as juxtaposed to the slow cardiac response at the vagal junction (see below), and, in such cases, he could be, according to Dale, "obstinate" (Geison, 1973). His intuitive capacity and "sensitivity" which, as stressed by Walter Cannon (1945), underlied Loewi's talent for discovery, were ready, at strategic moments, to break loose of his cautiousness.

Altogether, the discovery mode exploited by Loewi may be defined as involving conscious and subconscious mastery of available evidence, creative search for parsimonious mechanisms and appropriate methodology, and a subconscious associative attempt at the resolution of the problem at hand. In the conclusion of this paper I will try to ascertain whether or not this "Aha" or "Eureka" mode of creativity may still apply to certain modern, indeed, XXI Century problems of the cholinergic field and of neurosciences.

## Expansion of Loewi's discovery and the current status of cholinergic transmission

Only three-quarters of a century divide us from Loewi's first "Mitteilung". Yet, this discovery and the simultaneous thought of Sir Henry Dale were so heuristic in nature and brought about so many accomplishments that we might be as well separated from the decade of their findings by what Thomas Mann (1948) termed an "abyss of time". Let me turn now to the expansion in question, including the advances presented at the International Symposia on Cholinergic Mechanisms (ISCMs), including the IX Symposium.

### Expansion over the last seventy years: immediate consequences of Loewi's work

I will first briefly enumerate the main advances in the field which originated directly with Loewi's discovery; I will stress certain subjects that are particularly relevant in the context of this presentation.

### Cholinergic sites and pathways

Loewi's demonstration at the vagocardiac site of chemical transmission, driven by ACh and therefore cholinergic, to use Dale's nomenclature (Dale, 1933a, 1938), incited the search for other cholinergic sites, and, by extension, for cholinergic pathways. This search was helped by the concept developed by Gaskell, Langley and Dale of nicotinic and muscarinic cholinergic sites. In fact, Loewi's research established for the future the criteria needed for accepting the notion of cholinergic transmission – or any other chemical transmission – at any particular site (Volle, 1966); and, the particular heuristic jump which originated with the

establishment of these criteria resulted in several decades of research concerning quantitative determination of neurotransmitters via first bioassays and then appropriate chemical methods, and the demonstration of their presence at pertinent sites.

Loewi himself assumed, following his work with the vagocardial junction, that ACh-driven transmission obtains at other autonomic peripheral sites, and in his own laboratory instigated Engelhart's studies of the sites in question. Thus, Engelhart, Henderson, Gibbs, Roepke, Dale, Feldberg and others established the presence of cholinergic transmission at parasympathetic (such as salivary glands and the bladder) and certain sympathetic (as in the case of sweat glands) nerve endings; Dale, Feldberg, Vogt and Brown demonstrated this presence with respect to the voluntary muscle; Kibjakow, Feldberg, Gaddum, Barsoum, Emmelin and Vartiainen, with regard to the sympathetic and parasympathetic ganglia (Rosenblueth, 1950; Holmstedt and Liljestrand, 1963; Karczmar, 1967, 1970, 1986); and Feldberg, Dale, Eccles, Curtis and Krnjevic with respect to the CNS (Karczmar, 1967, 1990) .

An additional comment should be made in the context of the demonstration of cholinergic transmission at the fast nicotinic response sites, such as the sympathetic ganglia and the neuromyal junction, these sites to be juxtaposed with the slow response sites exemplified by the cardiovagal synapse. As already mentioned, Loewi was opposed to the notion of the chemical transmission at these sites; he and other investigators such as Sir John Eccles (Karczmar, 1991) doubted at that time that chemical transmission may obtain in the case of fast transmission processes such as those at the voluntary muscle, the ganglia and in the brain, as they did not see any mechanism that would lead to a rapid termination of the transmitter's action; in the absence of such a mechanism a transmitter would clog the junction! Neither Eccles nor Loewi – in spite of his own and Dale's pioneering work concerning ChEs – did realize in the 1930s and 1940s how rapidly one of the ChEs, acetylcholinesterase (AChE), can hydrolyze ACh, although this was also the time of the studies of Nachmansohn, Stedman, Mendel and Rudney (Karczmar,

1967, 1993) which brought to light the unique kinetics of AChE action.

The sites of particular importance today and of special significance in the framework of this presentation are the cholinergic synapses of the CNS. While the studies of these synapses were based on the notion of chemical transmission, Loewi, as already indicated, was skeptical with respect to this matter, and he did not seem to comment in print on the pertinent work during his life time, of Eccles, Dale and Feldberg. In two unpublished manuscripts entitled "Meaning of Life" and "The Organism as a Unit" (Lembeck and Giere, 1968) he refers to the CNS in the context of adaptive behavior, regulatory and information processes, and Pavlovian reflexology, but he does not refer to the possibility of the presence of chemical transmission in the CNS even once!

The CNS studies in question dealt particularly with the demonstration of the presence of cholinoceptive and cholinergic neurons in the CNS and the description of their pathways. The first line of these studies concerned electrophysiological and pharmacological analysis carried out at the neuronal level. This analysis was made possible by the development by Ralph Gerard and the Canberra investigators such as Eccles, Krnjevic, Curtis and Phillis, of the microelectrode and multibarrel micropipette methodology (Eccles et al., 1954; Karczmar, 1967, 1993; Krnjevic, 1969). Their work established the presence of cholinoceptive and cholinergic neurons in most parts of the CNS and of the pertinent synaptic potentials, including the muscarinic, nicotinic and mixed potentials.

The second line, intuited by Dale and initiated by Feldberg (1945) concerned the presence and differential distribution of ACh in the CNS, and their relevance for the central cholinergic pathways (Karczmar, 1967, 1993). This work, followed by the biochemical and histochemical determination in the CNS of AChE and choline acetyltransferase, the enzyme discovered by Nachmanson (Nachmansohn and Machado, 1943) led Gerebtzoff, Koelle, Shute and Lewis, and the McGeers with Kimura to define the central cholinergic pathways; this particular line of work was followed by several participants of the ISCMs,

including Larry Butcher, Bruce Wainer and Marsel Mesulam (Karczmar, 1990, 1993a).

The third line of the studies of the cholinergic CNS pathways concerned the central cholinoceptive and cholinergic receptors. Following particularly the development of appropriate pharmacological and molecular methods, the presence of nicotinic and muscarinic receptors in the CNS was demonstrated and their CNS location identified; in fact, several subtypes of either receptor are present in the brain (Cuello, 1993). In this context, George Koelle, Syogoro Nishi, John Szerb and Steve Polak (Schuetze and Role, 1987; Karczmar, 1990) made the important discovery of the presence of nicotinic and muscarinic receptors at nerve terminals of both cholinergic (autoreceptors) and non-cholinergic (heteroreceptors) neurons; these receptors regulate the release of ACh and other transmitters. The matter of the characterization, molecular biology and other aspects of cholinergic receptors is discussed below.

These three lines of approach and the pertinent findings of the last three decades which include the contributions of the ISCMs including the present Symposium yielded results that are mutually consistent and that led to an essentially definitive identification of the central cholinergic pathways, although there is a place for further refinements, particularly with respect to the spinal cord and the spinocerebral ascending and descending pathways (De Groat, 1976). There are two important aspects of this matter.

First, it appears that the central cholinergic pathways of rodents, felines and primates including man are basically similar, as discussed at this Symposium by Marsel Mesulam (Chapter 28), particularly with respect to the important, dense cholinergic radiation emanating for the CH4 cholinergic neurons of the nucleus basalis of Meynert. A second important aspect of this matter is the ubiquity of central cholinergic pathways and of their interactions with other transmitter systems (see below).

### Cholinesterases

Still another advance of the last 75 years concerns ChEs; the pioneering work of Loewi and

Dale in this area as well as the need for understanding the characteristics of the action of AChE in establishing cholinergic transmission at the "fast" cholinergic sites were already emphasized. Today, the important notion is that of multiplicity of biochemical and physical forms of ChEs and of their functions, which, as we know today, may not concern solely the termination of the action of ACh at the cholinergic synapses. Again, the pertinent findings were made by the past and present participants in the ISCMs such as Massoulié, Sussman and Soreq (Cuello, 1993; Karczmar, 1993). Thus, the human genes for AChE and BuChE were mapped by Mona Soreq and her associates (Chapter 26; Ben Aziz-Aloya et al., 1993); these investigators identified the AChE mRNA exons specific for the brain (and neuromyal junction) AChE as well as the exons concerned with the formation of non-synaptic AChE. And, Jean Massoulié, Joel Sussman and their associates (Chapter 4; Massoulié et al., 1993) distinguished at least six molecular forms of AChE which include symmetric, globular, either amphiphilic or nonamphiphilic forms on the one hand and asymmetric, collagen tailed forms on the other. They described the anchoring of these structurally different AChE proteins on the neuronal membrane, their molecular biology and associated these forms with two catalytic subunits. Finally, employing crystallographic methods, they posited the three-dimensional models of the pertinent structures.

*Anticholinesterases*

As already indicated, Loewi established the antiChE mechanism of action of physostigmine; this notion jived well with Dale's emphasis on the evanescent action of ACh, and served to explain the results of the Edinburgh investigators obtained with the extracts of the calabar bean. Our understanding of antiChE action augmented, even before the work of Dale and Loewi, with the isolation and crystallization in the 1860s by Jobst with Hesse and Vee with Leven of the pure alkaloid from the extract (Karczmar, 1970; Holmstedt, 1972). Subsequent progress was due to the establishment of the structure of physostigmine by Max

and Michel Polonowski in 1923, and the synthesis in the 1920s and 1930s of physostigmine and its derivatives, including neostigmine, by Stedman and by my Oak Park, Illinois neighbor, Percy Julian (Karczmar, 1963, 1970).

Two boosts propelled the knowledge of antiChEs. The first concerned the organophosphorus (OP) antiChEs; their studies started modestly enough with the 1850s and 1870s synthesis and studies of tetraethyl pyrophosphate and methylphosphoryldichloride by Moschnine and De Clermont, and von Hofman, respectively (Holmstedt, 1959, 1963; Karczmar, 1970), but an explosion of this particular knowledge occurred as the German, USA and British scientists raced in the 1930s and 1940s to develop both OP war gases and their antidotes (Holmstedt, 1959, 1963; Koelle, 1981). The second boost relates to the use of antiChEs in several disease states, beginning with myasthenia gravis, dystonias, etc., and culminating in the last two decades in their use in senile dementia of Alzheimer type (SDAT), as emphasized particularly by Ezio Giacobini, a participant at these Symposia (Chapter 30; Giacobini, 1990). The matter of the use of antiChEs in SDAT will be raised again in this presentation.

Today, the antiChE knowledge includes good understanding of the mechanism of their action and of the radicals involved in the latter (Usdin, 1970; Ishihara et al., 1991). Furthermore, it was shown that antiChEs may have effects which relate only indirectly to the inhibition of ChEs and which could not be foreseen by Loewi or Dale, namely their teratologic, ontogenetic and neurotoxic actions (Karczmar, 1963; 1984; Abou-Donia et al., 1979; Buznikov, 1984; Abou-Donia and Lapadula, 1990); their direct neuronal effects that include their sensitizing actions (Karczmar, 1984); and their delayed, chronic EEG and behavioral effects (Karczmar, 1984).

*Receptors*

Now, to return to the story of receptors as such, rather than as they relate to the cholinergic pathways. As already mentioned, Loewi was one of the pioneers in this area, along with the pathfinders such as Bernard, Langley and Ehrlich (Holmstedt and Liljestrand, 1963). Actually, as Clark (1926, 1937) applied mathematical formulation for their quantitative description, the receptor concept became somewhat abstract, as exemplified by the pertinent pharmacological work (Furchgott, 1955; Ariens and Simonis, 1967). Accordingly, several of us who are present at this Symposium in Mainz, were amazed when, at the 1959 Symposium organized by Carlos Chagas (Chagas and Carvalho, 1961), Chagas, Sy Ehrenpreis and David Nachmansohn approached the matter directly and attempted to isolate the nicotinic receptor as just any another substance! Ultimately, this line of work, jointly with the development of molecular biology, particularly cloning techniques, led to dramatic chemical, structural, pharmacological and molecular identification of multiple subtypes of both nicotinic and muscarinic receptors and of their amino acid sequences; several participants of the ISCMs such as Schimerlik, Barnard, Burgen, Birdsall, Karlin, Ladinsky, Lindstrom, Patrick, Waser, Changeux (Changeux, 1993) and Nordberg (Brown, 1989) contributed much to this lore.

The nicotinic receptors (Chapters 8–10) are composed of basically five subunits; their variations and the pattern of their assembly leads to the generation of very many receptor subtypes and to the formation of their transmembrane and channel or pore domains. As posited by Jean-Pierre Changeux (1993), these subtypes, their distribution and their contribution to the formation of synapses (Changeux and Danchin, 1976; Bertrand and Changeux, 1995) are resultants of the expression of a multigenic function.

Similarly, the muscarinic receptor subtypes (Chapters 11–16) are an expression of a multigenic system (Brown, 1989; Schimerlik, 1990; Brann et al., 1993). These subtypes share a general constitution consisting of several transmembrane helices or loops and several extra- and intracellular domains. Subtle differences in the pattern of these components and in their amino acid composition underlie the genetically controlled formation of their subtypes. Similarly genetically controlled is their nerve terminal and postsynaptic distribution in the CNS and elsewhere (Vilaro et al., 1993). Until recently it was thought that mainly M2 re-

ceptors characterize the CNS nerve terminal autoreceptors; it is opined today that all five subtypes of the muscarinic receptors may serve as autoreceptors (Vilaro et al., 1990).

*Indirect consequences of Loewi's work*

There are a number of modern concepts of the cholinergic lore which are not directly related to Loewi's, Dale's and Feldberg's findings. All these concepts have this in common that they underlie the flexible, plastic and multifactorial nature of the cholinergic system.

*ACh turnover and related phenomena*

First, there is the story of metabolism, turnover, release and storage of ACh, pioneered by William Feldberg, Frank MacIntosh, Victor Whittaker and Eduardo de Robertis (Karczmar, 1970, 1993; Whittaker, 1988; Cuello, 1993). This is a most dynamic and flexible phenomenon, as it embraces choline uptake, ChAT, various parameters of the synthesis and availability of choline and acetate, release phenomena, recycling of synaptic vesicles and blood–brain barrier events.

The understanding of ACh synthesis and the role of choline acetyltransferase, CoA and the ACh precursors, particularly choline is due to the efforts of the past and current participants in the ISCMs, such as Stanislav Tucek, the late Brian Ansell, Konrad Löffelholz, Giancarlo Pepeu, Lynn Wecker and Richard Wurtman (see below).

Then, the notion of the flexibility and multifactorial regulation of release phenomena which was initiated by Bernard Katz's discovery of the $Ca^{2+}$ role in the release and of the quantal or elementary events, was expanded subsequently by members of the ISCMs such as Giancarlo Pepeu, Maurice Israel, Brian Collier, Mario Marchi and Yves Dunant. Among the factors in question are the mechanisms involving cytoplasmic as well as vesicular (pace Victor!) release of ACh and the multiple mechanisms of presynaptic regulation of release. These mechanisms involve cholinoceptive autoreceptors as well as non-ACh transmitters (Chapters 5 and 21–23); vesamicol-sensitive proteins which control the uptake of ACh into synaptic vesicles (Chapter 7); transport proteins concerned with the movement and recycling of synaptic vesicles (Parsons, 1993); and protein regulators of the release of ACh (Greengard, 1987). It must be stressed that specific phosphorylation mechanisms and phosphokinases are involved in most of these processes (Greengard, 1987; Karczmar, 1993). Obviously, phenomena which are as multifactorial in nature as these are subject to very complex, point-to-point regulation processes, particularly since, as demonstrated by Brian Collier (Chapter 24), the processes in question are sensitive to and change with, the neuronal activity.

*Signal transduction, receptors, channels, elementary events, second messengers and G-proteins*

Here, we deal with interrelated phenomena that culminate, in a complex way, with signal transduction and effective synaptic transmission.

This story begins with the most heuristic discovery of Sir Bernard Katz and Ricardo Miledi concerning the postsynaptic "noise" which results from the application of ACh or current, and which Katz and Miledi (1970) identified as due to the interaction of small packets – several molecules – of ACh with ionophores or channels; they called these phenomena, most appropriately, "elementary events". The channels in turn control, via active mechanisms, ionic movements that lead to the signals, namely pre- and post-synaptic potentials and, via additional mediation of second messengers and related proteins, to effective transmission.

The studies of the many forms of ionic channels were pioneered by Sir John Eccles and Chris Krnjevic in the 1950s and 1960s, and pursued particularly by Brown and Adams (1980) (see Karczmar, 1993a,b); the pertinent data were presented at the past ISCMs by Krnjevic, Caulfield, McCormick and Brown (Aquilonius and Gillberg, 1990; Cuello, 1993), as well as at this Symposium. In the context of this presentation the significance of these findings lies in the flexibility of the responses of the channels ("Nachschlag" hypothesis; Colquhoun and Sakmann, 1985; Albuquerque et al., 1986, Chapter 9), as well as in their multiple modulatory role in modifying the pre- and post-

synaptic responses to whether ACh or other transmitters (Chapters 8–10).

The function of receptors and channels relates to still another multifactorial phenomenon, i.e., the multiplicity of postsynaptic cholinergic potentials which, in fact, include both excitatory and inhibitory potentials. The pertinent work was initiated by Rosamuad Eccles and Ben Libet (1961) with respect to sympathetic ganglia and by the Kurume-Loyola team of investigators, led by Syogoro Nishi, Kyozo Koketsu (1960) and the Gallaghers (Shinnick-Gallagher et al., 1986); this work was extended by Alan North (1982) (see Karczmar et al., 1986) to the enteric plexus and by Eccles, Koketsu, Cole, Nicoll, McCormick and Adams (McCormick, 1993; Karczmar, 1993a) to the CNS. Again, the emphasis here should be on the modulatory capacity of the potentials to regulate the global neuronal response, including the spike; this vectorial interaction between the synaptic potentials was first demonstrated by Sir John Eccles (1963), who emphasized its modulatory significance.

The receptors and the channels relate to the second and third messengers. The concept of messengers that intervene and, in fact, translate the activation of the receptors into a channel and an ionic effect, is indeed novel and could not be predicted in the 1920s. In fact, as we became aware in the 1960s how concentrated is ACh in the synaptic vesicles (Nishi et al., 1967; Karczmar, 1967b) – we knew from our own experience that this concentration of ACh causes burns when in contact with skin – we could not imagine that ACh's postsynaptic effect requires intermediaries. That it is so, however, was demonstrated by Earl Sutherland (Robison et al., 1971), Paul Greengard (Greengard, 1976, 1987), Michell (1975) and Lowell Hokin (Hokin and Dixon, 1993); basically, the messenger effects arise via activation of phosphoproteins and phosphorylation mechanisms, leading to changes in permeability and channel phenomena. Today, the second and related messengers include several nucleotides, phospholipid products, and nitric oxide (Dawson et al., 1992); their effect requires still another intermediary, the G proteins (Gilman, 1987). Besides Hokin, other participants at this (Chapters 14–20) as well as earlier ISCM Symposia, contributed to the knowledge in question.

There are specific linkages between the phenomena in question, as structurally specific links connect receptors, G proteins, channels and second messengers. This important notion was emphasized at this Symposium (Chapters 14–20). Thus, Nathanson, Löffelholz, Klein, Mullaney and their associates, and Jacobs demonstrated specific couplings between muscarinic receptor subtypes on the one hand, and G protein subtypes, certain phospholipases and components of phospholipid cascade on the other. Altogether, we deal here with biochemical pathways that link the early, more simple processes, such as those involving the muscarinic receptor, phospholipases and $Ca^{2+}$ channels, with, subsequently, deeper cellular compartments, such as the Golgi system, cytoskeleton and the nucleus.

In the context of this paper, the particular relevance of these multifactorial phenomena, their couplings and their receptor-to-cytoskeleton pathways is that they subserve the flexibility and plasticity which characterize the generation of the synaptic signal transduction – a matter that we will return to below.

*Modulations and sensitizations*

I referred above to modulations and interactions. These arise, as already stated, from vectoral interactions between postsynaptic potentials and from the presynaptic phenomena, at either auto- or hetero-receptors that regulate the release of ACh and other transmitters (Karczmar et al., 1972; Karczmar, 1993b). Another, well-known modulatory transmitter action arises when the catabolic enzymes of these transmitters are inhibited; in the case of ACh, these particular effects arise following the use of antiChEs.

A related but independent effect is that of sensitization; as already pointed out, Loewi (1952) used this term to describe the synaptic action of fluoride. It should be stressed that sensitization cannot be identified, causally, with the effects described in the preceding paragraph; in particular, it is not a direct consequence of any changes in the membrane potential or interactions between membrane potentials. It appears today that fluoride-induced sensitization may be due to $Ca^{2+}$-dependent changes of the membrane fluidity and permeability which occur via second messenger mechanism. Thus, Julius Axelrod (Hirata and Axelrod, 1978; Karczmar, 1990) referred to cyclic AMP actions of fluoride, and Brown and Birnbaumer (Codina and Birnbaumer, 1994), to its actions on heterotrimeric G-proteins. Allosteric configuration changes, described by Jean-Pierre Changeux (1993) may also be classified as instances of sensitization.

Desensitization may constitute an opposite, membrane-directed action, and, again, second messenger, G-proteins and related phosphorylation phenomena may be involved, as shown with respect to presynaptic transmitter release phenomena, by Paul Greengard (1976, 1987).

The significance, which cannot be exaggerated, of modulations, sensitizations and desensitizations is that they provide a most subtle point-to-point control of synaptic phenomena; the problems in computability that are involved in this type of control, will be stressed subsequently.

*Cholinergic correlates of behavior*

Although central effects of physostigmine were known since the work in the 1820s of the Edinburgh pharmacologists and although central actions of other antiChEs and cholinergic agonists were described since the 1920s and 1930s (see above; Holmstedt and Liljestrand, 1963; Karczmar, 1967, 1970), Loewi never envisaged the notion of central chemical – or cholinergic – transmission, even though he referred, in an unpublished manuscript (Lembeck and Giere, 1968), to "psychic regulations" and to their relation to "adaptations". On the other hand, Dale speculated on the subject of central cholinergic and non-cholinergic transmission in 1935, as did Feldberg a few years later (Feldberg, 1950); neither investigator did, however, refer to behavioral correlates of this notion, nor could they envisage the explosion of the pertinent findings of the last five decades and the variety of the central roles of ACh that are known today.

It is convenient to consider the behavioral cholinergic effects as either "organic" or psychological (Karczmar, 1978, 1979, 1990, 1995); for example, respiratory, convulsive, appetitive, thermostatic or reflexogenic effects may be classified as "organic", while effects on memory, REM sleep or aggression as psychological or cognitive in nature. Some phenomena, such as the EEG events including evoked potentials, defy classification. Again, the participants in these Symposia such as Pepeu, Dunnett, Steriade, Fibiger, Gillin, Hobson, Bigl, Giacobini and others (Karczmar, 1990, 1993; Pepeu, 1993), contributed significantly to this area.

Particularly important advances were made in the complex area of memory and learning. The cholinergic facilitation of memory and learning and their attenuation by anticholinergics is known since the investigations in the 1940s of William Funderburk and Karl Pfeiffer (Karczmar, 1967; 1995) concerning several conditioning paradigms. Subsequently, Drachman (1978) extended this work to the area of memory debilitation in aging; his studies form the basis for the use of cholinergic agonists, ACh precursors and antiChEs in Alzheimer's disease (Chapters 29 and 30).

What neuronal mechanisms are concerned with the phenomena in question? It is generally assumed that the facilitation of transmission along pertinent circuits is involved here. While this facilitation may depend on dendritic sprouting (Eccles, 1979), it may involve also electrophysiological processes which exhibit cholinergic correlates. In fact, this concept, due mainly to Ben Libet, is based on Libet's work with cholinergic transmission at the sympathetic ganglia; on the basis of this work, Libet suggested that certain slow potentials induce long term enhancement (LTE) of certain types of synaptic transmission, leading to the processes of memory and memory storage; long term potentiation (LTP) represents a related phenomenon (Libet, 1978). Besides ACh, other transmitters, such as dopamine and peptides, as well as second messengers are involved in these phenomena.

Three considerations should be raised here. First, cholinergic agonists and antagonists are very potent modifiers of the EEG patterns and of the evoked potentials, such as recruitment and several anticipatory potentials ("Bereitschaft" potentials). There is an interesting concatenation between these cholinergic effects and their action on mood, alertness, learning and awareness (Karczmar, 1979, 1990, 1995); this concatenation amounts to a behavioral syndrome referred to as cholinergic alert non-mobile behavior (CANMB) (Karczmar, 1979). Second, there is no known central function, whether animal or man, that does not show significant, frequently very marked, cholinergic correlates! Third, the pertinent phenomena are, again, multifactorial, as more than one central cholinergic pathway as well as non-cholinergic pathways, participate in these events (Karczmar, 1978). This complex, multifactorial character of these phenomena, particularly as it relates to their cognitive and conscious aspects, will be stressed subsequently.

*Trophic phenomena*

The antecedents for the recognition of tropisms were laid down by Ramon y Cajal, Paul Weiss and Victor Hamburger (I cannot resist it: also Karczmar, 1946), but it was the demonstration of the trophic action of the nerve growth factor (NGF) and its identification by Rita Levi-Montalcini (Levi-Montalcini and Angeletti, 1980) that led to notable developments in this area. Several participants in this and earlier ISCMs, such as Hans Thoenen and Micheal Sendtner (Thoenen et al., 1987; Chapter 35), Claudio Cuello (Cuello, 1993; Chapter 33) and Lars Olson (Ebendal et al., 1991) contributed significantly to this field; they emphasized that the trophic factors are effective both in the course of ontogenesis and beyond, and that they relate to central function and behavior, both in animals and perhaps in man (Lars Olson, pers. commun.). Their effort and that of others led to the recognition at this time of some 30 trophic factors that affect the cholinergic neurons and such components of the cholinergic system as choline acetyltransferase and ChEs (Koelle, 1988; Karczmar, 1990, 1993). Also, these investigators accentuated the role of the trophic factors in plasticity and flexibility of central cholinergic action and in the responses of cholinergic and non-cholinergic neurons.

**Problems in the cholinergic field that remain to be resolved in this and the next century**

There must be a healthy future for a knowledge that is growing as fast as the cholinergic knowledge and which, therefore, must be so heuristic. The proof, if needs be, of such a future is that a number of questions readily arise, begging for their resolution.

I would like to differentiate two categories of such questions. First, there are those that argue logically from the present status of the cholinergic lore; to answer those, we need ingenuity, technical competence, and, perhaps, significant technical advances, but no particular inspiration – we may resolve many of such problems still within this and the beginning of the next century. And then, there are questions the resolution of which is not readily apparent, which are speculative in nature and of immense importance. These are the problems for the next century; for their resolution we need more than ingenuity – their resolution requires creativity and a Loewi- or Dale-like modus operandi.

*Immediate problems*

*Synthesis of acetylcholine and related matters*

The following, unresolved problems are, at this time, associated with ACh synthesis:

(a) This synthesis involves ATP, CoA, citrate, mitochondrial dehydrogenase and the appropriate transport systems, leading to the formation of acetyl-CoA. "How is acetyl-CoA, generated primarily from mitochondrial dehydrogenase, transported to the cytosol where choline and most choline acetyltransferase are localized?" (Cooper, 1994). Are there any other mechanisms for the formation of acetyl-CoA in the cytosol (Szutowicz et al., 1994), and how does the whole process proceed in the cholinergic nerve terminal as compared to the neuronal soma cytosol?

(b) "What is the source of choline for ACh synthesis other than that released by ACh hydrolysis?" (Cooper, 1994; see also Köppen et al., 1993; Löffelholz et al., 1993). At this time, it appears that there may be two sources of choline to consider. First, dietary choline may come into the brain via the blood, whether as free and bound choline. It was thought that (lyso-)phosphatidylcholine cannot diffuse through the blood–brain barrier (BBB) and that more choline leaves than enters the brain (Dross and Kewitz, 1972). This problem is compounded by the finding that there is no de novo synthesis of choline in the brain. Löffelholz et al. (1993) may have provided the answer to the problem in question by showing that when blood's dietary choline is high, choline is taken up by the brain, as the BBB transport has no saturation limit. Ultimately, the free extracellular brain choline depends on the BBB transport in question and synaptically released choline as well as on the release of choline from the membrane phospholipids via receptor activation (Wurtman, 1992; Löffelholz et al., 1993). Homeostastic processes are involved here to keep free choline at a relatively low level; further clarification is needed as to the messenger and transporter systems that are involved in this homeostasis (Rylett and Schmidt, 1993). Altogether, one of the immediate problems in this area is to put in balance free choline derived from its various putative sources and to define the homeostatic mechanisms involved.

(c) A related question is that of the uptake of ACh into the vesicles as well as of the regulation of the distribution of cytosolic and vesicular ACh (Chapters 5–8).

(d) Two matters that concern the role of choline acetyltransferase (ChAT) in ACh synthesis also need definition. Thus, we should decide "which ChAT is physiologically relevant, the cytosolic form or the minor amount that is particulate and bound to terminal membranes?" (Cooper, 1994); and there is the question of the identification of the limiting factor of ACh synthesis: is it choline acetyltransferase? Is it choline (Tucek, 1990; Collier et al., 1993; Löffelholz et al., 1993; Rylett and Schmidt, 1993; Cooper, 1994)?

*Cholinesterases*

While there was dramatic progress with respect to identification of several biochemical and structural forms of ChEs and of their molecular origin, due particularly to the present and past participants in these Symposia such as Mona Soreq, Victor

Whittaker, Jean Massoulié and their associates, many questions remain. Thus, there needs to be further identification of the catalytic sites, their forms and the mechanisms of their hydrolytic activity (Whittaker, 1990), and their molecular biology (Chapters 4 and 26). Then, the participation of ChEs in the formation of synapses and the pertinent molecular mechanisms must be further described (Chapter 26). Another problem – an old one, to boot – is that of the significance and function of ChEs which are present in non-neuronal tissues (Karczmar, 1963 a,b; Koelle, 1963; Heilbronn, 1993; Whittaker, 1993); a related problem is that of the role of ChEs, particularly AChE, which are released from cholinergic and, possibly, noncholinergic nerve terminals (Appleyard, 1992).

*Anticholinesterases*

The basic structure of whether carbamate, oxamide or organophosphorus antiChEs is such (Holmstedt, 1963; Long, 1963; Usdin, 1970) that it is easy to synthesize new antiChE compounds; literally thousands of those are available today and new ones arrive on the scene daily. This new synthesis is directed at developing an antiChE compound that would be effective, kinetics and distribution-wise, in SDAT; this matter was discussed amply at the recent ISCMs including this one (Chapters 29 and 30). As believed by this author (Karczmar and Dun, 1988; Karczmar, 1991b, 1993), a successful outcome of such work is, for theoretical and evidential reasons, doubtful, and somewhat similar views were expressed at this symposium by Giancarlo Pepeu and Leon Thal (Chapter 29).

As already mentioned, antiChEs exhibit ontogenetic, neurotoxic and teratological actions as well as effects on the blood–brain barrier; these effects may actually depend both on their actions on ChEs as well as on their direct, noncholinesterasic effects (Karczmar, 1967, 1984); this aspect of antiChE and cholinergic action is not studied very much today. Furthermore, these compounds, particularly the organophosphorus drugs, may exhibit effect on EEG, mood and behavior that may extend, in man, for many years, well past the regeneration of ChEs (Duffy and Burchfiel, 1980;

Karczmar, 1984). Sometimes, these and related actions take the form of delayed neurotoxicity (Karczmar, 1967, 1993; Abou-Donia and Lapadula, 1990). The current interest in these actions of antiChEs is that these effects as well as those of their antidotes, such as pyridostigmine and atropine may underlie the so-called Persian Gulf Syndrome and certain phenomena observed in Bosnia-Herzegovina.

*Receptor subtypes, their channels and messengers*

Several matters concerning the cholinergic receptors are investigated currently, whether via molecular biology and cloning methods or pharmacological analysis (Chapters 8–13). The definitive answers are not in as yet. How many, actually, muscarinic and nicotinic receptor subtypes exist (Brown, 1989; Brann et al., 1993; Patrick et al., 1993; Sargent, 1993; Bertrand and Changeux, 1995)? What are their molecular structures, binding characteristics, desensitizing capacities, currents that they generate, their molecular biology, their function and distribution? To answer these questions is important both from the basic and clinical viewpoint.

A related question is that of receptor-channel and receptor-second messengers relations (Chapters 9, 17–19). What is the specific relationship between the receptor subtypes, channels and second and third messenger systems, as well as G proteins (Caulfield et al., 1993)? How many types of channels and post- or pre-synaptic potentials there are (Caulfield et al., 1993)? May a receptor subtype activate a number of channels? Do the channels, the second and third messengers, and the pertinent phosphokinases interact (Nishizuka, 1986)?

Two related matters also require further studies. First, there is the question of modulatory, allosteric, sensitization and desensitization phenomena (Bertrand and Changeux, 1955; Thesleff, 1955, 1990; Karczmar, 1957; Karczmar et al., 1972; Changeux, 1993). Many of these events depend on second messenger and phosphorylation mechanisms. In fact, there may be several "affinity" states, depending on the kinetics of the action of ACh and the state of gene expression at these receptors (Colquhoun et al., 1990; Changeux, 1993). Second, the postsynaptic as well as presynaptic cholinergic and noncholinergic potentials interact, as known since Eccles' demonstration (1963); the LTP phenomena belong here. But, except for the work of Chris Krnjevic, Ben Libet, David Brown and Mark Caulfield (Krnjevic, 1969, 1993; Ben Libet, 1970; Karczmar, 1993a), this area is insufficiently explored; this important matter will be returned to subsequently.

*Trophic phenomena*

As already mentioned, trophic phenomena received ample recognition at these Symposia (Karczmar, 1990b), and the pertinent research continues unabated, particularly due to the efforts of Cuello, Hefti, Thoenen, Olson and others (Chapters 31–35; Ebendal et al., 1991). Pertinent evidence concerns the possible trophic role of neurotransmitters, including ACh; trophic effects on neuronal activity; and trophic role of second messengers, including nitric oxide, and peptides such as ACTH (Racagni et al., 1994). The important aspect of tropisms is that they contribute to the phenomenon of brain plasticity (Rasmusson, 1993; Sillito, 1993), a novel concept that replaces the classical view of relative statism of the CNS and inflexibility of brain function. Creative insight into all this evidence promises rapid progress with respect to the treatment of aging and senile dementia of Alzheimer type (SDAT).

*Central cholinergic pathways, their interaction
with non-cholinergic pathways, and their
behavioral correlates*

Following the successful investigations of the last 40 years, only subtle refinements need to be added to the identification of the central cholinergic pathways whether in animals or in man (Chapter 28).

An important, related subject is that of joint topography of the cholinergic and noncholinergic pathways and of the related multitransmitter interactions. There is some pertinent knowledge of this interaction with respect to nerve terminal release, whether of ACh or other transmitters (Chapters 21–23; Consolo et al., 1990). And, certain specific

behaviors or functions were discussed from the viewpoint of what Michel Jouvet (1972) refers to as "Monoamine game"; this is particularly true for phases of sleep, including REM sleep. On the whole, however, the interactions at the postsynaptic site are relatively neglected. Indeed, there were very few Symposia devoted to this topic (Deniker et al., 1978; Karczmar, 1978), and only sporadically are these matters discussed in individual research communications (Gerfen and Keefe, 1994; Turek, 1994; Karczmar, 1996). And yet, whenever the central functions and behaviors were studied from this particular viewpoint, their multitransmitter nature was clearly illustrated (Karczmar, 1978). Accordingly, the identification of links between pathways of various transmitters and transmitter interactions are of prime importance for our understanding of the regulation of behavior and function; furthermore, these studies are needed to clarify the pre- and post-synaptic potential interactions and the generation of the synaptic output, that is, the evocation of the neuronal signal.

Now, as to the matter of the cholinergic nature of central functions and behaviors. As already stressed, the pertinent list is long, much was accomplished in this area over this entire Century, important pertinent insights were presented at the ISCM symposia (Karczmar, 1990, 1993a,b) and elsewhere (Napier et al., 1991), and new findings continue to accrue; for example, at the VIII ISCM Allan Hobson demonstrated the existence of a previously not identified form of REM sleep and localized its origin in the peribrachial region (Hobson et al., 1993). Of course, further delineation of the cholinergic correlates of such behaviors or functions as learning and memory processes, aggression, cholinergic alert non-mobile behavior (CANMB), sexual activity or REM sleep is still needed. The phenomena of memory and learning may be particularly difficult to clarify, as they are multifactorial and complex. First, we confront here again the problem of interaction in the pertinent processes between the cholinergic and other transmitter systems as well as non-transmitter, modulatory and trophic phenomena, and of the brain localization of these interactions. Then, this interaction and its localization must be related to several components of learning, such as a number of storage "compartments", consolidation, retrieval, extinction, and attention (McGaugh and Herz, 1982; Woody, 1982; Rose, 1992; Dunnett and Fibiger, 1993; Karczmar, 1993, 1995). Finally, there are the neuronal mechanisms, such as sprouting, LTE and LTP (see above). All these components of the phenomena of learning and memory must be integrated before their full understanding can be reached. Certain methodological and experimental difficulties are inherent; for example, it is difficult to relate LTE and LTP processes to pertinent transmitters and brain sites, and thence, to memory and learning processes and their components; technical advances may be needed, such as the capacity of carrying out appropriate recordings in conscious animals, in conjunction with measurements of memory processes. Yet, while the definitive progress may take time, the pertinent concepts were posited, the course is set and the pertinent methodology available.

*Clinical aspects*

The cholinergic agents do not have the clinical importance that is commensurate with the immense functional and basic importance of the cholinergic, whether peripheral or central systems. Yes, cholinergic drugs are employed in the treatment of myasthenia (less importantly with the advent of the notion of the auto-immunological nature of myasthenia); intestinal, bladder and asthmatic disease; certain ocular conditions; familiar dysautonomia and hypertension (where their value is essentially of historic interest), and as anesthetic adjuncts and antagonists (Cooper, 1994); the potential use of antiChEs and cholinergic agonists in SDAT was already referred to. Yet, all these uses do not appear to be very dramatic – as of today! However, further work in the area of receptor subtypes and their functional identity, and the employment of genetic manipulation and cloning techniques (Jenden, 1990; Brann et al., 1993) should lead in the near future to the development of effective cholinergic therapy in a number of diseases, already referred to, as well as motoneuron disease; pain regulation (Gillberg et al., 1990; Karczmar, 1993b); mental disease (as in the case

of nootropic agents, Pepeu and Spignoli, 1989) and psychosis (Karczmar, 1995); and dystonias, atonias and amyotrophic lateral sclerosis (Askmark et al., 1990).

*Problems that need long range approach: "computable" and "non-computable" problems*

I have stressed frequently in this paper the complexity and the multifactorial nature of cholinergic phenomena, such as, for example, learning and memory. We were concerned then with macromolecular management, yet, several sequential levels of complexity underlie these parameters. Thus, the phenomena in question begin at the neuronal level of channels and receptors, their multiple affinity and conductance states, and the modulatory and sensitization or desensitization effects (Colquhoun et al., 1990; Changeux, 1993; Chapters 10 and 15); then, there is the level, at each neuron, of interactions between the transmitters and the second and third messengers. How do the interactions in question result in the ultimate synaptic output or signal? And then we reach the macromanagement level, also complex, as it concerns the topography of the transmitter pathways and the conjunction between the pathways of the various transmitters (some 30 transmitters may be recognized at this time!), the linkage between these pathways, the modulatory systems and the trophic factors, and, finally, the behavioral outcome. It may be that all that is needed to predict the outcome of the processes occurring at each level is a computer program. Of course, the sophistication of the program must be increased at each subsequent hierarchical level, particularly when the third level – that of the "overt" behavioral outcome – is reached. In 1993, I made a "plea that a general conceptual framework ... and a computer program ... are needed to give a finalistic meaning to the bewildering number of messages transmitted by the cholinergic neurons ...", and Giancarlo Pepeu (1993) supported my plea ... Today, that is a few years later, I am not sure whether this conceptual framework is computational in nature; the total system is complex enough to become internally inconsistent and not amenable to complete description (Goedel, 1931). Actually, the Universal Turing Machine, that is, a two-dimensional computer capable of computing any computable function and testing validity of any mathematical solution (Turing, 1981) may not be able, paradoxically, to give an answer to the problem in question in a finite time, and, what may be particularly vexing, it may be impossible to decide whether or not such a solution would be ultimately forthcoming (Penrose, 1989, 1994).

Now comes the real crux of the matter. As already emphasized, cholinergic system relates particularly to the phenomena of awareness, cognition and organism-environment interaction (Karczmar, 1995). The obvious question for the XXI Century is then, how does the cholinergic system relate to the mind and such special mental functions as "free will" and creativity? Or, to state this problem more generally, how do the concepts of chemical transmission and related "organic" phenomena relate to the mind? And, are we today in the position to transcend perennial Cartesian (Descartes, 1637) notion of dualistic nature of the mind-brain relationship (for modern proposals concerning the dualism or "triadism" of this relationship, see Popper and Eccles, 1977; Penrose, 1994); to the contrary, should we become reductionists (Changeux and Monnes, 1995)? Even though the past and present approaches to this dilemma may not always address directly the cholinergic system, they must be briefly alluded to, to clarify the nature of this dilemma.

There were several attempts to bridge drastically the gap via a reductionist approach. The approach represented since the 1920s by J. Watson and his student B.F. Skinner is based on the notion that only the overt or operant processes are real (Skinner, 1957). Today, Skinner's approach is generally looked askance at; for example, his attempt at describing cognitive processes – such as those exhibited in the course of human conversation – as reducible to an operant behavior analysis appears to be a skinnerian self-parody rather than a defensible view (Koestler, 1967; Rose, 1992).

Presumably more sophisticated, more recent reductionist or semi-reductionist approaches relate mental and "free will" phenomena to certain EEG patterns or rhythms analyzed via multichannel and coherence power spectrum EEG analysis and tomographic EEG and magnetoencephalography (John, 1972; Deeke et al., 1976; Thatcher et al., 1995). Ben Libet (1985) introduced a related notion which has to do with the slow evoked potentials generically referred to as anticipatory or readiness potentials; he suggested that the significant delay that separates these potentials from a voluntary overt act is indicative of conscious and unconscious phenomena including the acts of "free will"; altogether, Libet seems to define consciousness as the delay in question. It is of interest that Libet's speculation is subject, to a certain degree, to experimental verification (Libet, 1985, and the discussion of Libet's paper by Eccles, Jung, Jasper and others).

Then, Penfield and Hebb (Rose, 1992) emphasized that conscious phenomena may be evoked by stimulation of certain brain parts, such as temporal lobe. In this context, Ben Libet (1994) speculated the brain exhibits "conscious mental fields" (CMFs) which are "... not describable in terms of any externally observable physical process or any known physical theory ... the CMF would be detectable only in terms of subjective experience accessible only to the individual who has the experience". He proposed further that the CMF may be represented by an isolated cortical slab which could be obtained, in human subjects, under certain therapeutic conditions; this paradigm would allow to test the subjective response – or a consciousness process – of the subjects to chemical or electric stimulation of the slab and the resulting CMF.

To this author at least, these approaches do not resolve the problem of mind–brain dichotomy as they always seem to contain a black box as the seat of transduction of an organic phenomenon into a mental occurrence. A glimmer of hope was offered recently, as Roger Penrose (1994) speculated that, as the neuronal events may be basically quantal in nature, the transduction in question occurs as quantal phenomena transit into classical events in the course of the process of "objective reduction". At least as to the matter of the quantal nature of neuronal phenomena Penrose may be not far from the truth. In this context, I am happy to recall that my Ph.D. mentor, the late Selig Hecht (Hecht et al., 1941) was the first investigator to tread the territory in question as he demonstrated that retinal cells respond to a few photons; as shown recently, that number may be actually one (Baylor et al., 1979). And, as is well known to cholinergikers, the components of the postsynaptic "elementary events" of Katz and Miledi (1970), the single channel responses and their "multiple conductance states" (Colquhoun et al., 1990) may express quantal events (Pollard et al., 1994); so do certain phenomena concerning the synaptic boutons and their paracrystalline structures or grids (Akert et al., 1975), and Eccles (1994) believes that this "quantal site" controls the probability of emission of ACh from synaptic vesicles. This matter of probability and quantal indeterminancy was actually associated by many investigators with the free will phenomenon (Eccles, 1994; see however Penrose, 1994). A related quantal level may be present at the neuronal microtubules (Lahoz-Beltra et al., 1993; Penrose, 1994).

Obviously, as of this time, the brain–mind or brain–consciousness problem remains unresolved. As pointed out above, even the partial approaches to this problem may be not computable; it is even more important that a "black box" seems to be always inherent in the approaches attempted and that the transduction of the pertinent phenomena into mentation such as involved in creativity or "free will" cannot be defined at this time.

### Is the Loewian mode of discovery still valid for resolving he problems of consciousness?

At the meetings of the NIH Study Groups it is frequently stated with respect to certain grant support applications that "the experiments proposed stem logically from the past work of the PI (principle investigator)", or words to this effect; the phenomena and the problems that we are concerned with here need an entirely different approach! It is frequently posited today that a number of mental processes cannot be readily defined on a reductionist or monist basis. Thus, certain linguistic processes and matters of syntax are not computable,

even though genetic mechanism may be involved in their generation and ontogenesis (Gardner, 1982; see however Chomsky, 1968); thus, certain sentences may make sense compared to other sentences, but the syntactical difference between them cannot be incorporated into a software program. Nor can a poetic sentence be differentiated, computer-wise, from prose; neither can a good piece of music. In other words, there cannot be "a Muse in the machine" (Rothstein, 1955; Pinsky, 1995), and a creative association between words or notes must be intuited.

The key word here seems to be "creativity", whether the phenomena in question relate to music, poetry, or, indeed, science and scientific discovery. It is my contention that Loewi's mode of discovery should be included within the category of cognitive creativity and that this mode, rather than a computable process, is needed today with respect to several unsolved biological problems, including the cholinergic correlates of consciousness and understanding (Karczmar, 1991).

Penrose's (1994) notion is pertinent in this context. Employing the Goedel-Turing theorem and the concept of universal Turing machine (Turing, 1937) Penrose sought to prove analytically that mathematical understanding, i.e., mathematical creativity and capacity for discovery are processes which are not computational in nature. By extension, he proposed that the same is true for non-mathematical understanding and non-mathematical creativity. Penrose juxtopposes his notion to that of Artificial Intelligence (AI) as he does not agree with the prediction that, with time, the computer and universal Turing machine processes will be capable not only of algorithmic computerization of any problem, but also of creative thinking and discovery (Churchland and Sejnowsky, 1992; Churchland, 1995; Freedman, 1995; Gelernter, 1995; Hofstadter, 1995). Actually, Dennet (1995) does not accept Penroses's (1994) proof of non-computational nature of discovery process. It must be emphasized that as many or more investigators of cognitive processes agree with Penroses's notion, beginning with Koestler (1964, 1967) and continuing with Searle (1980) and Dyson (1988).

Let me now complete the cycle and return to Loewi's discovery of chemical, cholinergic transmission. As already described, appropriate background was available at the time, and Loewi himself employed, in a different context, the method that he used ultimately to demonstrate the existence of the theorem in question. The stage and the actors were set (Dale, 1938), but the play with its plot had to be written and produced or, to put it differently, the words were available, but the play had still to be created. It may not be easy to prove that indeed a non-algorithmic, creative process was involved in what happened to Loewi at the time, perhaps in the form of a dream, but to me it appears that "something quite new ... was ... needed" (Penrose, 1994); a "Eureka" process was required.

Since, as already pointed out, many consciousness-related parameters exhibit marked cholinergic correlates, it is appropriate to predict that a XXI Century Loewi will employ the Loewian mode, create a solution, and execute the definitive experiment.

## Acknowledgements

Some of the research from this and Dr. N.J. Dun's laboratory referred to in this paper was supported by NIH grants NS6455, NS15858, RR05368, NS16348 and GM77; VA grant 4380; grants from Potts, Fidia and M.E. Ballweber Foundations; and Guggenheim (1969) and Fullbright (1985) Fellowships. In addition, help afforded by CARES, Chicago, is gratefully acknowledged. The author wishes also to thank Ms. J. Mixter and the staff of the LUMC Library for their expert help. Special gratitude is extended to my friend, Konrad Löffelholz, for providing me with pertinent documents and for his help with analyzing Otto Loewi's person, as well as for his many constructive suggestions with respect to ACh metabolism and second messenger phenomena.

## References

Abou-Donia, M.D., Graham, D.G. and Komcil, A.A. (1979) Delayed neurotoxicity of O-(2,4-dichlorophenyl)-O-ethyl phosphonothioate. Effect of a single dose on hens. *Toxicol. Appl. Pharmacol.*, 49: 293–303.

Abou-Donia, M.B. and Lapadula, D.M. (1990) Mechanisms of organophosphorus ester-induced neurotoxicity: Type I and Type II. *Annu. Rev. Pharmacol. Toxicol.*, 30: 405–440.

Akert, K., Peper, K. and Sandri, C. (1975) Structural organization of motor end plate and central synapses. In: P.G. Waser (ed.), *Cholinergic Mechanisms*, Raven Press, New York, pp. 43–57.

Albuquerque, E.X., Allen, C.N., Aracava, Y., Akaike, A., Shaw, K.P. and Rickett, D.L. (1986) Activation and inhibition of the nicotinic receptor: actions of physostigmine, pyridostigmine and meprodifen. In: I. Hanin (ed.), *Dynamics of Cholinergic Function*, Plenum Press, New York, pp. 677–695.

Antony, H. and Rotmann, M. (1968) Zum Mechanismus der negativ inotropen Acetylcholin-Wirkung auf das isolierte Froschmyokard. *Pflügers Arch.*, 300: 67–86.

Appleyard, M. (1992) Secreted acetylcholinesterase: Non-classical aspects of a classical enzyme. *Trends Neurosci.*, 15: 485–490.

Aquilonius, S.-M and Gillberg, P.-G (eds.) (1990) *Cholinergic Neurotransmission: Functional and Clinical Aspects*, Elsevier, Amsterdam, 493 pp.

Ariens, E.J. and Simonis, A.M. (1967) Cholinergic and anticholinergic drugs, do they act on common receptors? *Ann. N. Y. Acad. Sci.*, 144: 842–867.

Arnsberg, P. (1973) Mitten im Satz vom Tod überrascht. Erinnerungen an Otto Loewi. *Frankfurter Allg. Ztg.*, 2: 6–9.

Arnsberg, P. (1983) Die Geschichte der Frankfurter Juden seit der Französischen Revolution. In: H.-O. Schembs (ed.), *Biographisches Lexikon der Juden in den Bereichen: Wissenschaft, Kultur, Bildung, Öffentlichkeitsarbeit in Frankfurt am Main*, Eduard Roether Verlag, Darmstadt, pp. 287–290.

Askmark, H., Aquilonius, S.-M and Gillberg, P.-G. (1990) Neuropharmacology of amyotrophic lateral sclerosis. In: S.-M. Aquilonius and P.-G. Gillberg (eds.), *Cholinergic Neurotransmission: Functional and Clinical Aspects*, Elsevier, Amsterdam, pp. 371–380.

Baylor, D.A., Lamb, T.D. and Yau, K.W. (1979) Responses of retinal rods to single photons. *J. Physiol.*, 288: 589–611.

Ben Aziz-Aloya, R., Sternfeld, M. and Soreq, H. (1993) Promoter elements and alternative splicing in the human ACHE gene. In: A.C. Cuello (ed.), *Cholinergic Function and Dysfunction. Progress in Brain Research*, Vol. 98, Elsevier, Amsterdam, pp. 147–159.

Bernal, J.D. (1956) *Science and Industry in the Nineteenth Century*, Rutledge, London, pp. 230.

Bernard, C. (1850) Action de curare et de la nicotine sur le systeme nerveux et sur le systeme musculaire. *C. R. Soc. Biol.*, 2: 195.

Bernard, C. (1858) *Lecons sur la Physiologie et la Pathologie du Systeme Nerveux*, J.-B. Bailliere, Paris, pp. 524.

Bertrand, D. and Changeux, J.-P. (1995) Nicotinic receptor: an allosteric protein specialized for intercellular communication. *Semin. Neurosci.*, 7: 75–90.

Brann, M.R., Ellis, J., Jorgensen, H., Hill-Eubanks, D. and Jones, S.V.P. (1993) Muscarinic acetylcholine receptor subtypes: localization and structure/function. In: A.C. Cuello (Ed.), *Cholinergic Function and Dysfunction. Progress in Brain Research*, Vol. 98, Elsevier, Amsterdam, pp. 121–127.

Brazier, M.A.B. (1959) The historical development of neurophysiology. In: J. Field (ed.), *Handbook of Physiology. Section I: Neurophysiology*, American Physiological Society, Washington, DC, pp. 1–58.

Brown, D.A. and Adams, P.R. (1980) Muscarinic suppression of a novel voltage-sensitive $K^+$ current in a vertebrate neuron. *Nature*, 283: 673–676.

Brown, J.H. (ed.) (1989) *The Muscarinic Receptors*, Humana Press, New Jersey, 478 pp.

Buznikov, G.A. (1984) The actions of neurotransmitters and related substances on early embryogenesis. *Pharmacol. Ther.*, 25: 23–59.

Cannon, W.B. (1945) *The Way of an Investigator*, Norton, New York, 229 pp.

Caulfield, M.P., Robbins, J., Higashida, H. and Brown, D.A. (1993) Postsynaptic actions of acetylcholine: the coupling of muscarinic receptor subtypes to neuronal channels. In: A.C. Cuello (ed.), *Cholinergic Function and Dysfunction*, Progress in Brain Research, Vol. 98, Elsevier, Amsterdam, pp. 293–308.

Chagas, C. and Carvalho, A.P. de (1963) *Bioelectrogenesis*, Elsevier, Amsterdam, 413 pp.

Changeux, J.-P. (1993) Chemical signaling in the brain. *Sci. Am.*, November, 58–62.

Changeux, J.-P. and Danchin, A. (1976) Selective stabilization of developing synapses as mechanisms for specification of neuronal networks. *Nature*, 264: 705–712.

Changeux, J.-P. and Monnes, A. (1995) *Conversations on Mind, Matter and Mathematics*, Princeton University Press, Princeton, NJ, pp. 260.

Chomsky, N. (1968) *Language and Mind*, Harcourt Brace & World, New York, 88 pp

Churchland, P.M. (1995) *The Engine of Reason, the Seat of the Soul*, MIT Press, Cambridge, MA, 329 pp

Churchland, P.S. and Sejnowsky, T.J. (1992) Perspectives on computational neuroscience. *Science*, 242: 741–745.

Clark, A.J.(1926) The reaction between acetylcholine and muscle cells. *J. Physiol.*, 61: 530–546.

Clark, A.J. (1937) *General Pharmacology.* Heffter's Handb. exp. Pharmakol., Vol. 4, Springer-Verlag, Berlin, 228 pp.

Codina, J. and Birnbaumer, L. (1994) Requirement for intramolecular domain interaction in activation of G protein alpha subunit by aluminum fluoride and GDP but not GTP gamma S. *J. Biol. Chem.*, 269: 29339–29342.

Collier, B., Tandon, A., Prado, M.A.M. and Bachoo, M. (1993) Storage and release of acetylcholine in a sympathetic ganglion. In: A.C. Cuello (ed.), *Cholinergic Function and Dysfunction*, Progress in Brain Research, Vol. 98, Elsevier, Amsterdam, pp. 183–189.

Colquhoun, D. and Sakmann, B. (1985) Fast events in single-channel currents activated by acetylcholine and its analogues at the frog muscle end-plate. *J. Physiol.*, 369: 501–557.

Colquhoun, D., Cachelin, A.B., Marshall, C.G., Mathie, A. and Ogden, D.C. (1990) Function of nicotinic receptors. In: S.-M. Aquilonius and P.-G. Gillberg (eds.), *Cholinergic Neurotransmission: Functional and Clinical Aspects*, Elsevier, Amsterdam, pp. 43–50.

Consolo, S., Palazzi, E., Bertorelli, R., Fisone, G., Crawley, J. Hökfelt, T. and Bartfai, T. (1990) Functional aspects of acetylcholine-galanine coexistence in the brain. In: S.-M. Aquilonius and P.-G. Gillberg (eds.), *Cholinergic Neurotransmission: Functional and Clinical Aspects*, Elsevier, Amsterdam, pp. 279–287.

Cooper, J.R. (1994) Unsolved problems in the cholinergic nervous system. *J. Neurochem.*, 63: 395–399.

Cuello, A.C. (ed.) (1993) *Cholinergic Neurotransmission: Function and Dysfunction*, Proc. 8th Int. Cholinergic Symp., Elsevier, Amsterdam, pp. 462.

Dale, H.H. (1914) The action of certain esters and ethers of choline, and their relation to muscarine. *J. Pharmacol. Exp. Ther.*, 6: 147–190.

Dale, H.H. (1933a) Nomenclature of fibres in the autonomic system and their effects. *J. Physiol.*, 82: 10.

Dale, H.H. (1933b) Progress in autopharmacology; survey of present knowledge of the chemical regulation of certain functions by natural constituents of the tissues. Introduction. *John Hopkins Hosp. Bull.*, 53: 297–312.

Dale, H.H. (1933c) Progress in autopharmacology; survey of present knowledge of chemical regulation of certain functions by natural constituents of tissues. Acetylcholine. *John Hopkins Hosp. Bull.*, 53: 312–329.

Dale, H.H. (1934) Chemical transmission of nerve impulses. *Br. Med. J.*, 1–20.

Dale, H.H. (1935) Pharmacology and nerve endings. *Proc. Soc. Med.*, 28: 319–322.

Dale, H.H. (1938) Acetylcholine as a chemical transmitter of the effects of nerve impulses. I. History of ideas and evidence. Peripheral autonomic actions. Functional nomenclature of nerve fibres. *J. Mt. Sinai Hosp.*, 4: 401–429.

Dale, H.H. (1962) Otto Loewi 1873–1961. Biographical *Memoirs Fellows R. Soc.*, 8: 67–89.

Dawson, T.M., Dawson, J.l. and Snyder, S.H. (1992) A novel neuronal messenger molecule in brain: the free radical, nitric oxide. *Ann. Neurol.*, 32: 297–311.

Descartes, R. (1637) *Discours de la Methode Pour Bien Conduire sa Raison, et Chercher la Verite Dans les Sciences.* A. Leyde, Paris, 413 pp.

De Groat, W.G. (1976) Mechanisms underlying recurrent inhibition in the sacral parasympathetic outflow to the urinary bladder. *J. Physiol.*, 257: 503–513.

Deniker, P., Radouco-Thomas, C. and Villeneuve, A. (Eds.) (1978) *Neuro-psychopharmacology*, Proc. 10th Cong. CINP, Vols. 1–2, Pergamon Press, Oxford, UK.

Dennet, D.C. (1995) *Darwin's Dangerous Idea.* Simon and Schuster, New York, 586 pp.

Dieterich, H.A., Löffelholz, K. and Pompetzki, H. (1977) Acetylcholine overflow from isolated perfused hearts of various species in the absence of cholinesterase inhibition.

*Naunyn-Schmiedeberg's Arch. Pharmacol.*, 296: 149–152.

Dieterich, H.A., Lindmar, R. and Löffelholz, K. (1978) The role of choline in the release of acetylcholine in isolated hearts. *Naunyn-Schmiedeberg's Arch. Pharmacol.*, 301: 207–215.

Dixon, W.E. (1906) Vagus inhibition. *Br. Med. J.*, II: 1807.

Dixon, W.E. (1907) On the mode of action of drugs. *Med. Mag. London*, 16: 454–457.

Drachman, D.S. (1978) Central cholinergic system and memory. In: M.A. Lipton and K.F. Killam (Eds.), *Psychopharmacology; A Generation of Progress*, Raven Press, New York, pp. 651–652.

Dross, K. and Kewitz, H. (1972) Concentration and origin of choline in the rat brain. *Naunyn-Schmiedeberg's Arch. Pharmacol.*, 274: 91–106.

Duffy, F.H. and Buchfiel, J.L. (1980) Long term effects of the organophosphate sarin on EEG in monkeys and humans. *Neurotoxicology*, 1: 667–689.

Dunnett, S.B. and Fibiger, H.C. (1993) Role of forebrain cholinergic systems in learning and memory: relevance to the cognitive deficits of aging and Alzheimer's dementia. In: A.C. Cuello (Ed.), *Cholinergic Function and Dysfunction*, Progress in Brain Research, Vol. 98, Elsevier, Amsterdam, pp. 413–420.

Dyson, F.J. (1988) *Infinite in All Directions*. Harper and Row, New York, pp. 321.

Ebendal, T., Persson, H. and Olson, L. (1991) New directions in NGF experimental therapy of Alzheimer's disease. In: R. Becker and E. Giacobini (Eds.), *Cholinergic Basis for Alzheimer Therapy*, Birkhäuser, Boston, MA, pp. 468–473.

Eccles, J.C. (1963) *The Physiology of Synapses*. Springer-Verlag, Berlin, 316 pp.

Eccles, J.C. (1979) *The Human Mystery*. Springer-Verlag, Berlin, 255 pp.

Eccles, J.C. (1994) *How the Self Controls Its Brain*. Springer-Verlag, Berlin, 197 pp.

Eccles, J. C., Katz, B. and Koketsu, K. (1954) Cholinergic and inhibitory synapses in a pathway from motor-axon collaterals to motoneurones. *J. Physiol.*, 216: 524–562.

Eccles, R.M. and Libet, B. (1961) Origin and blockade of synaptic responses of curarized sympathetic ganglia. *J. Physiol.*, 157: 484–503.

Elliott, T.R. (1904) On the action of adrenalin. *J. Physiol.*, 31: 20–21.

Feldberg, W. (1945) Present views on the mode of action of acetylcholine in the central nervous system. *Physiol. Rev.*, 25: 596–642.

Feldberg, W. and Krayer, O. (1933) Das Auftreten eines azetylcholinartigen Stoffes im Herzvenenblut von Warmblütern bei Reizung der Nervi vagi. *Naunyn-Schmiedebergs Arch. exp. Path. Pharmak.*, 172: 170–193.

Fletcher, W.M. (1926) John Newport Langley. In memoriam. *J. Physiol.*, 61: 1–27.

Freedman, A. (1995) *Brainmakers*. Simon and Schuster, New York, 214 pp.

Friedman, A. (1972) Circumstances influencing Otto Loewi's discovery of chemical transmission in the nervous system. *Pflügers Arch.*, 325: 85–86.

Furchgott, R. (1955) The pharmacology of vascular smooth muscle. *Pharmacol. Rev.*, 7: 183–265.

Gardner, H. (1982) *Art, Mind and Brain*. Basic Books, New York, 380 pp.

Geison, G.H. (1973) Otto Loewi. In: C.C. Gillispie (Ed.), *Dictionary of Scientific Biography*, Vol. VIII, Charles Scribner's Sons, New York, pp. 451–457.

Gelernter, D. (1995) *The Muse in the Machine*. The Free Press, New York, 203 pp.

Gerfen, C.F. and Keefe, K.A. (1994) Neurostriatal receptors. *Trends Neurosci.*, 17: 2–3.

Giacobini, E. (1993) Pharmacotherapy of Alzheimer disease: new drugs and novel strategies. In: A.C. Cuello (Ed.), *Cholinergic Function and Dysfunction*, Progress in Brain Research, Vol. 98, Elsevier, Amsterdam, pp. 447–454.

Gillberg, P.-G., Askmark, H. and Aquilonius, S.-M. (1990) Spinal cholinergic mechanisms. In: S.-M. Aquilonius and P.-G. Gillberg (Eds.), *Cholinergic Neurotransmission: Functional and Clinical Aspects*, Elsevier, Amsterdam, pp. 361–370.

Gilman, A. (1987) G proteins: transducers of receptor-generated signals. *Annu. Rev. Biochem.*, 55: 505–649.

Godel, K. (1931) Über formal unentscheidbare Sätze per Principia Mathematica und verwandter Systeme. *Monatshefte f. Mathem. u. Phys.*, 3: 173–198.

Greengard, P. (1976) Possible role for cyclic nucleotides and phosphorylated membrane proteins in postsynaptic actions of neurotransmitters. *Nature*, 260: 101–108.

Greengard, P. (1987) Neuronal phosphoproteins. *Neurobiology*, 1: 81–119.

Hecht, S., Shlaer, S. and Pirenne, M.H. (1941) Energy, quanta and vision. *J. Gen. Physiol.*, 25: 891–940.

Held, I. W. (1948) Professor Otto Loewi. In: S.R. Kagan (Ed.), *Victor Robinson Memorial Volume*, Froben Press, New York, pp. 123–128.

Hirata, F. and Axelrod, J. (1978) Enzymatic synthesis and rapid translocation of phosphatidylcholine by two methyltransferases in erythrocyte membrane. *Proc. Natl. Acad. Sci. USA*, 75: 2348–2352.

Hobson, J.A., Datta, S., Calvo, J.M. and Quatrochi, J. (1993) Acetylcholine as a brain state modulator: triggering and long-term regulation of REM sleep. In: A.C. Cuello (Ed.), *Cholinergic Function and Dysfunction*, Progress in Brain Research, Vol. 98, Elsevier, Amsterdam, pp. 389–404.

Hofmann, F.B. (1917) Über die Einheitlichkeit der Herzhemmungsfasern und über die Abhängigkeit ihrer Wirkung vom Zustande des Herzens. *Zeitsch. f. Biol.*, 67: 427–452.

Hofstadter, D. (1995) *Computer Models of the Fundamental Mechanisms of Thought*. Basic Books, New York, 518 pp.

Hokin, L.E. and Dixon, J.F. (1993) I. Historical background. II. Effects of lithium on the accumulation of second messenger inositol 1,4,5-trisphosphate in brain cortex slices. In:

A.C. Cuello (Ed.), *Cholinergic Function and Dysfunction.* Progress in Brain Research, Vol. 98, Elsevier, Amsterdam, pp. 309–315.

Holmstedt, B. (1959) Pharmacology of organophosphorus cholinesterase inhibitors. *Pharmacol. Rev.*, 11: 567–688.

Holmstedt, B. (1963) Pharmacology of organophosphorus anticholinesterase agents. In: G.B. Koelle (Ed.), *Cholinesterases and Anticholinesterase Agents*, Handb. exp. Pharmakol., Vol. 15, Springer-Verlag, Berlin, pp. 428–485.

Holmstedt, B. (1972) The ordeal bean of old calabar: the pageant of Physostigma venenosum in medicine. In: T. Swain (Ed.), *Plants in the Development of Modern Medicine*, Cambridge University Press, Cambridge, MA, pp. 303–360.

Holmstedt, B. and Liljenstrand, G. (1963) Readings in Pharmacology. Pergamon Press, Oxford, 395 pp.

Hunt, R. (1901) Further observations on the blood pressure lowering body in the suprarenal gland. *Am. J. Physiol.*, 5: 6–7.

Hunt, R. and Taveau, R. de M. (1906) On the physiological action of certain choline derivatives and new methods for detecting choline. *Br. Med. J.*, II: 1788–1791.

Ishihara, Y., Kato, K. and Goto, G. (1991) Central cholinergic agents. I. Potent acetylcholinesterase inhibitors. *Chem. Pharmacol. Bull. (Tokyo)*, 39: 3225–3235.

Jenden, D.J. (1990) Achievements in cholinergic research, 1969–1989: drug development. In: S.-M. Aquilonius and P.-G. Gillberg (Eds.), *Cholinergic Neurotransmission: Functional and Clinical Aspects*, Elsevier, Amsterdam, pp. 479–486.

John, E.R. (1972) Neural correlates of learning and memory. In: A.G. Karczmar and J.C. Eccles (Eds.), *Brain and Human Behavior*, Springer-Verlag, Berlin, pp. 291–301.

Karczmar, A.G. (1946) The role of amputation and nerve resection in the regressing limbs of urodele larvae. *J. Exp. Zool.*, 103: 401–127.

Karczmar, A.G. (1957) Antagonism between a bis-quaternary oxamide, WIN8078, and depolarizing and competitive blocking agents. *J. Pharmacol. Exp. Ther.*, 119: 39–47.

Karczmar, A.G. (1963) Ontogenetic effects. In: G.B. Koelle (Ed.), *Cholinesterases and Anticholinesterase Agents*, Handb. exp. Pharmakol., Vol. 15, Springer-Verlag, Berlin, pp. 799–832.

Karczmar, A.G. (1967a) Pharmacologic, toxicologic and therapeutic properties of anticholinesterase agents. In: W.S. Root and F.G. Hoffman (Eds.), *Physiological Pharmacology*, Vol. 3, Academic Press, New York, pp. 163–322.

Karczmar, A.G. (1967b) Discussion of the paper of S. Ehrenpreis. *Ann. N. Y. Acad. Sci.*, 144: 734–735.

Karczmar, A.G. (1970) History of the research with anticholinesterase agents. In: A.G. Karczmar (Ed.), *Anticholinesterase Agents*, Section 13, Vol. 1: pp. 1–44, Int. Encyclop. Pharmacol. Ther., Pergamon Press, Oxford, UK.

Karczmar, A.G. (1978) Multitransmitter mechanisms underlying selected functions, particularly aggression, learning and sexual behavior. In: P. Deniker, C. Radouco-Thomas and A. Villeneuve (Eds.), *Neuropsychopharmacology*. Proc. Xth

Cong. CINP, Vol. 1, Pergamon Press, Oxford, UK, pp. 581–608.

Karczmar, A.G. (1979) Brain acetylcholine and animal electrophysiology. In: K.L. Davis and P.A. Berger (Eds.), *Brain Acetylcholine and Neuropsychiatric Disease*, Plenum Press, New York, pp. 265–310.

Karczmar, A.G. (1984) Acute and long lasting central actions of organophosphorus agents. *Fund. Appl. Toxicol.*, 4: S1–S17.

Karczmar, A.G. (1986) Historical development of concepts of ganglionic transmission. In: A.G. Karczmar, K. Koketsu and S. Nishi (Eds.), *Autonomic and Enteric Ganglia*, Plenum Press, New York, pp. 3- 26.

Karczmar, A.G. (1990) Physiological cholinergic functions in the CNS. In: S.-M. Aquilonius and P.-G. Gillberg (Eds.), *Cholinergic Neurotransmission: Functional and Clinical Aspects*, Elsevier, Amsterdam, pp. 437–466.

Karczmar, A.G. (1991a) Historical aspects of the cholinergic transmission. In: T.C. Napier, P.W. Kalivas and I. Hanin (Eds.), *The Basal Forebrain*, Plenum Press, New York, pp. 453–476.

Karczmar, A.G. (1991b) SDAT models and their dynamics. In: R. Becker and E. Giacobini (Eds.), *Cholinergic Basis for Alzheimer Therapy*, Birkhäuser, Boston, MA, pp. 141–152.

Karczmar, A.G. (1993a) Comments to session on electrophysiological aspects of cholinergic mechanisms. In: A.C. Cuello (Ed.) *Cholinergic Function and Dysfunction*, Progress in Brain Research, Vol. 108, Elsevier, Amsterdam, pp. 279–284.

Karczmar, A.G. (1993b) Brief presentation of the story and the present status of studies of the vertebrate cholinergic system. *Neuropsychopharmacology*, 9: 181–199.

Karczmar, A.G. (1995) Cholinergic substrates of cognition and organism-environment interaction. *Prog. Neuro-Psychopharmacol. Biol. Psychiatry*, 19: 187–211.

Karczmar, A.G. (1996) Loewi's discovery and the future of our notions of central cholinergic transmission and its behavioral correlates. *Perspect. Biol. Med.*, in press.

Karczmar, A.G. and Dun, N.J. (1988) Effects of anticholinesterases pertinent for SDAT treatment but not necessarily underlying their clinical effectiveness. In: E. Giacobini and R. Becker (Eds.), *Current Research in Alzheimer Therapy*, Taylor and Francis, New York, pp. 15–29.

Karczmar, A.G., Nishi, S. and Blaber, L.C. (1972) Synaptic modulations. In: A.G. Karczmar and J.C. Eccles (Eds.), *Brain and Human Behavior*, Springer-Verlag, Berlin, pp. 63–92.

Katz, B. and Miledi, R. (1970) Membrane noise produced by acetylcholine. *Nature*, 226: 962–963.

Kewitz, H., Pleul, O., Dross, K. and Schwartzkopff (1975) The supply of choline in rat brain. In: P.G. Waser (Ed.), *Cholinergic Mechanisms*, Raven Press, New York, pp. 131–135.

Koelle, G.B. (1981) Organophosphorus poisoning - an overview. *Fund. Appl. Toxicol.*, 1: 129–137.

Koelle, G.B. (1986) Otto Loewi. *Trends Pharmacol. Sci.*, 7: 290–291.

Koelle, G.B. (1988) Enhancement of acetylcholinesterase synthesis by glycyl-l-glutamine: an example of a small peptide that regulates differential transcription? *Trends Pharmacol. Sci.*, 9: 318–321.

Koestler, A. (1964) *The Act of Creation*. Macmillan, New York, 751 pp.

Koestler, A. (1967) *The Ghost in the Machine*. Macmillan, New York, 384 pp.

Koketsu, K. (1966) Restorative action of fluoride on synaptic transmission blocked by organophosphorus anticholinesterases. *Int. J. Neuropharmacol.*, 5: 254–257.

Koketsu, K. (1969) Cholinergic synaptic potentials and the underlying ionic mechanisms. *Fed. Proc. Am. Soc. Exp. Biol.*, 28: 101–112.

Koketsu, K. and Karczmar, A.G. (1966) Action of NaF at various cholinergic synapses. *Fed. Proc.*, 25: 627.

Köppen, A., Klein, J., Holler, T. and Löffelholz, K. (1993) Synergistic effect of nicotinamide and choline administration on extracellular choline levels in the brain. *J. Pharm. Exp. Ther.*, 266: 720–725

Krnjevic, K. (1969) Central cholinergic pathways. *Fed. Proc.*, 28: 115–120.

Krnjevic, K. (1993) Central cholinergic mechanisms and function. In: A.C. Cuello (Ed.), *Cholinergic Function and Dysfunction*, Progress in Brain Research, Vol. 98, Elsevier, Amsterdam, pp. 285–292.

Kutscher, F. (1903) Beiträge zur Kenntnis der Eiweisskörper. *Ztschft. physiol. Chem.*, 28: 111–134.

Lahoz-Beltra, R., Hameroff, S. and Dayhoff, J.E. (1993) Cytoskeletal logic: a model for molecular computation via Boellean operations in microtubules and microtubule-associated proteins. *Biosystems*, 19: 1–23.

Langley, J.N. (1905) On the reaction of cells and of nerve endings to certain poisons, chiefly as regards the reaction of striated muscle to nicotine and to curare. *J. Physiol.*, 33: 374–413.

Langley, J.N. and Dickinson, W.L. (1889) On the local paralysis of peripheral ganglia, and on the connection of different classes of nerve fibres with them. *Proc. R. Soc.*, 46: 423–431.

Lembeck, F. and Giere, W. (1968) *Otto Loewi. Ein Lebensbild in Dokumenten*. Springer-Verlag, Berlin, 241 pp.

Libet, B. (1970) Generation of slow inhibitory and excitatory potentials. *Fed. Proc.*, 29: 1945–1956.

Libet, B. (1979) Slow postsynaptic actions in ganglionic functions. In: C.M. Brooks, K. Koizumi and A. Sato (Eds.), *Integrative Functions of the Autonomic Nervous System*, University of Tokyo Press, Tokyo, pp. 199–222.

Libet, B. (1985) Unconscious cerebral initiative and the role of conscious will in voluntary action. *Behav. Brain Sci.*, 8: 529–566.

Libet, B. (1993) The neural time factor in conscious and unconscious events. In: T. Nagel (Ed.) *Experimental and Theoretical Studies of Consciousness*, CIBA Foundation Symposium No. 174, CIBA, New York, pp. 123–146.

Libet, B. (1994) A testable field theory of mind-brain interaction. *J. Consciousness Studies*, 1: 119–126.

Levi-Montalcini, R. and Angeletti, P.L. (1968) Nerve growth factor. *Physiol. Rev.*, 48: 534–569.

Loewi, O. (1902) Über Eiweiszsynthese in Thierkorper. *Arch. exp. Path. Pharmakol.*, 48: 303–330.

Loewi, O. (1918) Über Beziehungen zwischen Herzmittel- und Kationenwirkung. II. Mitteilung. Über den Zusammenhang zwischen Digitalis- und Kalziumwirkung. *Arch. exp. Path. Pharmakol.*, 82: 131–158.

Loewi, O. (1921) Über humorale Übertragbarkeit der Herznervenwirkung. I. Mitteilung. *Pflügers Arch. ges. Physiol.*, 189: 239–242.

Loewi, O. (1922) Über humorale Übertragbarkeit der Herznervenwirkung. II. Mitteilung. *Pflügers Arch. ges. Physiol.*, 193: 201–213.

Loewi, O. (1924) Über humorale Übertragbarkeit der Herznervenwirkung. III. Mitteilung. *Pflügers Arch. ges. Physiol.*, 203: 408–412.

Loewi, O. (1938) Über den Mechanismus der Stoffwechselwirkung der Hypophysen-vorderlappens. *Fiziol. Zh. SSSR*, 24: 241–144.

Loewi, O. (1952) On the action of fluoride on the heart of Rana pipiens; preliminary note. *J. Mt. Sinai Hosp.*, 19: 1–3.

Loewi, O. (1960) An autobiographic sketch. *Perspect. Biol. Med.*, 4: 1–25.

Loewi, O. and Mansfeld, G. (1911) Über den Wirkungsmodus des Physostigmins. *Arch. exp. Pathol. Pharmakol.*, 62: 180–185.

Loewi, O. and Navratil, E. (1924) Über humorale Übertragbarkeit der Herznervenwirkung. VII. Mitteilung. *Pflügers Arch. ges. Physiol.*, 206: 135–140.

Loewi, O. and Navratil, E. (1926a) Über humorale Übertragbarkeit der Herznerven-wirkung. X. Mitteilung. Über das Schicksal des Vagusstoffes. *Pflügers Arch. ges. Physiol.*, 214: 678–688.

Loewi, O. and Navratil, E. (1926b) Über humorale Übertragbarkeit des Herznerven-wirkung. XI. Mitteilung. Über den Mechanismus der Vaguswirkung von Physostigmin und Ergotamin. *Pflügers Arch. ges. Physiol.*, 214: 689–696.

Löffelholz, K. (1989) Receptor regulation of choline phospholipid hydrolysis. *Biochem. Pharmacol.*, 38: 1543–1549.

Löffelholz, K., Klein, J. and Köppen, A. (1993) Choline, a precursor of acetylcholine and phospholipids in the brain. In: A.C. Cuello (Ed.), *Cholinergic Function and Dysfunction*, Progress in Brain Research, Vol. 98, Elsevier, Amsterdam, pp. 197–200.

Long, J.P. (1963) Structure-activity relationships of the reversible anticholinesterase agents. In: G.B. Koelle (Ed.), *Cholinesterases and Anticholinesterase Agents*, Handb. exp. Pharmakol., Vol. 15, Springer-Verlag, Berlin, pp. 374–427.

Mann, T. (1948) *Joseph and His Brothers*. (English translation by H.T. Lowe-Porter). A.A. Knopf, New York, 1207 pp.

Massoulié, J., Bon, S., Duval, N., Legay, C., Krejci, E. and Coussens, F. (1993) Expression of Torpedo and rat acetyl-cholinesterase forms in transfected cells. In: A.C. Cuello (Ed.), *Cholinergic Function and Dysfunction*, Progress in Brain Research, Vol. 98, Elsevier, Amsterdam, pp. 157–153.

McCormick, D.A. (1993) Actions of acetylcholine in the cerebral cortex and thalamus and implications for function. In: A.C. Cuello (Ed.), *Cholinergic Function and Dysfunction*, Progress in Brain Research, Vol. 98, Elsevier, Amsterdam, pp. 303–308.

McGaugh, J.L. and Herz, M.J. (1972) *Memory Consolidation*. Albion Publishing Co., San Francisco, CA, 204 pp.

Michell, R.H. (1975) Inositol phospholipids and cell surface receptor function. *Biochim. Biophys. Acta*, 415: 81–147.

Nachmansohn, D. and Machado, A.L. (1943) The formation of acetylcholine. A new enzyme "choline acetylase". *J. Neurophysiol.*, 6: 397–404.

Napier, T.C., Kalivas, P.W. and Hanin, I. (Eds.) (1991) *The Basal Forebrain*. Plenum Press, New York, 489 pp.

Nishi, S., Soeda, H. and Koketsu, K. (1967) Release of acetylcholine from sympathetic preganglionic nerve terminals. *J. Neurophysiol.*, 30: 114–134.

Nishizuka, Y. (1986) Studies and perspectives of protein kinase C. *Science*, 233: 305–311.

North, R.A. (1982) Electrophysiology of the enteric neurons. In: G. Bertaccini, (Ed.), *Mediators and Drugs in Gastrointestinal Motility*, Vol. 59, Springer-Verlag, Berlin, pp. 145–179.

Parsons, S.M., Bahr, B.A., Rogers, G.A., Clarkson, E.D., Noremberg, K. and Hicks, B.W. (1993) Acetylcholine transporter - vesamicol receptor pharmacology and structure. In: A.C. Cuello (Ed.), *Cholinergic Function and Dysfunction*, Progress in Brain Research, Vol. 98, Elsevier, Amsterdam, pp. 175–181.

Patrick, J., Sequela, P., Vernino, S., Amador, M., Luetje, C. and Dani, J.A. (1993) Functional diversity of neuronal nicotinic acetylcholine receptors. In: A.C. Cuello (Ed.) *Cholinergic Function and Dysfunction*, Progress in Brain Research, Vol. 98, Elsevier, Amsterdam, pp. 113–120.

Peacock, A. (1982) The relationship between the soul and the brain. In: F.C. Rose and W.F. Bynum (Eds.), *Historical Aspects of the Neurosciences*, Raven Press, New York, pp. 83–98.

Penrose, R. (1989) *The Emperor's New Mind*. Oxford University Press, Oxford, UK, 466 pp.

Penrose, R. (1994) *Shadows of the Mind*. Oxford University Press, Oxford, UK, 457 pp.

Pepeu, G. (1993) Overview and future directions of CNS cholinergic mechanisms. In: A.C. Cuello(Ed.), *Cholinergic Function and Dysfunction*, Progress in Brain Research, Vol. 98, Elsevier , Amsterdam, pp. 455–458.

Pepeu, G. and Spignoli, G. (1989) Nootropic drugs and brain cholinergic mechanisms. *Prog. Neuro-Psychopharmacol. Biol. Psychiatry*, 13: S77–S88.

Pepeu, G., Casamenti, F., Giovannini, M.G., Vannucchi, M.G.

and Pedata, F. (1990) Principal aspects of the regulation of acetylcholine release in the brain. In: S.-M. Aquilonius and P.-G. Gillberg (Eds.), *Cholinergic Neurotransmission: Functional and Clinical Aspects*, Elsevier, Amsterdam, pp. 273–278.

Pinsky, R. (1955) *The Muse in the Machine: or, the Poetics of Zork*. Book Review Section, The New York Times, March 19.

Pollard, J.R., Arispe, N. Rojas, E. and Pollard, H.B. (1994) A geometric sequence that accurately describes allowed multiple conductance levels of ion channels: the "Three-Halves" rule. *Biophysics J.*, 67: 647–655.

Popper, K.R. and Eccles, J.R. (1977) *The Self and its Brain*. Springer-Verlag, Berlin.

Protas, L.L. and Leontieva, G.R. (1992) Ontogeny of cholinergic and adrenergic mechanisms in the frog (Rana temporaria) heart. *Am. J. Physiol.*, 262: R150–R161.

Racagni, G., Brunello, N. and Langer, S.Z. (Eds.) (1994) *Recent Advances in the Treatment of Neurodegenerative Disorders and Cognitive Dysfunction*. Karger, Basel, 264 pp.

Rasmusson, D.D. (1993) Cholinergic modulation of sensory information. In: A.C. Cuello (Ed.), *Cholinergic Function and Dysfunction*, Progress in Brain Research, Vol. 98, Elsevier, Amsterdam, pp. 357–364.

Robeson, G.A., Butcher, R.W. and Sutherland, E.W. (Eds.) (1971) *Cyclic AMP*. Academic Press, New York, 531 pp.

Rose, S. (1992) *The Making of Memory*. Anchor Books, New York, 355 pp.

Rothstein, E. (1955) *Where Melody Merges With Moonlight*. New York Times, Arts Section, June 4, 1995, p. 17.

Rylett, R.J. and Schmidt, B.M. (1993) Regulation of the synthesis of acetylcholine. In: A.C. Cuello (Ed.), *Cholinergic Function and Dysfunction*, Progress in Brain Research, Vol. 98, Elsevier, Amsterdam, pp. 161–166.

Sargent, P.B. (1993) The diversity of neuronal nicotinic acetylcholine receptors. *Annu. Rev. Neurosci.*, 16: 403–443.

Schuetze, S.M. and Role, L.W. (1987) Developmental regulation of nicotinic acetylcholine receptors. *Annu. Rev. Neurosci.*, 10: 403–457.

Schimerlik, M.I. (1990) Structure and function of muscarinic receptors. In: S.-M. Aquilonius and P.-G. Gillberg (Eds.), *Cholinergic Neurotransmission: Functional and Clinical Aspects*, Elsevier, Amsterdam, pp. 11–19.

Searle, J. (1980) Minds, brains and programs. *Behav. Brain Sci.*, 3: 417–453.

Shinnick-Gallagher, P., Gallagher, J.P. and Yoshimura, P. (1986) The pharmacology of sympathetic preganglionic neurons. In: A.G. Karczmar, K. Koketsu and S. Nishi (Eds.), *Autonomic and Enteric Ganglia*, Plenum Press, New York, pp. 409–424.

Sillito, A.M. (1993) Cholinergic neuromodulatory system: an evaluation of its functional roles. In: A.C. Cuello, (Ed.), *Cholinergic Function and Dysfunction*, Progress in Brain Research, Vol. 98, Elsevier, Amsterdam, pp. 371–378.

Skinner, B.F. (1957) *Verbal Behavior*. Appleton-Century-Crofts, New York, 478 pp.

Szutowicz, A., Bielarczyk, H. and Skulimowska, H. (1994) Effects of dichloroacetate on acetyl-CoA content and acetylcholine synthesis in rat brain synaptosomes. *Neurochem. Res.*, 19: 1107–1112.

Thesleff, S. (1955) The mode of neuromuscular block caused by acetylcholine, nicotine, decamethonium and succinylcholine. *Acta Physiol. Scand.*, 34: 218–231.

Thesleff, S. (1990) Functional aspects of quantal and nonquantal release of acetylcholine at the neuromuscular junction. In: S.-M. Aquilonius and P.-G. Gillberg (Eds.), *Cholinergic Neurotransmission: Functional and Clinical Aspects*, Elsevier, Amsterdam, pp. 93–99.

Tucek, S. (1990) The synthesis of acetylcholine: twenty years of progress. In: S.-M. Aquilonius and P.-G. Gillberg (Eds.), *Cholinergic Neurotransmission: Functional and Clinical Aspects*, Elsevier, Amsterdam, pp. 467–477.

Turek, F.W.(1994) Circadian rhythms. *Rec. Prog. Horm. Res.*, 49: 43–90.

Turing, A.M. (1937) On computable numbers, with an application to the Entscheidungsproblem. *Proc. London Math. Soc., Ser. 2*, 43: 544–546.

Usdin, E. (1970) Reactions of cholinesterases with substrates, inhibitors and reactivators. In: A.G. Karczmar (Ed.), *Anticholinesterases Agents*, Int. Encyclop. Pharmacol. Therap., Vol. 1, Pergamon Press, Oxford, pp. 47–354.

Van der Kloot, W.R. (1972) *Letter to G.L. Geison*, June 2, 1972.

Vilaro, M.T., Mengod, G. and Palacios, J.M. (1993) Advances and limitations of the molecular neuroanatomy of cholinergic receptors: the example of multiple muscarinic receptors. In: A.C. Cuello (Ed.), *Cholinergic Function and Dysfunction*, Progress in Brain Research, Vol. 98, Elsevier, Amsterdam, pp. 95- 101.

Volle, R.L. (1966) Muscarinic and nicotinic stimulant actions at autonomic ganglia. In: A.G. Karczmar (Ed.), *Ganglionic Blocking and Stimulating Agents*, Int. Encyclop.. Pharmacol. Therap., Vol. 1, Pergamon Press, Oxford, pp. 1–106.

Whittaker, V.P. (Ed.) (1988) *The Cholinergic Synapse*. Handb. exp. Pharmacol., Vol. 86, Springer-Verlag, Berlin, 762 pp.

Whittaker, V.P. (1992) *The Cholinergic Neuron and Its Target: The Electromotor Innervation of the Electric Ray "Torpedo" as a Model*. Birkhäuser, Boston, MA, 572 pp.

Woody, C.D. (1982) *Memory, Learning and Higher Function. A Cellular View*. Springer-Verlag, Berlin, 483 pp.

Wurtman, R.F. (1992) Choline metabolism as a basis for selective vulnerability of cholinergic neurons. *Trends Neurosci.*, 15: 117–122.

# Section I

# Neurobiology of Cholinergic Transmission

J. Klein and K. Löffelholz (Eds.)
*Progress in Brain Research*, Vol. 109
© 1996 Elsevier Science B.V. All rights reserved.

CHAPTER 1

# Molecular properties and cellular distribution of cholinergic synaptic proteins

## H. Zimmermann, W. Volknandt, A. Hausinger and Ch. Herrmann

*Biozentrum der J.W. Goethe-Universität, AK Neurochemie, Zoologisches Institut, Marie-Curie-Str. 9,
D-60439 Frankfurt am Main, Germany*

## Introduction

Synaptic vesicles govern the fast exocytotic release of neurotransmitters. Uptake storage and stimulus-dependent release but also axonal transport and recycling are controlled by specific proteins associated with the synaptic vesicle membrane (Zimmermann et al., 1993; Pevsner and Scheller, 1994; Südhof, 1995; Volknandt, 1995). According to our present understanding cholinergic neurons contain a set of synaptic vesicle proteins identical to that of neurons releasing other neurotransmitters. The only exception represents presumably the transporters that mediate the lumenal accumulation of neurotransmitter, the vesicular transporters for acetylcholine (Chapters 5–7; Erickson et al., 1994; Roghani et al., 1994; Varoqui et al., 1994) and ATP (Schläfer et al., 1994). At present the only source for the isolation of purely cholinergic synaptic vesicles is the electric organ of electric rays (Zimmermann, 1988). Major molecular protein constituents of cholinergic vesicles that have been identified also in synaptic vesicles of unspecified transmitter type include the proton pumping vacuolar ATPase, synapsins, synaptic vesicle protein 2 (SV2), synaptophysin, synaptotagmin, VAMP/synaptobrevin, small G-proteins of the rab family and possibly a calcium transporter (Damer and Creutz, 1994; Volknandt, 1995).

Synaptic vesicles appear to belong to a neuronal vesiculo/endosomal membrane compartment that occurs in a similar form also in endocrine cells and in a more distantly related form also in other cells (Kelly, 1993; Régnier-Vigouroux and Huttner, 1993). A major difference between neural and non-neural cells is the large distance between the site of generation of membrane compartments for the secretory pathway (the cell body) and the site of secretion (generally the axon terminal). There is increasing evidence that the mature form of the synaptic vesicle membrane compartment may not be formed at the trans-Golgi network. Rather vesicular precursor compartments may be synthesized first that are transported into the axon terminal by fast axonal transport where the bona fide synaptic vesicles are formed by additional sorting processes. After fulfilling its task in the nerve terminal and by mechanisms not yet understood the synaptic vesicle membrane compartment is sorted into the axon and returned to the cell body by fast retrograde axonal transport. In axons of the rat sciatic nerve integral membrane proteins of synaptic vesicles such as synaptophysin, are returned to the cell body to a large extent whereas surface associated proteins such as synapsin I or the small G-protein rab3A appear to be retained in the axon terminal where they may be degraded (Dahlström and Li, 1994). The mechanisms that sort the vesicular membrane compartment into the retrograde transport pathway are not yet understood. Experiments with extracellular volume markers and immunoelectron-microscopic studies suggest that vesicles fuse into a retrogradely directed endosomal membrane compartment that is represented by multivesicular bodies, vesiculo-tubular structures and lamellar bodies (Zimmermann et al.,

1989, 1993). In the axons of the *Torpedo* cholinergic electromotoneurons these membrane compartments can be labeled by an antibody directed against SV2 (Janetzko et al., 1989; Wittich et al., 1994).

### Axonal distribution of the synaptic vesicle proteins SV2 and o-rab3

The density of synaptic vesicle membrane constituents is much higher in the nerve terminals than in the axon. The synaptic vesicle membrane compartment predominates in the nerve terminal. Multivesicular bodies or vesiculotubular structures are enriched in the axon. On immunofluorescence analysis using antibodies against integral synaptic vesicle membrane proteins axonal membrane compartments reveal a dotted distribution. When the peripheral nerve is ligated immunofluorescence is accumulated proximally and distally to the ligature (Dahlström and Li, 1994). In addition, synaptic vesicle membrane proteins such as SV2 and synaptophysin where found to accumulate at consecutive nodes of Ranvier that represent natural constrictions of the axon lumen (Janetzko et al., 1989; Zimmermann and Vogt, 1989; Zimmermann, 1996).

In order to compare the axonal localization of membrane integral and membrane associated vesicle proteins we studied the distribution of the vesicle proteins SV2 and o-rab3 in the cholinergic nerve terminals of the *Torpedo* electromotor system and in the electromotor axons. SV2 is a membrane integral protein whose polypeptide chain spans the synaptic vesicle membrane 12 times and which presumably represents a transporter (Bajjalieh et al., 1992; Feany et al., 1992). In contrast, the polypeptide chain of the small G-protein o-rab3 does not traverse the synaptic vesicle membrane. Like other members of the rab3 subfamily of small GTP-binding proteins it is expected to be attached to synaptic vesicles by geranylgeranylated prenylation of its C-terminus (Volknandt et al., 1991a) and to play a role in targeting synaptic vesicles to sites of exocytosis (Fischer von Mollard et al., 1994). Electromotor axons have a diameter of about 7 $\mu$m and are myelinated. The thick elec-

tromotor nerves branch when entering the electric organs and bundles of electromotor axons proceed through the electric organ. Finer bundles approach individual columns of electroplaque cells (Whittaker and Zimmermann, 1976; Whittaker, 1992). When they enter the space between electroplaque cells they are still myelinated. Myelin is lost several micrometers before they branch intensively and make multiple synaptic contacts with the ventral surface of the electroplaque cell. When a monoclonal antibody against SV2 (Wittich et al., 1994) or a polyclonal antibody against o-rab3 (Volknandt et al., 1993) and indirect immunofluorescence are applied there is an intense labeling of the multiple synaptic contacts formed by the terminal branches of the electromotor axons that cover the ventral surface of the electroplaque cells (Fig. 1a,b,g,h). In contrast, the terminal axon segments running between electroplaque cells, whether unmyelinated or myelinated lack significant immunofluorescence suggesting that vesicular membrane compartments are cleared very efficiently from this preterminal axon compartment. This has previously escaped our attention (Volknandt and Zimmermann, 1990; Volknandt et al., 1993).

There is a clear difference between the density of vesicle protein containing membrane compartments in the preterminal axon compartment and in the axon segments more proximal to the cell soma. Axonal profiles outside the columns of electroplaque cells (Fig. 1c,d) or within the electromotor nerve (Fig. 1e,f) reveal significant and punctuate immunofluorescence for SV2. Dots are very fine in the distal axon segments and are rather coarse within the electromotor nerve. The nature of the particles is not clear from the immunofluorescence images. Our previous immunoelectron microscopic investigations suggest that they represent both anterogradely moving organelles as well as retrogradely moving organelles of the endosomal pathway such as multivesicular bodies (Janetzko et al., 1989). Immunofluorescent membrane particles are further accumulated at nodes of Ranvier that attenuate the cross sectional area of the electromotor axons by about 80% (Zimmermann and Vogt, 1989; Zimmermann, 1996). Accumulation occurs

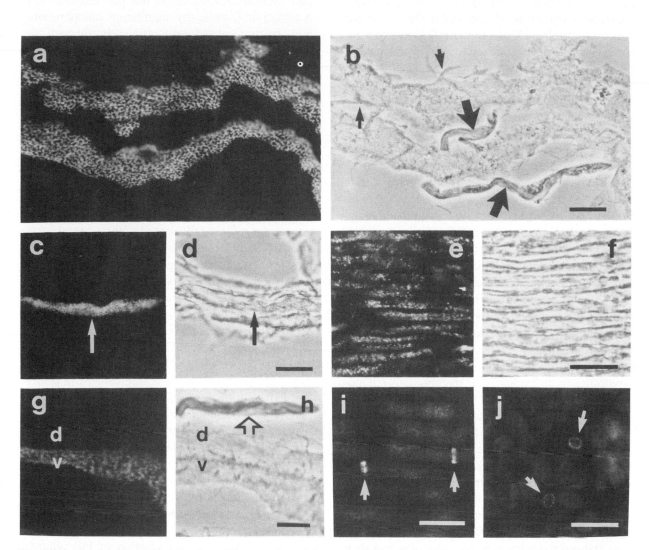

Fig. 1. Immunocytochemical localization of the synaptic vesicle proteins SV2 and o-rab3 in the axons of the *Torpedo marmorata* electromotor system. (a) Indirect immunofluorescence (Cy3-labeled second antibody) after application of the monoclonal SV2 antibody to cryostat sections of the electric organ with selective labeling of the synaptic contact area at the ventral surface of the electroplaque cells. (b) Corresponding phase contrast image revealing the preterminal ramifications of the electromotor axons. Neither large myelinated (large arrows) nor small unmyelinated axon ramifications (small arrows) are immunolabeled (cf. (a)). (c,d) Localization of SV2 (Cy3-labeled second antibody) in an electromotor axon running outside a column of electroplaque cells. In this case significant dotted immunofluorescence can be observed within the axon lumen (arrow). (e,f) Punctuate localization of SV2 (Cy3-labeled second antibody) within the electromotor nerve (longitudinal section) outside the electric organ. (g) Indirect immuofluorescence after application of the polyclonal o-rab3 antibody (FITC-labeled second antibody) to cryostat sections of the electric organ. As for SV2 there is a selective labeling of the synaptic terminals but preterminal axons (open arrow) remain unlabeled. (h) Corresponding phase contrast image. Preterminal axons (open arrow) remain unlabeled (cf. (g)). v, d, ventral and dorsal side of the electroplaque cell respectively. (i) Immunolocalization of o-rab3 (FITC-labeled second antibody) in the axons of the electromotor nerve (corresponding to (e,f)). There is weak punctuate labeling of the axoplasm and significant labeling of a band in the nodal region of nodes of Ranvier (arrows). (j) Cross-section corresponding to (i) revealing ring-like o-rab3 immunopositive nodal structures. The technical procedures are as described by Volknandt et al. (1993). Bars correspond to 25 µm.

both proximal and distal to the constriction. This can be observed in longitudinal and transversal sections of the axon. The situation differs for the small G-protein o-rab3. Immunofluorescence for o-rab3 in the electromotor axon has a much finer punctuate structure and is much weaker than that for SV2 (Fig. 1i). This suggests that o-rab3 is not associated with the bulk of structures that are labeled by the SV2 antibody. A peculiar result is the immunolabeling of a ring like structure within the nodal compartment at the nodes of Ranvier (Fig. 1i,j). The position of this structure appears to be outside the axonal lumen and corresponds to that of filamentous actin observed after labeling with fluoresceinated phalloidin (Zimmermann and Vogt, 1989, Zimmermann, 1996). Phalloidin labeling was assigned to the terminal processes of the Schwann cell lamellae surrounding the nodal axon. Since we have no reason to assume that our antibody against o-rab3 (Volknandt et al., 1993) is not specific we consider the possibility that o-rab3 is also contained in the Schwann cell lamellae or that the protein is transported from the axon lumen to the surrounding Schwann cell compartment. The possibility of an exchange of membranous components between the axon lumen and the nodal Schwann cell compartment has indeed been suggested (Gatzinsky et al., 1988, 1991).

The observation that the preterminal axon compartments of the electromotor neuron contain a very low density of immunolabeled membrane compartments (in contrast to the remainder of the axon and the synaptic contacts) could be indicative of the presence of an additional and effective axonal transport mechanism. Microtubules and their associated motors drive the bulk of fast axonal transport (Hirokawa, 1993; Brady, 1995). But vesicular organelles can also be transported along axonal actin filaments (Kuznetsov et al., 1992; Bearer et al., 1993). Actin filaments are present within the terminal axon compartment where they may serve as anchoring structures for reserve vesicles (Hirokawa et al., 1989; Benfenati and Valtorta, 1993). The recent observation that myosin II is involved in neurotransmitter release by a mechanism acting upstream exocytosis (Mochida et al., 1994) further supports the notion that the

terminal axon segment may use a membrane transport system in addition to the microtubular one. Alternatively, vesicular membrane compartments may be transported less efficiently through the bulk of the axon or the large amount of material to be transported may cause a backlog and axonal accumulation.

## Endogenous proteolysis of synaptic vesicle proteins

At present the mechanism that terminate the life cycle of a synaptic vesicle inside the nerve terminal is unknown. The elucidation of the molecular action of the clostridial neurotoxins in blocking transmitter release (Blasi et al., 1994; Montecucco and Schiavo, 1994) has provided an example of how proteolytic cleavage of identified synaptic vesicle proteins makes vesicles incapable of further participating in release. The integral synaptic vesicle protein VAMP/synaptobrevin is a substrate of the clostridial neurotoxins tetanus toxin (TeTx) and botulinum toxin B, D, F, and G whose light chains all represent $Zn^{2+}$-containing metalloproteases. These toxins cleave a fragment from the $N$-terminal and the cytoplasm-facing end of the protein. This is thought to represent the mechanism responsible for the observed blockade of synaptic transmission. Other synaptic protein substrates for clostridial neurotoxins include syntaxin, a protein associated with both the synaptic vesicle membrane and the plasma membrane and the soluble synaptic protein SNAP-25 (synaptosomal-associated protein of $M_r$ 25 kDa). All these proteins participate in the formation of the synaptic vesicle docking and fusion complex (Scheller, 1995; Südhof, 1995).

Like its relative in mammalian vesicles VAMP/synaptobrevin of the cholinergic synaptic vesicles of the *Torpedo* electric organ is a substrate of tetanus toxin and botulinum neurotoxin B and is resistant to botulinum neurotoxin A (Fig. 2, see also Herreros et al., 1995). After 60 min of incubation with tetanus toxin and botulinum neurotoxin B immunoreactivity for VAMP/synaptobrevin can no longer be detected.

Using synaptosomal fractions from rat cerebral

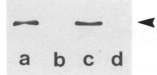

Fig. 2. Proteolytic cleavage of the synaptic vesicle protein VAMP/synaptobrevin by clostridial neurotoxins. Synaptic vesicles were isolated from the electric organ of the electric ray *Torpedo marmorata* by density gradient centrifugation as previously described (Wittich et al., 1994). Vesicles were sonified, incubated with clostridial neurotoxins (10 μg/ml) for 1 h at 37°C, and taken up in sample buffer for immunoelectrophoresis. Synaptobrevin (arrow head) remained stable in the absence of neurotoxins (a) or on addition of botulinum neurotoxin A (c). The immunopositive band at 18 kDa disappeared completely on addition of tetanus toxin (b) or botulinum toxin B (d).

cortex we investigated the question whether synaptic vesicle proteins could be substrates also of endogenous proteases (Hausinger et al., 1995). Fractions containing intact synaptosomes were incubated for up to 30 min in the presence of the calcium ionophore ionomycin and in the presence of calcium or the calcium chelators EGTA or EDTA. No hydrolysis of synaptic vesicle proteins could be observed on subsequent analysis by immunoblotting when calcium chelators were present in the medium. However, in the presence of extracellular calcium (150 μM) we observed the time dependent appearance of a proteolytic breakdown product of the synaptic vesicle proteins VAMP/synaptobrevin and synaptotagmin (Fig. 3a,b). The immunopositive proteolytic fragment derived from VAMP/synaptobrevin with a molecular mass of 8 kDa was detectable within 10 min of incubation with ionomycin and calcium and was further enhanced after 15 min (Fig. 3a). This time course is comparable to the appearance of hydrolysis products on addition of clostridial neurotoxins. Longer periods of incubation did not result in a further degradation of the synaptosomal pool of VAMP/synaptobrevin. The molecular mass of the endogenous proteolysis product is in the size range of the cleavage fragments obtained after addition of clostridial neurotoxins. Whereas on addition of clostridial neurotoxins to a fraction of synaptic vesicles synaptobrevin is completely degraded (cf.

Fig. 2) the fragment generated by the endogenous protease appears to be stable.

The synaptic vesicle protein synaptotagmin, a further constituent of the synaptic vesicle docking complex, may act as a calcium sensor in triggering exocytosis (Südhof, 1995). It is not a reported substrate of clostridial neurotoxins. In fractions of freshly prepared and untreated synaptosomes the antibody against synaptotagmin recognized the expected band at 65 kDa as well as a presumptive proteolytic breakdown product of 44 kDa that is generally observed in standard preparations of subcellular fractions (Fig. 3b). In the presence of ionomycin and calcium a further immunopositive band at 37 kDa appeared within the same time course as the proteolysis product derived from VAMP/synaptobrevin. When tested under the same experimental conditions other synaptic vesicle proteins such as synapsin I, synaptophysin, the membrane associated syntaxin, the cytoskeletal proteins tubulin and actin or lactate dehydrogenase revealed no immunopositive degradation products.

The specificity of the endogenous proteolysis was further defined in an additional series of ex-

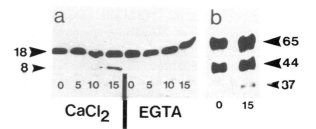

Fig. 3. Time- and calcium-dependence of the endogenous synaptosomal proteolysis of the synaptic vesicle protein VAMP/synaptobrevin (a) or synaptotagmin (b). A crude synaptosomal fraction (fraction $P_2$ according to Whittaker et al., 1964) was isolated in the presence of the protease inhibitor phenanthroline (20 mM) and incubated for up to 15 min in sucrose solution (Hausinger et al., 1995) in the presence of the calcium ionophore ionomycin (30 μM) and $CaCl_2$ (150 μM) or EGTA (3 mM) followed by immunoelectrophoresis. (a) VAMP/Synaptobrevin: Only in the presence of calcium ions a immunopositive proteolytic product of VAMP/synaptobrevin (8 kDa) appears. (b) Synaptotagmin: after 15 min and in the presence of calcium ions an immunopositive band of 37 kDa makes its appearance in addition to the two bands at 65 kDa and 44 kDa normally recognized by the antibody.

36

periments. Synaptosomes were hypoosmotically
shocked and incubated for 16 h at room temperature in the presence of various inhibitors. Incubation of the osmotically shocked synaptosomal
fraction in the presence of protease inhibitors such
as E-64, PMSF, pepstatin, or captopril, phenanthroline and phosphoramidon (potential inhibitors
of $Zn^{2+}$-containing proteases) could not prevent
proteolysis of VAMP/synaptobrevin (Fig. 4a).
Phenanthroline and phosphoramidon could, however, attenuate the proteolysis. In contrast, addition
of EGTA, EDTA or $Zn^{2+}$ completely prevented the
hydrolysis of the vesicle protein. When the synaptosomal fraction was first prepared in the presence
of EDTA and subsequently incubated in the presence of $CaCl_2$ hydrolysis occurred (Fig. 4b). This
suggests that the endogenous enzyme is not a
$Zn^{2+}$-containing but rather a calcium-dependent
protease. Addition of a heptapeptide (AA 74–80:
ASQFETS, 1 mM, Alomone Labs, Jerusalem, Israel) that contains the tetanus toxin cleavage site
Gln76-Phe77 of synaptobrevin 2 to osmotically
shocked synaptosomes (16 h room temperature)
did not prevent the formation of the 8 kDa peptide.
Thus, the endogenous proteolysis occurs at a
cleavage site different from that of tetanus toxin.

Our experiments imply that the production of
cleavage products of VAMP/synaptobrevin and
synaptotagmin depend on the entry of $Ca^{2+}$ into the
synaptosomal compartment. Only a relatively
small pool of the two vesicle proteins is accessible
to the endogenous protease. Most of the vesicle
protein appears to be protected from the protease
and becomes accessible only in a specific molecular environment or at a specific stage of the synaptic vesicle life cycle. In this connection it is of interest that formation of the fusion complex appears
to prevent hydrolysis of VAMP/synaptobrevin by
tetanus toxin (Pellegrini et al., 1994). $Ca^{2+}$-
dependent cleavage of synaptic vesicle proteins
may represent a physiological mechanism that
controls the life cycle of the synaptic vesicle. Since
the $Ca^{2+}$-signal is short lived under conditions of
moderate synaptic transmission activation of the
protease would be very low and transient. Prolonged synaptic activation leads to increased presynaptic $Ca^{2+}$-levels. It has long been known that

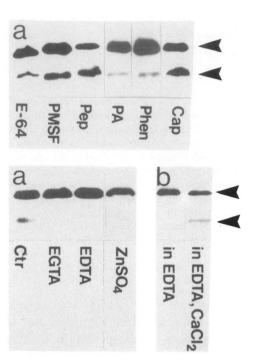

Fig. 4. Studies on the inhibition of the endogenous synaptosomal proteolysis of VAMP/synaptobrevin by immunoblot
analysis. (a) Synaptosomal fractions were prepared as for Fig.
3 except that no protease inhibitor was present during the
preparation and that the fraction was hypoosmotically
shocked prior to application of substances (Hausinger et al.,
1995). Lysed synaptosomes were incubated at room temperature for 16 h in the absence (control, Ctr) or in the presence of
the protease inhibitors E-64 (3 mg/ml), PMSF (1 mM), pepstatin (Pep, 4 $\mu$g/ml), phosphoramidon (PA, 100 $\mu$M) or captopril (Cap, 1 mM) or of the calcium chelators EGTA (1 mM)
or EDTA (1 mM) or of $ZnSO_4$ (50 $\mu$M). Only the calcium
chelators and $ZnSO_4$ inhibit the formation of the proteolysis
product at 8 kDa. (b) In an additional series of experiments
synaptosomes were prepared in the presence of EDTA
(1 mM) and after osmotic lysis incubated for 16 h either in the
absence (in EDTA) or presence of $CaCl_2$ (3 mM) (in EDTA,
$CaCl_2$). Readdition of calcium ions overcomes the inhibitory
effect of EDTA and causes the appearance of the 8 kDa band.

high frequency stimulation causes a rapid decre-
ment in transmitter release, loss of synaptic vesicle
numbers and the formation of vacuoles and cisternae (Zimmermann, 1979). The $Ca^{2+}$-dependent
truncation of specific synaptic vesicle proteins
would prevent the immediate reuse of the vesicle
for exocytosis and could act as a signal terminating
its life cycle inside the presynaptic compartment.

## The major vault protein MVP100, a novel constituent of cholinergic electromotor neurons

Cholinergic synaptic vesicles from the electric organ of the electric ray *Torpedo* are routinely isolated by sucrose density gradient centrifugation followed by chromatography on a sizing column (Bonzelius and Zimmermann, 1990). When analyzing the protein composition of the synaptic vesicle enriched fractions by SDS polyacrylamide gel electrophoresis we noticed the presence of a protein with the apparent $M_r$ of 100 000. It is associated with synaptic vesicle containing fractions on sucrose density gradient centrifugation (step gradient or continuous gradients). But it separates from the bulk of synaptic vesicles on subsequent column chromatography using Sephacryl 1000. Material containing the 100 kDa protein is delayed and its elution pattern only partially overlaps with that of synaptic vesicles. However, the 100 kDa protein elutes well before the salt peak. Like synaptic vesicles the material containing the 100 kDa protein can be sedimented by high speed centrifugation suggesting that it is contained in a particle. When a synaptic vesicle fraction isolated by density gradient centrifugation is subjected to glycerol velocity gradient centrifugation a considerable amount of the 100 kDa protein is contained within the synaptic vesicle fraction but the remainder sediments with lighter fractions.

By direct microsequencing of the 100 kDa protein derived from the glycerol gradient six stretches of amino acid residues were obtained. Polyclonal antibodies were generated from the intact protein and from one of the peptides derived. Furthermore an oligonucleotide derived from one of the sequences was used as a probe for screening a cDNA library constructed using mRNA derived from the electric lobe of the electric ray *Discopyge ommata*, a species related to *Torpedo marmorata* (Volknandt et al., 1991b). A full length clone derived had a short 5′-untranslated region without a stop codon and an open reading frame encoding 852 amino acids (Herrmann et al., 1996). This is followed by a short untranslated flanking region containing a polyadenylation signal and the poly-A tail. A computer based search revealed a high homology of the 100 kDa protein of the electric ray with cDNAs coding for a protein of similar molecular mass in the lower eucaryote *Dictyostelium discoideum* and the rat (Rome et al., 1991; Kickhoefer and Rome, 1994; Vasu and Rome, 1995). Previous investigations on this protein isolated from a variety of species (rat, cow, chicken, the bullfrog *Rana catesbeiana*, and the South African clawed frog *Xenopus laevis*) had revealed that it is associated with a particle. Interestingly, the rat protein of 104 kDa was first identified in subpopulations of clathrin coated vesicles using a preparative agarose gel electrophoresis techniques. Refined isolation procedures allowed to separate the 104 kDa containing material from the coated vesicles (Kedersha and Rome, 1986a). The protein was found to be associated with ovoid particles of about 35 × 65 nm named "vaults". The term was chosen because negative staining revealed a particle morphology reminiscent of the multiple arches that form cathedral vaults (Kedersha and Rome, 1996b; Rome et al., 1991). Vaults were found to be highly polymeric structures containing more than one protein. In the rat proteins of 210, 192, 104, 54 kDa were identified. The 104 kDa is the major constituent of the particle accounting for 75% of the particle mass. It is thus referred to as the major vault protein (MVP). Accordingly we name the homologous protein from the electric ray MVP100.

Vaults from rat and bullfrog were found to contain a unique small RNA contributing to about 5% by weight and thus represent a novel type of ribonucleoprotein particle (Kickhoefer et al., 1993). At present the function of vaults is unknown. By immunofluorescence evidence for a partial association with the nuclear pore complex has been derived (Chugani et al., 1993). Disruption of major vault proteins in *Dictyostelium* reveals a mild growth defect under conditions of nutritional stress (Vasu and Rome, 1995).

The cDNA of the electric ray MVP100 reveals several phosphorylation sites which are motifs for protein kinase C, casein kinase II and tyrosine protein kinase (Fig. 5). In fractions of isolated synaptic vesicles MVP100 becomes phosphorylated in the presence of divalent cations without addition of exogenous protein kinase. Secondary structure

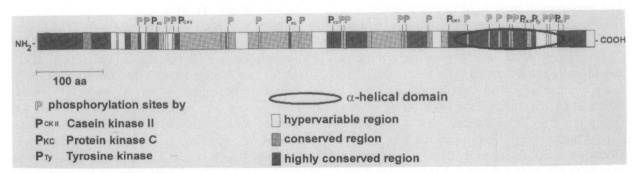

Fig. 5. Domain structure of MVP100 from electric ray. The primary structure of the protein (852 amino acid residues, aa) is compared by alignment with its mammalian homologue in rat (Kickhoefer and Rome, 1994) and two isoforms of the slime mold *Dictyostelium discoideum* (Vasu and Rome, 1995). The overall amino acid identity between the proteins from different species of 60% indicates a high degree of phylogenetic conservation. Dark areas depict stretches of amino acid residues with more than 80% amino acid identity. Conserved regions (more than 60% identity) are in gray and stretches with completely differing amino acids or where gaps had to be introduced (hypervariable regions) are in white. Phosphorylation sites for several types of protein kinases are indicated. Those that are evolutionary conserved are in bold. Secondary structure analysis predicts a long $\alpha$-helical domain near the C-terminus of all major vault proteins although this is only partially reflected by high conservation at the amino acid level.

analysis predicts a very long (ca. 150 amino acids) $\alpha$-helical domain towards the C-terminus. This suggests that the protein contains an elongated tail structure. An immunocytochemical analysis using the colloidal gold method confirms the high contents of MVP100 in the cholinergic nerve terminals. Colloidal gold particles are often found in close association with synaptic vesicles and labeling is absent from plasma membranes and mitochondria. At present it is not clear whether the apparent association with synaptic vesicles is due to the close packing of the two organellar structures or a direct structural linkage.

Our results suggest that the cholinergic electromotoneurons contain large and stable proteinaceous particles with MVP100 as their major constituent. Nerve terminals are rich not only in synaptic vesicles but also in these particles. The functional role of the particles needs further elucidation.

## Summary and outlook

Cholinergic synaptic vesicles share their major protein constituents with synaptic vesicles from non-specified transmitter type. The origin of the cholinergic synaptic vesicle membrane compartment and its recycling can be studied in the electromotor axon of the electric ray. Immunocytochemical evidence suggests that the synaptic vesicle integral membrane proteins SV2 and the vesicle associated small GTP-binding protein o-rab3 are associated with particulate yet differing organelles within the axon. As compared to the remainder of the axon the preterminal axon compartment appears to be cleared very effectively of vesicular membrane constituents. Calcium influx into synaptosomes is associated with the endogenous proteolysis of a pool of the synaptic vesicle proteins VAMP/synaptobrevin and synaptotagmin. Although the physiological function of this hydrolysis is not clear it may play a role in inactivating the synaptic vesicle within the presynaptic compartment. In addition to synaptic vesicles cholinergic nerve terminal have a high content in proteinaceous particles that contain the major vault protein MVP100 as their major constituent and that may be physically linked to vesicles. The presence of the MVP containing protein particle in the nerve terminal implies new and yet unrecognized functional properties of this highly specialized compartment of the nerve cell.

## Acknowledgements

The present study was supported by a grant from the Deutsche Forschungsgemeinschaft (SFB 169/A10) and by a grant to A.H. from the Boehringer Ingelheim Stiftung.

# References

Bajjalieh, S.M., Peterson, K., Shinghal, R. and Scheller, R.H. (1992) SV2, a brain synaptic vesicle protein homologous to bacterial transporters. *Science*, 257: 1271–1273.

Bearer, E.L., De Giorgis, J.A., Bodner, R.A., Kao, A.W. and Reese, T.S. (1993) Evidence for myosin motors on organelles in squid axoplasm. *Proc. Natl. Acad. Sci. USA*, 90: 11252–11256.

Benfenati, F. and Valtorta, F. (1993) Synapsins and synaptic transmission. *News Physiol. Sci.*, 8: 18–23.

Blasi, J., Binz, T., Yamasaki, S., Link, E., Niemann, H. and Jahn, R. (1994) Inhibition of neurotransmitter release by clostridial neurotoxins correlates with specific proteolysis of synaptosomal proteins. *J. Physiol. Paris*, 88: 235–241.

Bonzelius, F. and Zimmermann, H. (1990) Recycled synaptic-vesicles contain vesicle but not plasma membrane marker, newly synthesized acetylcholine, and a sample of extracellular medium. *J. Neurochem.*, 55: 1266–1273.

Brady, S.T. (1995) A kinesin medley: biochemical and functional heterogeneity. *Trends Cell Biol.*, 5: 159–164.

Chugani, D.C., Rome, L.H. and Kedersha, N.L. (1993) Evidence that vault ribonucleoprotein particles localize to the nuclear pore complex. *J. Cell Sci.*, 106: 23–29.

Dahlström, A.B. and Li, J.Y. (1994) Fast and slow axonal transport-different methodological approaches give complementary information: contributions of the stop-flow/crush approach. *Neurochem. Res.*, 19: 1413–1419.

Damer, C.K. and Creutz, C.E. (1994) Secretory and synaptic vesicle membrane proteins and their possible roles in regulated exocytosis. *Prog. Neurobiol.*, 43: 511–536.

Erickson, J.D., Varoqui, H., Schäfer, M.K.H., Modi, W., Diebler, M.F., Weihe, E., Rand, J., Eiden, L.E., Bonner, T.I. and Usdin, T.B. (1994) Functional identification of a vesicular acetylcholine transporter and its expression from a "cholinergic" gene locus. *J. Biol. Chem.*, 269: 21929–21932.

Feany, M.B., Lee, S., Edwards, R.H. and Buckley, K.M. (1992) The synaptic vesicle protein SV2 is a novel type of transmembrane transporter. *Cell*, 70: 861–867.

Fischer von Mollard, G., Stahl, B., Li, C., Südhof, T.C. and Jahn, R. (1994) Rab proteins in regulated exocytosis. *Trends Biochem. Sci.*, 19: 164–168.

Gatzinsky, K.P., Berthold, C.-H. and Corneliuson, O. (1988) Acid phosphatase activity at nodes of Ranvier in the alpha-motor and dorsal root ganglion neurons of the cat. *J. Neurocytol.*, 17: 531–544.

Gatzinsky, K.P., Berthold, C.-H. and Rydmark, M. (1991) Axon-Schwann cell networks are regular components of nodal regions in normal large nerve fibres of cat spinal roots. *Neurosci. Lett.*, 124: 264–268.

Hausinger, A., Volknandt, W. and Zimmermann, H. (1995) Calcium-dependent endogenous proteolysis of the vesicle proteins synaptobrevin and synaptotagmin. *NeuroReport*, 6: 637–641.

Herreros, J., Blasi, J., Arribas, M. and Marsal, J. (1995) Teta-nus toxin mechanism of action in *Torpedo* electromotor system: a study on different steps in the intoxication process. *Neuroscience*, 65: 305–311.

Herrmann, C., Volknandt, W., Wittich, B., Kellner, R. and Zimmermann, H. (1996) The major vault protein (MVP100) is contained in cholinergic nerve terminals of electric ray electric organ. *J. Biol. Chem.*, 271 (1996) 13908–13915.

Hirokawa, N. (1993) Mechanism of axonal transport – identification of new molecular motors and regulations of transports. *Neurosci. Res.*, 18: 1–9.

Hirokawa, N., Sobue, K., Kanda, K., Harada, A. and Yorifuji, H. (1989) The cytoskeletal architecture of the presynaptic terminal and molecular structure of synapsin 1. *J. Cell Biol.*, 108: 111–126.

Janetzko, A., Zimmermann, H. and Volknandt, W. (1989) Intraneuronal distribution of a synaptic vesicle membrane protein: antibody binding sites at axonal membrane compartments and *trans*-Golgi network and accumulation at nodes of Ranvier. *Neuroscience*, 32: 65–77.

Kedersha, N.L. and Rome, L.H. (1986a) Preparative agarose gel electrophoresis for the purification of small organelles and particles. *Anal. Biochem.*, 156: 161–170.

Kedersha, N.L. and Rome, L.H. (1986b) Isolation and characterization of a novel ribonucleoprotein particle: large structures contain a single species of small RNA. *J. Cell Biol.*, 103: 699–709.

Kelly, R.B. (1993) Storage and release of neurotransmitters. *Cell*, 72: 43–53.

Kickhoefer, V.A. and Rome, L.H. (1994) The sequence of a cDNA encoding the major vault protein from *Rattus norvegicus*. *Gene*, 151: 257–260.

Kickhoefer, V.A., Searles, R.P., Kedersha, N.L., Garber, M.E., Johnson, D.L. and Rome, L.H. (1993) Vault ribonucleoprotein particles from rat and bullfrog contain a related small RNA that is transcribed by RNA polymerase III. *J. Biol. Chem.*, 268: 7868–7873.

Kuznetsov, S.A., Langford, G.M. and Weiss, D.G. (1992) Actin-dependent organelle movement in squid axoplasm. *Nature*, 356: 722–725.

Mochida, S., Kobayashi, H., Matsuda, Y., Yuda, Y., Muramoto, K. and Nonomura, Y. (1994) Myosin II is involved in transmitter release at synapses formed between rat sympathetic neurons in culture. *Neuron*, 13: 1131–1142.

Montecucco, C. and Schiavo, G. (1994) Mechanism of action of tetanus and botulinum neurotoxins. *Mol. Microbiol.*, 13: 1–8.

Pellegrini, L.L., O'Connor, V. and Betz, H. (1994) Fusion complex formation protects synaptobrevin against proteolysis by tetanus toxin light chain. *FEBS Lett.*, 353: 319–323.

Pevsner, J. and Scheller, R.H. (1994) Mechanisms of vesicle docking and fusion: Insights from the nervous system. *Curr. Opin. Cell Biol.*, 6: 555–560.

Régnier-Vigouroux, A. and Huttner, W.B. (1993) Biogenesis of small synaptic vesicles and synaptic-like microvesicles. *Neurochem. Res.*, 18: 59–64.

Roghani, A., Feldman, J., Kohan, S.A., Shirzadi, A., Gundersen, C.B., Brecha, N. and Edwards, R.H. (1994) Molecular cloning of a putative vesicular transporter for acetylcholine. *Proc. Natl. Acad. Sci. USA*, 91: 10620–10624.

Rome, L., Kedersha, N. and Chugani, D. (1991) Unlocking vaults: organelles in search of a function. *Trends Cell Biol.*, 1: 47–50.

Scheller, R.H. (1995) Membrane trafficking in the presynaptic nerve terminal. *Neuron*, 14: 893–897.

Schläfer, M., Volknandt, W. and Zimmermann, H. (1994) Putative synaptic vesicle nucleotide transporter identified as glyceraldehyde-3-phosphate dehydrogenase. *J. Neurochem.*, 63: 1924–1931.

Südhof, T.C. (1995) The synaptic vesicle cycle: a cascade of protein-protein interactions. *Nature*, 375: 645–653.

Varoqui, H., Diebler, M.F., Meunier, F.M., Rand, J.B., Usdin, T.B., Bonner, T.I., Eiden, L.E. and Erickson, J.D. (1994) Cloning and expression of the vesamicol binding protein from the marine ray *Torpedo* – homology with the putative vesicular acetylcholine transporter unc-17 from *Caenorhabditis elegans*. *FEBS Lett.*, 342: 97–102.

Vasu, S.K. and Rome, L.H. (1995) *Dictyostelium* vaults: disruption of the major proteins reveals growth and morphological defects and uncovers a new associated protein. *J. Biol. Chem.*, 270: 16588–16594.

Volknandt, W. (1995) The synaptic vesicle and its targets. *Neuroscience*, 64: 277–300.

Volknandt, W. and Zimmermann, H. (1990) Identical properties of transmembrane synaptic vesicle protein $M_r$ 100,000 in *Torpedo* and $M_r$ 86,000 in bovine brain. *Neurochem. Int.*, 4: 539–547.

Volknandt, W., Pevsner, J., Elferink, L.A., Schilling, J. and Scheller, R.H. (1991a) A synaptic vesicle specific GTP-binding protein from ray electric organ. *Mol. Brain Res.*, 11: 283–290.

Volknandt, W., Vogel, M., Pevsner, J., Misumi, Y., Ikehara, Y. and Zimmermann, H. (1991b) 5′-Nucleotidase from the electric ray electric lobe; primary structure and relation to mammalian and procaryotic enzymes. *Eur. J. Biochem.*, 202: 855–861.

Volknandt, W., Hausinger, A., Wittich, B. and Zimmermann, H. (1993) The synaptic vesicle associated G-protein o-rab3 is expressed in subpopulations of neurons. *J. Neurochem.*, 60: 851–857.

Whittaker, V.P. (1992) *The Cholinergic Neuron and its Target: The Electromotor Innervation of the Electric Ray Torpedo as a Model*, Birkhäuser, Boston, MA.

Whittaker, V.P. and Zimmermann, H. (1976) The innervation of the electric organ of *Torpedinidae*: a model cholinergic system. In D.C. Malins and J.R. Sargent (Eds.), *Biochemical and Biophysical Perspectives of Marine Biology. 3*, Academic Press, London, pp. 67–116.

Whittaker, V.P., Michaelson, I.A. and Kirkland, R.J.A. (1964) The separation of synaptic vesicles from nerve-ending particles ('synaptosomes'). *Biochem. J.*, 90: 293–303.

Wittich, B., Volknandt, W. and Zimmermann, H. (1994) SV2 and o-rab3 remain associated with recycling synaptic vesicles. *J. Neurochem.*, 63: 927–937.

Zimmermann, H. (1979) Vesicle recycling and transmitter release. *Neuroscience*, 4: 1773–1804.

Zimmermann, H. (1988) Cholinergic synaptic vesicles. *Handb. Exp. Pharmacol.*, 86: 349–382.

Zimmermann, H. (1996) Accumulation of synaptic vesicle proteins and cytoskeletal specializations at the peripheral node of Ranvier. *Microsc. Res. Techn.*, 34: 462–473.

Zimmermann, H. and Vogt, M. (1989) Membrane proteins of synaptic vesicles and cytoskeletal specializations at the node of Ranvier in electric ray and rat. *Cell Tissue Res.*, 258: 617–629.

Zimmermann, H., Volknandt, W., Henkel, A., Bonzelius, F., Janetzko, A. and Kanaseki, T. (1989) The synaptic vesicle membrane: origin, distribution, protein components, exocytosis and recycling. *Cell Biol. Int. Rep.*, 13: 993–1006.

Zimmermann, H., Volknandt, W., Wittich, B., and Hausinger, A. (1993) Synaptic vesicle life cycle and synaptic turnover. *J. Physiol. Paris*, 87: 159–170.

J. Klein and K. Löffelholz (Eds.)
*Progress in Brain Research*, Vol. 109
© 1996 Elsevier Science B.V. All rights reserved.

CHAPTER 2

# Redistribution of clathrin and synaptophysin at the frog neuromuscular junction triggered by nerve stimulation: immunocytochemical studies of vesicle trafficking

Andreas W. Henkel and William J. Betz

*Department of Physiology, Box C240, University of Colorado Health Sciences Center, 4200 East Ninth Avenue, Denver, CO 80262, USA*

## Introduction

Nerve terminals release neurotransmitter into the synaptic cleft by fusion of synaptic vesicles with the plasma membrane. The vesicle membrane is retrieved by endocytosis without intermixing of vesicle and plasma membrane proteins (Chapter 1; Bonzelius and Zimmermann, 1990; Sudhof and Jahn, 1991). This process is triggered by depolarization of the synaptic plasma membrane and the subsequent influx of calcium ions (Katz and Miledi, 1968). For a short period of time the vesicles become incorporated into the synaptic plasma membrane and are retrieved subsequently via coated vesicles, including most or probably all synaptic vesicle proteins (Heuser, 1989; Smythe and Warren, 1991; Maycox et al., 1992). The mechanism of membrane retrieval by coated vesicles has been intensively studied in cells that use receptor mediated endocytosis for the uptake of receptor-bound proteins or hormones. Soluble clathrin molecules aggregate in the vicinity of the target receptor and form a coated pit. This process also depends on the presence of G-proteins and adaptor proteins (Carter et al., 1993; van der Bliek et al., 1993). Then the pit buds off and produces a coated vesicle. Besides this model, other exo- endocytotic mechanisms have been proposed. One important alternative hypothesis, often referred to as the "kiss and run" model, is the formation of a fusion pore that connects the lumen of a vesicle with the extracellular space and allows the release of transmitter (Ceccarelli and Hurlbut, 1980). The existence of such transient connections has been shown by capacitance measurements in several cell types, including mast cells and chromaffin cells. These fusion pores may consist of proteins, which form a bottle neck between vesicle and plasma membrane (Almers and Tse, 1990), or they may be composed of lipids (Monck and Fernandez, 1992; Nanavati et al., 1992). It has been hypothesized that both mechanisms may be acting at nerve terminals, depending on the rate of stimulation (Fesce et al., 1994). Low frequency stimulation favors the 'kiss and run' model but higher frequency leads to the complete fusion of vesicles with the plasma membrane. The situation at the neuromuscular junction appears to be more complicated because electron micrographs show that coated vesicles are not very abundant even after intense stimulation, when large vacuoles appear (Miller and Heuser, 1984; Torri-Tarelli et al., 1987).

In order to study the dynamics of synaptic vesicles during nerve stimulation, we used antibodies against clathrin, the major protein of coated vesicles, and synaptophysin, an integral membrane protein of synaptic vesicles, to monitor the distribution of synaptic vesicles and clathrin.

## Transient relocation of synaptic vesicles during nerve stimulation

The frog (*Rana pipiens*) neuromuscular junction of

the *Cutaneous pectoris* (Cp) muscle is an elongated (up to 500 $\mu$m length) , multi-branched synapse, and synaptic vesicles are organized in distinct clusters with average diameters between 1 and 2 $\mu$m. These anatomical properties enable light- and fluorescence microscopy observations in order to monitor vesicle translocations inside the terminal. Changes in vesicle distribution can be triggered by the exocytotic agent $\alpha$-latrotoxin, the active component in black widow spider venom. The toxin elicits synaptic vesicle exoytosis independently of extracellular calcium ions but blocks endocytosis only in the absence of calcium (Ceccarelli and Hurlbut, 1980). Exposure of neuromuscular preparations to $\alpha$-latrotoxin without extracellular calcium leads to a permanent incorporation of vesicle proteins into the plasma membrane, indicating a selective block of endocytosis. This has been demonstrated by immunostaining with a polyclonal antibody, raised against frog, synaptophysin (Valtorta et al., 1988).

We used the same antibody (a gift from Dr. M. Browning, UCHSC) to study the relocation of synaptic vesicles after intensive nerve stimulation. In Western blots, the antibody labeled frog synaptophysin (apparent MW = 39 kDa), derived from brain, with a high specificity, but bound only weakly to rat (brain) and bovine (chromaffin cell) synaptophysin (MW = 38 kDa). Synaptophysin was not labeled in chick brain samples (Fig. 1). Immunostaining of Triton X-100 permeabilized neuromuscular preparations revealed a characteristic punctate staining pattern under resting conditions (Fig. 2A). Each spot represented a cluster of synaptic vesicles. When the preparation was electrically stimulated via the nerve at a frequency of 20 Hz for 12 min, and immediately fixed, the anti-synaptophysin staining was no longer arranged in bright spots but it was evenly spread over the whole terminal (Fig. 2B). This observation could be interpreted as an incorporation of synaptic vesicles in the synaptic plasma membrane although we could not rule out by light microscopy that vesicles are dispersed within the nerve terminal.

The translocation of synaptic vesicles appeared to be totally reversible. Stimulation of preparations under the same conditions resulted in a very dif-

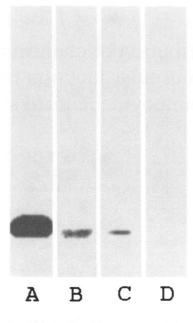

Fig. 1. Antibody specificity for anti-frog synaptophysin. All samples (A–D) were homogenized and centrifuged for 10 min at 1000 × $g$. The supernatants were mixed with the same volume of 2× sample buffer, containing mercaptoethanol. The protein concentration was determined as described by Henkel and Bieger (1994). Five micrograms of protein were separated by SDS-gel electrophoresis (10% gels), blotted onto nitrocellulose and probed with the antibody. The detection system used was a biotinylated secondary antibody, avidine conjugated horseradish peroxidase and enhanced chemoluminescence (ECL, Amersham). (A) Frog brain, the band labeled is 39 kDa; (B) rat brain, 38 kDa; (C) bovine chromaffin cells, 38 kDa. (D) chick brain, not labeled.

ferent staining pattern, when the muscle was given some time for recovery. If the preparations were fixed 45 min after the end of the stimulation, the observed immunostaining pattern were virtually identical to these of a resting, unstimulated terminal (Fig. 2C). The most likely explanation for this result is that the vesicles were recycled from the plasma membrane by endocytosis and rearranged in clusters again.

### Ultrastructural effects of nerve stimulation

In order to answer the question, if the homogenous distribution of vesicles immediately after stimula-

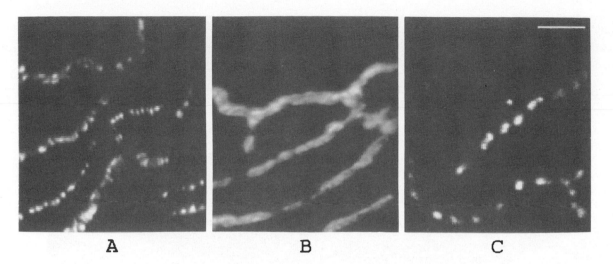

A      B      C

Fig. 2. Immunofluorescence of frog neuromuscular junctions with anti frog synaptophysin. The preparations were fixed with 4% paraformaldehyde in phosphate buffer, permeabilized with 0.25% Triton X-100 and incubated overnight with one antibody (concentration: 1:500) at 4°C. The detection system was a biotinylated secondary antibody and a avidine-Cy3-fluorophore conjugate (Vector). (A) A unstimulated control preparation shows a cluster-like staining pattern. (B) A stimulated (12 min at 20 Hz) and immediately fixed preparation shows a homogenous staining of the nerve terminal membrane. (C) The preparation was stimulated, rested 45 min and was then fixed. The cluster-like staining pattern become visible again. Scale bar = 5 $\mu$m.

tion (Fig. 2B) was actually due to their incorporation in the synaptic plasma membrane or because of a spreading inside the nerve terminal, frog neuromuscular preparation were stimulated for 12 min at 20 Hz, immediately fixed, and processed for electron microscopy. Cross sections through these synapses showed a tremendous reduction in vesicle number inside the nerve terminals. This confirmed early observations that vesicles remain incorporated into the synaptic plasma membrane for a transient period of time (Heuser and Reese, 1973), and are not just dispersed inside the nerve terminal upon nerve stimulation. Furthermore, several endocytotic vacuoles became visible (Fig. 3). Therefore, the primary mechanism for endocytosis under intense stimulation (12 min at 20 Hz) appeared to be the retrieval of synaptic vesicle membrane by endocytotic vacuoles (Miller and Heuser, 1984). These observations led to the conclusion that the "kiss and run" mechanism is not active under these stimulation conditions. Synaptic vesicles completely collapse into the plasma membrane, and become endocytosed, creating endocytotic vacuoles.

## Clathrin at the neuromuscular junction

One proposed mechanism for recycling of fused synaptic vesicles involves clathrin coated vesicles, similar to receptor mediated endocytosis of specific receptor-bound ligands. The first step of this process involves the binding of adaptor proteins (AP-2, AP 180, auxillin and adaptins) to the cytoplasmic domain of the synaptic vesicle protein synaptotagmin (Zhang et al., 1994). Soluble clathrin molecules bind to adaptor proteins, and form coated pits, consisting of triscelion-shaped clathrin aggregates. Then a coated vesicle buds of the coated pit. Finally, an uncoating ATPase (Schlossman et al., 1984) strips the vesicle of its clathrin coat, and the coat then disintegrates into single, soluble clathrin molecules.

If there is a big fluctuation of the clathrin distribution during exo- and endocytosis cycles, it should be possible to monitor the aggregation and disaggregation of clathrin molecules by immunocytochemical techniques.

We used a monoclonal anti-clathrin antibody that labeled frog clathrin with a high affinity.

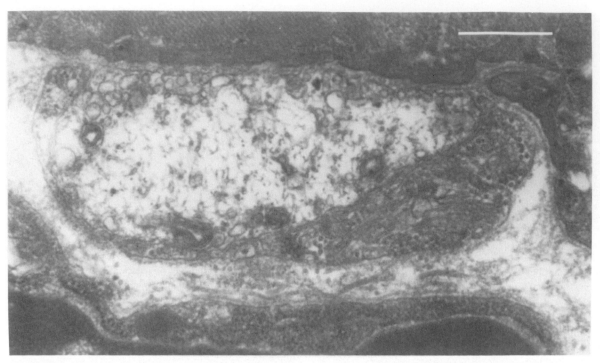

Fig. 3. Ultrastructural effects of nerve stimulation. Frog motor nerve terminals were stimulated (12 min at 20 Hz), immediately fixed (2% glutaraldehyde, 2% paraformaldehyde, 100 mM sodium phosphate buffer, pH 7.2) for 1 h at 21°C and examined with an electron microscope. This cross section through a nerve terminal appears to be depleted of most synaptic vesicles and some endocytotic vacuoles are visible. Scale bar = 0.5 µm.

Samples of frog brain (Fig. 4A1), frog nerve (Fig. 4A2), synapse rich region of frog Cp-muscle (Fig. 4A3), and rat brain (Fig. 4A4) were probed with the antibody (0.6 µg/ml) in a Western blot. Clathrin was detected in rat and frog brain, and to a much lesser extent in frog Cp, but it was absent in nerve.

Then we did immunofluorescence on Triton X-100 permeabilized, unstimulated neuromuscular junctions. The staining pattern resembled the distribution of vesicle clusters (Fig. 4B, top picture). However, some terminals showed a more diffuse clathrin distribution (Fig. 4B, lower picture). Schwann cells also showed punctuated clathrin staining but muscle fibers were not labeled (data not shown).

Strong nerve stimulation for 12 min at 20 Hz and immediate fixation, caused a concentration of immunoreactivity in small, very bright dots and a strong overall decrease of brightness within the terminal (Fig. 4C). The vesicle cluster-like staining

pattern, as observed in unstimulated preparations, totally disappeared. It has not become clear yet, if the tiny bright dots represent individual coated vesicles, or if they are active sites of endocytosis where clathrin molecules aggregate (coated pits). However, in several evaluated electron micrographs of stimulated nerve terminals, no increase in coated vesicle number was observed. Hence the accumulation of clathrin molecules in small areas could not be directly interpreted as the formation of coated vesicles from previously soluble clathrin monomers.

## Summary and outlook

Exo- and endocytosis at the frog neuromuscular junction are accompanied by major changes in synaptic vesicle and clathrin distribution inside the nerve terminal. According to the most widely accepted model, synaptic vesicles fuse completely with the plasma membrane and release their con-

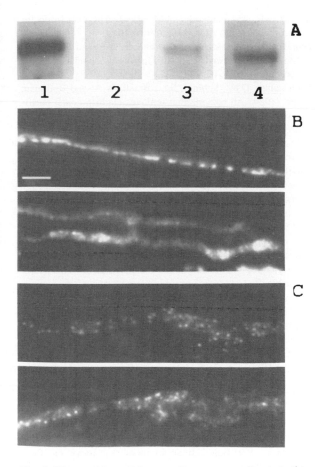

Fig. 4. Western blot and immunofluorescence of anti-clathrin antibody. (A) Western blot: the samples were treated as described above (see Fig. 1) and probed with anti-clathrin (CHC 5.9 MAb from ICN Pharmaceuticals Inc.). The detection system was the same as described in Fig. 1. The band labeled in A1 and A3 has a apparent molecular weight of 185 kDa, that in A4 is 180 kDa. (A1) frog brain, 2.5 µg protein/lane; (A2) frog nerve (10 µg/lane), no band labeled; (A3) synapse-rich region of frog *Cutaneous pectoris* muscle (10 µg/lane); (A4) rat brain, 2.5 µg/lane. (B,C) Immunofluorescence with anti clathrin (CHC 5.9 MAb) at the frog neuromuscular junction. (B) The unstimulated control terminals show a cluster-like (upper picture) or a rather diffuse (lower picture) staining pattern. (C) After electrical nerve stimulation (12 min at 20 Hz) multiple small, bright spots do appear and the total absolute brightness decreases. Scale bar = 5 µm.

tents into the synaptic cleft. Vesicles remain part of the plasma membrane for a transient period of time, then become internalized, and are subsequently rearranged in vesicle clusters. Our im-

munofluorescence studies with anti-synaptophysin, serving as a marker for synaptic vesicles, support this hypothesis, at least under conditions of very intense nerve stimulation.

There are currently three models for synaptic vesicle recycling under debate (Schweizer et al., 1995): (i) recycling via clathrin coated vesicles (Heuser and Reese, 1973); (ii) recycling via endocytotic cisternae (Miller and Heuser, 1984) and (iii) the "kiss and run" mechanism, that does not postulate the total fusion of the vesicle with the plasma membrane, but rather proposes a short opening of a "fusion pore" (Ceccarelli et al., 1973).

In the present study, we found that clathrin, the major component of coated vesicles, is accumulated in small intra-synaptic spots, if the terminal is highly stimulated. In a resting terminal this protein shows a more or less widespread distribution inside the synapse with a trend towards clustering. These clusters resemble synaptic vesicle clusters regarding shape and distribution, but future double-labeling experiments have to be done to show the identity of clathrin- and vesicle clusters. It has not become clear yet, if the stimulation-dependent clathrin aggregation is directly related to the formation of coated vesicles, especially because no increased number of coated vesicles have been observed in our electron microscopic images.

### Acknowledgements

We thank Dr. M. Browning for the α-synaptophysin antibody. Furthermore, we like to thank Mr. Steven Fadul for outstanding technical assistance in all light and fluorescence microscopy experiments and Ms. Dot Dill for excellent electron microscopic assistance. The present study was supported by a long term fellowship from the Human Frontier Science Program Organization.

### References

Almers, W. and Tse, F.W. (1990) Transmitter release from synapses: does a preassembled fusion pore initiate exocytosis? *Neuron*, 4: 813–818.
Bonzelius, F. and Zimmermann, H. (1990) Recycled synaptic vesicles contain vesicle but not plasma membrane marker,

newly synthesized acetylcholine and a sample of extracellular medium. *J. Neurochem.*, 55: 1266–1273.

Carter, L.L., Redelmeier, T.E., Woollenweber, L.A. and Schmid, S.L. (1993) Multiple GTP-binding proteins participate in clathrin coated vesicle-mediated endocytosis. *J. Cell. Biol.*, 120: 37–45.

Ceccarelli, B. and Hurlbut, W.P. (1980) Vesicle hypothesis of the release of quanta of acetylcholine. *Physiol. Rev.*, 60: 396–441.

Ceccarelli, B., Hurlbut, W.P. and Mauro, A. (1973) Turnover of transmitter and synaptic vesicles at the frog neuromuscular junction. *J. Cell. Biol.*, 57: 499–527.

Fesce, R., Grohovaz, F., Valtorta, F. and Meldolesi, J. (1994) Neurotransmitter release: fusion or 'kiss and run'? *Trends Cell. Biol.*, 4: 1–4.

Henkel, A.W. and Bieger, S.C. (1994) Quantification of proteins dissolved in an electrophoresis sample buffer. *Anal. Biochem.*, 223: 329–331.

Heuser, J.E. (1989) The role of coated vesicles in recycling of synaptic vesicle membrane. *Cell. Biol. Int. Rep.*, 13: 1063–1076.

Heuser, J.E. and Reese, T.S. (1973) Evidence for recycling of synaptic vesicle membrane during transmitter release at the frog neuromuscular junction. *J. Cell. Biol.*, 57: 315–344.

Katz, B. and Miledi, R. (1968) The role of calcium in neuromuscular facilitation. *J. Physiol.(London)*, 195: 481–492.

Maycox, P.R., Link, E., Reetz, A., Morris, S.A. and Jahn, R. (1992) Clathrin-coated vesicles in nervous tissue are involved primarily in synaptic vesicle recycling. *J. Cell. Biol.*, 118: 1379–1388.

Miller, T.M. and Heuser, J.E. (1984) Endocytosis of synaptic vesicle membrane at the frog neuromuscular junction. *J. Cell. Biol.*, 98: 685–698.

Monck, J.R. and Fernandez, J.M. (1992) The exocytotic fusion pore. *J. Cell. Biol.*, 119: 1395–1404.

Nanavati, C., Markin, V.S., Oberhauser, A.F. and Fernandez, J.M. (1992) The exocytotic fusion pore modeled as a lipidic pore. *Biophys. J.*, 63: 1118–1132.

Schlossman, D.M., Schmid, S.L., Braell, W.A. and Rothman, J.E. (1984) An enzyme that removes clathrin coats: purification of an uncoating ATPase. *J. Cell. Biol.*, 99: 723–733.

Schweizer, F.E., Betz, H. and Augustine, G.J. (1995) From vesicle docking to endocytosis: Intermediate reactions of exocytosis. *Neuron*, 14: 689–696.

Smythe, E. and Warren, G. (1991) The mechanism of receptor-mediated endocytosis. *Eur. J. Biochem.*, 202: 689–699.

Sudhof, T.C. and Jahn, R. (1991) Proteins of synaptic vesicles involved in exocytosis and membrane recycling. *Neuron*, 6: 665–677.

Torri-Tarelli, F., Haimann, C. and Ceccarelli, B. (1987) Coated vesicles and pits during enhanced quantal release of acetylcholine at the neuromuscular junction. *J. Neurocytol.*, 16: 205–214.

Valtorta, F., Jahn, R., Fesce, R., Greengard, P. and Ceccarelli, B. (1988) Synaptophysin (p38) at the frog neuromuscular junction: its incorporation into the axolemma and recycling after intense quantal secretion. *J. Cell. Biol.*, 107: 2717–2727.

Van der Blieck, A.M., Redelmeier, T.E., Damke, H., Tisdale, E.J., Meyerowitz, E.M. and Schmid, S. (1993) Mutations in human dynamin block an intermediate stage in coated vesicle formation. *J. Cell. Biol.*, 123: 553–563.

Zhang, J.Z., Davletov, B.A., Sudhof, T.C. and Anderson (1994) Synaptotagmin I is a high affinity receptor for clathrin AP-2: Implications for membrane recycling. *Cell*, 78: 751–760.

J. Klein and K. Löffelholz (Eds.)
*Progress in Brain Research*, Vol. 109
© 1996 Elsevier Science B.V. All rights reserved.

# CHAPTER 3

# Transcriptional regulation of the human choline acetyltransferase gene

Louis B. Hersh, Hiroyasu Inoue and Yi-Ping Li

*Department of Biochemistry, University of Kentucky, Lexington, KY, USA*

## Introduction

There are three proteins which are specifically expressed in cells which synthesize and store the neurotransmitter acetylcholine. These are the high affinity choline transporter, which brings extracellular choline into the cell, choline acetyltransferase (ChAT), the enzyme which utilizes choline and acetylcoenzyme A to synthesis acetylcholine, and the vesicular acetylcholine transporter (VAChT), which transfers cytosolic acetylcholine into storage vesicles (Chapters 5–7). To date studies on this cholinergic system have focused primarily on the biosynthetic enzyme choline acetyltransferase (EC 2.3.1.6), due in large part to the availability of specific antisera and cDNA probes with which to study this protein. Choline acetyltransferase exhibits a discreet expression pattern limited to specific subpopulations of neurons within the mammalian brain. Studies on the mechanism by which the cell type-specific expression of ChAT, and the vesicular acetylcholine transporter are controlled at the transcriptional level are now emerging (Chapters 5 and 6). Studies in this laboratory have focused on the human ChAT gene and its transcriptional regulation.

## Structure of the cholinergic gene

A human ChAT genomic fragment was first isolated in this laboratory in 1989 (Kong et al., 1989). This was followed by the isolation of the rat ChAT gene in which it was found that the coding sequence is contained within 13 or 14 exons spanning more than 64 kb of DNA (Hahn et al., 1992). Studies by Deguchi and his colleagues (Misawa et al., 1992; Kengaku et al., 1993) demonstrated that in both rat and mouse there are multiple ChAT mRNA transcripts differing only in their 5′ untranslated sequence, being derived from three non-coding exons. These 5′ non-coding exons were referred to as the R, N, and M exons (Misawa et al., 1992; Kengaku et al., 1993). The structure of the ChAT gene is depicted in Fig. 1. Within the human ChAT gene sequences homologous to the rodent R and M exons have been identified, however the functionality of these exons has yet to be established. The presence of an N type exon in the human ChAT gene has, to date, not been demonstrated.

We noted that the AUG used as the initiation codon in the rat, mouse, and porcine ChAT genes, is changed to an ACG in the human gene (Kong et al., 1989). Since the N-terminus of the human gene is blocked (Hersh et al., 1988), its N-terminal sequence cannot be determined. Thus there are three possible translational start sites in the human ChAT gene; the ACG located at the same position as the first AUG in other species, an AUG sequence which lies 30 nucleotides downstream of the ACG codon, which if utilized would yield a protein 10 amino acids smaller than other mammalian ChAT species, or an in-frame AUG found 324 nucleotides upstream within the M exon, which would yield a protein 108 amino acids larger than other mammalian species. These are

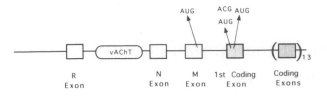

Fig. 1. Schematic illustration of the cholinergic gene. The three non-coding exons, R, N and M as well as the position of the coding exons is shown for the rodent gene. The AUG in the rodent gene is replaced by an ACG in the human gene. Two in-frame AUG sequences are shown for the human gene. Within the first intron of the ChAT gene is found the gene for the vesicular acetylcholine transporter. The coding sequence for this transporter is found within a single exon.

noted in Fig. 1. If the start site in the M exon were utilized, it would make this exon the first coding exon, rather than the third non-coding exon as it is in other species.

Subsequent studies have shown this gene to actually represent a cholinergic gene since the coding sequence for the vesicular acetylcholine transporter lies within the first intron of the mammalian ChAT gene (Bejanin et al., 1994; Erickson et al., 1994) (Fig. 1). This unusual gene structure appears to be conserved evolutionarily as it was first described in C. elegans (Alfonso et al., 1993). Whether the regulatory elements which control transcription of the ChAT gene also transcriptionally regulate the vesicular acetylcholine transporter has yet to be established.

## Identification of an enhancer in the human ChAT gene

We have examined the 5′ flanking region of the human ChAT gene for transcriptional regulatory elements. The technique employed was to clone fragments of the human ChAT gene into a promoterless vector containing the luciferase gene as a reporter gene. We then tested for promoter activity of such constructs by measuring luciferase activity following transient transfection expression in various cell lines. If promoter elements are present in the test fragment luciferase activity will be detectable in the transfected cells. In the absence of promoter activity no luciferase activity will be de-

tectable. Furthermore, comparison of the expression of these constructs in cholinergic versus non-cholinergic cells can be used to test for the presence of elements which may be responsible for cholinergic specific expression of the gene.

Utilizing such transient expression assays, basal promoter activity has been localized to a region just upstream of the M exon, between nucleotides −163 to +94 of the human ChAT gene (Li et al., 1993a). This region of the gene does not contain a classical "TATA" box, but is rather GC rich, typical of many housekeeping genes. Further analysis of the gene revealed the presence on an enhancer element initially localized between nucleotides −929 and −1190, which significantly increased the transcriptional activity of the gene (Fig. 2) (Hersh et al., 1993). However, this element did not act as a cell specific element since it was active in both cholinergic and non-cholinergic cell lines (Fig. 3).

More detailed analysis of this enhancer element localized it to nucleotides −970 to −941 and showed it to be composed of three elements acting in concert. These are two GC-box like sequences which are similar to sequences which bind the transcription factor Sp1, which flank a third unique sequence (Inoue et al., 1993) (Fig. 4). Footprint analysis shows this entire region to bind nuclear

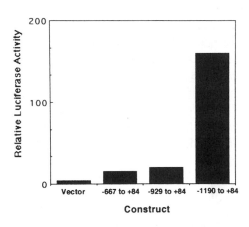

Fig. 2. Identification of an enhancer element in the human ChAT gene. Enhancer activity of the indicated fragments from the 5′ region of the human ChAT gene were cloned into the pXP luciferase reporter vector and transfected into PC12 cells. Enhancer activity was assessed by measuring luciferase activity produced from the transfected cells.

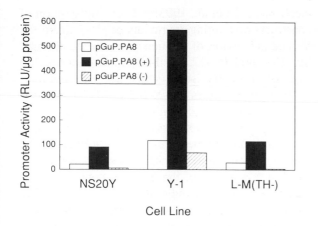

Fig. 3. Specificity of the human ChAT enhancer element. Constructs containing nucleotides −1190 to −929 (+) from the human ChAT gene in the vector pGUP.PA8, which contains a heat shock promoter driving the luciferase gene, were transfected into the indicated cell lines. Promoter activity was assessed by measuring luciferase activity. pGuP.PA8 is the control vector, pGuP.PA8 (+) contains the insert in the normal or + orientation, while pGuP.PA8 (−) contains the insert in the reverse or − orientation, NS20Y is a cholinergic cell line, Y-1 and L-M(TK-) are non-cholinergic cell lines.

factors, and gel shift analysis demonstrated that each of the three elements can bind nuclear derived factors, presumably transcription factors. The GC-boxes appear to bind transcription factors related to Sp1, while the third element binds a unique transcription factor. Mutation of any of these three elements reduces, but does not abolish, enhancer activity, demonstrating that all three elements are required to obtain maximal enhancer activity (Fig. 4).

## The human ChAT enhancer requires protein kinase A for activity

Insight into the function of this enhancer came from studies with a mutant PC-12 cell line, A123.7, which has been engineered to express a defective regulatory subunit of protein kinase A, and thus has very low protein kinase A activity (Correll et al., 1989). We found that the mutant A123.7 PC12 cell line exhibited less than 4% of the ChAT activity of its parental cell line. Furthermore, examination of the steady-state ChAT

mRNA levels by Northern blot analysis, showed that in contrast to the 4.4 kb mRNA detected in the parental PC-12 cell line, ChAT mRNA could not be detected in the protein kinase A deficient mutant cell line (Inoue et al., 1995). Because of the relatively low expression of ChAT mRNA in the parental cell, RT-PCR analysis was used to demonstrate the presence of a low, but detectable level of ChAT mRNA in the protein kinase A deficient cell line (Fig. 5). Thus the low level of ChAT mRNA in the protein kinase A deficient cell line correlates with the inability to detect ChAT activity in these cells.

Transient transfection analysis using luciferase as a reporter gene and the human ChAT enhancer

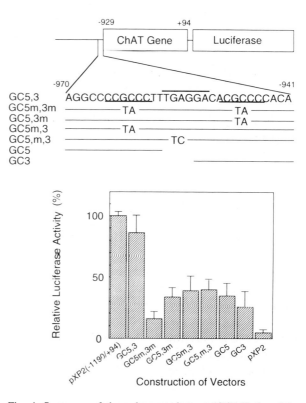

Fig. 4. Sequence of the enhancer element (GC5,3) found in the human ChAT gene and the effect of mutations on its activity. The GC boxes are underlined, while the middle element has a line above it. Mutations made in these sequences are shown below them. Activity of these sequences compared to the vector pXP were determined in NS20Y cells as described in Fig. 3 and are expressed relative to the activity of a construct containing nucleotides −1190 to +94 of the human ChAT gene.

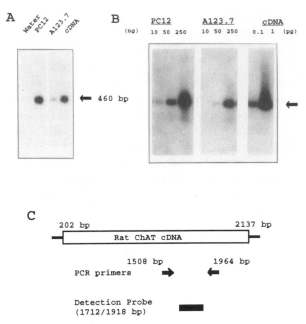

Fig. 5. PCR analysis of ChAT mRNA expressed in wild-type and PKA-deficient PC12 cells. Poly(A)$^+$ RNA from the indicated cells were reverse transcribed by MMLV reverse transcriptase (A) or thermostable reverse transcriptase (B) and then analyzed by PCR with primers as shown in (C). The PCR product was hybridized with a $^{32}$P-labeled detection probe (C). Solid arrows indicate the length (460 basepairs) of hybridization bands.

to drive basal promoter activity, showed that the activity of the enhancer element was barely detectable in the mutant cell line. However, when the catalytic subunit $\beta$ (C$\beta$) of protein kinase A was co-expressed with the enhancer, enhancer activity increased 280% (Inoue et al., 1995). These experiments demonstrate a role for protein kinase A in regulating the enhancer activity of the human ChAT gene, and furthermore show the necessity of this enhancer element for efficient expression of the human ChAT gene.

## Cholinergic cell specific expression is controlled by repressor elements

Studies on the cell specific expression of the human ChAT gene, using transient transfection analysis with luciferase as a reporter gene, led to the conclusion that cholinergic specific expression is brought about by a differential repression

mechanism (Li et al., 1993b). Thus, transfection of constructs containing fragments of DNA from the 5′ flanking region of the gene into the cholinergic MC-IXC and PC-12 cell lines were compared to expression of the same constructs in the non-cholinergic cell lines SK-N-BE(2)-M17 and SHS-Y5Y. Constructs containing ~1 or 2 kb of 5′ flanking sequence expressed luciferase at similar levels in both cholinergic and non-cholinergic cell lines, indicating the absence of a cell specific element in this region of the gene. In contrast extension to ~3.3 kb showed higher expression in the cholinergic cell lines as compared to the non-cholinergic cell lines, and further repression of expression was observed if ~6.6 kb of 5′ flanking sequence was used. These results suggest the presence of two silencers residing in the 5′-flanking sequence of the human ChAT gene, which work cooperatively to repress the expression of the gene in non-cholinergic cells. One of these silencers is located between −2 and −3.3 kb while the other is located between −3.3 and −6.6 kb.

The proximal silencer element was further studied by testing a series of deletion fragments of the human ChAT gene 5′ sequence, subcloned into the expression vector pGUP.PA8. This vector contains a minimal promoter from the human heat shock gene and the luciferase reporter gene. We added to it the enhancer sequence from the human ChAT gene (Fig. 6), to increase expression from this relatively weak basal promoter. Expression in the human cholinergic neuroblastoma cell line MC-IXC was compared to that in the human adrenergic neuroblastoma cell line SK-N-BE(2)-M17. The fragment between nucleotides −2043 to −2665 reduced reporter gene expression in the adrenergic line by ~4 fold (Fig. 6). When this sequence was further divided into two fragments, only one, nucleotides −2043 to −2409 exhibited repression. Further deletion to a 215 basepair fragment comprised of nucleotides −2195 to −2409, retained the silencer activity (Fig. 6).

## The cell specific repressor is composed of two E-boxes

The 215 basepair repressor sequence contained

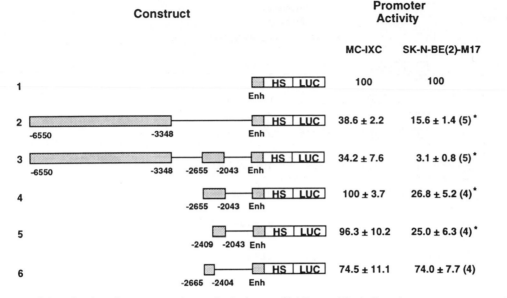

Fig. 6. Localization of a repressor element in the human ChAT gene. The indicated constructs were transfected into the cholinergic cell line MC-IXC and the adrenergic cell line SK-N-BE(2)-M17. Expression of luciferase activity was used to measure promoter activity and its repression in the non-cholinergic cell line. HS, heat shock promoter; LUC, luciferase. The symbol * indicates a significant difference ($P < 0.05$) between the two cell lines.

three consensus E box sequences (CANNTG), referred to as Ea, Eb and Ec (Fig. 7). The E box sequence CANNTG is recognized by a family of protein transcription factors with basic helix-loop-helix (bHLH) structures, some of which have been shown to be involved in tissue-specific gene expression. Further deletions from the 5′ or 3′ end, such that Ea or Ec was removed, resulted in a total loss of silencer activity showing that E boxes Ea and Ec are important for the silencer activity (Li et al., 1995). It is interesting to point out that these regulatory sequences lie within the coding sequence of the acetylcholine transporter.

To further study the importance of the E boxes for silencer activity, each was mutated. In agreement with the deletion constructs, mutations made in E-box Ea or Ec resulted in a loss of silencing activity (Fig. 7). In contrast a mutation made in E-box Eb had no effect on silencer activity. These experiments demonstrate that E-box Ea and Ec are necessary for cholinergic-specific silencer activity, while Eb is not involved.

If, as proposed the E-boxes Ea and Ec act as a silencer elements, there should be specific protein

transcription factors differentially present in cholinergic and non-cholinergic cells capable of binding to these DNA sequences. Gel retardation experiments were thus conducted using DNA probes containing each of the E boxes. An oligonucleotide containing E-box A showed strong binding of nuclear proteins prepared from both the cholinergic MC IXC as well as the adrenergic SK-N-BE(2)-M17 cell line. Gel retardation experiments performed with a DNA fragment containing E-box Eb showed no significant binding, a result which supports the conclusion that Eb is not functional.

Gel retardation experiments using a DNA fragment containing Ec showed a different binding pattern between the cholinergic and adrenergic cell lines (Fig. 8). There was strong binding when a nuclear protein extract from SK-N-BE(2)-M17 cells was used while the binding of nuclear proteins from MC-IXC cells to Fragment C was barely detectable. These results indicate that there is a specific protein binding to Ec in the adrenergic cell line which is greatly reduced or absent in the cholinergic cell line. The lack of binding to Ec

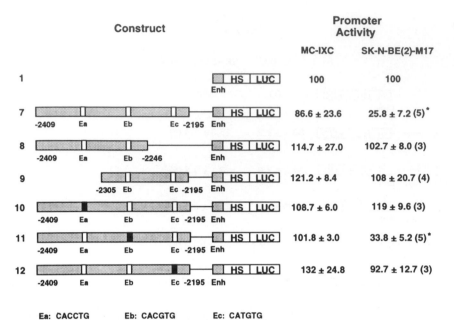

Ea: CACCTG     Eb: CACGTG     Ec: CATGTG

Fig. 7. Schematic illustration of constructs containing the 215 basepairs sequence with silencer activity and the effect of deletions or mutations of this sequence on repressor activity. Repressor activity was compared between MC-IXC and SK-N-BE(2)-M17 cell lines. Open boxes represent E boxes and closed boxes represent mutated E boxes. Data are presented as the mean ± SE values of percentage of the control, construct #1. The symbol * indicates a significant difference ($P < 0.05$) between the two cell lines.

could explain the inactivity of the silencer in the cholinergic cell line.

In order to further verify the specificity of protein binding to Ea and Ec, two mutant 18 mers (Am and Cm) containing mutated Ea or Ec sequences were tested for their ability to compete for protein binding to E-box A or C. The two 18 mers with a mutated E box did not compete with protein binding while the same amount of wild type 18 mers did compete. These results along with the transfection data on mutant Ea and Ec provide evidence for a critical role of Ea and Ec in the cholinergic cell type-specific silencer activity. The lack of protein binding to Ec in cholinergic cells appears to be pivotal in cholinergic phenotype determination.

**Summary and conclusions**

These studies demonstrate the presence of two important regulatory elements in the human ChAT gene. An enhancer element, which is composed of two GC-box sequences plus an intervening sequence, which is required for efficient transcription of the gene, and a repressor element which is composed of two E-box sequences, and is in part responsible for cell specific expression of the ChAT gene. These regulatory elements are located downstream of the most distal exon, exon R. Studies of the rat ChAT gene also indicate the presence of a cell specific repressor element; however in this case the repressor element is located upstream of the distal R exon. Preliminary studies in this laboratory confirm the presence of a cell specific element in this region of the human ChAT gene. Thus, there appears to be at least two, and possibly three cell specific regulatory elements in the ChAT gene. Whether these regulatory elements work in a cooperative fashion, or whether they serve to regulate different groups of cholinergic neurons, has yet to be established.

Cell-type specific silencers have been identified

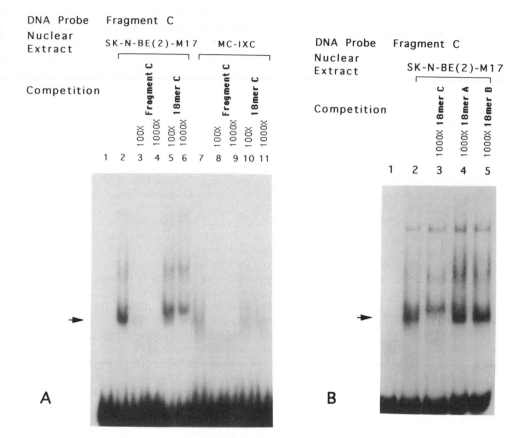

Fig. 8. (A) Gel retardation assays identifying an E-box Ec-binding protein in SK-N-BE(2)-M17 nuclear extracts, indicated by the arrow, but absent in MC-IXC nuclear extracts. Lane 1, radiolabeled 67 mer probe containing E-box C only; lane 2, same with nuclear extract from SK-N-BE(2)-M17 cells; lane 3, same with 100 fold excess of unlabeled 67 mer probe containing E-box C; lane 4, same with 1000 fold excess of unlabeled 67 mer probe containing E-box C, lane 5, same with 100 fold excess of unlabeled 18 mer containing E-box C; lane 6, same with 1000 fold excess of unlabeled 18 mer containing E-box C; lanes 7–11, same as lanes 2–6 except nuclear extract from MC-IXC cells was used. (B) Demonstration of specificity of E box binding proteins. Lane 1, radiolabeled probe, a 67 mer probe containing E-box C only; lane 2, same with nuclear extract from SK-N-BE(2)-M17 cells; lane 3, same with 1000-fold excess of unlabeled 18 mer containing E-box C; lane 4, same with 1000 fold excess of unlabeled 18 mer containing E-box A; lane 5, same with 1000-fold excess of unlabeled 18 mer containing E-box B.

in some other neuron-specific genes. The SCG10 gene and the type II $Na^+$ channel gene both contain a common silencer element which is responsible for the neuron-specific expression of these genes (Maue et al., 1990; Mori et al., 1992). Functionally, this cholinergic neuron-specific silencer is different in that it restricts the expression of the ChAT gene, not just to neurons, but to a subset of neurons.

An E box has been found to be a part of the adrenergic neuron-specific enhancer identified in the rat tyrosine hydroxylase gene (Yoon and Chikaraishi, 1992). That enhancer requires synergy between the E box and an overlapping AP1 motif. We demonstrated here that the two interactive E boxes in the ChAT gene (CACCTG and CATGTG) are necessary for the cholinergic neuron-specific repressor. It thus appears that there are a variety of bHLH protein factors involved in the control of neuronal subtype-specific gene expression.

54

## Acknowledgements

This research was supported in part by a grant from the National Institute on Aging (AG05893) to LBH.

## References

Alfonso, A., Grundahl, K., Duerr, J.S., Han, H.-P. and Rand, J.B. (1993) The *Caenorhabditis elegans unc-17* gene: a putative vesicular acetylcholine transporter. *Science,* 261: 617–619.

Bejanin, S., Cervini, R., Mallet, J. and Berrard, S. (1994) A unique gene organization for two cholinergic markers, choline acetyltransferase and a putative vesicular transporter of acetylcholine. *J. Biol. Chem.,* 269: 21944–21947.

Correll, L.A., Woodford, T.A., Corbin, J.D., Mellon, P.L. and McKnight, G.S. (1989) Functional characterization of cAMP-binding mutations in type I protein kinase. *J. Biol. Chem.,* 264: 16672–16678.

Erickson, J.D., Varoqui, H., Shafer, M.K.-H., Modi, W., Diebler, M.-F., Weihe, E., Rand, J., Eiden, L., Bonner, T.I. and Usdin, T.B. (1994) Functional identification of a vesicular acetylcholine transporter and its expression from a cholinergic locus. *J. Biol. Chem.,* 269: 21929–21932.

Hahn, M., Hahn, S.L., Stone, D.M. and Joh, T.H. (1992) Cloning of the rat gene encoding choline acetyltransferase, a cholinergic neuron-specific marker. *Proc. Natl. Acad. Sci. USA,* 89: 4387–4391.

Hersh, L.B., Takane, K., Gylys, K., Moomaw, C. and Slaughter, C. (1988) Conservation of amino acid sequences between human and porcine choline acetyltransferase. *J. Neurochem.,* 51: 1843–1845.

Hersh, L.B., Kong, C.F., Sampson, C., Mues, G., Li, Y.-P., Fisher, A., Hilt, D. and Baetge, E.E. (1993) Comparison of the human and porcine choline acetyltransferase genes: identification of an enhancer-like element. *J. Neurochem.,* 61: 306–314.

Inoue, H., Baetge, E.E. and Hersh, L.B. (1993) Enhancer containing unusual GC Box-like sequences on the human choline acetyltransferase gene. *Mol. Brain Res.,* 3: 299–304.

Inoue, H., Li, Y.-P., Wagner, J.A. and Hersh, L.B. (1995) Expression of the choline acetyltransferase gene depends on protein kinase A activity. *J. Neurochem.,* 64: 985–990.

Kengaku, M., Misawa, H. and Deguchi, T. (1993) Multiple mRNA species of choline acetyltransferase from rat spinal cord. *Mol Brain Res.,* 18: 71–76.

Kong, C.F., Hilt, D. and Hersh, L.B. (1989) Isolation of a genomic clone of human choline acetyltransferase. In H. Kewitz, T. Thomsen and U. Bickel (Eds.), *Pharmacological Interventions on Central Cholinergic Mechanisms in Senile Dementia (Alzheimer's Disease).* Zuckschwerdt-Verlag, Berlin, pp. 15–19.

Li, Y.-P., Baetge, E.E. and Hersh, L.B. (1993a) Cyclic AMP regulation of the human choline acetyltransferase gene. *Neurochem. Res.,* 18: 271–275.

Li, Y.-P., Baskin, F., Davis, R. and Hersh, L.B. (1993b) Cholinergic neuron-specific expression of the human choline acetyltransferase gene is controlled by silencer elements. *J. Neurochem.,* 61: 748–751.

Li, Y.-P., Baskin, F., Davis, R. Wu, D. and Hersh, L.B. (1995) A cell type-specific silencer in the human choline acetyltransferase gene requiring two distinct and interactive E boxes. *Mol. Brain Res.,* 30: 106–114.

Maue, R.A., Kraner, S.D., Goodman, R.H. and Mandel, G. (1990) Neuron-specific expression of the rat brain type II sodium channel gene is directed by upstream regulatory elements. *Neuron,* 4: 223–231.

Misawa, H., Ishii, K. and Deguchi, T. (1992) Gene expression of mouse choline acetyltransferase. Alternative splicing and identification of a highly active promoter region. *J. Biol. Chem.,* 267: 20392–20399.

Mori, N., Schoenherr, C., Vandenbergh, D.J. and Anderson, D.J. (1992) A common silencer element in the SCG 10 and type II $Na^+$ channel genes binds a factor present in non neuronal cells but not in neuronal cells. *Neuron,* 9: 45–54.

Yoon, S.O. and Chikaraishi, D.M. (1992) Tissue-specific transcription of the rat tyrosine hydroxylase gene requires synergy between an AP-1 motif and an overlapping E box-containing dyad. *Neuron,* 9: 55–67.

J. Klein and K. Löffelholz (Eds.)
*Progress in Brain Research*, Vol. 109
© 1996 Elsevier Science B.V. All rights reserved.

CHAPTER 4

# Biosynthesis and integration of acetylcholinesterase in the cholinergic synapse

Jean Massoulié, Claire Legay, Alain Anselmet, Eric Krejci,
Françoise Coussen and Suzanne Bon

*Laboratoire de Neurobiologie Moléculaire et Cellulaire, CNRS URA 1857, Ecole Normale Supérieure, 46, rue d'Ulm, 75005 Paris, France*

## Introduction

After its release in cholinergic synapses, acetylcholine is rapidly hydrolyzed by acetylcholinesterase (AChE, EC 3.1.1.7). This enzyme possesses an extremely high catalytic efficiency, and also presents a variety of molecular forms that may be anchored in various ways in the synaptic architecture. In this chapter, we focus our attention on the latter aspect.

## Molecular forms of AChE

### Alternative splicing and post-translational processes

Vertebrates possess a single gene for AChE, but multiple transcripts generated by alternative splicing may produce several types of subunits in which the catalytic domain is associated with distinct C-terminal peptides (for a review, see Massoulié et al., 1993). In mammals, for example, the last exon encoding the catalytic domain may be followed by three distinct coding sequences in mRNAs, called R ("readthrough"), H (hydrophobic) and T (tailed) (Fig. 1).

In the case of R mRNAs, the following sequence is retained, without splicing. This type of mRNA is generally minor (Legay et al., 1993b); it produces a soluble monomeric form of AChE in transfected cells (Li et al., 1993), but this enzyme has not yet been characterized in vivo.

The "H" sequence encodes a C-terminal peptide that contains a cysteine and a signal allowing post-translational cleavage and addition of a glyco-phosphatidylinositol (GPI) anchor, so that the derived H subunits produce disulfide-linked GPI-anchored dimers (Duval et al., 1992a; Legay et al., 1993).

The "T" sequence encodes a 40 amino acid peptide that is remarkably conserved throughout evolution, from invertebrates such as *Caenorhabditis* to mammals. This sequence again contains a cysteine that may form intersubunit disulfide bonds, and may partially organize as an amphiphilic $\alpha$-helix (Massoulié et al., 1993; Bon et al., unpublished). It is able to mediate amphiphilic interactions with phospholipidic or detergent micelles, but also quaternary interactions, forming oligomers of catalytic subunits or heteromeric associations with structural subunits. The T subunits thus produce a variety of quaternary combinations (Duval et al., 1992a; Legay et al., 1993a). The homo-oligomeric forms include amphiphilic monomers ($G_1$) and dimers ($G_2$), which are called amphiphilic forms of type II, as opposed to the GPI-anchored dimers (Bon et al., 1988a,b, 1991), as well as soluble tetramers ($G_4$). The hetero-oligomeric forms are the asymmetric or collagen-tailed tetramers and the hydrophobic-tailed tetramers. In both cases, a collagenic subunit or a hydrophobic 20 kDa subunit is associated with a tetramer of T subunits, in such a way that two of these subunits are disulfide-linked together, while

Genomic structure (3')

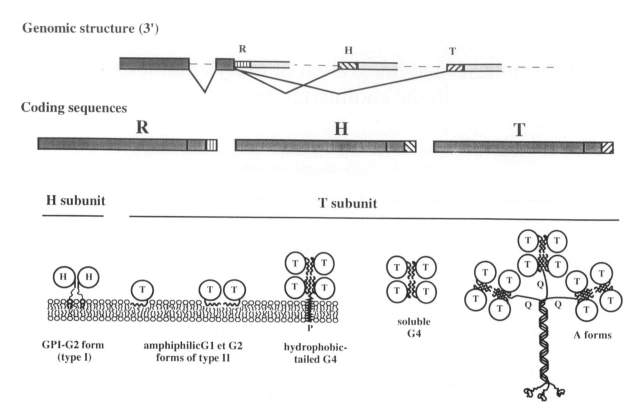

Fig. 1. Splicing pattern of the 3′ region of the acetylcholinesterase gene, and quaternary structure of the molecular forms generated from H and T subunits.

the other two are disulfide-linked to the structural anchoring subunit (for a review, see Massoulié et al., 1993). When co-expressed in transfected COS cells, the *Torpedo* or mammalian T subunits readily assemble with *Torpedo* collagenic subunits to produce asymmetric forms (Krejci et al., 1991; Legay et al., 1993a). T subunits may also combine with a chimeric protein, $Q_N/H_C$, in which the N-terminal domain of the collagenic tail subunit ($Q_N$) is fused to C-terminal peptide from the H subunit ($H_C$), generating GPI-anchored tetramers (Duval et al., 1992b). This demonstrates that the $Q_N$ domain is sufficient to bind one catalytic tetramer.

Thus, the diverse molecular forms of acetylcholinesterase arise from alternative splicing, which specifies the nature of a small C-terminal domain, and from post-translational processes, i.e. addition of GPI anchor in the case of H subunits, oligomerization and association with anchoring proteins in the case of T subunits.

*Analysis of GPI addition and heteromeric associations*

Because the site of cleavage of the C-terminal peptide of H subunits has not been directly determined, we have attempted to analyze the corresponding GPI-addition signal by site-directed mutagenesis. This analysis was directly based on previous extensive studies performed mostly on the GPI-anchored decay acceleration factor (Moran and Caras, 1991a,b, 1994) and alkaline phosphatase (Micanovic et al., 1990; Gerber et al., 1992; Kodukula et al., 1992, 1993). From these studies, it appeared that the GPI addition signal consists of a C-terminal hydrophobic sequence, preceded by a cleavage/addition site. The last amino acid of the mature protein, $\omega$, to which the GPI anchor is linked with an ethanolamine through an amide bond, is a small residue (S, G, A, D, N, or more rarely C). The two groups, however, differed in their conclusions regarding the following residues: whereas the group of Caras initially emphasized the importance of a small residue at position $\omega + 1$, the group of Underfriend found that only proline was prohibited at this site, but that $\omega + 2$ must be a small residue. These rules did not unambiguously define an $\omega$ position in the H sequences of either *Torpedo* or rat acetylcholinesterases, because of the existence of multiple sites which followed one of these conditions, or both. Because both groups agreed that the presence of proline at positions $\omega$, $\omega + 1$, or $\omega + 2$ was incompatible with the process of glypiation, we introduced prolines at various positions, in the hope of identifying the functional $\omega$ site. We expressed the modified molecules in COS cells, and analyzed the efficiency of processing (Coussen et al., unpublished).

We also modified residues around potential $\omega$ sites, in order to assess their importance. We thus found (1) that multiple potential $\omega$ sites may coexist in a glypiation signal sequence, (2) that the restrictions originally proposed on the $\omega + 1$ and $\omega + 2$ residues are neither sufficient nor necessary, (3) that the presence of a proline at $\omega - 1$ reduces the efficiency of GPI addition. In fact, the glypiation process appears to be remarkably robust: the very sensitive assay of acetylcholinesterase activity allowed us to detect low levels of GPI-anchored molecules, even in the absence of any $\omega$ site, as defined by conditions on the $\omega + 1$ or $\omega + 2$ residues.

In order to analyze the GPI signal of rat acetylcholinesterase, we expressed the H subunit in COS cells, and examined the enzyme activity associated with the cells and released in the medium. Depending on the mutations introduced in the signal sequence, we observed considerable variations in the total yield of activity and in the quantity released into the medium, as well as in the proportion of GPI-linked dimers, as characterized by their sensitivity to the specific phospholipase PI-PLC. It is interesting that a high level of GPI-anchored form is not always associated with a high rate of release into the medium. A soluble form may be produced either during processing, by hydrolysis of the cleavage intermediate instead of transamidation to the GPI ethanolamine (Maxwell et al., 1995), or after glypiation, by the action of an endogenous phospholipase or protease. A protease appears more likely, as it would explain the importance of the terminal residues of the mature protein. The release of acetylcholinesterase has been reported in many cases and may play an important role in some physiological processes. Immunofluorescence of intact living cells, with or without previous treatment with PI-PLC, showed that only mature GPI-anchored molecules were exposed at the cell surface, even when they were produced in very low proportion. In contrast, the non-glypiated molecules were entirely contained in intracellular vesicular bodies, that could be visualized after permeabilization. These molecules probably retained the C-terminal hydrophobic sequence because their sedimentation was sensitive to the presence of detergents.

This observation is in excellent agreement with previous studies, which concluded that an abortive GPI addition sequence, in which the cleavage/addition site had been neutralized by mutagenesis, served as a retention signal, and that the unprocessed precursors were eventually degraded in a post-ER compartment (Moran and Caras, 1992; Delahunty et al., 1993; Kodukula et al., 1993).

In order to analyze the GPI addition signal from *Torpedo* acetylcholinesterase, we chose not to express the H subunit itself, because this protein cannot be correctly folded at 37°C. Lowering the temperature of transfected cultures to 27°C only allows a small proportion of molecules to become active, and might alter the post-translational processes. We therefore analyzed the glypiation of the chimeric $Q_N/H_C$ protein, which uses GPI addition signal of the *Torpedo* collagenic subunit. Coexpression of this protein with rat T subunits in COS cells, at 37°C, generates GPI-anchored tetramers which may be conveniently characterized by their sensitivity to PI-PLC.

In this study, we found that the rat T subunits, when expressed alone, mostly formed monomers and dimers which were partly retained within the cells, and largely released into the medium, together with tetramers. These molecules were not associated with the cell outer membrane.

When they were co-expressed with the $Q_N/H_C$ protein, however, we observed the formation of tetramers which were GPI-anchored and exposed at the cell surface. It is interesting that $Q_N/H_C$ mutants, in which the cleavage/addition site was mutated, were still able to associate with T subunits, but were mostly cleaved without addition of a GPI anchor, thus producing secreted tetramers.

## Acetylcholinesterase in muscles: alternative splicing and focalization of transcription during development

The GPI-anchored form of acetylcholinesterase, generated from H subunits, exists in the muscles of lower vertebrates. It is for example the only form of enzyme in the dorsal muscle of *Torpedo* (S. Bon, unpublished result) and has also been characterized in *Xenopus* muscle (Inestrosa et al., 1988). In higher vertebrates, however, adult muscles exclusively express T subunits, that generate amphiphilic monomers and dimers of type II, as well as hydrophobic tailed tetramers and collagen-tailed forms. A number of studies have focussed on the latter forms, because their presence is clearly related to the synaptic interaction between nerve and muscle.

In the fast muscles of adult rats, for example, the $A_{12}$ form is concentrated at the neuromuscular junctions, whereas extrajunctional regions contain only globular forms (essentially the intracellular $G_1$ form and the membrane-bound $G_4$ form). The contrast is less marked in slow muscles such as the soleus, since asymmetric forms are also present in extrajunctional regions; in addition, this muscle contains a lower level of $G_1$ and $G_4$ forms, and a significant proportion of $A_8$ form (Sketelj, 1994). These characteristics may be partly imposed upon the muscle by its pattern of activity, as shown by electrical stimulation of the rat denervated soleus (Lømo et al., 1985).

It must be emphasized that the distribution of molecular forms of AChE does not depend only on the physiological characteristics of the muscles, but also varies dramatically between different species. In human muscles, particularly, the collagen-tailed molecules are present in non-innervated as well as in innervated regions (Carson et al., 1979; Sketelj and Brzin, 1985). In addition, denervation may induce a decrease of the overall acetylcholinesterase activity, as in rat, or an increase, as in guinea-pig, rabbit and chicken. When total activity increases, collagen-tailed forms may either decrease, as in the pectoral muscle of chicken or in the fast semi-membranous muscle of the rabbit, or increase, as in the slow part of the rabbit semi-membranous muscle (Bacou et al., 1982).

The diaphragm of rodents allows an exceptionally convenient analysis of the junctional distribution of AChE forms. In fact, the localization of collagen-tailed forms at neuromuscular junctions was first demonstrated in the rat diaphragm by Hall (1973). We studied the splicing of AChE mRNAs, in the mouse diaphragm, during development, as well as in cultures of the myogenic cell lines C2 and Sol8. We found that while adult muscle only contains T transcripts, minor proportions of R and H transcripts also exist in the developing diaphragm, e.g. at embryonic days 14 or 16, and also in the myogenic cultures (Legay et al., 1995). Their proportion did not decrease upon fusion of myoblasts to myotubes or maturation in the presence of insulin (Pinset et al., 1991). It is interesting that the R transcript was more abundant than the H transcript, both in the embryonic diaphragm and in the myogenic cultures, because its presence had only been reported as a minor component, in tissues expressing the H transcript, such as the rat fetal liver (Legay et al., 1993b), or in the mouse erythroleukemia cell line, MEL (Li et al., 1993). In these conditions, it could be considered to result from a defect of the splicing mechanism. The fact that it is more abundant than H in the developing muscle raises the question of its physiological significance. The presence of the corresponding AChE form remains to be established.

Since R and H transcripts persist in differentiated myotubes in culture, their disappearance during muscle development may be related to innervation. They did not, however, reappear in the adult diaphragm after denervation, at least up to 3 days.

We also studied the localization of AChE transcripts in the mouse diaphragm by in situ hybridization. Their distribution was rather diffuse, with a broad concentration along the midline of the diaphragm at embryonic day 13 (E13), and they became progressively concentrated around the junctional nuclei, at the time of birth, as previously observed in the case of the mRNA encoding the $\alpha$ and $\gamma$ subunits of the nicotinic acetylcholine receptor (Piette et al., 1993). It thus appears that the presence of the nerve is not directly related to the splicing of the AChE transcripts towards T mRNAs, but rather promotes a differentiation state that determines this mechanism.

A comparison with the evolution of mRNAs encoding the subunits of the acetylcholine receptor suggests that the synaptic expression in the junctional nuclei may be controlled by similar factors, such as those which particularly specify an exclusively synaptic localization of the e subunit, after birth. Nevertheless, it is clear that the regulation of AChE transcripts differs in many respects from that of AChR subunits. First, it has been shown by the group of P. Taylor that the increase in mRNA levels observed upon fusion of myoblasts into myotubes results from a higher rate of transcription in the case of AChR, and from a stabilization of transcripts in the case of AChE (Fuentes and Taylor, 1993). Secondly, while muscle denervation invariably induces a reexpression of AChR transcripts (except those encoding the $\varepsilon$ subunit), it leads to highly variable effects, depending on the muscle and on the species, in the case of AChE. It seems likely that, in contrast to the greater susceptibility to re-innervation resulting from extra junctional reexpression of AChR, the changes observed in AChE expression do not represent an adaptive response, but rather reflect the dysfunction of the system.

## Cellular differentiation and post-translational processing of AChE subunits

The different molecular forms of AChE which may be generated from H and T subunits, as described above, have been extensively studied in transiently transfected COS cells. In these cells, transfection with vectors expressing the two types of subunits of rat AChE produces roughly equivalent levels of catalytic activity. In our culture conditions, about half of the activity generated from H subunit consisted of cell-associated GPI-anchored dimers (GPI-$G_2$) which were largely exposed at the external cell surface, and the other half was released in soluble form in the medium. On the other hand, about 80% of the monomers, dimers and tetramers generated from T subunits were released in the medium.

We also studied the production of AChE in stably transfected rat basoleukemia (RBL) cell lines (Coussen et al., 1995). The molecular forms produced from each type of subunit were the same as in COS cells. The GPI-$G_2$ form appeared to be nearly exclusively associated with the cell surface, where immunofluorescence showed a punctate pattern, probably representing its accumulation in caveolae. In cells expressing the T subunits, the activity was entirely intracellular, it was about 10-fold lower than for H subunits and the protein could not be labelled by immunostaining, even after permeabilization. The rate of protein synthesis, however, appeared to be equivalent in both cases, as judged by metabolic labeling. The difference is most likely due to post-translational maturation and stability, because the apparent mass of H subunits increased during a chase period, indicating maturation of its glycosylation, while there was no change in the T subunits.

The fact that H and T subunits do not follow the same maturation route may be illustrated in vivo by differences in their glycosylation: in *Torpedo* electric organs, only the H subunits carry the distinct carbohydrate epitope that binds Elec-39 or HNK-1 monoclonal antibodies (Bon et al., 1987).

A striking contrast between the fate of the different AChE subunits in different tissues was observed by the group of H. Soreq in transiently transfected *Xenopus* embryos. These embryos had been injected at the two-cell stage with synthetic mRNA, so that tissue-specificity could not arise at the transcription level. In these experiments, the T subunits produced abundant activity in muscles, specifically at endplates, while R transcripts were only expressed in epidermal ciliated cells, which secreted a soluble form of the enzyme in the tank water (Seidman et al., 1995). It is thus clear that the alternatively spliced C-terminal peptides which characterize the R, H and T subunits, and perhaps other types of subunits which have not yet been identified, do not only determine the mode of anchoring and the quaternary associations of the subunits, but at the same time play a major role in the efficiency with which different cell types produce, process and eventually stabilize the enzyme.

## Acetylcholinesterase during development of the avian central nervous system

### Acetylcholinesterase mRNAs

Unlike *Torpedo* and mammals, the AChE gene of chicken or quail does not seem to possess alternative splicing of the 3′ coding exons, since the only cDNAs which were cloned belonged to the T type, and no potential exon H could be found in the genomic sequence extending between the last exon of the catalytic domain and exon T, where it exists in other species (A. Anselmet, unpublished results). Northern blots in fact reveal the existence of multiple transcripts, differing in their size between 4.5 kb and 6 kb, and RT-PCR analysis showed that they all contain the T coding sequence.

Although the coding sequence thus appears invariable, development of the quail brain is accompanied by marked changes in the pattern of transcripts: the longer transcripts are abundant in the embryo, and show distinct evolutions in different brain regions, but disappear in the adult which only expresses the shorter 4.5 kb transcript (Anselmet et al., 1994). These variations most likely reflect the size of the 3′ untranslated regions, as in mammals (Fuentes and Taylor, 1993), and may be related to the stability of the transcripts.

### Active and inactive acetylcholinesterase

We also observed important variations at the protein level during development. The specific activity generally increased, but went through a maximum in some brain regions. In addition we observed large variations in the proportion of inactive AChE. In birds, a fraction of AChE subunits is catalytically inactive, or perhaps possesses a very low activity (Chatel et al., 1993a). The inactive subunits may be quantified, using a two-site immunoassay (Chatel et al., 1994). They represent a significant proportion of the protein in the embryonic brain, e.g. 40% in the chicken at embryonic day 11. They exist as monomers and dimers which do not differ in their molecular properties (sedimentation, interaction with detergents) from the corresponding active molecular forms. They also react with the organophosphate inhibitor MTP, indicating the presence of an active serine.

Besides their lack of catalytic activity, these subunits differ from active ones by the fact that they do not form tetramers, and that they are not recognized by a conformation-dependent monoclonal antibody, C-131 (Chatel et al., 1993b). Active and inactive subunits thus probably differ in their conformation.

Variations in the abundance of inactive subunits was revealed by comparison of the intensity of immunological staining with monoclonal antibodies, in Western blots of samples containing identical catalytic activities: depending on the brain region, this intensity passed through a maximum, at different stages of development. It was particularly marked at embryonic day 16, in the quail optic lobes (Anselmet et al., 1994).

The existence of inactive subunits is puzzling. Such an inactive component has not been detected in rat or in human tissues (Brimijoin et al., 1987; Hammond and Brimijoin, 1988), although the kinetics of biosynthesis indicate that catalytic activity is only acquired after a lag period of about 15 min (Lazar et al., 1984). It is therefore unlikely that they represent precursors of active subunits. Instead, they appear to result from an alternative folding pathway; the observed variations suggest that the probability of entering the active or inactive conformation does not reflect an intrinsic property of the polypeptide chain, but largely depends on the differentiation state of the cell, possibly on the cellular equipment in chaperon proteins such as BiP and calnexin (Kim and Arvan, 1995). The possibility of an inactive folding may be related to the presence, in the primary structure of chicken and quail AChEs, of a peptide that has no equivalent in *Torpedo* and mammalian enzymes (Randall et al., 1994; A. Anselmet et al., unpublished). This inserted peptide contains 237 residues in the $\alpha$-allele of quail AChE; it is very hydrophilic, it contains several repeated motifs, and its position suggests that it may form an extended loop on the outside of the catalytic subunit, according to the three-dimensional structure determined by X-ray crystallography for *Torpedo* AChE (Sussman et al., 1991). It is possible that this peptide may perturb the folding of the catalytic domain.

Recent studies by the group of Silman have suggested that heat-shock proteins may participate in the biosynthesis of AChE, in cultures of chicken myotubes (Eichler et al., 1991). These studies also showed that recovery of activity after a heat-shock involved a re-activation of thermally inactivated molecules, but no activation of the pre-existing inactive component (Eichler and Silman, 1995).

## Molecular forms of acetylcholinesterase during development of the quail brain

The development of the mammalian and avian brain is accompanied by an increase in the proportion of the tetrameric $G_4$ form, and a correlative decrease of the dimeric $G_2$ and monomeric $G_1$ forms. In the quail, the proportion of $G_4$ continuously increases from about 50% at embryonic day 13 (E13), to more than 85% in the adult. In addition, an analysis of the sedimentation patterns observed in the presence of different detergents (Triton X-100 and Brij-96) indicates that while $G_4$ is entirely amphiphilic in the adult brain, it consists of approximately equal amphiphilic and non-amphiphilic components at E13. The proportion of non-amphiphilic $G_4$ continuously decreases during development. This change may reflect the availability of an associated hydrophobic subunit which anchors the tetramers of T subunits in plasma membranes, as demonstrated in the case of the mammalian brain (Gennari et al., 1987; Inestrosa et al., 1987; Roberts et al., 1991). Thus, the development of the brain is accompanied by a progressive increase in the proportion of the physiologically active, membrane-bound form of AChE, with a transient abundance of inactive molecules resulting from inappropriate folding, and a decrease in the proportion of intracellular monomeric and dimeric precursors, and of non-anchored tetramers.

## Significance and regulation of the hetero-oligomeric forms

In muscles, development is also accompanied by an increase in the activity of the more complex molecular forms, in which T subunits are associated with hydrophobic or collagenic anchoring subunits. It seems likely that this evolution reflects the availability of the anchoring proteins. An increased capacity of mature muscle cells to correctly assemble the hetero-oligomeric structures may perhaps also play a role, although the co-expression of collagenic Q subunits and catalytic T subunits is sufficient to produce collagen-tailed forms in the non myogenic COS cells.

The collagen-tailed forms are more clearly associated with neuromuscular endplates than globular forms, e.g. in rat fast muscles (but not in human muscles, as discussed above). These molecules may be tethered in the synaptic basal lamina through ionic interactions with glycosaminoglycans (Bon et al., 1978; Brandan et al., 1985). The collagenic sequence of the tail (Krejci et al., 1991) contains basic residues that appear to constitute heparin-binding sites (Deprez and Inestrosa, 1995). The collagen-tail forms may thus be localized at close proximity of the presynaptic sites of acetylcholine release and of the postsynaptic receptors. They are considered to play a major role in the inactivation of acetylcholine, preventing multiple activation of the receptors.

Strong experimental evidence, however, suggests that the hydrophobic-tailed tetramers, $G_4$, may also play an important role in muscle physiology. The $G_4$ form is specifically deficient in various dystrophic conditions, e.g. in the 129/ReJ or mdx mice (Gisiger and Stephens, 1984; Oliver et al., 1992). The observed defect in $G_4$ is probably not causally related to the disease, but rather represents an index of muscle malformation, since these dystrophies are of different genetic origins.

In fact, muscular exercise, in normal physiological conditions, specifically affects the level of the $G_4$ form (Fernandez and Donoso, 1988; Gisiger et al., 1991, 1994). This regulation seems to involve the release of CGRP by motoneurons (Hodges-Savola and Fernandez, 1995). Careful analysis has shown that this form is localized in a perijunctional zone, immediately surrounding the neuromuscular endplate (Gisiger and Stephens, 1988). This enzyme may thus be responsible for scavenging acetylcholine molecules that might escape hydrolysis by the collagen-tailed forms and diffuse out of the synaptic cleft.

Because the collagen-tailed and hydrophobic-tailed molecules contain the same catalytic T subunits, the distinct regulation of their expression must largely rely on the expression of their associated structural subunits. The collagenic subunits have been cloned in *Torpedo* (Krejci et al., 1991) and more recently in mammals (E. Krejci et al., unpublished), and the cloning of the hydrophobic subunit is under way. It will thus be possible to examine the expression of these proteins and their regulation. The significance of the different forms of AChE in vivo, will be assessed by inactivating the genes encoding each type of anchoring protein in transgenic mice.

### Non-conventional roles of cholinesterases

Both acetylcholinesterase and butyrylcholinesterase are expressed very early during embryogenesis, with distinct temporal and spatial patterns. This has led embryologists to suggest that these proteins may play a role during morphogenesis. The implication of cholinesterases in non-synaptic functions may imply their catalytic activity on acetylcholine or possibly on other substrates. It should be emphasized, in this respect, that the possibility of a peptidasic activity, which has long been discussed, has now been conclusively ruled out (Checler and Vincent, 1989; Michaelson and Small, 1993; Checler et al., 1994). On the other hand, non-conventional functions of cholinesterases might be related exclusively to their protein structure. For example, secreted acetylcholinesterase has been proposed to exert neuromodulatory effects (Greenfield, 1991; Appleyard, 1992), or activate macrophages (Klegeris et al., 1994), independently of its catalytic activity. During embryogenesis, cholinesterases may play a structural role in cellular adhesion or signalling (Layer et al., 1993).

An implication in cellular interactions would be consistent with the fact that some acetylcholinesterase molecules carry the HNK-1 epitope, which is considered a hallmark of adhesion proteins, and more convincingly that they are homologous with non enzymatic adhesion molecules, i.e. the *Drosophila* neurotactin (Barthalay et al., 1990) and the recently described mammalian neuroligin (Ichtchenko et al., 1995). Neuroligin was identified by its $Ca^{2+}$-dependent attachment to the extracellular domain of a subset of neurexins, neuron-specific membrane molecules which exhibit considerable diversity through the existence of three genes, possessing two promoters each and many splice variants (Ullrich et al., 1995). It is conceivable that cholinesterases may specifically interact with some neurexin species.

### Acknowledgements

This work was supported by grants from the Centre National de la Recherche Scientifique (CNRS), the Direction de la Recherche et de la Technologie (DRET) and the Association Française contre les Myopathies (AFM).

### References

Anselmet, A., Fauquet, M., Chatel, J.-M., Maulet, Y., Massoulié, J. and Vallette, F.-M. (1994) Evolution of acetylcholinesterase transcripts and molecular forms during development in the central nervous system of the quail. *J. Neurochem.*, 62: 2158–2165.

Appleyard, M.E. (1992) Secreted acetylcholinesterase: non-classical aspects of a classical enzyme. *Trends Neurosci.*, 15: 485–490.

Bacou, F., Vigneron, P. and Massoulié, J. (1982) Acetylcholinesterase forms in fast and slow rabbit muscle. *Nature*, 296: 661–664.

Barthalay, Y., Hipeau-Jacquotte, R., De la Escalera, S., Jiménez, F. and Piovant, M. (1990) *Drosophila* neurotactin mediates heterophilic cell adhesion. *EMBO J.*, 9: 3603–3609.

Bon, S., Cartaud, J. and Massoulié, J. (1978) The dependence of acetylcholinesterase aggregation at low ionic strength upon a polyanionic component. *Eur. J. Biochem.*, 85: 1–14.

Bon, S., Méflah, K., Musset, F., Grassi, J. and Massoulié, J. (1987) An immmunoglobulin M monoclonal antibody, recognizing a subset of acetylcholinesterase molecules from electric organs of *Electrophorus* and *Torpedo*, belongs to the HNK-1 anti-carbohydrate family. *J. Neurochem.*, 49: 1720–1731.

Bon, S., Toutant, J.P., Méflah, K. and Massoulié, J. (1988a) Amphiphilic and nonamphiphilic forms of *Torpedo* cholinesterases: I. Solubility and aggregation properties. *J. Neurochem.*, 51: 776–785.

Bon, S., Toutant, J.P., Méflah, K. and Massoulié, J. (1988b) Amphiphilic and nonamphiphilic forms of *Torpedo* cholinesterases: II. Existence of electrophoretic variants and of phosphatidylinositol phospholipase C-sensitive and -insensitive forms. *J. Neurochem.*, 51: 786–794.

Bon, S., Rosenberry, T.L. and Massoulié, J. (1991) Amphiphilic, glycophosphatidyl-inositol-specific phospholipase C (PI-PLC)-insensitive monomers and dimers of acetylcholinesterase. *Cell. Mol. Neurobiol.*, 11: 157–172.

Brandan, E., Maldonado, M., Garrido, J. and Inestrosa, N.C. (1985) Anchorage of collagen-tailed acetylcholinesterase to the extracellular matrix is mediated by heparan sulfate proteoglycans. *J. Cell Biol.*, 101: 985–992.

Brimijoin, S., Hammond, P. and Rakonczay, Z. (1987) Two-site immunoassay for acetylcholinesterase in brain, nerve, and muscle. *J. Neurochem.*, 49: 555–562.

Carson, S., Bon, S., Vigny, M., Massoulié, J. and Fardeau, M. (1979) Distribution of acetylcholinesterase molecular forms in neural and non-neural sections of human muscle. *FEBS Lett.*, 97: 348–352.

Chatel, J.M., Grassi, J., Frobert, Y., Massoulié, J. and Vallette, F.M. (1993a) Existence of an inactive pool of acetylcholinesterase in chicken brain. *Proc. Natl Acad. Sci. USA*, 90: 2476–2480.

Chatel, J.M., Vallette, F.M., Massoulié, J. and Grassi, J. (1993b) A conformation-dependent monoclonal antibody against active chicken acetylcholinesterase. *FEBS Lett.*, 319: 12–15.

Chatel, J.M., Eichler, J., Vallette, F.M., Bon, S., Massoulié, J. and Grassi, J. (1994) Two-site immunoradiometric assay of chicken acetylcholinesterase: active and inactive molecular forms in brain and muscle. *J. Neurochem.*, 63: 1111–1118.

Checler, F. and Vincent, J.P. (1989) Peptidasic activities associated with acetylcholinesterase are due to contaminating enzymes. *J. Neurochem.*, 53: 924–928.

Checler, F., Grassi, J. and Vincent, J.P. (1994) Cholinesterases display genuine arylacylamidase activity but are totally devoid of intrinsic peptidase activities. *J. Neurochem.*, 62: 756–763.

Coussen, F., Bonnerot, C. and Massoulié, J. (1995) Stable expression of acetylcholinesterase and associated collagenic subunits in transfected RBL cell lines: production of GPI-anchored dimers and collagen-tailed forms. *Eur. J. Cell Biol.*, 67: 254–260.

Delahunty, M.D., Stafford, F.J., Yuan, L.C., Shaz, D. and Bonifacino, J.S. (1993) Uncleaved signals for glycosylphosphatidylinositol anchoring cause retention of precursor proteins in the endoplasmic reticulum. *J. Biol. Chem.*, 268: 12017–12027.

Deprez, P.N. and Inestrosa, N.C. (1995) Two heparin-binding domains are present on the collagenic tail of asymmetric acetylcholinesterase. *J. Biol. Chem.*, 270: 11043–11046.

Duval, N., Massoulié, J. and Bon, S. (1992a) H and T subunits of acetylcholinesterase from *Torpedo*, expressed in COS cells, generate all types of globular forms. *J. Cell Biol.*, 118: 641–653.

Duval, N. Krejci, E., Grassi, J., Coussen, F., Massoulié, J. and Bon, S. (1992b) Molecular architecture of acetylcholinesterase collagen-tailed forms; construction of a glycolipid-tailed tetramer. *EMBO J.*, 11: 3255–3261.

Eichler, J. and Silman, I. (1995) The activity of an endoplasmic reticulum-localized pool of acetylcholinesterase is modulated by heat shock. *J. Biol. Chem.*, 270: 4466–4472.

Eichler, J., Toker, L. and Silman, I. (1991) Effect of heat shock on acetylcholinesterase activity in chick muscle cultures. *FEBS Lett.*, 293: 16–20.

Fernandez, H.L. and Donoso, J.A. (1988) Exercise selectively increases $G_4$ AChE activity in fast-twitch muscle. *J. Appl. Physiol.*, 65: 2245–2252.

Fuentes, M.E. and Taylor, P. (1993) Control of acetylcholinesterase gene expression during myogenesis. *Neuron*, 10: 679–687.

Gennari, K., Brunner, J. and Brodbeck, U. (1987) Tetrameric detergent-soluble acetylcholinesterase from human caudate nucleus: subunit composition and number of active sites. *J. Neurochem.*, 49: 12–18.

Gerber, L.D., Kodukula, K. and Udenfriend, S. (1992). Phosphatidylinositol glycan (PI-G) anchored membrane proteins. *J. Biol. Chem.*, 267: 12168–12173.

Gisiger, V. and Stephens, H. (1984) Decreased $G_4$ (10 S) acetylcholinesterase content in motor nerves to fast muscles of dystrophic 129/ReJ mice: lack of a specific compartment of nerve acetylcholinesterase? *J. Neurochem.*, 43: 174–183.

Gisiger, V. and Stephens, H.R. (1988) Localization of the pool of $G_4$ acetylcholinesterase characterizing fast muscles and its alteration in murine dystrophy. *J. Neurosci. Res.*, 19: 62–68.

Gisiger, V., Sherker, S. and Gardiner, P.F. (1991) Swimming training increases the $G_4$ acetylcholinesterase content of both fast ankle extensors and flexors. *FEBS Lett.*, 278: 271–273.

Gisiger, V., Bélisle, M. and Gardiner, P.F. (1994) Acetylcholinesterase adaptation to voluntary wheel running is proportional to the volume of activity in fast, but not slow, rat hindlimb muscles. *Eur. J. Neurosci.*, 6: 673–680.

Greenfield, S. (1991) A non-cholinergic action of acetylcholinesterase (AChE) in the brain: from neuronal secretion to the generation of movements. *Cell. Mol. Neurobiol.*, 11: 55–78.

Hall, Z.W. (1973) Multiple forms of acetylcholinesterase and their distribution in endplate and non-endplate regions of rat diaphragm muscle. *J. Neurol.*, 4: 343–361.

Hammond, P. and Brimijoin, S. (1988) Acetylcholinesterase in Huntington's and Alzheimer's diseases: simultaneous enzyme assay and immunoassay of multiple brain regions. *J. Neurochem.*, 50: 1111–1116.

Hodges-Savola, C.A. and Fernandez, H.L. (1995) A role for calcitonin gene-related peptide in the regulation of rat skeletal muscle $G_4$ acetylcholinesterase. *Neurosci. Lett.*, 190: 117–120.

Ichtchenko, K., Hata, Y., Nguyen, T., Ullrich, B., Missler, M., Moomaw, C. and Südhof, T.C. (1995) Neuroligin 1: a splice site-specific ligand for $\beta$-neurexins. *Cell*, 81: 435–443.

Inestrosa, N.C., Roberts, W.L., Marshall, T. and Rosenberry, T.L. (1987) Acetylcholinesterase from bovine caudate nucleus is attached to membranes by a novel subunit distinct from those of acetylcholinesterase in other tissues. *J. Biol. Chem.*, 262: 4441–4444.

Inestrosa, N.C., Fuentes, M.E., Anglister, L., Futerman, A.H. and Silman, I. (1988) A membrane-associated dimer of acetylcholinesterase from Xenopus skeletal muscle is solubilized by phosphatidylinositol-specific phospholipase C. *Neurosci. Lett.*, 90: 186–190.

Kim, P.S. and Arvan, P. (1995) Calnexin and BiP act as sequential molecular chaperones during thyroglobulin folding in the endoplasmic reticulum. *J. Cell Biol.*, 128: 29–38.

Klegeris, A., Budd, T.C. and Greenfield, S.A. (1994) Acetylcholinesterase activation of peritoneal macrophages is independent of catalytic activity. *Cell. Mol. Neurobiol.*, 14: 89–98.

Kodukula, K., Cines, D., Amthauer, R., Gerber, L.D. and Udenfriend, S. (1992) Biosynthesis of phosphatidylinositol-glycan (PI-G)-anchored membrane proteins in cell-free systems: cleavage of the nascent protein and addition of the PI-G moiety depend on the size of the COOH-terminal signal peptide. *Proc. Natl. Acad. Sci. USA*, 89: 1350–1353.

Kodukula, K., Gerber, L.D., Amthauer, R., Brink, L. and Udenfriend, S. (1993) Biosynthesis of glycosylphosphatidylinositol (GPI)-anchored membrane proteins in intact cells: specific amino acid requirements adjacent to the site of cleavage and GPI attachment. *J. Cell Biol.*, 120: 657–664.

Krejci, E., Coussen, F., Duval, N., Chatel, J.M., Legay, C., Puype, M., Vandekerckhove, J., Cartaud, J., Bon, S. and Massoulié, J. (1991) Primary structure of a collagenic tail subunit of *Torpedo* acetylcholinesterase: co-expression with catalytic subunit induces the production of collagen-tailed forms in transfected cells. *EMBO J.*, 10: 1285–1293.

Layer, P.G., Weikert, T. and Alber, R. (1993) Cholinesterases regulate neurite growth of chick nerve cells *in vitro* by means of a non-enzymatic mechanism. *Cell Tissue Res.*, 273: 219–226.

Lazar, M., Salmeron, E., Vigny, M. and Massoulié, J. (1984) Heavy isotope labeling study of the metabolism of monomeric and tetrameric acetylcholinesterase forms in the murine neuronal-like T28 hybrid cell line. *J. Biol. Chem.*, 259: 3703–3713.

Legay, C., Bon, S., Vernier, P., Coussen, F. and Massoulié, J. (1993a) Cloning and expression of a rat acetylcholinesterase subunit; generation of multiple molecular forms, complementarity with a *Torpedo* collagenic subunit. *J. Neurochem.*, 60: 337–346.

Legay, C., Bon, S. and Massoulié, J. (1993b) Expression of a cDNA encoding the glycolipid-anchored form of rat acetylcholinesterase. *FEBS Lett.*, 315: 163–166.

Legay, C., Huchet, M., Massoulié, J. and Changeux, J.P. (1995) Developmental regulation of acetylcholinesterase transcripts in the mouse diaphragm: alternative splicing and focalization. *Eur. J. Neurosci.*, 7: 1803–1809.

Li, Y., Camp, S. and Taylor, P. (1993) Tissue-specific expression and alternative mRNA processing of the mammalian acetylcholinesterase gene. *J. Biol. Chem.*, 268: 5790–5797.

Lømo, T., Massoulié, J. and Vigny, M. (1985) Stimulation of denervated rat soleus muscle with fast and slow activity patterns induce different expression of acetylcholinesterase molecular forms. *J. Neurosci.*, 5: 1180–1187.

Massoulié, J., Pezzementi, L., Bon, S., Krejci, E. and Vallette, F.M. (1993) Molecular and cellular biology of cholinesterases. *Prog. Neurosci.*, 41: 31–91.

Maxwell, S.E., Ramalingam, S., Gerber, L.D. and Udenfriend, S. (1995) Cleavage without anchor addition accompanies the processing of a nascent protein to its glycophosphatidylinositol-anchored form. *Proc. Natl. Acad. Sci. USA*, 92: 1550–1554.

Micanovic, R., Gerber, L., Berger, J., Kodukula, K. and Udenfriend, S. (1990) Selectivity of the cleavage/attachment site of phosphatidylinositol-glycan-anchored membrane proteins determined by site-specific mutagenesis at Asp-484 of placental alkaline phosphatase. *Proc. Natl. Acad. Sci. USA*, 87: 157–161.

Michaelson, S. and Small, D.H. (1993) A protease is recovered with a dimeric form of acetylcholinesterase in fetal bovine serum. *Brain Res.*, 611: 75–80.

Moran, P. and Caras, I.W. (1991a) A nonfunctional sequence converted to a signal for glycophosphatidylinositol membrane anchor attachment. *J. Cell Biol.*, 115: 329–336.

Moran, P. and Caras, I.W. (1991b) Fusion of sequence elements from non-anchored proteins to generate a fully functional signal for glycophosphatidylinositol membrane anchor attachment. *J. Cell Biol.*, 115: 1595–1600.

Moran, P. and Caras, I.W. (1992) Proteins containing an uncleaved signal for glycophosphatidylinositol membrane anchor attachment are retained in a post-ER compartment. *J. Cell Biol.*, 119: 763–772.

Moran, P. and Caras, I.W. (1994) Requirements for glycophosphatidylinositol attachment are similar but not identical in mammalian cells and parasitic protozoa. *J. Cell Biol.*, 125: 333–343.

Oliver, L., Chatel, J.M., Massoulié, J., Vigny, M. and Vallette, F.M. (1992) Localization and characterization of the molecular forms of acetylcholinesterase in dystrophic (mdx) mouse tissues. *Neuromusc. Disord.,* 2: 87–97.

Piette, J., Huchet, M., Houzelstein, D. and Changeux, J.-P. (1993) Compartmentalized expression of the $\alpha$- and $\gamma$-subunits of the acetylcholine receptor in recently fused myofibers. *Dev. Biol.,* 157: 205–213.

Pinset, C., Mulle, C., Benoit, P., Changeux, J.-P., Chelly, J., Gros, F. and Montarras, D. (1991) Functional adult acetylcholine receptor develops independently of motor innervation in Sol 8 mouse muscle cell line. *EMBO J.,* 10: 2411–2418.

Randall, W.R., Rimer, M. and Gough, N.R. (1994) Cloning and analysis of chicken acetylcholinesterase transcripts from muscle and brain. *Biochim. Biophys. Acta,* 1218: 453–456.

Roberts, W.L., Doctor, B.P., Foster, J.D. and Rosenberry, T.L. (1991) Bovine brain acetylcholinesterase primary sequence involved in intersubunit disulfide linkages. *J. Biol. Chem.,* 266: 7481–7487.

Seidman, S., Sternfeld, M., Ben Aziz-Aloya, R., Timberg, R., Kaufer-Nachum, D. and Soreq, H. (1995) Synaptic and epidermal accumulations of human acetylcholinesterase are encoded by alternative 3′-terminal exons. *Mol. Cell. Biol.,* 15: 2993–3002.

Sketelj, J. (1994) Neural regulation of acetylcholinesterase in skeletal muscles. *Basic Appl. Myology,* 4: 281–291.

Sketelj, J. and Brzin, M. (1985) Asymmetric molecular forms of acetylcholinesterase in mammalian skeletal muscles. *J. Neurosci. Res.,* 14: 95–103.

Sussman, J.L., Harel, M., Frolow, F., Oefner, C., Goldman, A., Toker, L. and Silman, I. (1991) Atomic structure of acetylcholinesterase from *Torpedo californica*: a prototypic acetylcholine-binding protein. *Science,* 253: 872–879.

Ullrich, B., Ushkaryov, Y.A. and Südhof, T.C. (1995) Cartography of neurexins: more than 1000 isoforms generated by alternative splicing and expressed in distinct subsets of neurons. *Neuron,* 14: 497–507.

# Section II

# Vesicular Acetylcholine Transporter

J. Klein and K. Löffelholz (Eds.)
*Progress in Brain Research*, Vol. 109
© 1996 Elsevier Science B.V. All rights reserved.

CHAPTER 5

# The VAChT/ChAT "cholinergic gene locus": new aspects of genetic and vesicular regulation of cholinergic function

Jeffrey D. Erickson[1], Eberhard Weihe[2], Martin K.-M. Schäfer[2], Elaine Neale[3], Lura Williamson[3], Tom I. Bonner[4], Jung-Hwa Tao-Cheng[5] and Lee E. Eiden[1]

[1]*Section on Molecular Neuroscience, Laboratory of Cell Biology, NIMH, NIH, Bethesda, MD, USA,* [2]*Department of Anatomy and Cell Biology, Philipps University, Marburg, Germany,* [3]*Laboratory of Developmental Neurobiology, NICHD, NIH, Bethesda, MD, USA,* [4]*Laboratory of Cell Biology, NIMH, NIH, Bethesda, Maryland, USA and* [5]*Laboratory of Neurobiology, NINDS, NIH, Bethesda, MD, USA*

## Introduction

Chemical coding in the nervous system requires cell-specific expression of several classes of neuronal proteins. Neuropeptide precursors, their processing enzymes, neurotransmitter biosynthetic enzymes, and transporters for neurotransmitter uptake and accumulation into synaptic vesicles are all expressed in cells at precise neuroanatomical locations to allow central and peripheral neurotransmission to occur.

In the last 4 years, cDNAs have been cloned and characterized that encode the vesicular monoamine transporters responsible for vesicular storage of dopamine, norepinephrine, epinephrine, serotonin and histamine, which act as neurotransmitters, hormones and autacoids in neuroendocrine cells (Erickson et al., 1992; Liu et al., 1992; Erickson and Eiden, 1993; Krejjci et al., 1993; Peter et al., 1993; Surratt et al., 1993; Howell et al., 1994). In 1993, Rand and colleagues identified the cDNA encoding the product of the *unc-17* gene in C. elegans (Alfonso et al., 1993). Based on its ubiquitous expression in the nematode cholinergic nervous system, the interruption of cholinergic neurotransmission by abrogating expression of the *unc-17* gene in *Caenorhabditis elegans*, and in particular the homology of UNC-17 to the vesicular monoamine transporters (VMATs), Rand and colleagues suggested that UNC-17 was the nematode vesicular acetylcholine transporter (Alfonso et al., 1993).

Complementary DNAs encoding the putative *Torpedo*, rat and human vesicular acetylcholine transporters have since been cloned based on their homology to UNC-17 (Erickson et al., 1994; Roghani et al., 1994; Varoqui et al., 1994). The rat protein has been functionally characterized and thus proved to be a vesicular acetylcholine transporter. It mediates accumulation of ACh after expression in a mammalian non-neuronal cell line that is dependent on vacuolar ATPase activity, on maintenance of an intracellular proton gradient, and is inhibited by vesamicol (Erickson et al., 1994). This protein has been named VAChT (vesicular **ACh t**ransporter) (Erickson et al., 1994). VAChT appears to be expressed in all known major cholinergic neuronal systems in rat brain (Chapter 7; Erickson et al., 1994; Roghani et al., 1994; Schäfer et al., 1994) and peripheral nervous system (Schäfer et al., 1994), and is therefore likely to be the sole protein required for vesicular ACh accumulation in cholinergic neurons in the rat (Erickson et al., 1994; Schäfer et al., 1994).

The nematode gene for VAChT (*unc-17*) maps to a shared gene locus with *cha-1*, which encodes the cholinergic biosynthetic enzyme choline acetyltransferase (ChAT) (Rand and Russell, 1984; Rand, 1989; Alfonso et al., 1994). The human VAChT gene has been mapped to chromosome 10q11.2, the chromosomal localization of human ChAT (Cohen-Haguenauer et al., 1990; Erickson et al., 1994). The nested organization of the VAChT and ChAT genes (discussed below) and the fact that they encode the major proteins responsible for expression of the cholinergic phenotype, has led to the designation of the VAChT/ChAT genes as the "cholinergic gene locus" (Erickson et al., 1994).

Here we review structural features of VAChT deduced from its primary sequence that give insight into its transporter function. We describe the distribution of VAChT in the rat and primate nervous systems and its intracellular targeting to synaptic vesicles in neuronal cells. We also describe the ontogeny of VAChT and ChAT expression in an in vitro model for developing spinal cord motor neurons and discuss the implications of the phylogenetic conservation of the VAChT/ChAT "cholinergic gene locus" for the developmental and evolutionary ontogenesis of the cholinergic nervous system.

## Structure, transport properties, and pharmacology of VAChT: comparison with vesicular monoamine transporters (VMATs)

The biogenic amine and acetylcholine vesicular transporters share important structural and functional similarities and have important differences. Both VAChT and the VMATs are significantly glycosylated integral membrane proteins with 12 putative hydrophobic membrane-spanning domains. Both types of transporters are responsible for the accumulation of amine as the positively charged species (Henry et al., 1987; Parsons et al., 1993), and accumulate amine by exchanging protons within the vesicle for amines in the cytoplasm (amine/proton antiport) (Johnson, 1988; Parsons et al., 1993). Since the interior of the vesicle is positively charged, motive force for amine accumulation arises solely from the proton electrochemical gradient, the relative impermeability of the vesicular membrane to protons, and the coupling of proton passage from the vesicle with amine entry into the vesicle through the transporter molecule (Njus et al., 1986; Maycox et al., 1990). Both the cholinergic and the biogenic amine-containing vesicle generate a proton gradient via energy-dependent

proton pumping into the vesicle catalyzed by a vacuolar ATPase on the vesicular membrane (Toll and Howard, 1980; Johnson, 1988). Finally, amine transport by both VMATs and VAChT are blocked by potent and specific inhibitors: reserpine for VMAT1 and VMAT2, tetrabenazine for VMAT2 (and less potently VMAT1), and vesamicol for VAChT (Henry et al., 1987; Erickson et al., 1992; Liu et al., 1992; Parsons et al., 1993; Peter et al., 1994; Weihe et al., 1994).

There are also significant differences among the transporters. Although VMATs and VAChT both transport positively charged amines, biogenic amines are not substrates for VAChT (Clarkson et al., 1993; Erickson et al., 1994). Vesamicol does not inhibit monoamine uptake by VMATs nor is VAChT-mediated ACh transport inhibited by reserpine or tetrabenazine (Erickson et al., 1992, 1994). VAChT mediates accumulation of ACh against a gradient of approximately 100-fold in cholinergic vesicles from *Torpedo*, whilst the chromaffin granule monoamine transporter mediates accumulation of biogenic amines as much as 10 000-fold from the extravesicular medium (Carmichael and Winkler, 1985; Parsons et al., 1993).

These general properties of amine vesicular transporters are maintained upon expression of either VMATs or VAChT against a common non-neuronal cell background, and are therefore intrinsic to the transporters themselves, and not imparted by interaction with other neuron-specific accessory factors in adrenergic or cholinergic cells. Both VAChT and VMATs, upon expression in fibroblasts, can mediate uptake of their respective substrates, which is blocked by specific inhibitors, into a cellular compartment that contains a vacuolar ATPase and maintains a proton gradient required for transport. Accumulation of biogenic amines mediated by VMATs occurs to a significantly greater extent than ACh accumulation mediated by VAChT (Erickson et al., 1992, 1994). These properties all generally parallel the transport properties established for ACh transport into *Torpedo* vesicles and PC12 cells, and biogenic amine uptake into biogenic amine-containing storage vesicles isolated either from chromaffin cells or

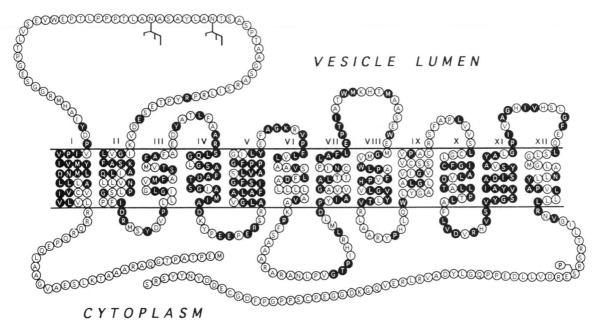

Fig. 1. Sequence and putative structure of rat VAChT, and conservation of its amino acid sequence among mammalian and non-mammalian species. The primary sequence of human VAChT is given in single-letter amino acid code, with residues conserved among unc-17, *Torpedo* VAChT, rat VAChT and human VAChT in black background.

from the central nervous system of mammals (Philippu and Beyer, 1973; Toll and Howard, 1978; Toll and Howard, 1980; Johnson, 1988; Parsons et al., 1993).

Expression of the recombinant vesicular acetylcholine transporter in non-neuronal cells has provided experimental proof that a single polypeptide is responsible for both ACh transport and high-affinity vesamicol binding in cholinergic neurons (Erickson et al., 1994; Varoqui et al., 1994). Previous hypotheses have included the formation of a multi-protein complex consisting in separate polypeptides subserving vesamicol binding and acetylcholine transport (Parsons et al., 1993). Demonstration that the vesamicol receptor and acetylcholine transporter are a single molecule has clarified the mechanism of vesamicol action at the neuromuscular junction and at autonomic cholinergic synapses, re-emphasizing the importance of vesicular accumulation in the quantal release of acetylcholine.

Both similar and divergent properties of VAChT and the VMATs, summarized in Table 1, ought to be reflected in conserved and divergent

amino acid sequences within the primary structures of the corresponding proteins. VAChTs and VMATs in general share significant sequence homology in the transmembrane domains of the proteins (Alfonso et al., 1993; Erickson et al., 1994). Single aspartic acid residues in transmembrane domains I, VI, X and XI of all VAChTs and VMATs may be critical for amine/proton antiport. An aspartic acid in transmembrane domain IV of all VAChTs, lacking in VMATs, may be a determinant of amine substrate specificity of the acetylcholine transporter (Usdin et al., 1995). The large intraluminal loop between transmembrane domains 1 and 2 in both VAChTs and VMATs is poorly conserved between VMATs and VAChTs, among the VAChTs of nematode, *Torpedo*, rat and human, and even between rat and human VAChTs. Conservation of multiple potential glycosylation sites among all transporters in this domain however, may indicate a role for this post-translational modification in addressing or function of the amine vesicular transporters. The N- and C-termini are poorly conserved between VMATs and VAChT and within VAChTs across species as well, although an aromatic amino

TABLE 1

Properties and structures of VAChTs and VMATs

| | VAChT | VMAT1/2 |
|---|---|---|
| Substrate(s) | Acetylcholine | DA, NE, Epi, 5-HT (both) histamine (VMAT2) |
| Transported species | Quaternary amine | Protonated amine |
| Accumulation of transported species | ~100-fold (*Torpedo* SSVs) | ~10000-fold (bovine chromaffin granules) |
| Countertransported species | Proton (2 per amine transport cycle calculated) | Proton (2 per amine transport cycle) |
| Inhibitor | Vesamicol | Reserpine (both) tetrabenazine (VMAT2) |
| Secondary structure | Putative "12-TMD" | Putative "12-TMD" |
| Conserved negatively charged residues | D, TMD I, VI, X, XI D, TMD IV | D, TMD I, VI, X, XI |
| Intraluminal loop | 2–3 potential glycosylation sites | 3–4 potential glycosylation sites |

"TMD", transmembrane domain; DA, dopamine; NE, norepinephrine; Epi, epinephrine; 5-HT, serotonin.

acid-rich portion of the C-terminal tail of the VAChTs is somewhat conserved. The C-terminal domain of all of the vesicular transporters, however, is generally rich in acidic amino acids, which could serve to "anchor" this portion of the transporter in the cytoplasm. In general the divergence and the hydrophilicity of the C-terminus of the amine transporters have made this region ideal for the generation of anti-peptide antibodies for the further study of their cellular and intracellular distribution, described for VAChT below.

## Distribution of VAChT in the rat and primate nervous systems

We originally investigated the distribution of VAChT mRNA in the central and peripheral nervous systems of the rat by in situ hybridization histochemistry with a riboprobe complementary to the 3′ untranslated region of the VAChT message (Erickson et al., 1994; Schäfer et al., 1994). The rationale for employing this probe was its high specificity for VAChT and the unlikelihood that it would cross-hybridize to transcripts from other VAChT isoform genes, should they exist. Two isoforms of the vesicular biogenic amine transporter, designated VMAT1 and VMAT2, are expressed throughout the neuroendocrine system, with VMAT1 expressed mainly in endocrine, and VMAT2 predominantly in neuronal, cells (Weihe

et al., 1994). In contrast, all of the major cholinergic cell groups of the central and peripheral nervous system of the rat, identified as such either by classical neuroanatomical criteria or staining for choline acetyltransferase (ChAT) mRNA or protein, appear to express VAChT mRNA or protein (Fig. 2) (Erickson et al., 1994; Roghani et al., 1994; Schäfer et al., 1994). Of particular interest, VAChT mRNA-positive cells have been identified in the rat cerebral cortex, confirming the existence of cortical cholinergic interneurons (Eckenstein et al., 1988; Schäfer et al., 1994), and in the medial habenular and hypothalamic arcuate nuclei, previously reported to contain cholinergic neurons based on staining for ChAT immunoreactivity (Schäfer et al., 1994 and references therein).

In the primate central and peripheral nervous systems as well, VAChT mRNA and protein are present in the basal forebrain, spinal motor neurons, parasympathetic and sympathetic preganglionic neurons, and in parasympathetic principal ganglion cells, as well as a subpopulation of cells in the stellate ganglion, presumably corresponding to the cholinergic sudomotor neurons of this sympathetic ganglion (Fig. 3) (Schäfer et al., unpublished). Examination of the human nervous system confirms the ubiquitous expression of a single isoform of VAChT in all major groups of cholinergic neurons in both central and peripheral nervous systems (Schäfer et al., 1995).

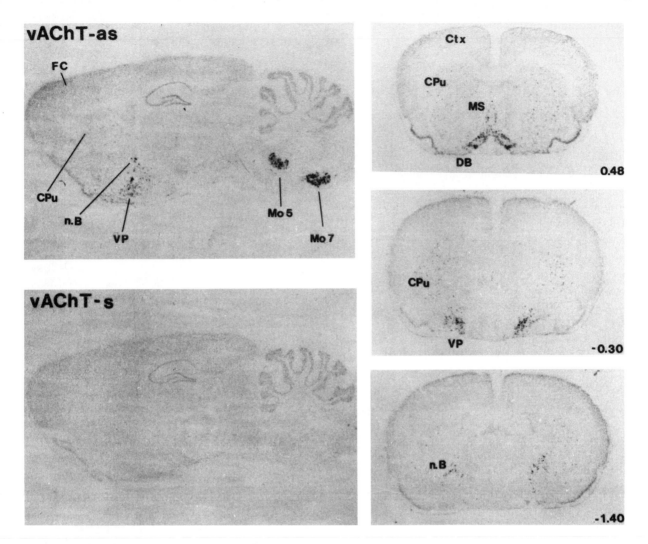

Fig. 2. Sagittal section of rat brain hybridized with 3′-UT VAChT probe. Distribution of rat vesicular acetylcholine transporter mRNA in the cholinergic nervous system using a $^{35}$S-labeled VAChT cRNA antisense (as) probe. Rat brain sections were hybridized using the 3′ VAChT probe, and the hybridization conditions, described previously (Schäfer et al., 1994). Sections were exposed to emulsion for 14 days. Hybridization in cholinergic areas of forebrain, i.e. ventral pallidum (VP), basal nucleus of Meynert (n.b.), caudate-putamen (CPu), medial septal nucleus (MS), nuclei of the horizontal and vertical limb of the diagonal band (DB) as shown on a sagittal section (upper left) and coronal sections (right). Specific VAChT mRNA hybridization is seen in low abundance throughout cerebral cortex, in moderate abundance in striatum and in ventral forebrain, and in great abundance in motor nuclei of the trigeminal (Mo5) and facial (Mo7) nerves. Hybridization with the complementary sense (s) probe demonstrates specificity of VAChT labeling in regions described above. Weak labeling over cerebellum and hippocampus is nonspecific. Numbers in right-hand panels indicate rostral (+)/caudal (−) distance of section planes from Bregma.

Thus, in contrast to the VMATs, multiple isoforms of VAChT do not appear necessary for transport of ACh in different divisions of the nervous system, either central versus peripheral, or motor versus autonomic. This may reflect the monolithic nature of neurons that express VAChT, all expressing the single classical neurotransmitter ACh, compared to those that express VMATs and also mutually exclusively express biosynthetic enzymes for catecholamines, serotonin, or histamine.

## The VAChT/ChAT "cholinergic gene locus"

Since VAChT and the ACh biosynthetic enzyme choline acetyltransferase (ChAT) are apparently invariably co-expressed in the mammalian nervous system, we have sought a mechanism for co-regulation at the level of the genes encoding each protein. Following the discovery of the genetic linkage between the gene encoding UNC-17 and that encoding choline acetyltransferase (ChAT), in the nematode C. elegans, the physical relationship between the ChAT and VAChT gene loci in the mammalian genome was examined. A fluorescent cDNA probe for human VAChT was used to locate the VAChT gene to human chromosome 10q11.2, the locus at which human ChAT was previously found (Cohen-Haguenauer et al., 1990).

Fig. 4 shows the detailed structure of the rat VAChT/ChAT gene. The VAChT gene in both human and rat is contained within the first intron of the "R-type" ChAT primary transcript (reviewed by Usdin et al., 1995). This genomic arrangement is identical to that for the *C. elegans unc-17/cha-1* gene locus (Alfonso et al., 1994). Conservation of this structure from nematode to human is particularly remarkable in that the nematode VAChT gene is itself composed of three separate coding exons in addition to the common VAChT/ChAT upstream non-coding exon ("R"

exon), while rat and human VAChTs comprise but a single coding exon within their respective genomes.

As described by Alfonso (1994), the nematode *unc-17/cha-1* gene locus appears to express a sepa

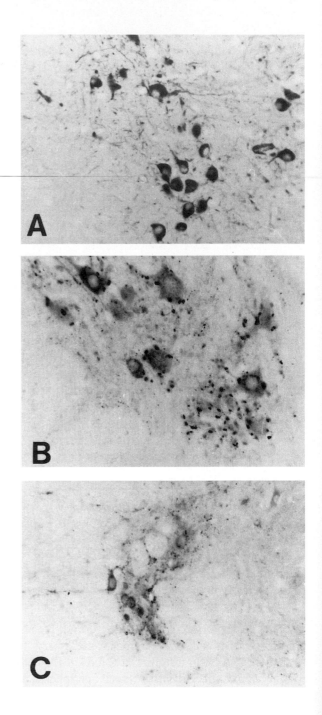

Fig. 3. Immunohistochemical localization of VAChT in the primate central nervous system. Tissue blocks from brain of *Macaca mulatta* were post-fixed in Bouin–Hollande after perfusion in situ with 10% formalin/1% methanol in phosphate-buffered saline. Six micrometer sections of Bouin–Hollande fixed deparaffinized tissue was stained with an anti-human anti-VAChT rabbit antiserum raised against a C-terminal VAChT dodecapeptide used at 1:1000 dilution with nickel-enhanced peroxidase-catalyzed diaminobenzidine deposition to afford visualization of the antigen-antibody complex. All micrographs are at identical magnification. (A) nucleus basalis of Meynert; (B) spinal motor neurons; (C) sympathetic pre-ganglionic cholinergic neurons of the intermediolateral cell column (IML). See Schafer et al. (1995) for additional details.

rate mRNA for each protein via alternative splicing following transcription beginning upstream of the single common 5′ exon (designated "R" in the mammal) of *unc-17* and *cha-1* (VAChT and ChAT). Indirect evidence (reviewed by Usdin et al., 1994) exists for a common utilization of this exon to generate alternative transcripts for VAChT and ChAT in the rat. Thus, Bejanin and co-workers have used reverse transcription-polymerase chain reaction (PCR) with rat spinal cord mRNA as template to generate DNA species containing both R-exon and VAChT sequences that imply splicing of the R exon to the VAChT coding exon. The cDNA initially cloned by Erickson et al. however, and demonstrated to encode functional VAChT, extends beyond this putative

splice junction (as shown in Fig. 4), suggesting that a second transcription initiation point for VAChT may exist downstream of the putative VAChT/ChAT common transcription start site that includes the R exon. Misawa et al. have directly cloned spinal cord cDNAs which contain R-exon sequences spliced to N and M exons, in addition to cDNAs which do not contain R-exon sequences (see Fig. 4). Indeed, Hersh and co-workers (Chpater 3; Li et al., 1995) have identified a cholinergic cell-specific promoter/enhancer domain for human ChAT gene transcription that is contained within the VAChT gene itself, corresponding to 3′ coding and untranslated regions of VAChT. Our detailed analysis of the rat VAChT/ChAT gene indicates that this ChAT promoter/enhancer in the rat (Kengaku et al., 1993) would also contain sequences within the VAChT gene itself (see Fig. 4).

Thus, the mammalian VAChT/ChAT gene locus may initiate transcription of VAChT and ChAT (i) from a common transcriptional start site located upstream of the R exon yielding separate transcripts by alternative splicing with a common 5′ exon, and (ii) from separate transcripts for VAChT (initiating within or downstream of the R

Fig. 4. Structure of the rat VAChT/ChAT "cholinergic" gene locus and its transcripts. The structure of the rat VAChT and ChAT genes and the putative VAChT and ChAT transcripts from the VAChT/ChAT gene locus are shown. Non-coding ChAT exons R, N and M are denoted as described in the literature (see Usdin et al., 1995 and references therein). The single VAChT coding exon, and the first coding exon of ChAT are indicated, with the asterisk denoting the initiating methionine codon. The rat VAChT/ChAT "cholinergic gene locus" was mapped using known restriction sites in both VAChT and ChAT cDNAs and the rat ChAT gene from a genomic clone obtained by PCR screening of a P1 rat genomic clone bank using rat VAChT oligonucleotides from the known sequence of the cDNA (Erickson et al., 1994). The rat VAChT mRNA appears to be represented within the rat genome as a single uninterrupted exon from 400 bases upstream of the initiating methionine through the polyadenylation signal in the 3′ untranslated region. As such, the structure of the rat VAChT/ChAT gene locus is essentially identical to the human VAChT/ChAT gene locus (see Erickson et al., 1994; Bejanin et al., 1994; Usdin et al., 1995). It is not yet known if the initiation of transcription of the rat and human VAChT gene is primarily at the R exon, with read-through to the VAChT coding region, primarily at the R exon with an intron allowing splicing to the VAChT coding region, or downstream of the R exon at an initiation site separate from R exon-intiated transcription of the ChAT gene (see Usdin et al., 1995). Note the starts of transcription of the V1a and V1b putative VAChT transcripts are depicted as indeterminate, since these transcripts have been deduced from RT-PCR amplification experiments using primers within the R exon (Bejanin et al., 1994). Figure adapted from Usdin et al. (1995) and modified based on additional rat structural gene information as described above.

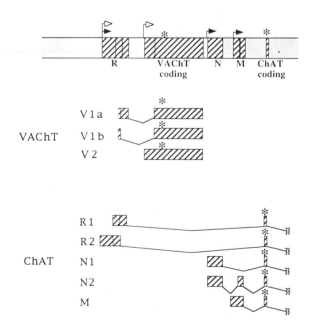

exon region) and ChAT (initiating downstream of both the R-exon and the VAChT coding region, and probably within the 3′ end of the VAChT gene itself). Work is in progress in a number of laboratories to delineate the detailed transcriptional pathways that generate VAChT and ChAT transcripts in both CNS and periphery. The cholinergic nervous system in mammals is considerably more complex than that of C. elegans. The latter lacks an autonomic nervous system so that the cholinergic phenotype is restricted to motor neurons of the head and body, whereas mammalian cholinergic gene regulation must occur not only in motor neurons (and not non-motor neurons and non-neuronal cells) but in parasympathetic (and not sympathetic) autonomic neurons.

We have partially sequenced the 5′ flanking sequences of the human VAChT/ChAT gene locus. Approximately 2 kb upstream of the 5′ border of the R-exon, a consensus RE-1/NRSE sequence exists (Chong et al., 1995; Schoenherr and Anderson, 1995) (see Table 2). This element binds a protein (REST, or NRSF) expressed in non-neuronal cells, and apparently absent from all neuronal cells, which suppresses gene transcription both in genes in which the RE-1/NRSE element is normally found upstream of the transcriptional start site (see Table 2) and in chimeric genes in which RE-1/NRSE is fused to the constitutive promoters of heterologous genes (Chong et al., 1995; Schoenherr and Anderson, 1995). The role of the RE-1/NRSE cis-acting locus of the VAChT/ChAT gene locus in imparting cholinergic-specific transcriptional regulation is currently under investigation. Recently, Ibanez and coworkers have produced transgenic mice containing a reporter gene under the control of 2 kb of the 5′ flank of the rat VAChT/ChAT gene (i.e. 2 kb upstream of the "R" exon) which is correctly expressed in the nervous system, within presumptive motor neurons, and appropriately up-regulated upon axotomy in vivo (Lönnerberg et al., 1994). The element also mediates correct expression in cholinergic versus non-neuronal and non-cholinergic cells in vitro. Although expression in the autonomic nervous system and the effects of deletions of this construct on expression in vivo

TABLE 2

The RE-1-like regulatory domain of human VAChT

| HVAChT | CAGCACCCTGGACAG |
|---|---|
| SCG10 | CAGCACCTTGGACAG |
| Synapsin | CAGAACCACGGACAG |
| Na II | CAGAACCACGGACAG |
| BDNF | CAGCACCACGGAGAG |
| Consensus | CAG-ACC--GGA-AG |

Abbreviations: SCG10, superior cervical ganglion-specific gene product, clone 10; NaII, sodium channel type II subunit; BDNF, brain-derived neurotrophic factor; RE-1 restrictive element-1, also called non-neuronal silencer response element (NRSE).

were not checked, it seems likely that the rat VAChT/ChAT gene also contains an RE-1-like element that may mediate this pattern of expression in transgenic mice.

It is noteworthy that Hersh and co-workers have identified a domain of the human gene well downstream of the VAChT RE-1 (Chapter 3; Li et al., 1995) that suppresses transcriptional activity in non-cholinergic neuronal cells. Clearly, an element in addition to RE-1 (which presumably restricts expression to neuronal cells, but not among subtypes of neuronal cells), must exist in order to properly address VAChT/ChAT expression not only to neurons, but specifically to cholinergic neurons. A Pou-like protein has been identified in Drosophila which binds to the VAChT/ChAT gene to mediate cholinergic neuron-specific expression (Kitamoto and Salvaterra, 1995). It may be that combinations of cis-acting elements act together to bind sets of proteins that mediate correct expression of the VAChT/ChAT gene locus in motor, sympathetic, parasympathetic and central cholinergic neurons in mammals.

## Co-expression of VAChT and ChAT in developing nervous system

Given the potential complexities of transcription from the VAChT/ChAT gene locus cited above, a number of laboratories have focused on the differential regulation of VAChT and ChAT in neuronal systems in culture in which VAChT and ChAT expression are regulated by neurotrophins, retinoids, and other signalling pathways. Berse and Blusztajn (1995) have suggested that VAChT and ChAT messages may be independently regulated, since the ratio of VAChT to ChAT mRNA species can be shown to vary as much as two-fold under stimulation by different agents, including cyclic AMP and retinoids, in a murine septal cholinergic cell line. It has not been determined if this differential regulation is wholly transcriptional or has components of post-transcriptional regulation as well. Misawa et al. have reported on up-regulation of VAChT and ChAT transcripts amplified by PCR in cultured superior cervical ganglia of the rat treated with cholinergic differentiation factor (CDF/LIF), and this has been confirmed by Berrard et al. in the same culture system stimulated with CDF/LIF as well as retinoids (Berrard et al., 1995; Misawa et al., 1995).

We have examined VAChT expression in the embryonic mouse spinal cord in culture to determine its developmental scheduling relative to that of choline acetyltransferase and other neuronal markers in this in vitro model for central nervous system development. As shown in Fig. 5, VAChT expression is discernible in cultured spinal cord removed from 12-day embryos by day 6 (corresponding to embryonic day 18) in culture, but is found only in cell bodies of neurons with short or absent processes. Cholinergic process development can be traced by the intense staining of varicosities with anti-VAChT in the following 3 weeks of culture. Expression of ChAT, as measured in extracts of cells by enzymatic assay, is likewise undetectable in the first week of culture, and increases exponentially, following a plateau phase around day 13–17, from 2 to 4 weeks in culture (Brenneman and Warren, 1983). Further quantitative analysis at the level of mRNA expression will be required to substantiate co-expression of VAChT and ChAT from the cholinergic gene locus during development, but it appears at this point unlikely that either VAChT or ChAT is rate-limiting with respect to the other component of the cholinergic phenotype in development of patent cholinergic cells in vivo.

## Subcellular targeting of VAChT to small synaptic vesicles (SSVs) in PC12 cells

Neuroanatomically precise chemical coding of neurotransmission requires cell-appropriate expression of genes encoding neurotransmitter biosynthetic enzymes and vesicular transporters, e.g. cholinergic neuron-specific expression of the VAChT/ChAT gene locus. Chemical coding is also dependent on the types of vesicle which contain a given neurotransmitter substance. For example, the contents of large dense-cored and small synaptic vesicles in mammalian cholinergic parasympathetic neurons are heterogeneous, and are also apparently preferentially released both from different sites at the nerve terminal, and under different conditions of nerve stimulation (Agoston et al., 1988).

Targeting of the vesicular acetylcholine transporter to specific vesicle subpopulations within cholinergic neurons would provide a second level of regulation in addition to cholinergic neuron-specific expression of the VAChT/ChAT gene. Cholinergic neurons contain many more small synaptic vesicles (SSVs) than large dense-cored vesicles (LDCVs), and adrenergic neurons contain large and small dense-cored vesicles (LDCVs and SDCVs) with few SSVs. Thus, it is difficult to address the question of whether VAChT and VMATs are targeted to, or excluded from, specific vesicle populations in cholinergic and adrenergic neurons. The PC12 rat pheochromocytoma cell line contains both LDCVs and small synaptic-like microvesicles (see Regnier-Vigouroux and Huttner, 1993). Following stimulation with nerve growth factor (NGF) PC12 cells differentiate into neurons, which contain abundant complements of both large dense-core (LDCV) and small synaptic (SSV) vesicles (Greene and Tischler, 1976; Tischler and Greene, 1978) and also express both VMAT1 and VAChT. These cells are thus ideal for exploring differential intracellular sorting of VMAT1 and VAChT to large and small secretory vesicles.

Specific antibodies against VAChT and VMAT1 were used to investigate the subcellular localization of each transporter in PC12 cells dif-

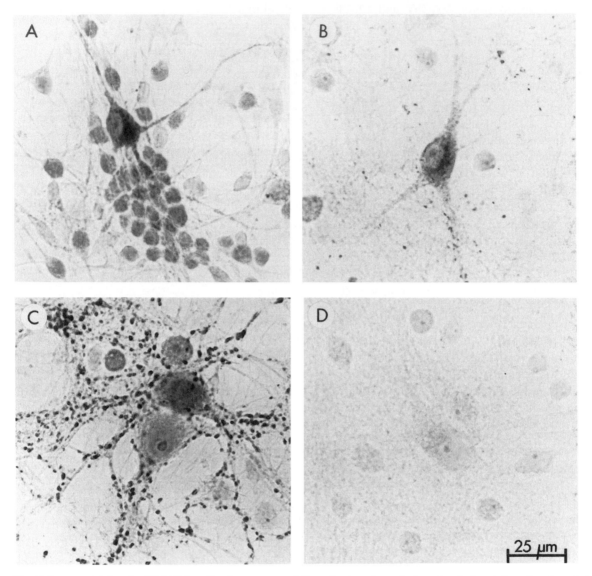

Fig. 5. Immunostaining for VAChT in cell cultures prepared from fetal murine spinal cord at 1, 2 and 4 weeks after plating. The primary rabbit polyclonal antibody is directed against a C-terminal peptide contained in rat VAChT conjugated to KLH, and visualized using biotin-avidin-peroxidase reagents (Vectastain). (A) After 1 week in culture, perinuclear immunoreactivity is seen in a few neurons. (B) After 2 weeks, delicate sprays of bouton-like structures stain for the transporter. (C) Between 3 and 4 weeks, dense accumulations of reactive boutons encrust neuronal somata and dendrites and are prominent in areas of the culture containing neurite networks. (D) A 4-week-old culture reacted with antibody pre-absorbed for 48 h at 4°C with 25 $\mu$M of the C-terminal VAChT peptide used to generate the antiserum.

ferentiated with NGF by immune electron microscopy, using pre-embedding immunohistochemical staining followed by immunogold labeling and silver enhancement. VAChT is contained predominantly on the membrane of SSVs within neurites of NGF-treated PC12 cells (Fig. 6A), and is largely absent from LDCV membranes. VMAT1, on the other hand, is much more abundantly localized to LDCVs than to SSVs in these cells (Fig. 6B), as suggested previously by Liu et al. (1994). The preferential localization of VAChT on SSVs in a single cell containing both SSVs and LDCVs

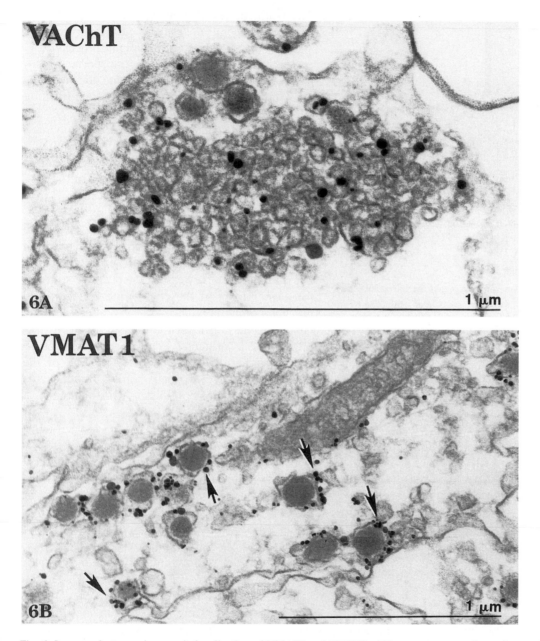

Fig. 6. Immunoelectron microscopic localization of VMAT1 and VAChT within secretory granules and synaptic vesicles of PC12 cells. Pre-embedding immune electron microscopical visualization of the VAChT and VMAT1 antigens was afforded using anti-rat VAChT and VMAT1 anti-peptide rabbit polyclonal antisera specific for each transporter at a dilution of 1:500, using methods described by Tao-Cheng and Tanner (1994). A. VAChT immunoreactivity confined mainly to small synaptic vesicles (SSVs) in neurites of PC12 cells grown on Matrigel and treated with nerve growth factor (NGF) for 12 days. B. VMAT1 immunoreactivity confined mainly to large dense-core vesicles (LDCVs) in neurites of NGF-treated PC12 cells (see Wiehe at al., 1996).

clearly demonstrates that VAChT inherently contains the structural information required for targeting to a specific vesicle subtype. Thus SSVs and LDCVs are created not merely as different-

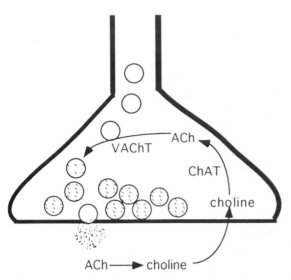

Fig. 7. The "cholinergic operon". Depicted are the metabolic and sequestration steps, mediated by ChAT and VAChT, respectively, that comprise the pathway leading to establishment of the cholinergic phenotype in mammalian neurons.

sized organelles with different storage capacities for acetylcholine, but with clearly different amine transport capabilities as well.

VAChT may be directed to SSVs due to the absence of an LDCV-directing signal found in VMAT1, or the presence of an SSV-directing signal not found in VMAT1. Further work involving examination of SSV/LDCV targeting of VAChT/VMAT chimeric proteins will be required to address this important "cholinergic" question.

## Summary and concluding remarks

VAChT encodes a vesicular acetylcholine transporter expressed specifically in cholinergic neurons and targeted specifically to small synaptic vesicles, responsible for accumulation of acetylcholine into those vesicles to allow generation of the neurotransmitter "quanta" that underlie cholinergic neurotransmission. VAChT has significant structural and functional similarity to the vesicular monoamine transporters, which should improve our current understanding of the molecular mechanisms of vesicular amine transport by combining data from studies of both types of transporters to develop a general model for the bioenergetics and allostery of the amine/proton antiport process

The vesicular acetylcholine transporter is, together with ChAT a part of a gene locus that is nominally sufficient to confer cholinergic function on a neuronal cell. As such (Fig. 7), the VAChT/ChAT gene locus can be looked upon as a "cholinergic operon" responsible for assembling a "metabolic/sequestration" pathway in cholinergic cells. The extracellular signals that are ultimately responsible for the regulation of this "operon" may include both trophic factors and other neurotransmitters. Recent work suggests that adrenergic neurotransmission, by stimulating production of cholinergic differentiation factor(s) from presumptive cholinergic target tissue, is a decisive stimulus for development of the cholinergic phenotype in the autonomic nervous system, and cholinergic neurotransmission itself may be necessary to stabilize the peripheral cholinergic phenotype (Habecker and Landis, 1994). Regulation of the VAChT/ ChAT gene locus during these events will likely be a subject of future investigation.

## Acknowledgements

The laboratory of Jim Rand, Oklahoma Medical Research Foundation, catalyzed much of the work described here through their pioneering studies of UNC-17 in *C. elegans*. The cloning and sequencing of *Torpedo* VAChT was a collaborative effort with Hélène Varoqui and Marie-François Diebler, Laboratoire de Neurobiologie Cellulaire, Gif-sur-Yvette. In particular, we wish to acknowledge the contribution of Hélène Varoqui and the Diebler laboratory in providing key experimental insights that led to the verification of the transport function of VAChT in vitro. Anna Iacangelo is thanked for analysis and mapping of the rat VAChT/ChAT gene locus. We thank Virginia Tanner for her important technical contributions to the immune electron microscopic work on the subcellular targeting of VAChT and VMAT1. M.K.-H.S. and E.W. were supported by grants from the German Research Foundation and the Volkswagen Foundation.

# References

Agoston, D.V., Conlon, J.M. and Whittaker, V.P. (1988) Selective depletion of the acetylcholine and vasoactive intestinal polypeptide of the guinea-pig myenteric plexus by differential mobilization of distinct transmitter pools. *Exp. Brain Res., 72:* 535–542.

Alfonso, A., Grundahl, K., Duerr, J.S., Han, H.-P. and Rand, J.B. (1993) The *Caenorhabditis elegans* unc-17 gene: a putative vesicular acetylcholine transporter. *Science, 261:* 617–619.

Alfonso, A., Grundahl, K., McManus, J.R., Asbury, J.M. and Rand, J.B. (1994) Alternative splicing leads to two cholinergic proteins in C. elegans. *J. Mol. Biol., 241:* 627–630.

Bejanin, S., Cervini, J., Mallet, J. and Berrard, S. (1994) A unique gene organization for two cholinergic markers, choline acetyltransferase and a putative vesicular transporter of acetylcholine. *J. Biol. Chem., 269:* 21944–21947.

Berrard, S., Varoqui, H., Cervini, R., Israël, M., Mallet, J. and Diebler, M.-F. (1995) Coregulation of two embedded gene products, choline acetyltransferase and the vesicular acetylcholine transporter. *J. Neurochem., 65:* 939–942.

Berse, B. and Blusztajn, J.K. (1995) Coordinated upregulation of choline acetyltransferase and vesicular acetylcholine transporter gene expression by the retinoic acid receptor $\alpha$, cAMP, and leukemia inhibitory factor/ciliary neurotrophic factor signaling pathways in a murine septal cell line. *J. biol. Chem., 270:* 22101–22104.

Brenneman, D.E. and Warren, D. (1983) Induction of cholinergic expression in developing spinal cord cultures. *J. Neurochem., 41:* 1349–1356.

Carmichael, S.W. and Winkler, H. (1985) The adrenal chromaffin cell. *Sci. Am., 253:* 39–49.

Chong, J.A., Tapia-Ramirez, J., Kim, S., Toledo-Arai, J.J., Zheng, Y., Boutros, M.C., Altshuller, Y.M., Frohman, M.A., Kraner, S.D. and Mandel, G. (1995) REST: a mammalian silencer protein that restricts sodium channel gene expression to neurons. *Cell, 80:* 949–957.

Clarkson, E.D., Bahr, B.A. and Parsons, S.M. (1993) Classical noncholinergic neurotransmitters and the vesicular transport system for acetylcholine. *J. Neurochem., 61:* 22–28.

Cohen-Haguenauer, O., Brice, A., Berrard, S., Nguyen, V.C., Mallet, J. and Frezal, J. (1990) Localization of the choline acetyltransferase (ChAT) gene to human chromosome 10. *Genomics, 6:* 374–378.

Eckenstein, F.P., Baughman, R.W. and Quinn, R.W. (1988) An anatomical study of cholinergic innervation in rat cerebral cortex. *Neuroscience, 25:* 457–474.

Erickson, J.D. and Eiden, L.E. (1993) Functional identification and molecular cloning of a human brain vesicle monoamine transporter. *J. Neurochem., 61:* 2314–2317.

Erickson, J.D., Eiden, L.E. and Hoffman, B. (1992) Expression cloning of a reserpine-sensitive vesicular monoamine transporter. *Proc. Natl. Acad. Sci. USA, 89:* 10993–10997.

Erickson, J.D., Varoqui, H., Schäfer, M., Diebler, M.-F., Weihe, E., Modi, W., Rand, J., Eiden, L.E., Bonner, T.I. and Usdin, T. (1994) Functional characterization of the mammalian vesicular acetylcholine transporter and its expression from a "cholinergic" gene locus. *J. Biol. Chem., 269:* 21929–21932.

Greene, L.A. and Tischler, A.S. (1976) Establishment of a noradrenergic clonal cell line of rat adrenal pheochromocytoma cells which respond to nerve growth factor. *Proc. Natl. Acad. Sci. USA, 73:* 2424–2428.

Habecker, B.A. and Landis, S.C. (1994) Noradrenergic regulation of cholinergic differentiation. *Science, 264:* 1602–1604.

Henry, J.-P., Gasnier, B., Roisin, M.P., Isambert, M.-F. and Scherman, D. (1987) Molecular pharmacology of the monoamine transporter of the chromaffin granule membrane. *Ann. N. Y. Acad. Sci., 493:* 194–206.

Howell, M., Shirvan, A., Stern-Bach, Y., Steiner-Mordoch, S., Strasser, J.E., Dean, G.E. and Schuldiner, S. (1994) Cloning and functional expression of a tetrabenazine sensitive vesicular monoamine transporter from bovine chromaffin granules. *FEBS Lett., 338:* 16–22.

Johnson, Jr., R.G. (1988) Accumulation of biological amines into chromaffin granules: a model for hormone and neurotransmitter transport. *Physiol. Rev., 68:* 232–307.

Kengaku, M., Misawa, H. and Deguchi, T. (1993) Multiple mRNA species of choline acetyltransferase from rat spinal cord. *Mol. Brain Res., 18:* 71–76.

Kitamoto, T. and Salvaterra, P.M. (1995) A POU homeo domain protein related to dPOU-19/pdm-1 binds to the regulatory DNA necessary for vital expression of the Drosophila choline acetyltransferase gene. *J. Neurosci., 15:* 3509–3518.

Krejjci, E., Gasnier, B., Botton, D., Isambert, M.-F., Sagne, C., Gagnon, J., Massoulié, J. and Henry, J.-P. (1993) Expression and regulation of the bovine vesicular monoamine transporter gene. *FEBS Lett., 335:* 27–32.

Li, Y.-P., Baskin, F., Davis, R., Wu, D. and Hersh, L.B. (1995) A cell type-specific silencer in the human choline acetyltransferase gene requiring two distinct and interactive E boxes. *Mol. Brain Res., 30:* 106–114.

Liu, Y., Peter, D., Roghani, A., Schuldiner, S., Prive, G.G., Eisenberg, D., Brecha, N. and Edwards, R.H. (1992) A cDNA that suppresses MPP+ toxicity encodes a vesicular amine transporter. *Cell, 70:* 539–551.

Liu, Y., Schweitzer, E.S., Nirenberg, M.J., Pickel, V.M., Evans, C.J. and Edwards, R.H. (1994) Preferential localization of a vesicular monoamine transporter to dense core vesicles in PC12 cells. *J. Cell Biol., 127:* 1419–1433.

Lönnerberg, P., Lendahl, U., Funakoshi, H., Arhlund-Richter, L., Persson, H. and Ibanez, C.F. (1994) Positive and negative regulatory elements in the choline acetyltransferase gene cooperate to direct developmental and tissue-specific expression in transgenic mice. *Soc. Neurosci. Abstr., 20:* 45.

Maycox, P.R., Hell, J.W. and Jahn, R. (1990) Amino acid eurotransmission: spotlight on synaptic vesicles. *Trends Neurosci., 13:* 83–87.

Misawa, H., Takahashi, R. and Deguchi, T. (1995) Coordinate expression of vesicular acetylcholine transporter and choline acetyltransferase in sympathetic superior cervical neurones. *NeuroReport,* 6: 965–968.

Njus, D., Kelley, P.M. and Harnadek, G.J. (1986) Bioenergetics of secretory vesicles. *Biochim. Biophys. Acta,* 853: 237–265.

Parsons, S.M., Prior, C. and Marshall, I.G. (1993) Acetylcholine transport, storage, and release. *Int. Rev. Neurobiol.,* 35: 279–390.

Peter, D., Finn, J.P., Klisak, I., Liu, Y., Kojis, T., Heinzmann, C., Roghani, A., Sparkes, R.S. and Edwards, R.H. (1993) Chromosomal localization of the human vesicular amine transporter genes. *Genomics,* 18: 720–723.

Peter, D., Jimenez, J., Liu, Y., Kim, J. and Edwards, R.H. (1994) The chromaffin granule and synaptic vesicle amine transporters differ in substrate recognition and sensitivity to inhibitors. *J. Biol. Chem.,* 269: 7231–7237.

Philippu, A. and Beyer, J. (1973) Dopamine and noradrenaline transport into subcellular vesicles of the striatum. *Naunyn-Schmiedeberg's Arch. Pharmacol.,* 278: 387–402.

Rand, J.B. (1989) Fine-structure genetic analysis of the *cha-1-unc-17* gene complex in *Caenorhabditis. Genetics,* 122: 73–80.

Rand, J.B. and Russell, R.L. (1984) Choline acetyltransferase-deficient mutants of the nematode *Caenorhabditis elegans. Genetics,* 106: 227–248.

Regnier-Vigouroux, A. and Huttner, W.B. (1993) Biogenesis of small synaptic vesicles and synaptic-like microvesicles. *Neurochem. Res.,* 18: 59–64.

Roghani, A., Feldman, J., Kohan, S.A., Shirzadi, A., Gundersen, C.B., Brecha, N. and Edwards, R.H. (1994) Molecular cloning of a putative vesicular transporter for acetylcholine. *Proc. Natl. Acad. Sci. USA,* 91: 10620–10624.

Schäfer, M.K.-H., Weihe, E., Varoqui, H., Eiden, L.E. and Erickson, J.D. (1994) Distribution of the vesicular acetylcholine transporter (VAChT) in the central and peripheral nervous systems of the rat. *J. Mol. Neurosci.,* 5: 1–18.

Schäfer, M.K.-H., Weihe, E., Erickson, J.D. and Eiden, L.E. (1995) Human and monkey cholingergic neurons visualized in paraffin-embedded tissues by immunoreactivity for VAChT, the vesicular acetylcholine transporter. *J. Mol. Neurosci.,* 6: 225–236.

Schoenherr, C.J. and Anderson, D.J. (1995) The neuron-restrictive silencer factor (NRSF): a coordinate repressor of multiple neuron-specific genes. *Science,* 267: 1360–1363.

Surratt, C., Persico, A., Yang, X., Edgar, S., Bird, G., Hawkins, A., Griffin, C., Li, X., Jabs, E. and Uhl, G. (1993) A human synaptic vesicle monoamine transporter cDNA predicts posttranslational modifications, reveals chromosome 10 gene localization and identifies Taq 1 RFLPs. *FEBS Lett.,* 318: 325–330.

Tao-Cheng, J.H. and Tanner, V.A. (1994). A modified method of pre-embedding EM immunocytochemistry which improves specificity and simplifies the process for *in vitro* cells. *Proceedings of the Microscopy Socoety of America,* San Francisco Press, San Francisco, CA.

Tischler, A.S. and Greene, L.A. (1978) Morphologic and cytochemical properties of a clonal cell line of rat adrenal pheochromocytoma cells which respond to nerve growth factor. *Lab. Invest.,* 39: 77–89.

Toll, L. and Howard, B.D. (1978) Role of $Mg^{2+}$-ATPase and a pH gradient in the storage of catecholamines in synaptic vesicles. *Biochemistry,* 17: 2517–2523.

Toll, L. and Howard, B.D. (1980) Evidence that an ATPase and a protonmotive force function in the transport of acetylcholine into storage vesicles. *J. Biol. Chem.,* 255: 1787–1789.

Usdin, T., Eiden, L.E., Bonner, T.I. and Erickson, J.D. (1995) Molecular biology of vesicular acetylcholine transporters (VAChTs). *Trends Neurosci.,* 18: 218–224.

Varoqui, H., Diebler, M.-F., Meunier, F.-M., Rand, J.B., Usdin, T.B., Bonner, T.I., Eiden, L.E. and Erickson, J.D. (1994) Cloning and expression of the vesamicol binding protein from the marine ray *Torpedo.* Homology with the putative vesicular acetylcholine transporter UNC-17 from Caenorhabditis elegans. *FEBS Lett.,* 342: 97–102.

Weihe, E., Schäfer, M.K.-H., Erickson, J.D. and Eiden, L.E. (1994) Localization of vesicular monoamine transporter isoforms (VMAT1 and VMAT2) to endocrine cells and neurons in rat. *J. Mol. Neurosci.,* 5: 149–164.

Weihe, E., Tao-Cheng, J.-H., Schäfer, M.K.-H., Erickson, J.D. and Eiden, L.E. (1996) Visualization of the vesicular acetylcholine transporter in cholinergic nerve terminals and its targeting to a specific population of small synaptic vesicles. *Proc. Natl. Acad. Sci. USA* 93: 3547–3552.

J. Klein and K. Löffelholz (Eds.)
*Progress in Brain Research*, Vol. 109
© 1996 Elsevier Science B.V. All rights reserved.

CHAPTER 6

# Expression of the vesicular acetylcholine transporter in mammalian cells

H. Varoqui[1], F.-M. Meunier[1], F.A. Meunier[1], J. Molgo[1], S. Berrard[2], R. Cervini[2], J. Mallet[2], M. Israël[1] and M.-F. Diebler[1]

[1]*Laboratoire de Neurobiologie Cellulaire et Moléculaire, CNRS, Gif sur Yvette, France and* [2]*Laboratoire de Génétique Moléculaire de la Neurotransmission et des Processus Neurodégénératifs, CNRS, Gif sur Yvette, France*

## Introduction

As earliest as 30 years ago, synaptic vesicles isolated from the rat brain (De Robertis et al., 1962; Whittaker et al., 1964) and the *Torpedo* electric organ (Israël et al., 1968; Whittaker et al., 1972) were shown to store large amounts of acetylcholine (ACh). During the past two decades, an enormous effort has been made to understand the properties, function and pharmacology of vesicular ACh storage. A variety of evidence has been presented supporting the view that concentration of ACh within vesicles requires a specific transport activity using the inwardly acidic pH gradient generated by a vacuolar $H^+$-ATPase to drive transmitter uptake (Michaelson and Angel, 1981; Anderson et al., 1982, 1983; Diebler and Lazereg, 1985; Bahr and Parsons, 1986a; Yamagata and Parsons, 1989). Thereby, the vesicular ACh transporter (VAChT) takes up cytoplasmic ACh in exchange for internal protons. It thus appears that vesicular storage of all classical neurotransmitters (biogenic amines, ACh, glutamate, gamma- aminobutyric acid and glycine) is supported by specific proton exchanging transporters, which thus form a class of functionally related proteins.

Active transport of ACh was shown to be specifically inhibited by vesamicol in a mixed noncompetitive way (Chapter 7; Bahr and Parsons, 1986b; Kornreich and Parsons, 1988; Diebler and Morot Gaudry, 1989). Progress in the understanding of the molecular properties of VAChT has been achieved with the development of vesamicol and ACh analogs and several models for VAChT organization have been consequently proposed (Marshall and Parsons, 1987; Rogers and Parsons, 1989, 1992; Bahr et al., 1992a; Parsons et al., 1993). However, although much valuable information has been gained on the behaviour of the protein through the many attempts to purify VAChT using classical biochemical techniques, this strategy has not allowed to elucidate the molecular identity of VAChT.

The advent of new techniques within molecular biology has recently enabled rapid progress in the molecular identification of vesicular neurotransmitter transporters. The first members of the family to be cloned were the monoamine transporters VMAT1 and VMAT2 (Erickson et al., 1992; Liu et al., 1992 a,b; Erickson and Eiden, 1993; Krejci et al., 1993; Peter et al., 1993; for review, see Schuldiner, 1994). The breakthrough for the cholinergic counterpart is due to Alfonso et al. (1993) who identified in *Caenorhabditis elegans* mutants the *unc17* gene which encodes a protein showing a clear homology with the vesicular monoamine transporters. This protein was suggested to be the cholinergic vesicular transporter. Starting with this observation, we took advantage of the cholinergic nature of the *Torpedo* electric lobe motoneurons to identify the homologous protein. We showed that it was the vesamicol binding protein (Varoqui et al., 1994) and that it was able to accumulate ACh into intracellular organelles (Erickson et al., 1994).

This paper recalls the cloning of the *Torpedo* vesicular acetylcholine transporter and its functional identification by expression in mammalian cells. Moreover, the present work addresses the issue of the cellular biology of VAChT through its immunological characterization and with regard to the control of the expression of a cholinergic phenotype.

## Molecular cloning and expression of the vesicular acetylcholine transporter

### Molecular cloning

Due to its high content in cholinergic cell bodies, the electric lobe of *Torpedo* is a material of choice to generate a cDNA library from which to isolate a VAChT clone.

A VAChT cDNA clone was isolated by screening a *Torpedo marmorata* electric lobe lambda ZAP library with a randomly primed radiolabeled *unc17* probe (Varoqui et al., 1994). The open reading frame encodes a protein of 511 amino acids with a calculated molecular weight of 56 kDa. At this time, the sequence of homologous proteins has been deduced from rat and human cDNA libraries (Erickson et al., 1994; Roghani et al., 1994) or rat gene analysis (Béjanin et al., 1994). The comparison of the amino acid sequences of these proteins and the rat VMATs is shown on Fig. 1. The *Torpedo* VAChT is 50% and 66% identical to nematode and rat proteins, respectively, with an overall identity of 43% with the rat VMATs. Hydropathy analysis suggested a common structural motif in which the proteins

have 12 transmembrane domains (TM). As the sequence predicts potential sites of N-glycosylation in the hydrophilic loop between TM1 and TM2, which is shorter and has additional glycosylation sites in the *Torpedo* protein, this loop presumably resides in the vesicular lumen. The N- and C-terminus would then be located in the cytoplasm. The highest degree of amino acid conservation amongst all members of this family is observed in the TM regions while the amino and carboxy terminus are the most divergent. Overall, because of the species differences, no functional domain specific for a type of neurotransmitter can be predicted from the primary sequence. Interestingly, aspartate residues, which may be involved in positively charged substrate interaction and/or proton transport, are conserved in TM 1, 6, 10 and 11 of all proteins. Further work to elucidate the key domains involved in substrate and ligand recognition is still needed. Bearing no resemblance with the neurotransmitter plasma membrane transporters which use different ionic gradients, this new class of proteins displays some homology with the multidrug resistance proteins, particularly in the N-terminus half of the protein, which are also proton exchangers.

### Expression in mammalian cells

Expression of the cDNAs encoding the *Torpedo*, nematode and rat proteins in CV1 cells has permitted the unambiguous functional identification of these proteins. Fig. 2 shows that cells expressing the *Torpedo* or rat cDNAs, but not cells expressing the VMATs, exhibited both a saturable high affinity binding of L-vesamicol ($K_D \approx 6$ nM, similar to that observed on *Torpedo* isolated synaptic vesicles) and an ACh transport specifically

Fig. 1. Alignment of the amino acid sequences of vesicular neurotransmitter transporters. Amino acid sequences deduced from cDNA nucleotide sequences are depicted: four first lines, the acetylcholine transporters of *Torpedo marmorata* (TVAChT), rat (RVAChT), human (HVAChT) and *Caenorhabditis elegans* (UNC 17); last two lines, the two rat monoamine transporters (*RVMAT1 and RVMAT2*). Amino acid sequence of the *Torpedo* VAChT was aligned to the other transporter sequences using the PILEUP program of the GCG Sequence Analysis Software Package. Gaps (dots) were introduced to facilitate this alignment. Sequence identities are indicated by using white on dark lettering; homologies, by black on light gray lettering. Bars are drawn over the 12 extensive hydrophobic sequences that may correspond to transmembrane domains (TM). Ψ, potential sites of N-glycosylation sites on internal loops. ●, potential phosphorylation sites by protein kinase C on externally oriented structures.

TM1

```
TVAChT  . . . . M A V G Q A K A A M G K I S A G G E R S K R I S G A M N P R R K R I L L V V C I N M L L D N M L Y M V I   56
RVAChT  M E P T A P T G Q A R A A A T K L S E A V G . . . . . . . A A T Q S P Q R Q R R P L V L V V C V A L L L D N M L Y M V I   53
HVAChT  M E S A E P A G Q A R A A A T K L S E A V G . . . . . . . A A T Q S P R R Q R R C V L V V C V A I A L L L D N M L Y M V I   53
UNC 17  . . . . . M G F N V P V I N R D S E I L K A D A K W . . . L E Q Q D N Q K K C V L V V F V A L L L D N M L L T V V   51
RVMAT1  . . . . . . . . M L Q V M L G A P Q R L . . K E G R Q S R K L L V L V V F V A L L L D N M L L T V V   41
RVMAT2  . . . . . . . . M A L S D L V L . . L R W . . . L R D S R H S R K L L F L V F L A L L L D N M L L T V V   40
```

ψ

```
TVAChT  V T I P N V I . . . E T I R . . . . . . M Y K L V Y I T T P S . . . . . . N G . . . . . . . . .   81
RVAChT  V T D V I . . A H M R G G S E G P T L V S E V W E P T P P T L A N A S A V T A N T S . . . . . . . . .   98
HVAChT  V T D V I . . A H M R G G G E G P T R T P E V W E P T P L P T P A N A S A V T A N T S . . . . . . . . .   98
UNC 17  L I P K Y I . . R D I H N . . . . . . . Y Q V S E . . . . . E G I H N E T S . . . . . .   77
RVMAT1  V T S L T Y A T E F K D S N S S L H R G P S V S S Q Q A L T S P A P S T I F S F D N M T T T V E E H V P F R V T   101
RVMAT2  I S S Y Y S I K H E K N S T E I Q . . . . . . T T R P E L V V S T S E S I F S Y Y N N S T V L I T G N A T G T L P   95
```

ψ          ψ                                                        TM2

```
TVAChT  . . . . . . . . . . . T N G S L L N S T Q R A V L E S N P N A N S D I Q I G V L F A S K A I L Q L L S N P F T G T F I   129
RVAChT  . . . . . . . . . A S P T A A G S A R S I L R P A Y P T E S E D V K I G V L F A S K A I L Q L L V N P F S G P F I   146
HVAChT  . . . . . . . . . A S P T A A W P A G S A L R P A Y P T E S L D V K I G V L F A S K A I L Q L L V N P F S G P F I   146
UNC 17  . . . . . . . . . . . Q L A N G T Y L V . R E V G G R I N F L D S E L E L W V A S K A L I Q L L V N P F V G P L T   124
RVMAT1  . . W T N G T I P P P V T E A S S V P K N N C L Q G I E F L E E N V R I G I L F A S K A L I Q L L V N P F V G P L T   158
RVMAT2  G G Q S H K A T S T Q H T V A N T T V P S D C P S E D R D L L . N S N V Q V C L F A S K A T V G L L T N P F I G L L T   154
```

TM3                                          TM4

```
TVAChT  S M V G D L L I C T I V F S I T . . G E S . . . I L A A S V Q L G S A A D T S G I A M I A D K V I P   189
RVAChT  M S F D V L L I C I G V M F A G V M . F A E D I A T L A A S V Q L G S A A D T S G I A M I A D K Y P E   206
HVAChT  M S F D V L L I C I G V M F A G V L V T A E D I A T L A A S V Q L G S A A D T S G I A M I A D K Y P E   206
UNC 17  R I V E S E M I L G C T K F A I A I L G K S G V L L F A S R S L Q G F C A A D T S G L A M I A D R F I E   184
RVMAT1  N I I G H R M F V C F M I L E L M A S G T L L T V V I T Q S I G S S S S V A G L M L A S V I I D   218
RVMAT2  N I I P S M F A G F C I L S V M . S S S F A L I G I G S C S S V A G M G N L A S V I I D   214
```

TM5                                          TM6

```
TVAChT  L S E A T Q . . . . . . . . . . . G V Q F S V Q Q . . . . . A S F V C L D G I L L M M V T F F   249
RVAChT  P F S R R L V V V T S G S G G L A C L G A S V Q Y A A C R V . . A V S I F D A L I L L A V A K F F   266
HVAChT  P L S R R L V V T S G S G G L A C L G A S V Q Y A A C R V . . A V S I F D A L I L L A V A K F F   266
UNC 17  S N S R A T Y S S S L S S S L . . . S L S E S V P I I S L C P V I T I S Y V M A T V M I N T H   244
RVMAT1  N Y R G R L M G L G G L A L C L G A S V M F V S S L P L A L L D G A L Q I C I L W F .   277
RVMAT2  D E I G K P M G G L A M G V I V G S G F V V C R T A A L V I L D G A I Q T F V L Q F .   273
```

TM7                                                          ψ

```
TVAChT  . . A G K T R V N T L Q G S T I Y K G M I D P V V V A G T T C N I P . . S N V M K K G M . N A S E W   306
RVAChT  S A A A T A R A N L P V C G P I H R L M L D P V V A G T T C N I P . . A T W M K H I . . A A S E W   325
HVAChT  S A A A T A R A N L P V C G P I H R L M L D P F A A G T T C N I P . . A T W M K H I . . A A S E W   325
UNC 17  R R G T D S H G E K V Q G P M W R F M D F F A C C S G A L I M A V S L G P Z P L T T T W S E M M P D T P G W   304
RVMAT1  . . S K V S P E S A M G T S L L T F L K D F L V A A G S I C L A N M G V A I L E P A L P I M Q T M C . S P E W   333
RVMAT2  . . S R V Q P E S Q K G P L T T F L K D F L V A A G S I C F A N M G I A M L E P A L P I M E T M C . S R K W   329
```

TM8                                          TM9

```
TVAChT  Q M G I T F S F E I L G I S L A K Y N Y Q L Y G A V G V I I G A S C T I P A C R N F E E L I I   366
RVAChT  E M G M A F V P V G L G I L I V R A R Y G H L Q V L Y G A L G I A V I G V S C V V P A C R S F A P L V V   385
HVAChT  E M G M A F V P V G L G I L I V R L A R Y G H L Q V L Y G A L G I A V I G V S S C I V P A C R S F A P L V V   385
UNC 17  L V Q V I G F P F L V I G I S L K K L R A F P E H T A I A M V G L A M E G I A C F A I P Y T T S V M Q I V T   364
RVMAT1  Q L G L A F L S V A Y L I G T N L F G V N K M G R . M C S L V G M A V G I S L L C V P L A H N I F G L I G   391
RVMAT2  Q L G V A L S I S Y L I G T N I F G I L H K M G R . . M C A L L G M V I V G I S I L C I P F A K N I Y G L I A   387
```

TM10                                    TM11

```
TVAChT  P I C A L P F G I A L V D T A L L P T I A F I V D I R Y V S V Y G S V Y A I A D I S Y S V A Y A L G P I M A G Q I V H D   426
RVAChT  S L C G L P F G I A L V D T A L L P T I A F I V D V R H V S V Y G S V Y A I A D I S Y S V A Y A L G P I V A G H I V H S   445
HVAChT  S L C G L P F G I A L V D T A L L P T I A F I V D V R H V S V Y G S V Y A I A D I S Y S V A Y A L G P I V A G H I V H S   445
UNC 17  P L S F V C F C G A L I D A S L L P I M G H L V D L R H V S V Y G S V Y A I A D I S Y S V A Y A F G P I I A G W I V T N   424
RVMAT1  F N A C L G F A G M V D S S M P I M G Y L L R H T S V Y G S V Y A I A D V A F C V G F A I G P S T G G V L V Q V   451
RVMAT2  F N F C V G F A I F M V D S M M P I M G Y L L R H V V Y G S V Y A I A D V A I C M G Y A I G P S A G G A L A K A   447
```

TM12

```
TVAChT  L G F V Q S N L G M G L V N I L Y A P G L L F L N V C Q M . . . . . K P S L S E R N I L L E E G P K G L Y D T I I . .   479
RVAChT  L G F E Q L S L G M Q L A N L L L A P V L L L R I V G L L . . . T R S R S E R D V L L D E P P Q G L Y D A V R . .   498
HVAChT  L G F E Q I S L G M Q L A N L L L A P V L L L R I V G L L . . . . T R S R S T R D V L L D E P P Q G L Y D A V R .   498
UNC 17  W C Y T A L N I I I F A T N V T L A P V L F L Y H S Y D T L G A K G D T A E M T Q L N S S A P A G G Y N G K P E A   484
RVMAT1  I G F P W L M V I I G T I N I L Y A P L C C F L Q N . . . . . . P P A K E E K R A I L . S Q E C P T E T Q M Y T F   501
RVMAT2  I G F P W L M T I I G I I D I A Y A P L C F F L R S . . . . . . P P A K E E K M A L L M D H N C P I K R K M Y T .   497
```

```
TVAChT  . . . . . . . . M E E R K E A K E P H G T S S G N H S V H A V L S D Q E G Y S E . . . . . . . . 511
RVAChT  . . . . . . . L R E V Q . G K D G G E P C S P P G P F D G C E D D Y N . Y Y S R S . . . . . . 530
HVAChT  . . . . . . L R E R P V S G Q D G E P R S P P G P F D E C E D D Y N Y Y Y T R S . . . . . . 532
UNC 17  T T A E S Y Q G W E D Q Q S Y Q N Q A Q I P N H A V S F Q D S R P Q A E F P A G Y D P L N P Q W 532
RVMAT1  Q K P T K A F P L G E N S D D P S S G E . . . . . . . . . . . . . . . . . . . . . . . 521
RVMAT2  Q N N V Q S Y P I G D D E E S E S D . . . . . . . . . . . . . . . . . . . . . . . . . . 515
```

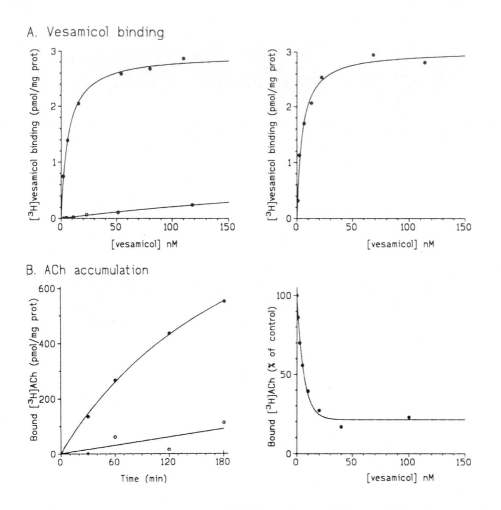

Fig. 2. Functional identification of VAChT. Expression of cDNAs in CV1 cells was performed using the vaccinia virus/bacteriophage T7 hybrid system (Fuerst et al., 1986). (A) Vesamicol binding on postnuclear cell membranes. Specific binding was measured as previously described (Varoqui et al., 1994). Left: ●, cells transfected with *Torpedo* VAChT; ○, cells transfected with rat VMAT2. Right: cells transfected with rat VAChT. (B) acetylcholine accumulation into cells transfected with rat VAChT. Intact cells grown in choline free medium were incubated in the presence of [$^3$H]ACh (0.4 mM) and esterase inhibitor. Bound ACh was measured after disruption of the cells as described by Erickson et al. (1994). Left: time course of ACh accumulation. Values were corrected for mock transfected background. Incubation was performed in the absence (●) or the presence (○) of $2\,\mu$M vesamicol. Right: vesamicol inhibition of [$^3$H]ACh uptake at 120 min. Mock transfected values (unaffected by any concentration of vesamicol) were subtracted from each data point. Values are the means of triplicate determinations from single experiments repeated at least once with similar results.

inhibited by L-vesamicol with an IC$_{50}$ similar to the $K_D$. Moreover, drugs which inhibit vacuolar-type H$^+$-ATPase (such as bafilomycin, *N*-ethyl-maleimide, tri-*n*-butyltin) or dissipate the proton gradient (FCCP), reduced by 80–95% the observed ACh uptake in transfected cells (Erickson et al., 1994). This result demonstrates that the expressed transporters, or at least a fraction of them, have been correctly inserted in membranes of intracellular organelles across which a proton gradient could sustain VAChT function. Taken together, these data make it clear that vesamicol binding and substrate recognition sites involved in ACh transport activity are situated on the same polypeptide. This does not at all rule out possible interactions of the transporter with other protein(s) or a proteoglycan (Bahr et al., 1992b; Parsons et al., 1993) in cholinergic neurons.

## Immunochemical characterization

Polyclonal antibodies against the *Torpedo* and rat VAChT were raised in rabbit and mice using a fusion protein with the last 66 amino acid C-terminus of the corresponding proteins. Using ultrastructural immunocytochemical techniques, we showed that immunoreactivity for VAChT at the presynaptic nerve terminals of the *Torpedo* electric organ is strictly associated with the synaptic vesicles (Fig. 3). Further analysis of purified *Torpedo* subcellular membrane fractions by immunoblotting (Fig. 4) showed that a signal is only detected in the purified vesicle preparation. It consisted of a diffuse staining between 70 and 200 kDa and a large banding between 50 and 64 kDa, showing the heterogenous mobility of VAChT in SDS-polyacrylamide gel electrophoresis. This result is in agreement with the previous observations of Parsons's group on *Torpedo* vesicles labelled with a photoreactive analog of ACh (Chapter 7; Rogers and Parsons, 1992; Parsons et al., 1993). A different VAChT immunoreactivity pattern was observed on synaptic vesicles purified from rat brain, where it was detected as a broad heterogenous band at 50–64 kDa. The significance of these differences between VAChT from rat and *Torpedo* vesicles remains to be elucidated.

### N-Glycosylation of VAChT

Amino acid sequence analysis of VAChT predicted several potential N-glycosylation sites. Fig. 5 shows that in *Torpedo* synaptic vesicle membranes, as well as in transfected cells expressing the *Torpedo* VAChT, the protein is indeed highly glycosylated. Whichever of these preparations was used, treatment with endoglycosidase F resulted in the decrease of immunoreactivity in the 50–64 kDa region and the appearance of a large doublet around 39 kDa with minor bands around 50 kDa. In CV1 cells expressing the rat protein, most of the VAChT immunoreactivity was detected at 39 kDa with a minor recognition at 50–64 kDa. The dis-

crepancy in the expression pattern of the rat and *Torpedo* proteins may reflect the differences in the number of potential N-glycosylation sites deduced from the respective sequences and/or different interactions with other proteins. By deglycosylation of polypeptides electroeluted from *Torpedo* vesicle proteins separated by SDS-gel, we ensured that the 39 kDa band did derive from deglycosylation of the higher molecular weight polypeptides recognized in control conditions (not shown). Furthermore, an antibody directed against the N-terminus of VAChT also recognized the 39 kDa form, showing that this peptide is not an artefactual proteolysis product. An electrophoretic mobility higher than expected from the calculated molecular mass has also been observed after deglycosylation of the monoamine transporter (Liu et al., 1994) and most likely reflects the highly hydrophobic characteristics of these proteins.

To examine the possible role of N-glycosylation on VAChT activity, we followed vesamicol binding on membranes of cells stably expressing the *Torpedo* VAChT at different times of exposure to endoglycosidase F, and have compared it to the VAChT immunoreactivity pattern by Western blotting (Fig. 6). Deglycosylation was almost complete at 90 min, and up to 6 h, vesamicol binding was similar to controls. Although the experiment was carried out at a saturating vesamicol concentration, and thus does not rule out possible conformational changes affecting the drug affinity, N-glycosylation does not appear crucial for the ligand binding.

## Targeting of the vesicular acetylcholine transporter in transfected cells

In the past few years, studies on the sorting of vesicular proteins expressed in transfected cells of neuronal and non neuronal origin, have proven a useful strategy to unravel synaptic vesicle biogenesis (Orci et al., 1987; Cameron et al., 1991; Lindstedt and Kelly, 1991; Feany et al., 1993; Leube et al., 1994). We report here a series of preliminary experiments addressing this issue.

Stable cell lines expressing the *Torpedo* VAChT were generated from two cell types by

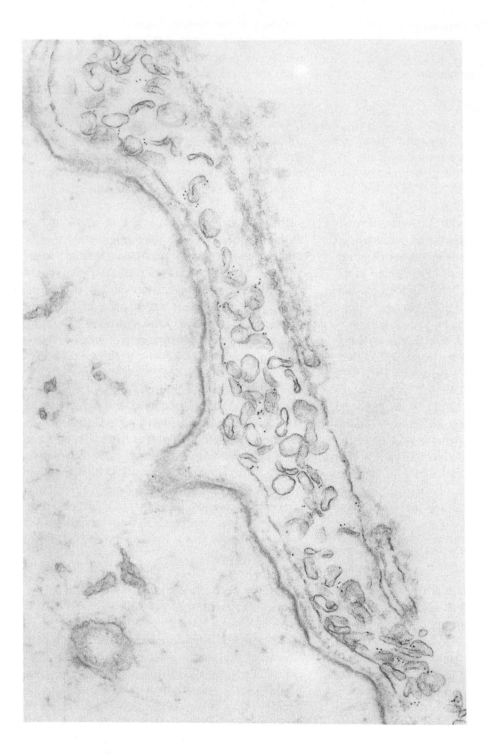

Fig. 3. Immunoelectron microscopy localization of VAChT in the electric organ of *Torpedo*. Immunogold detection of VAChT was performed as described by Brochier et al. (1993). Only synaptic vesicles are decorated with 5 nm gold particles. Magnification: ×74 000.

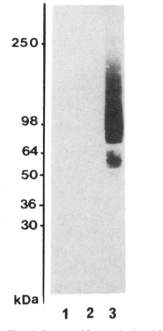

Fig. 4. Immunoblot analysis of VAChT distribution. Proteins (30 μg) were separated by SDS polyacrylamide gel electrophoresis (5–15%) in the presence of 50 mM dithiothreitol. Western blot was probed with a polyclonal affinity purified antibody against the C-terminus of *Torpedo* VAChT. Visualization was performed with an enhanced chemiluminescent procedure using peroxidase-labelled secondary antibody. Lane 1: postsynaptic membranes; lane 2, total synaptosomes; lane 3, purified synaptic vesicle membranes.

recovered in the light and heavy fractions depended on the preparation. Upon analysis of the fractions by Western blotting (Fig. 7B), VAChT immunoreactivity followed the profile of vesamicol binding but with different patterns. The expected mature form at 50–64 kDa was only detected in the light fractions. That the glycosylated protein was routed to separate organelles was further confirmed on cells treated for 4 days with 0.1 μg/ml tunicamycin. Treated cells displayed similar vesamicol binding as controls, but upon fractionation, the activity was shifted towards denser fractions. Concomitantly, on Western blots, the 50–64 kDa band disappeared and VAChT immunoreactivity was mostly present at 50 kDa with reinforcement of the bands detected below 36 kDa. These small peptides likely result from a proteolytic degradation of the protein, enhanced by the absence of protective glycosylation. Distribution of VAChT immunoreactivity was compared to known organelle markers. It appears that rab5, a marker of early endosomes (Chavrier et al., 1990), and rab6, a marker of the Golgi complex (Goud et al., 1990), partly coincide with VAChT containing membranes sedimenting in the light and dense fractions, respectively.

To further analyse VAChT targeting, we examined its subcellular distribution by immunofluores-

transfection with a CMV promoter expression construct: a neuronal cell line of N18TG2 lacking synaptic vesicle-like structures and the vesicular protein marker synaptophysin, and a neuroendocrine cell line AtT20 shown to contain ACTH secretory granules (Matsuuchi et al., 1988; Tooze et al., 1989) and expressing synaptophysin. We examined the intracellular distribution of VAChT in the transfected cells by biochemical and immunocytochemical methods.

Vesamicol binding and VAChT immunoreactivity patterns from subcellular fractions of transfected N18 cells separated by velocity sedimentation is illustrated on Fig. 7. Vesamicol binding activity was recovered in two regions of the gradient, with a peak in the light fractions at ~8% sucrose and a broad distribution in rapidly sedimenting membranes. The relative amount of activity

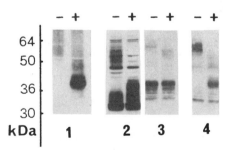

Fig. 5. N-Linked glycosylation of VAChT. Membranes of *Torpedo* synaptic vesicles (1), CV1 cells transiently expressing the *Torpedo* (2) or rat (3) VAChT, or N18TG2 stably transfected with the *Torpedo* cDNA (4), were treated (+) for 18 h at 37°C with endoglycosidase F/N glycosidase F (Boehringer). Controls in the absence of endoglycosidase (−) were run in parallel. Proteins were separated by SDS gel electrophoresis in the presence of 50 mM dithiothreitol. After Western blotting, VAChT immunoreactivity was detected as in Fig. 4.

## A. Immnunoblot analysis

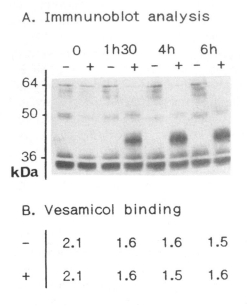

## B. Vesamicol binding

|   | 0 | 1h30 | 4h | 6h |
|---|---|---|---|---|
| – | 2.1 | 1.6 | 1.6 | 1.5 |
| + | 2.1 | 1.6 | 1.5 | 1.6 |

Fig. 6. Recovery of VAChT immunoreactivity and vesamicol binding activity after endoglycosidase F treatment. Membranes of N18 cells stably expressing the *Torpedo* VAChT were incubated in the presence (+) or the absence (–) of endoglycosidase F at 37°C for the exposure times indicated in the upper line. Samples were then probed for both VAChT immunoreactivity and vesamicol binding. (A) VAChT immunoreactivity. Samples were processed for gel electrophoresis and Western blotting as described in Fig. 4. (B) vesamicol binding. Specific binding at 80 nM L-[$^3$H]vesamicol was measured as described by Varoqui et al. (1994). Activity is expressed in pmol/mg protein.

cence microscopy. VAChT staining was primarily intracellular and was seen throughout the cell body and along the entire length of the processes (Fig. 8A). Immunoreactivity was detected as a coarsely granular staining particularly visible in the cell body (Fig. 8A, right panel), in addition to a smaller punctuate staining mainly seen in the processes. Mock transfected cells were not labelled (Fig. 8B). Using confocal laser scanning microscopy, we compared the overall distribution of VAChT with that of the following markers: medial and trans Golgi complex (Jasmin et al., 1989; Goud et al., 1990) (Fig. 8C,D); β-COP membrane traffic vesicles (Duden et al., 1991) (Fig. 8E); endoplasmic reticulum (Louvard et al., 1982) (Fig. 8F) and endocytosed transferrin (Fig. 8G). It appeared that only partial overlapping could be

seen with the Golgi marker rab6 and rare dots of endocytosed transferrin were also positive for VAChT.

Immunofluorescence studies performed on transfected AtT20 are illustrated on Fig. 9. In these neuroendocrine cells, VAChT staining appeared as an intracellular finely punctuate pattern, markedly different from that observed in transfected N18 cells. Accumulation of immunoreactivity was observed in the perinuclear region. VAChT staining was enriched in the peripheral cytoplasm and at the tips of the processes (Fig. 9A,B), a pattern characteristic of the distribution of the ACTH secretory granules of these cells (Orci et al., 1987; see Matsuuchi et al., 1988). Remarkably, some of the small punctuate structures were positive for both VAChT and synaptophysin (Fig. 9C,D). However, some other punctua were only positive for one or the other protein.

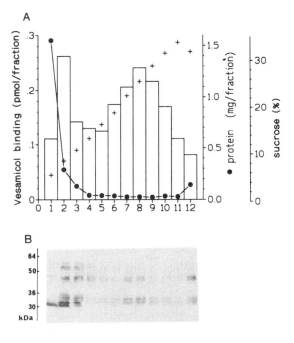

Fig. 7. Subcellular distribution of VAChT in stably transfected N18 cells. Postnuclear supernatants of N18 cells stably expressing the *Torpedo* VAChT were separated on a linear 5–30% sucrose gradient and centrifuged at 25 000 rev./min in a SW41 rotor for 15 min. Each fraction was assayed for vesamicol binding (A) and VAChT immunoreactivity (B). Experimental protocol as in Fig. 6.

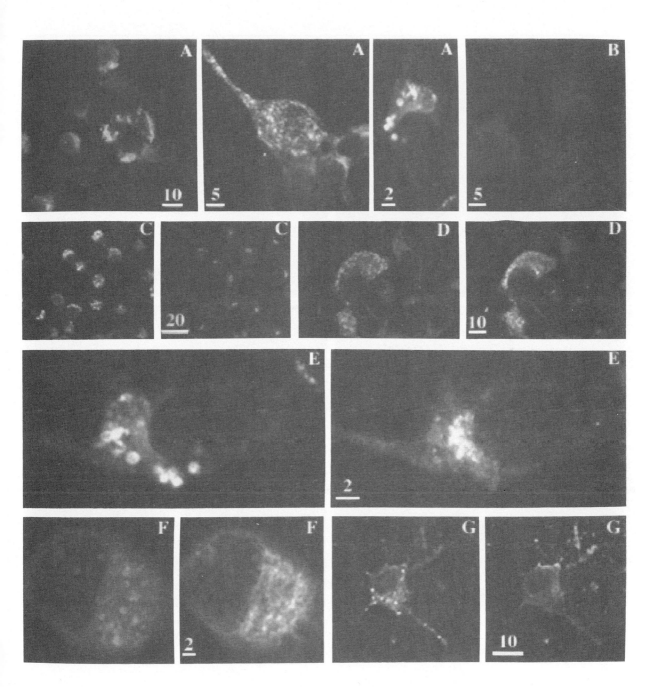

Fig. 8. Immunolocalization of VAChT in transfected N18 cells by confocal laser scanning microscopy. Acetone-fixed N18 cells stably expressing the *Torpedo* VAChT were incubated with primary antibody, followed by fluorescein or Texas red-conjugated secondary antibody. Cells exposed to fluorescein-conjugated transferrin for 1 h were fixed with paraformaldehyde, permeabilized with Triton X-100 and incubated with VAChT antibody, followed by Texas red-conjugated secondary antibody. (A,B) Immunolocalization of VAChT in transfected cells (A) and mock transfected cells (B); (C–G) Double-label immunolocalization of VAChT (left panels) with: CTR 433, C, right; rab6, (D), right; β-COP, (E) right; endoplasmic reticulum, (F), right; endocytosed transferrin (Sigma), (G), right. Scales in μm.

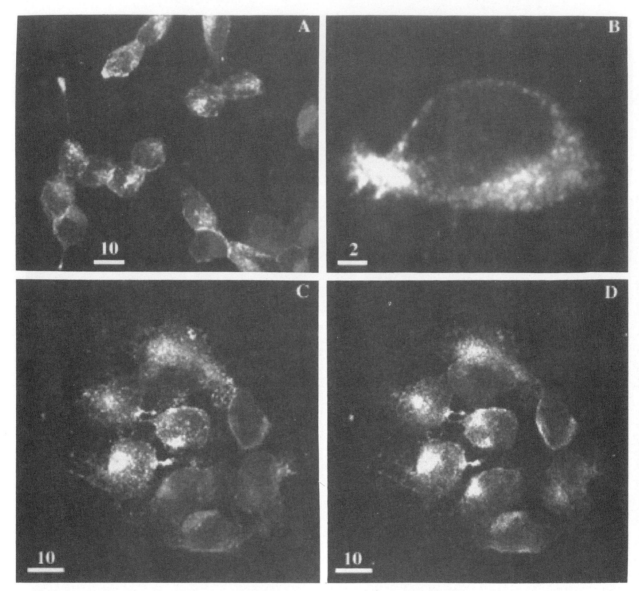

Fig. 9. Immunolocalization of VAChT in transfected AtT20 by confocal laser scanning microscopy. Transfected cells were processed for immunohistochemistry as described in Fig. 8. (A,B) VAChT. (C,D) Double labeling for VAChT (left) and synaptophysin (Boehringer) (right). Scales in $\mu$m.

Taken together, the present data indicate that in transfected cells devoid of synaptic vesicle markers, non glycosylated and glycosylated VAChT is seen in various intracellular organelles, part of them being endosomes. In neuroendocrine cells equipped with secretory machinery, the distribution pattern of VAChT supports the view that it is sorted along with endogenous vesicle proteins to the secretory vesicles. Confirmatory experiments are now in progress. It still remains to ascertain whether VAChT expressed in the absence of correct vesicle structures enters the general endosomal and lysosomal pathway, as shown for other vesicle proteins in transfected fibroblasts (Cameron et al.,

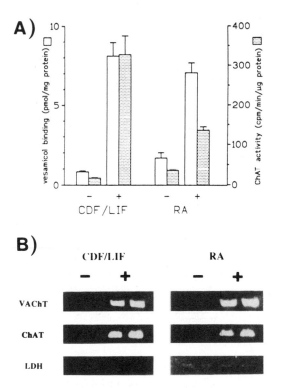

**A)**

**B)**

Fig. 10. Effect of CDF/LIF and retinoic acid on vesamicol binding, ChAT activity and the level of VAChT and ChAT mRNAs. Neurons from newborn rat sympathetic ganglia were grown in control conditions (−) or in the presence (+) of 10 ng/ml CDF/LIF or 5 $\mu$M retinoic acid for 11 days. (A) The binding of 100 nM L-[$^3$H] vesamicol was performed as described by Varoqui et al. (1994). ChAT activity was determined by the technique of Fonnum (1975). Results are means ± SEM values of triplicate independent samples from one experiment representative of two. (B) RT-PCR were performed in duplicate with primers specific for VAChT or ChAT coding sequences or, as a control, for LDH mRNA (for experimental details, see Berrard et al., 1995).

1991; Lindstedt and Kelly, 1991; Liu et al., 1994) or, as observed for synaptophysin in more competent epithelial cells by Leube et al. (1994), is sorted away in its mature form to a novel population of vesicles.

## Regulation of the expression of the vesicular ACh transporter

Both in nematode and mammals, the gene encoding VAChT was recently shown to be nested within the first intron of the gene that encodes choline acetyltransferase (ChAT), the biosynthetic enzyme of ACh (Chapter 5; Alfonso et al., 1993, 1994; Béjanin et al., 1994; Erickson et al., 1994). This unique gene organization thus forms a potentially remarkable template for a coordinated regulation of the expression of two proteins playing a key role in cholinergic function.

We addressed this issue on primary cultured sympathetic neurons from newborn rats where induction of both the multiple ChAT mRNA species and ChAT activity by several classes of factors has been well demonstrated (see Cervini et al., 1994 and references within). We have shown that treatment of these cells with the cytokine CDF/LIF (cholinergic differentiation factor/leukemia inhibitory factor) or with retinoic acid (RA) also induced VAChT mRNA and VAChT protein levels. VAChT amount was quantified by measuring the number of vesamicol binding sites (Fig. 10A) and immunodetection (Berrard et al., 1995). ChAT activity and vesamicol binding were increased by each factor in a parallel way: about 10- and 4-fold increase by CDF/LIF and RA, respectively. By reverse transcription (RT) PCR with specific primers, we demonstrated that the amount of both total VAChT and ChAT mRNAs was higher in CDF/LIF or RA treated cultures than in controls (Fig. 10B). Among the three VAChT mRNA species described by Béjanin et al. (1994), only the major type which does not share the R-exon with two ChAT mRNAs could be detected and was shown to be induced by these factors (not shown). Similar data with RT-PCR have been recently reported by Misawa et al. (1995) with CDF/LIF and the ciliary neurotrophic factor.

Our results demonstrate the coregulation of ChAT and VAChT mRNA and protein expression by two classes of factors which induce differentiation of noradrenergic neurons towards a cholinergic phenotype. Because of the conserved embedded organization of the VAChT and ChAT genes, it is tempting to envision that common enhancer elements permit regulatory factors to ensure the coordinated expression of both the enzyme responsible for the synthesis of acetylcholine and the protein allowing its storage in synaptic vesicles.

## Acknowledgements

We thank M. Synguelakis for skilful assistance in the preparation of cell cultures; Mrs M. Tomasi and R. Charré for technical assistance. Antibodies to CTR433, rab6, rab5, endoplasmic reticulum, $\beta$-COP were kindly provided by Drs M. Bornens, B. Goud, R. Jahn, D. Louvard, T. Kreis, respectively. We are indebted to J. Rand for the generous gift of the *unc17* clone. This work was supported by CNRS, INSERM, A.F.M., the Ministère de l'Enseignement Supérieur et de la Recherche, Rhone-Poulenc Rorer, the Institut de Recherche sur la Moëlle Epinière, DRET Grants 92/175 and 94/067, and fellowships from IFSBM (H.V.) and MERS (F.A.M.).

## References

Alfonso, A., Grundahl, K., Duerr, J.S., Han, H.-P. and Rand, J.B. (1993) The *Caenorhabditis elegans unc17* gene: a putative vesicular acetylcholine transporter. *Nature*, 261: 617–619.

Alfonso, A., Grundahl, K., McManus, J.R., Asbury, J.M. and Rand, J.B. (1994) Alternative splicing leads to two cholinergic proteins in *Caenorhabditis elegans*. *J. Mol. Biol.*, 241: 627–630.

Anderson, D.C., King, S.C. and Parsons, S.M. (1982) Proton gradient linkage to active uptake of $^3$H-acetylcholine by *Torpedo* electric organ synaptic vesicles. *Biochemistry*, 21: 3037–3043.

Anderson, D.C., King, S.C. and Parsons, S.M. (1983) Pharmacological characterization of the acetylcholine transport system in purified *Torpedo* electric organ synaptic vesicles. *Molec. Pharmacol.*, 24: 48–54.

Bahr, B.A. and Parsons, S.M. (1986a) Acetylcholine transport and drug inhibition kinetics in *Torpedo* synaptic vesicles. *J. Neurochem.*, 46: 1214–1218.

Bahr, B.A. and Parsons, S.M. (1986b) Demonstration of a receptor in *Torpedo* synaptic vesicles for the acetylcholine storage blocker L-trans-2-(4-phenyl[3,4$^3$H]piperidino) cyclohexanol. *Proc. Natl. Acad. Sci. USA*, 83: 2267–2270.

Bahr, B.A., Clarkson, E.D., Rogers, G.A., Noremberg, K. and Parsons, S.M. (1992a) A kinetic and allosteric model for the acetylcholine transporter-vesamicol receptor in synaptic vesicles. *Biochemistry*, 31: 5752–5762.

Bahr, B.A., Noremberg, K., Rogers, G.A., Hicks, B.A. and Parsons, S.M. (1992b) Linkage of acetylcholine transporter-vesamicol receptor to proteoglycan in synaptic vesicles. *Biochemistry*, 31: 5778–5784.

Béjanin, S., Cervini, R., Mallet, J. and Berrard, S. (1994) A unique gene organization for two cholinergic markers, cho-

line acetyltransferase and a putative vesicular transporter of acetylcholine. *J. Biol. Chem.*, 269: 21944–21947.

Berrard, S., Varoqui, H., Cervini, R., Israël, M., Mallet, J. and Diebler, M.-F. (1995) Coregulation of two embedded gene products: choline acetyltransferase and the vesicular acetylcholine transporter. *J. Neurochem.*, 65: 939–942.

Brochier, G., Israël, M. and Lesbats, B. (1993) Immunolabelling of the presynaptic membrane of *Torpedo* electric organ nerve terminals with an antiserum towards the acetylcholine releasing protein mediatophore. *Biol Cell*, 78: 145–154.

Cameron, P.L., Südhof, T.C., Jahn, R. and de Camilli, P. (1991) Colocalization of synaptophysin with transferrin receptors: implications for synaptic biogenesis. *J. Cell Biol.*, 115: 151–164.

Cervini, R., Berrard, S., Béjanin, S. and Mallet, J. (1994) Regulation by CDF/LIF and retinoic acid of multiple ChAT mRNAs produced from distinct promoters. *NeuroReport*, 5: 1346–1348.

Chavrier, P., Parton, R.G., Hauri, H.P., Simons, K. and Zerial, M. (1990) Localization of low molecular weight GTP binding proteins to exocytic and endocytic compartments. *Cell*, 62: 317–329.

De Robertis, E.D., Del Arnaiz, G.R. and De Iraldi, A.P. (1962) Isolation of synaptic vesicles from nerve endings of the rat brain. *Nature*, 194: 794–945.

Diebler, M.-F. and Lazereg, S. (1985) Mg-ATPase and cholinergic synaptic vesicles. *J. Neurochem.*, 44: 1633–1641.

Diebler, M.-F. and Morot Gaudry-Talarmain, Y. (1989) AH5183 and cetiedil: two potent inhibitors of acetylcholine uptake into isolated synaptic vesicles from *Torpedo* marmorata. *J. Neurochem.*, 52: 813–821.

Duden, R., Allan, V. and Kreis, T. (1991) Involvement of ß-COP in membrane traffic through the Golgi complex. *Trends Cell Biol.*, 1: 14–19.

Erickson, J.D. and Eiden, L.E. (1993) Functional identification and molecular cloning of a human brain monoamine vesicular transporter. *J. Neurochem.*, 61: 2314–2317.

Erickson, J.D., Eiden, L.E. and Hoffman, B.J. (1992) Expression cloning of a reserpine-sensitive vesicular monoamine transporter. *Proc. Natl. Acad. Sci. USA.*, 89: 10993–10997.

Erickson, J.D., Varoqui, H., Schäfer, M.K.H., Modi, W., Diebler, M.-F., Weihe, E., Rand, J., Eiden, L.E., Bonner, T.I. and Usdin, T.B. (1994) Functional identification of a vesicular acetylcholine transporter and its expression from a "cholinergic" gene locus. *J. Biol. Chem.*, 269: 21929–21932.

Feany, M.B., Yee, A.G., Delvy, M.L. and Buckley, K.M. (1993) The synaptic vesicle proteins SV2, synaptotagmin and synaptophysin are sorted to separate cellular compartments in CHO fibroblasts. *J. Cell Biol.*, 123: 575–584.

Fonnum, F. (1975) A rapid radiochemical method for the determination of choline acetyltransferase. *J. Neurochem.*, 24: 407–409.

Fuerst, T.R., Niles, E.G., Studier, F.W. and Moos, B. (1986) Eukaryotic transient-expression system based on recombinant vaccinia virus that synthesizes bacteriophage T7

RNA polymerase. *Proc. Natl. Acad. Sci. USA.*, 83: 8122–8126.

Goud, B., Zahroui, A., Tavitian, A. and Saraste, J. (1990) Small GTP-binding protein associated with Golgi cisternae. *Nature*, 345: 553–556

Israël, M., Gautron, J. and Lesbats, B. (1968) Isolement des vésicules synaptiques de l'organe électrique de la Torpille et localisation de l'acétylcholine à leur niveau. *C.R. Acad. Sci. (Paris)*, 266: 273–275.

Jasmin, B.J., Cartaud, J., Bornens, M. and Changeux, J.P. (1989) Golgi apparatus in chick skeletal muscle: changes in its distribution during end plate development and after denervation. *Proc. Natl. Acad. Sci. USA.*, 86: 7218–7222.

Kornreich, W.D. and Parsons, S.M. (1988) Sidedness and chemical and kinetics properties of the vesamicol receptor of cholinergic synaptic vesicles. *Biochemistry*, 27: 5262–5267.

Krejci, E., Gasnier, B., Isambert, M.-F., Sagné, C., Gagnon, J., Massoulié, J. and Henry, J.-P. (1993) Expression and regulation of the bovine vesicular monoamine transporter gene. *FEBS Lett.*, 1: 27–32.

Leube, R.E., Leimer U., Grund, C., Franke, W.W., Harth, N. and Wiedenmann, B. (1994) Sorting of synaptophysin into special vesicles in nonneuroendocrine epithelial cells. *J. Cell Biol.*, 127: 1589–1601.

Linstedt, A.D. and Kelly, R.B. (1991) Synaptophysin is sorted from endocytic markers in neuroendocrine PC12 cells but not transfected fibroblasts. *Neuron*, 7: 309–317.

Liu, Y., Peter, D., Roghani, A., Schuldiner, S., Prive, G.G., Eisenberg, D., Brecha, N. and Edwards, R.H. (1992a) A cDNA that suppresses MPP+ toxicity encodes a vesicular amine transporter. *Cell*, 70: 538–551.

Liu, Y., Roghani, A. and Edwards, R.H. (1992b) Gene transfer of a reserpine-sensitive mechanism of resistance to MPP+. *Proc. Natl. Acad. Sci. USA.*, 89: 9074–9078.

Liu, Y., Schweitzer, E.S., Nirenberg, M.J., Pickel, V.M., Evans, C.J. and Edwards, R.H. (1994) Preferential localization of a monoamine transporter to dense core vesicles in PC12 cells. *J. Cell Biol.*, 127: 1419–1433.

Louvard, D., Reggio, H. and Warren, G. (1982) Antibodies to the Golgi complex and the rough endoplasmic reticulum. *J. Cell Biol.*, 92: 92:107.

Marshall, I.G. and Parsons, S.M. (1987) The vesicular acetylcholine transport system. *Trends Neurosci.*, 10: 174–177.

Matsuuchi, L., Buckley, K.M., Lowe, A.W. and Kelly, R.B. (1988) Targeting of secretory vesicles to cytoplasmic domains in AtT-20 and PC12 cells. *J. Cell Biol.*, 106: 239–251.

Michaelson, D.M. and Angel, I. (1981) Saturable acetylcholine transport into purified cholinergic synaptic vesicles. *Proc. Natl. Acad. Sci. USA*, 78: 2048–2052.

Misawa, H., Takahashi, R. and Deguchi, T. (1995) Coordinate expression of vesicular acetylcholine transporter and choline acetyltransferase in sympathetic superior cervical neurones. *NeuroReport*, 6: 965–968.

Orci, L., Ravazzola, M., Amherdt, M., Perrelet, M., Powell, S.K., Quinn, D.L. and Moore, H.-P.H. (1987) The *trans*-most cisternae of the Golgi complex: a compartment for sorting of secretory and plasma membrane proteins. *Cell*, 51: 1039–1051.

Parsons, S.M., Bahr, B.A., Clarkson, E.D., Noremberg, K. and Hicks, B.W. (1993) Acetylcholine transporter-vesamicol receptor pharmacology and structure. In: A.C. Cuello (Ed.), *Cholinergic Function and Dysfunction, Progress in Brain Research, Vol. 98,* Elsevier, Amsterdam, pp. 175–181.

Peter, D., Finn, J., Klisak, I., Liu, Y., Kojis, T., Heinzmann, C., Roghani, A., Sparkes, R. and Edwards, R.H. (1993) Chromosomal localization of the human vesicular amine transporter genes. *Genomics*, 18: 720–723.

Rogers, G.A. and Parsons, S.M. (1989) Inhibition of acetylcholine storage by acetylcholine analogs in vitro. *Mol. Pharmacol.*, 36: 333–341.

Rogers, G.A. and Parsons, S.M. (1992) Photoaffinity labeling of the acetylcholine transporter. *Biochemistry*, 31: 5770–5777.

Roghani, A., Feldman, J., Kohan, S.A., Shirzadi, A., Gundersen, C.B., Brecha, N. and Edwards, R.H. (1994) Molecular cloning of a putative vesicular transporter for acetylcholine. *Proc. Natl. Acad. Sci. USA*, 91: 10620–10624.

Schuldiner, S. (1994) A molecular glimpse of vesicular monoamine transporters. *J. Neurochem.*, 62: 2067–2078.

Tooze, J., Hollingshead, M., Fuller, S.D., Tooze, S.A. and Huttner, W.B. (1989) Morphological and biochemical evidence showing neuronal properties in AtT20 cells and their growth cones. *Eur. J. Cell Biol.*, 49: 259–273.

Varoqui, H., Diebler, M.-F., Meunier, F.M., Rand, J.B., Usdin, T.B., Bonner, T.I., Eiden, L.E. and Erickson, J.D. (1994) Cloning and expression of the vesamicol binding protein from the marine ray *Torpedo. FEBS Lett.*, 342: 97–102.

Whittaker, V.P., Michaelson, I.A. and Kirkland, R.J. (1964) The separation of synaptic vesicles from nerve endings particles ("synaptosomes"). *Biochem. J.*, 90: 293–303.

Whittaker, V.P., Essman, W.B. and Dowe, G.H.C. (1972) The isolation of pure cholinergic synaptic vesicles from the electric organs of elasmobranch fish of the family Torpinidae. *Biochem. J.*, 128: 833–846.

Yamagata, S.K. and Parsons, S.M. (1989) Cholinergic synaptic vesicles contain a V-Type and a P-Type ATPase. *J. Neurochem.*, 53: 1354–1362.

J. Klein and K. Löffelholz (Eds.)
*Progress in Brain Research*, Vol. 109
© 1996 Elsevier Science B.V. All rights reserved.

CHAPTER 7

# Interactions of protons with the acetylcholine transporter of synaptic vesicles

Marie L. Nguyen and Stanley M. Parsons

*Neuroscience Research Institute and Department of Chemistry, University of California, Santa Barbara, CA 93106, USA*

## Introduction

Most biochemical studies of acetylcholine (ACh) active transport by synaptic vesicles have utilized intact vesicles isolated from the electric organ of the marine ray *Torpedo* and ATP to generate the required acidification of the vesicular interior (Parsons et al., 1993). The acidification generates a transmembrane pH gradient that induces protons to flow through the vesicular ACh transporter (VAChT; Chapters 5 and 6) to the cytoplasm, thus driving ACh uptake by proton antiport. A kinetic model for the VAChT was developed recently (Bahr et al., 1992). The outwardly oriented VAChT binds ACh with a dissociation constant of 20–50 mM. The complex then reorients to the inside of the vesicles with rate constant $k_1$, after which ACh is released. The empty, inwardly oriented VAChT reorients towards the outside with rate constant $k_2$. Rate constant $k_2$ is about 100-fold smaller than $k_1$. Thus, the rate determining step in the overall transport cycle is return of the empty ACh binding site to the outside of the vesicle. The approximately 100-fold ratio of $k_1$ to $k_2$ causes the apparent affinity of ACh during steady state transport to be greater than it is under equilibrium conditions. The Michaelis constant ($K_M$) is about 0.3 mM, which is within the physiological range of cytoplasmic ACh concentration. This kinetics model specifies neither the number of protons exchanged per ACh taken up nor the steps in the transport cycle in which the proton(s) flow through the VAChT.

Proton interaction with and flow through the VAChT have been studied very little. This is because a kinetics analysis of how protons interact with the VAChT requires that the pH be varied on each side of the membrane. However, variation of the external pH will affect the activity of the V-type ATPase that pumps protons into the vesicles, thus superimposing the pH dependence of the ATPase on the pH dependence of the VAChT. Also, with intact vesicles the experimenter has no control over the internal pH, which is set by the activity of the V-type ATPase. Thus, it has been necessary to develop a different approach to driving ACh uptake in order to study the interaction of protons with the VAChT.

*Torpedo* synaptic vesicles can be hyposmotically lysed, and they reseal right side out after lysis (Noremberg and Parsons, 1989). When lysis is carried out at a low pH and the resealed vesicles then are suddenly jumped to a basic pH, they take up ACh (Nguyen and Parsons, 1995). The uptake is inhibited by vesamicol and nigericin and exhibits a $K_M$ similar to that obtained in the ATP-driven uptake of ACh. Thus, it occurs through the VAChT because of the transmembrane proton gradient. Using this pH-jump technique, we have studied the effects of different internal and external pH values and electrical potential on the transport activity of the VAChT in isolated *Torpedo* synaptic vesicles. The observations cannot be explained by the simplest models of VAChT-mediated exchange of ACh for protons.

## The pH-jump method

Although Noremberg et al. (1989) showed that essentially all of the lysed vesicles reseal right side out when hyposmotic lysis is carried out at pH 7.8, we do not know the efficiency of resealing at other pH values. This was determined as follows. Vesicles were hyposmotically lysed at pH 4.3–8.1 in [$^{14}$C]ribitol. Ribitol is an uncharged, chemically unreactive sugar that passes non-specifically through the vesicular membrane. It equilibrates at the same concentration in the aqueous phases inside and outside of the vesicles. Equilibrated lysed vesicles were filtered and washed. The amount of [$^{14}$C]ribitol retained by the vesicles is shown in Fig. 1. The same amount was retained from pH 5.2 to 8.1, consistent with successful resealing. of single vesicles. Below pH 5.0, the amount of trapped [$^{14}$C]ribitol increased substantially. This is consistent with fusion of vesicles with each other such that the same amount of membrane enclosed a larger volume of solvent. Thus, the pH-jump technique can be used successfully to obtain resealed monomeric vesicular ghosts down to a pH

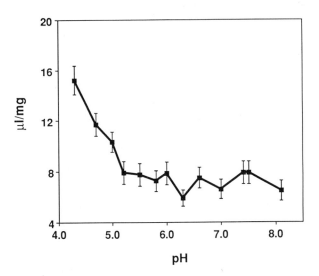

Fig. 1. Retention of [$^{14}$C]ribitol after lysis of vesicles at the indicated pH values. Vesicles were hyposmotically lysed in [$^{14}$C]ribitol and 0.1 M citrate–KOH buffer at various pH values. After a 30 min incubation at 23°C to allow for equilibration, portions of vesicles were filtered in triplicate and washed. The amounts of [$^{14}$C]ribitol bound were converted to μl internal solvent volume/mg vesicular protein.

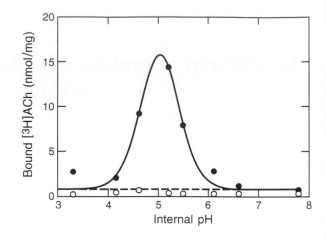

Fig. 2. Internal pH profile for uptake of [$^3$H]ACh with constant external pH. Vesicles were set to the indicated internal pH values by hyposmotic lysis in 0.1 M citrate–KOH and the external pH was jumped to 7.8 with HEPES–KOH. One hundred μl of ghosts containing 7.5 μg of vesicular protein were periodically filtered and washed. The amounts of [$^3$H]ACh (50 μl) transported at 5 min in the absence (●) and presence (○) of 2 μM vesamicol are shown. To fit the uptake data, two protons binding to a site of $pK_a$ 5.3 had to be assumed for the data above pH 5. As shown in Fig. 1, a lysis-resealing artifact inactivates uptake below pH 5.0. Reproduced with permission from Nguyen and Parsons (1995).

of about 5.0–5.2, but below this pH range the method might be subject to limitations.

## Effects of variable internal pH on uptake of ACh

Uptake of a subsaturating concentration of [$^3$H]ACh at external pH 7.8 was studied over the internal pH range 3.3–7.8 (Fig. 2). It exhibited a maximum at about internal pH 5.0 and decreased very steeply to zero at both higher and lower internal pH values. The data could be fit well in a determinant manner only by assuming that two internal sites on the VAChT of about $pK_a$ 5.3 ± 0.2 must be protonated in order to activate transport. The fit to the data below pH 5.0 assumed that protonation of two internal sites of $pK_a$ 4.7 ± 0.1 caused inactivation of transport. However, the data in Fig. 1 tell us that the inhibition of ACh transport at very low internal pH probably arises from a disrupted vesicular structure. Possibly, the vesicular

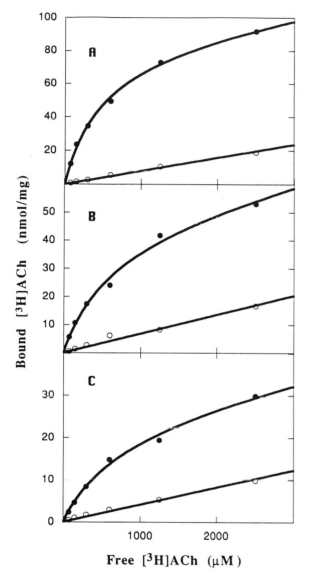

Fig. 3. Saturation of [³H]ACh uptake at different internal pH values. Vesicular ghosts were taken through the pH-jump in the presence of indicated concentrations of [³H]ACh. The amounts of uptake in the absence (total, ●) and presence (nonspecific, ○) of $2 \mu M$ vesamicol were determined at 5 min. Equations describing non-specific and total uptake were fit simultaneously to the two sets of data to estimate $K_M$ and $V_{max}$ values at different internal pH values Frames A, B, and C show the data and regression analyses at internal pH 5.2, 5.7, and 6.2, respectively.

ghosts formed at low internal pH are sealed against leakage of ribitol but not protons or are composed of inverted membranes. Inhibition of ACh trans-

port by very low internal pH probably is an artifact of the hyposmotic lysis-resealing procedure, and only the increase in transport occurring from internal pH 6.5 to 5.0 can be assigned to the VAChT per se.

The behavior of $K_M$ and $V_{max}$ values for transport of [³H]ACh over the internal pH range 6.2–5.2 was studied as follows. Synaptic vesicles were hyposmotically lysed at pH 5.2, 5.7 and 6.2 in the presence of various concentrations of [³H]ACh. The pH of the ghost suspension was then jumped to 7.8. The amounts of total and non-specific uptake are shown in Fig. 3. Simultaneous regression analysis of the two sets of data at each pH value yielded $K_M$ and $V_{max}$ estimates shown in Table 1. At lower internal pH, the $K_M$ decreased slightly but the major effect was a large increase in the $V_{max}$. To the extent that the structure of some of the vesicles might have been disrupted at the lowest internal pH value, the true $V_{max}$ would have been even higher than observed, but the true $K_M$ would not differ. Thus, the trends accurately reflect the dependence of $K_M$ and $V_{max}$ values on internal pH even if the vesicular structure was partially compromised at the lowest internal pH.

It is informative to analyze the expected effects of variable internal pH on $K_M$ and $V_{max}$ values for simple models of the interaction of internal protons with the VAChT. The analysis uses the previously developed kinetics model (Bahr et al., 1992). If a proton is carried by the VAChT in the $k_1$ step, decreased internal pH will increase $k_1$. This will have no effect on $V_{max}$ because $k_2$ is rate limiting. However, this will decrease $K_M$ by increasing the ratio of $k_1$ to $k_2$. If a proton is carried in the $k_2$ step, decreased internal pH will increase $k_2$. This will increase both $V_{max}$ and $K_M$ because the ratio of $k_1$ to $k_2$ will decrease. Neither pattern corresponds to the trends of the parameters in Table 1, suggesting that a more complicated model is required.

### Effects of variable external pH on uptake of ACh

A proton transported through the membrane by the VAChT will be bound on the outside of the VAChT before it dissociates into the cytoplasm.

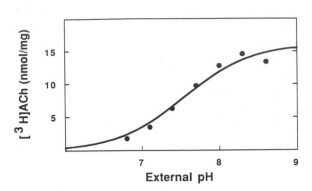

Fig. 4. Uptake of [³H]ACh at different external pH values. Vesicles were hyposmotically lysed in the absence and presence of vesamicol at pH 5.1, and the pH of the ghost suspension then was jumped to the indicated values in the presence of $50\,\mu$M [³H]ACh. A hyperbolic equation was fit to the amount of specific uptake at 7 min (obtained by subtracting the average of the amount of nonspecific uptake at each pH value from the amount of total uptake at 7 min) to estimate a p$K_a$ of $7.6 \pm 0.2$.

Such a proton binding site will exhibit a p$K_a$, and equilibrium protonation of a significant fraction of the externally oriented site will inhibit transport of ACh. To examine the effects of lower external pH values on ACh transport, vesicles at constant internal pH were subjected to a jump to different external pH values in the presence of subsaturating [³H]ACh. ACh uptake increased from 0 at low external pH to a maximum at high external pH (Fig. 4). A hyperbola consistent with the need to deprotonate a single external site to activate ACh transport fit the data well and resulted in an estimate for the p$K_a$ of $7.6 \pm 0.2$.

To determine whether the inactivation of transport by external protons arises from an effect on $K_M$ or $V_{max}$, [³H]ACh uptake was titrated at external pH 6.8 and 8.3. The amounts of total and nonspecific uptake at different concentrations [³H]-ACh are shown in Fig. 5. Simultaneous regression analysis of the two sets of data at each external pH value yielded the $K_M$ and $V_{max}$ estimates listed in Table 1. $K_M$ increased at lower internal pH but $V_{max}$ remained unchanged. This behavior is characteristic of competitive inhibition of ACh transport by protons and is consistent with the binding of either ACh or a proton to a common site in the externally oriented VAChT.

## Effects of electrical potential change on uptake of ACh

The steep dependence of ACh uptake on internal acidification and the complex trends in $K_M$ and $V_{max}$ for ACh uptake at different internal pH values suggest that two internal protons are involved in ACh uptake. Two models are possible. The first is that two protons are exchanged for one ACh. The second model is that one proton is exchanged for one ACh and the second proton binds to an allosteric site on the inside of the VAChT. If two protons are transported across the membrane, uptake of ACh will be electrogenic and will depend

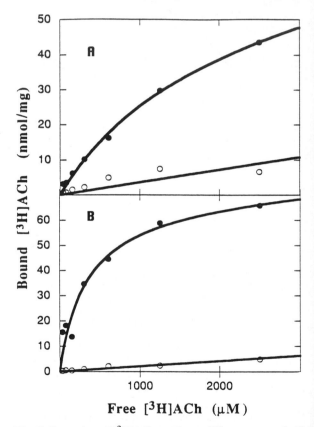

Fig. 5. Saturation of [³H]ACh uptake at different external pH values. Vesicular ghosts at internal pH 5.1 were taken through the pH-jump in the presence of indicated concentrations of [³H]ACh. The amounts of uptake in the absence (total, ●) and presence (non-specific, ○) of $2\,\mu$M vesamicol were determined at 7 min. Equations describing non-specific and total uptake were fit simultaneously to the two sets of data to estimate $K_M$ and $V_{max}$ values at each external pH. Frames A and B show the fits at external pH 6.8 and 8.3, respectively.

TABLE 1

Michaelis–Menten parameters at different internal and external pH values

|  | $K_M$ ($\mu$M) | $V_{max}$ (nmol/mg) |
|---|---|---|
| Internal pH[a] |  |  |
| 5.2 | 496 ± 34 | 86 ± 3 |
| 5.7 | 630 ± 104 | 46 ± 3 |
| 6.2 | 727 ± 109 | 25 ± 2 |
| External pH[b] |  |  |
| 6.8 | 1622 ± 426 | 57 ± 8 |
| 8.3 | 327 ± 78 | 69 ± 6 |

[a]Regression values obtained from Fig. 3.
[b]Regression values obtained from Fig. 5.

on the transmembrane electrical gradient. To test for a dependence of ACh transport on electrical potential, the following experiment was done. Synaptic vesicles were hyposmotically lysed in Na-based buffer containing [3H]ACh and [14C]-methylamine in the absence or presence of 300 nM valinomycin. Methylamine accumulates in acidic compartments. The external pH was jumped with a K+-based buffer. The valinomycin will mediate uptake of K+, producing a positive internal electrical potential in the vesicles. The amounts of [3H]ACh and [14C]methylamine taken up by the vesicles are shown in Fig. 6A and B, respectively. Valinomycin induced a much more rapid collapse of the pH gradient, as indicated by the faster loss of [14C]methylamine. This indicates that vesicles exposed to valinomycin indeed became more positively charged. Yet there was essentially no difference in the amounts of ACh taken up. The result indicates that increased positive transmembrane electrical potential can substitute for protons as a driving force for ACh uptake, consistent with the exchange of two protons for one ACh.

## A model for the interactions of protons with the VAChT

The results of these studies lead to the model shown in Fig. 7. If steps $k_1$ and $k_2$ both involve movement of a proton through the VAChT, lower

internal pH will speed both steps. If the $pK_a$s of the two internal protonation sites were identical to each other, lower internal pH would increase the $V_{max}$ but have no effect on the $K_M$ because the ratio of $k_1$ to $k_2$ would not change (although their values would increase). The fact that $K_M$ decreases slightly at lower internal pH implies that the $pK_a$ for the internal protonation site involved in the $k_1$ step is slightly lower than the $pK_a$ for the internal protonation site involved in the $k_2$ step. The regression analysis in Fig. 2 indicates that the internal protonation sites have $pK_a$s of approximately 5.3, suggesting that the $pK_a$ of the $k_1$ site is <5.3

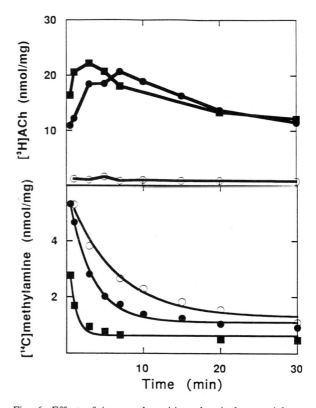

Fig. 6. Effect of increased positive electrical potential on [3H]ACh and [14C]methylamine uptake. Vesicles isolated in isosmotic buffers neutralized with NaOH were subjected to hyposmotic lysis with 0.1 M citrate–NaOH at pH 5.1 to yield final concentrations during transport of 0 nm (●) or 100 nm (■) valinomycin or 2 $\mu$M vesamicol (○). External pH was jumped with 0.2 M HEPES–KOH to pH 7.8. Dpms bound to the filters for each isotope were determined by double channel liquid scintillation spectroscopy. Frame A shows [3H]ACh and frame B shows [14C]methylamine uptake.

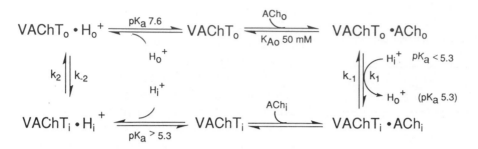

Fig. 7. A model for the interaction of protons and ACh with the VAChT. Cytoplasmic ACh ($ACh_o$) binds to the outwardly oriented transporter ($VAChT_o$) with an equilibrium dissociation constant ($K_{Ao}$) of 20–50 mM. Internal protons ($H_i^+$) bind to an internal site with a $pK_a$ somewhat lower than 5.3. Transporter loaded on both sites reorients in step $k_1$ after which it releases ACh to the inside and a proton to the outside. The $pK_a$ of the protonation site that moves in the $k_1$ step to the outside is assumed to remain unchanged at about 5.3. Internally oriented transporter ($VAChT_i$) protonates at the ACh site with a $pK_a$ somewhat higher than 5.3 and reorients to the outside in step $k_2$. The transported proton dissociates into the cytoplasm with a $pK_a$ of 7.6 to complete one cycle of ACh transport. Step $k_1$ occurs about 100-fold faster than step $k_2$, which accounts for a Michaelis constant that is about 100-fold lower than the dissociation constant for ACh.

and the $pK_a$ of the $k_2$ site is >5.3. After a proton translocates to the outside in the $k_2$ step, it dissociates with $pK_a$ 7.6 to create the competent ACh binding site. The requirement that this site deprotonate before binding ACh could account for the competitive inhibition of ACh transport by external protons. A shift in $pK_a$ of the ACh binding site from >5.3 to 7.6 upon reorientation from the inside to the outside of the vesicles is energetically reasonable. External ACh binds to the externally oriented substrate site with a dissociation constant of 20–50 mM. When a separate site on the inside of this form of the VAChT becomes protonated, step $k_1$ occurs with movement of ACh to the inside simultaneously with movement of the bound proton to the outside of the vesicle. We have no information about the $pK_a$ of this protonation site when it is outwardly oriented and merely assume that it remains unchanged from the value it has when oriented to the inside. Protonation of this site when it is outwardly oriented does not have a significant effect on uptake of ACh because uptake will already have been nearly completely inhibited by prior protonation of the more basic outwardly oriented ACh binding site.

This model is an elaboration of the simplest possible model for a proton antiporter. The simplest model has a single binding site that alternately binds a proton and a molecule of substrate on opposite sides of the membrane. Our model hypothesizes such a site (written on the right hand side of VAChT in Fig. 7). In addition, the VAChT has a second protonation site that drives the microscopic step for uptake of ACh. This step ($k_1$) would occur spontaneously in simple proton antiporters. Our model for the VAChT is similar to the model hypothesized for the vesicular monoamine transporter (Schuldiner, 1994).

## Acknowledgements

We thank Ricardo Garcia and Barry Sanchez for their excellent preparation of vesicles. This study was supported by grant NS15047 from the National Institute of Neurological Disorders and Stroke, United States Public Health Service, and a grant from the Muscular Dystrophy Association.

## References

Bahr, B.A., Clarkson, E.D., Rogers, G.A., Noremberg, K. and Parsons, S.M. (1992) A kinetic and allosteric model for the acetylcholine transporter-vesamicol receptor in synaptic vesicles. *Biochemistry*, 31: 5752–5762.

Nguyen, M.L. and Parsons, S.M. (1995) Effects of internal pH on the acetylcholine transporter of synaptic vesicles. *J. Neurochem.*, 64: 1137–1142.

Noremberg, K. and Parsons, S.M. (1989) Regulation of the vesamicol receptor in cholinergic synaptic vesicles by acetylcholine and an endogenous factor. *J. Neurochem.*, 52: 913–920.

Parsons, S.M., Prior, C. and Marshall, I.G. (1993) Acetylcholine transport, storage, and release. *Int. Rev. Neurobiol.*, 35: 277–390.

Schuldiner, S. (1994) A molecular glimpse of vesicular monoamine transporters. *J Neurochem.*, 62: 2067–2078.

# Section III

# Neuronal Nicotinic Receptors

J. Klein and K. Löffelholz (Eds.)
*Progress in Brain Research*, Vol. 109
© 1996 Elsevier Science B.V. All rights reserved.

CHAPTER 8

# Nicotinic receptors of the vertebrate CNS: introductory remarks

Alfred Maelicke

*Laboratory of Molecular Neurobiology, Institute of Physiological Chemistry and Pathobiochemistry,
Johannes-Gutenberg University Medical School, Duesbergweg 6 , D-55099 Mainz, Germany*

In vertebrates, nicotinic cholinergic neurotransmission is found in both the central nervous system (CNS) and the periphery (muscle endplate). Although muscle and neuronal nicotinic acetylcholine receptors (nAChR) have evolved from a common ancestor, it is striking that the muscle receptor has remained rather stable in evolution whereas the neuronal receptor has evolved to a wide diversity of subtypes (Bertrand and Changeux, 1995; Lindstrom, 1995). The latter is due to the fact that in the course of evolution the vertebrate CNS has undergone much larger structural and functional specialization, concomitant with topological diversification, than has muscle tissue. Clearly, since nicotinic receptor isoforms are widely and differentially distributed in the adult brain and in brain development (Wada et al., 1989; Seguela et al., 1993; Alkondon et al., 1994; Lobron et al., 1995) they must play important roles in brain function. Although neurons may express several nAChR subtypes simultaneously, their expression patterns are selective (Role, 1992; Alkondon et al., 1994; Lobron et al., 1995; Storch et al., 1995), indicating a correlation between subtype expression and neuronal function. Indeed, neuronal nAChR subtypes differ in their cation selectivities and permeabilities, in their activation and inactivation kinetics, in their affinities for ACh, and in their general pharmacology (Alkondon and Albuquerque, 1993; Alkondon et al., 1994; Montes et al., 1994). Nicotinic receptors have also been implicated in neurological diseases, in particular those associated with impairments in learning and memory (Nordberg, 1992; Schröder et al., 1995), with evidence emerging that particular nAChR subtypes may be involved (Lobron et al., 1995).

In Chapter 9 of this volume, Albuquerque and colleagues discuss the distinct cellular distribution, and the functional properties of the neuronal nicotinic receptor subtypes expressed in hippocampal neurons. They have succeeded in assigning three distinct whole-cell currents that are activated by nicotinic agonists to specific nAChR subtypes, and to characterize the functional properties of these nAChR channels to a level that is only matched by the muscle receptor and ectopically expressed nAChR subtypes.

In Chapter 10 of this volume, Lindstrom and colleagues summarize the molecular cloning and subunit composition data presently available on nicotinic receptors. In addition, their contribution addresses two functional aspects of neuronal nicotinic receptors that are also of practical interest: (i) the mechanism of upregulation by chronic exposure to nicotine of the most abundant neuronal nAChR subtype in the mammalian brain, and (ii) the pharmacological profile of neuronal nAChRs in regard to epibatidine, a natural compound initially isolated from the skin of Ecuadoran poison frogs.

My laboratory has contributed to this area by the discovery of a novel class of regulatory ligands, named "non-competitive agonists, NCA", which allosterically modulate the ACh-activated nAChR channel (Okonjo et al., 1991; Maelicke et al., 1993; Pereira et al, 1993; Schrattenholz et al.,

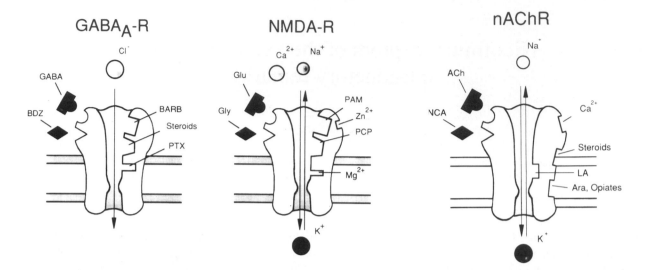

Fig. 1. Regulatory control by ligands of the GABA$_A$ receptor, the NMDA receptor, and the nAChR. Left: the GABA$_A$ receptor is a ligand-gated Cl$^-$ channel. Benzodiazepines (BDZ) increase the probability of GABA-induced channel opening; barbiturates (BARB), steroids and picrotoxin (PTX) affect the mean open times of the channel. Middle: The NMDA receptor is a ligand-gated cation channel with high Ca$^{2+}$ permeability. Glycine (Gly) acts as co-agonist of glutamate (Glu); polyamines (PAM) increase the probability of channel opening induced by the concerted action of Glu ad Gly; Mg$^{2+}$, phencyclidine (PCP), ketamine and MK-801 exert voltage-dependent blocks, whereas the block exerted by Zn$^{2+}$ is voltage-independent (Maelicke et al., 1995). Right: the nAChR is a ligand-gated nonspecific cation channel. NCAs modulate the ACh-sensitivity of the nAChR, and can overcome ACh-induced desensitization; Ca$^{2+}$ positively affects nAChR activation; local anaesthetics (LA) induce a voltage-dependent block of the channel; steroids, arachidonic acid (Ara) and opiates decrease the responsiveness to ACh of the receptor.

1996). (i) At low concentrations (submicromolar), they facilitate the conversion by ACh (and ACh-competitive agonists) of the resting to the open-channel state of the receptor, which results in a potentiation of the ACh-induced response (Schrattenholz et al., 1996). (ii) At higher concentrations, NCAs, by themselves and independently, can activate nAChRs, albeit with only very low efficacy (Maelicke et al., 1993; Pereira et al, 1993). (iii) At about the same concentrations, NCAs also produce a fast non-competitive block of the ACh-activated nAChR channel (Schrattenholz et al., 1996). The actions as sensitizing agent and as noncompetitive agonist are produced by the interaction of the NCAs with a binding site in the extracellular region of the nAChR $\alpha$-subunit that is distinct from that of ACh (Schrattenholz et al., 1993; Schröder et al., 1994). The action as non-competitive inhibitor probably is produced by the interaction with the local anesthetic binding site which is known to be accessible only in the open-channel conformation of the nAChR.

The functional properties of the novel class of nicotinic ligands are reminescent, in some respects, to those of glycine and polyamines with regard to the NMDA-subtype of glutamate receptor (Johnson and Ascher, 1987), and of benzodiazepines with regard to GABA$_A$ receptors (Rock and Macdonald, 1991). This has led us to postulate that modulatory control of channel activity of neurotransmitter-gated ion channels probably is a property that is intimately connected to the role these neuroreceptors play in the realm of CNS function. As the figure schematically summarizes, these three major neuroreceptors of the mammalian brain are subject to a multitude of ligand-controlled regulatory mechanisms, at least some of which may also be functional under physiological in situ conditions.

The fundamental properties of ligand-gated ion channels, such as ligand and ion selectivity, ion conductivity, mean open-channel time and binding and activation/inactivation kinetics, are mainly defined by the isoform composition of the particular subtype, i.e. on the molecular level (Kusama et al., 1993). Transcriptional regulation, including alternative splicing and RNA editing (Barnard, 1992; Hollmann and Heinemann, 1994), and post-translational modifications (Sommer et al., 1990) are the major mechanisms of modulatory control on the cellular level. They are under combined genetic and hormonal control, with their time range being from several minutes to days. Additional levels of control enable the CNS to selforganize during development and to function intelligently and in a "creative" fashion. The most prominent of these levels is the neuronal network, which produces stable pathways of communication by means of preformed synaptic contacts. Probably even more important, although hardly understood at present, are the "chemical networks", which transiently form in response to the constantly changing chemical environment to which the receptors of each and every synapse, dendritic spine and cell are exposed. The multiple modes of ligand-dependent control observed for typical CNS neuroreceptors (Fig. 1), including the action of NCAs on nicotinic receptors, are suggestive in this regard. As an attractive hypothesis, neurotransmitters and neurohormones may not only interact with their archetypic cognate receptors, but rather also with other neuroreceptors, albeit in a modulatory fashion. By modulating the sensitivity and response kinetics of neuroreceptors to their natural transmitters, the action of allosteric regulators could not only provide additional levels of integration of response, or coincidence detection, but distinct local combinations and concentrations of chemical messengers could also produce messages of a new quality. By their implementation on sets of different receptors in neighboring synapses and spines, and by linking many of these locally distinct incidences on the cellular network level, the kind of neuronal activity may be produced that could be associated with higher order brain activity, such as consciousness and mind. It is noteworthy in this respect that because chemical network control is achieved by ligand binding, its time

resolution is in the subsecond range, i.e. similar to that of thoughts and emotions. To summarize this discussion, we suggest that endogenous nicotinic NCAs are members of a much larger family of intracellular messengers, which are involved in higher order modes of control of CNS function.

## References

Alkondon, M. and Albuquerque, E.X. (1993) Diversity of nicotinic acetylcholine receptors in rat hippocampal neurons: I. Pharmacological and functional evidence for distinct structural subtypes. *J. Pharmacol. Exp. Ther.*, 265: 1455–1473.

Alkondon, M., Reinhardt-Maelicke, S., Lobron, C., Hermsen, B., Maelicke, A. and Albuquerque, E.X. (1994) Diversity of nicotinic acetylcholine receptors in rat hippocampal neurons: II. The rundown and inward rectification of agonist-elicited whole-cell currents and in situ hybridization studies. *J. Pharmacol. Exp. Ther.*, 271: 494–506.

Barnard, E.A. (1992) Receptor classes and transmitter-gated ion channels. *Trends Biochem. Sci.,* 17: 368–374.

Bertrand, D. and Changeux, J.-P. (1995) Nicotinic receptor: an allosteric protein specialized for intercellular communication. *Semin. Neurosci.*, 7: 75–90.

Hollmann, M. and Heinemann, S. (1994) Cloned glutamate receptors. *Annu. Rev. Neurosci.,* 17: 31–108.

Johnson, J.W. and Ascher, P. (1987) Glycine potentiates the NMDA response in cultured mouse brain neurons. *Nature,* 325: 529–531.

Kusama, T., Spivak, C.E., Whiting, P., Dawson, V.L., Schaeffer, J.C. and Uhl, G.R. (1993) Pharmacology of GABA receptors expressed in Xenopus oocytes and COS cells. *Br. J. Pharmacol.,* 109: 200–206.

Lindstrom, J. (1995) Nicotinic acetycholine receptors. In: A. North (Ed.), *CRC Handbook of Receptors*, CRC Press, Boca Raton, FL, pp. 153–175.

Lobron, C. Wevers, A., Dämgen, K., Jeske, A., Rontal, D., Birtsch, C., Heinemann, S., Reinhardt, S., Maelicke, A. and Schröder, H. (1995) Cellular distribution in the rat telencephalon of mRNAs encoding for the $\alpha 3$ and $\alpha 4$ subunits of the nicotinic acetylcholine receptor. *Mol. Brain Res.,* 30: 70–76.

Maelicke, A., Coban, T., Schrattenholz, A., Schröder, B., Reinhardt-Maelicke, S., Storch, A., Godovac-Zimmermann, J., Methfessel, C., Pereira, E.F.R. and Albuquerque, E.X. (1993) Physostigmine and neuromuscular transmission In: A.S. Penn, D.P. Richman, R.L. Ruff and V.A. Lennon (Eds), *Myasthenia Gravis and Related Disorders: Experimental and Clinical Aspects. Ann. N.Y. Acad. Sci.,* 681: 140–154.

Maelicke, A., Schrattenholz, A., Storch, A. and Schröder, B. (1995) Modulatory control by non-competitive agonist of nicotinic cholinergic neurotransmission in the central nervous system. *Semin. Neurosci.*, 7: 103–114.

Montes, J.G., Alkondon, M., Pereira, E.F.R., Castro, N. and Albuquerque, E.X. (1994) Electrophysiological methods for the study of neuronal nicotinic acetylcholine receptor ion channels. In: T. Narashi (Ed.), *Ion channels of excitable cells. Methods Neurosci.*, 19: 121–146.

Nordberg, A. (1992) Neuroreceptor changes in Alzheimer disease. *Cerebrovasc. Brain Metab. Rev.,* 4: 303–328.

Okonjo, K.O., Kuhlmann, J. and Maelicke, A.A. (1991) Second pathway of activation of the Torpedo acetylcholine receptor channel. *Eur. J. Biochem.*, 200: 671–677.

Pereira, E.F.R., Reinhardt-Maelicke, S., Schrattenholz, A., Maelicke, A. and Albuquerque, E.X. (1993) Identification and functional characterization of a new agonist site on nicotinic acetylcholine receptors of cultured hippocampal neurons. *J. Pharmacol. Exp. Ther.*, 265: 1474–1491.

Rock, D.M. and Macdonald, R.L. (1991) The polyamine spermine has multiple actions on *N*-methyl-D-aspartate receptor single-channel currents in cultured cortical neurons. *Mol. Pharmacol.*, 41: 83–88.

Role, L.W. (1992) Diversity in primary structure and function of neuronal nicotinic acetylcholine receptor channels. *Curr. Opin. Neurobiol.,* 2: 254–262.

Schrattenholz, A., Godovac-Zimmermann, J., Schäfer, H.-J., Albuquerque, E.X. and Maelicke, A. (1993) Photoaffinity labeling of Torpedo acetylcholine receptor by physostigmine. *Eur. J. Biochem.,* 216: 671–677.

Schrattenholz, A., Pereira, E.F.R., Methfessel, C., Roth, U., Weber, K.-H., Albuquerque, E.X. and Maelicke, A. (1996) Agonist responses of neuronal nicotinic acetylcholine receptors are potentiated by a novel class of allosterically active ligands. *Mol. Pharmacol.*, 49: 1–6.

Schröder, B., Reinhardt-Maelicke, S., Schrattenholz, A., McLane, K.E., Kretschmer, A., Conti-Tronconi, B.M. and Maelicke, A. (1994) Monoclonal antibodies FK1 and WF6 define two neighboring ligand binding sites on Torpedo acetylcholine receptors $\alpha$-polypeptide. *J. Biol. Chem.,* 269: 10407–10416.

Schröder, H., Lobron, C., Wevers, A., Maelicke, A. and Giacobini, E. (1995) Cellular acetylcholine receptor expression in the brain of patients with Alzheimer's and Parkinson's dementia. *Adv. Behav. Biol.,* 44: 63–67.

Seguela, P., Wadiche, J., Dinelly-Miller, K., Dani, J. and Patrick, J. (1993) Molecular cloning, functional properties and distribution of rat brain $\alpha7$: a nicotinic cation channel highly permeable to calcium. *J. Neurosci.*, 13: 596–604

Sommer, B., Keinanen, K., Verdoorn, T.A., Wisden, W., Burnashev, N., Herb, A., Kohler, M., Takagi, M., Sakmann, B. and Seeburg, P.H. (1990) Flip and flop: A cell specific functional switch in glutamate-operated channels of the CNS. *Science,* 249: 1580–1585.

Storch, A., Cooper, J.C., Gutbrod, O., Weber, K.-H., Reinhardt, S., Lobron, C., Hermsen, B., Soskic, V., Schrattenholz, A., Pereira, E.F.R., Albuquerque, E.X., Methfessel, C. and Maelicke, A. (1995) Physostigmine, galanthamine and codeine act as non-competitive nicotinic agonists on clonal rat pheochromocytoma cells. *Eur. J. Pharmacol.*, 290: 207–219.

Wada, E., Wada, K., Boulter, J., Deneris, E., Heinemann, S., Patrick, J. and Swanson, L.W. (1989) Distribution of $\alpha2$, $\alpha3$, $\alpha4$ and $\beta2$ neuronal nicotinic receptor subunit mRNAs in the central nervous system: a hybridization histochemical study in the rat. *J. Comp. Neurol.* , 284: 314–335.

J. Klein and K. Löffelholz (Eds.)
*Progress in Brain Research*, Vol. 109

CHAPTER 9

# Nicotinic acetylcholine receptors on hippocampal neurons: cell compartment-specific expression and modulatory control of channel activity

E.X. Albuquerque[1,2], E.F.R. Pereira[1], R. Bonfante-Cabarcas[1,2], M. Marchioro[1,2], H. Matsubayashi[1], M. Alkondon[1] and A. Maelicke[1,3]

[1]*Department of Pharmacology and Experimental Therapeutics, University of Maryland School of Medicine, Baltimore, MD 21201, USA,* [2]*Laboratory of Molecular Pharmacology II, Institute of Biophysics Carlos Chagas Filho, Federal University of Rio de Janeiro, Rio de Janeiro, RJ 21994, Brazil, and* [3]*Department of Physiological Chemistry and Pathobiochemistry, Johannes-Gutenberg University Medical School, Mainz D-6500, Germany*

## Introduction

Neuronal nicotinic receptors (nAChRs) are ACh-gated cationic channels and belong to a super-family of ligand-gated ion channels (Betz, 1990). The neuronal nAChRs are believed to be penta-oligomers made up of a single type of the agonist-binding $\alpha$ subunit or of a combination of $\alpha$ subunits with structural, $\beta$ subunits (Anand et al., 1991; Role, 1992; Karlin, 1993; Sargent, 1993). Screening of mRNA libraries from neurons of various brain regions of rats, chicks, and humans resulted in the isolation, identification, and cloning of eight $\alpha$ ($\alpha2$–$\alpha9$) and three $\beta$ ($\beta2$–$\beta4$) neuronal nAChR subunits (Heinemann et al., 1991; Lindstrom, 1995). According to studies of ectopic expression of a variety of combinations of nAChR subunits, out of the eight $\alpha$ subunits cloned to date, only three ($\alpha7$, $\alpha8$, and $\alpha9$) can form homooligomeric functional neuronal nAChRs; all the other $\alpha$ subunits have to be combined with $\beta$ subunits to give rise to functional nAChRs (Lindstrom, 1995). Because of the different possible combinations of the various neuronal nAChR subunits, neuronal nAChRs are diverse in their functional and pharmacological properties (Role, 1992; Albuquerque et al., 1995a).

Based on binding studies, neuronal nAChRs were classified into two broad classes: the $\alpha$-BGT-insensitive nAChRs (which include the nAChRs that bind nicotine with high affinity) and the $\alpha$-BGT-sensitive nAChRs (Romano and Goldstein, 1980; Lukas, 1984; Clarke et al., 1985). Electrophysiological studies carried out in CNS neurons suggested that various functional nAChR subtypes an be found in a single brain area, and that in some brain areas a single neuron can express more than one subtype of neuronal nAChR (Alkondon and Albuquerque, 1993, 1995; Alkondon et al., 1994). It is less clear whether there is segregation of the distinct subtypes of neuronal nAChRs on the apical and basal dendrites, cell body, and dendritic spines of the neurons, or whether rather than being segregated on these different regions, the various nAChR subtypes would be intermingled on the entire neuronal surface. In addition, it is unknown whether the nAChRs expressed on postsynaptic neurons in the CNS are synaptic or extrasynaptic receptors. However, the great diversity of neuronal nAChRs, in addition to the fast rate of activation/inactivation of some of these receptors in the CNS, still hinders their functional characterization in situ (Albuquerque et al., 1995b).

There is mounting evidence in support of the notion that ACh plays an important role in learning and memory, as well as in the control of neuronal development, partly by binding to neuronal nAChRs in the CNS. Given that the hippocampus is the brain structure most prominently associated with cognitive functions, our studies have been focused in the characterization of the properties of nAChRs expressed on hippocampal neurons.

Hippocampal neurons either in culture or acutely dissociated respond to nicotinic agonists with at least one of three pharmacologically and kinetically distinct ionic currents, which are referred to as type IA, II, and III currents (Alkondon and Albuquerque, 1993, 1995; Alkondon et al., 1994; Ishihara et al., 1995a). Type IA currents, the predominant responses of hippocampal neurons to nicotinic agonists, are fast-desensitizing currents sensitive to blockade by methyllycaconitine (MLA, 1 nM) or $\alpha$-bungarotoxin ($\alpha$-BGT, 10 nM) (Alkondon et al., 1992; Alkondon and Albuquerque, 1993). Type II currents, which can be recorded from 5% of the tested neurons, are slowly desensitizing currents sensitive to blockade by dihydro-$\beta$-erythroidine (DH$\beta$E) (10 nM) (Alkondon and Albuquerque, 1993, 1995). Type III currents, nicotinic responses recorded from 2.5% of the tested neurons, are slowly desensitizing currents that are sensitive to blockade by mecamylamine (1 $\mu$M) (Alkondon and Albuquerque, 1993). About 10% of the hippocampal neurons respond to nicotinic agonists with whole-cell currents that have two components. The fast-decaying component of these currents (which are referred to as type IB currents) has the same pharmacological and biophysical properties as type IA currents, whereas the slowly decaying component of these currents have the same characteristics as type II currents, in this way indicating that a single hippocampal neuron can express more than one nAChR subtype (Alkondon and Albuquerque, 1993, 1994; Alkondon et al., 1994). It is likely that an $\alpha$7-bearing nAChR gives rise to type IA currents, an $\alpha$4$\beta$2 nAChR gives rise to type II currents, and an $\alpha$3$\beta$4 nAChR gives rise to type III currents (Alkondon and Albuquerque, 1993). Indeed, hippocampal neurons express mRNAs coding for nAChR $\alpha$7, $\alpha$4, and $\beta$2 subunits (Alkondon et al., 1994), and the proportion of hippocampal neurons binding either nicotine with high affinity or $\alpha$-BGT corresponds to the proportion of neurons that respond to nicotinic agonists with type II or type IA currents, respectively (Barrantes et al., 1995). Also in agreement with the concept that $\alpha$7-bearing nAChRs account for the $\alpha$-BGT-sensitive nicotinic currents recorded from hippocampal neurons was the finding that the $\alpha$-BGT-sensitive hippocampal nAChRs have a high permeability to $Ca^{2+}$ as does the homomeric $\alpha$7-based nAChR ectopically expressed in *Xenopus* oocytes (Vijayaraghavan et al., 1992; Séguéla et al., 1993; Castro and Albuquerque, 1995).

Expanding these studies, we now show for the first time that neuronal nAChRs are differentially expressed on the somato-dendritic surface of hippocampal neurons. In addition, we demonstrate that various ions and drugs play a crucial role in modulating the activity of neuronal nAChRs.

## Neuronal nAChRs and cholinergic synapses in the hippocampus

Very recently, we have concentrated our efforts on the identification and characterization of the different types of nicotinic responses that can be recorded by focal application of nicotinic agonists to the dendrites and to the soma of hippocampal neurons that are visualized in slices obtained from 13–18-day-old rats or of cultured hippocampal neurons (Alkondon et al., 1995). Using infrared differential interference contrast videomicroscopy (Spruston et al., 1995) and a computerized system of micromanipulators, a drug-delivery pipette having a tip diameter of <2 $\mu$m can be positioned near the neuronal cell body, the apical or basal dendrites, or the dendritic regions surrounding the spines.(Alkondon et al., 1996). When rapidly and locally applied to the cell body and to various regions of the apical and basal dendrites of hippocampal neurons, either in slices or in cultures, ACh and other nicotinic agonists can activate nicotinic currents (Figs. 1 and 2). These studies demonstrate for the first time that functional nAChRs can be found on the entire surface of hippocampal neurons, including their dendrites and

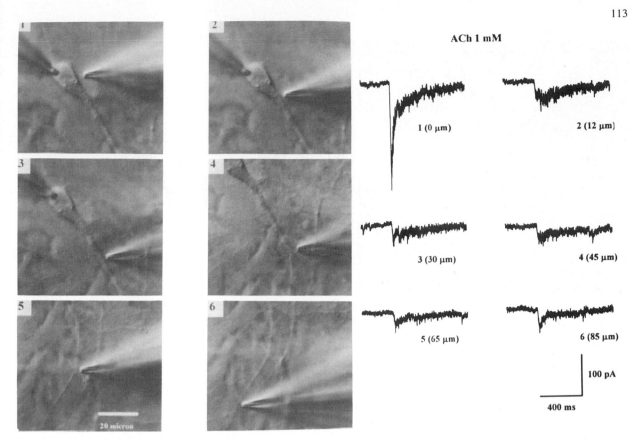

**ACh 1 mM**

1 (0 μm)    2 (12 μm)

3 (30 μm)    4 (45 μm)

5 (65 μm)    6 (85 μm)

100 pA

400 ms

20 micron

Fig. 1. Infrared images of a cultured hippocampal neuron and samples of the ACh-evoked currents from the same neuron. On the left: set of images of the neuron with the recording pipette on the cell soma and the ACh-delivery pipette positioned near the cell body or at various locations along the apical dendrite. All images have been contrast enhanced using a Hamamatsu Image Processor. On the right: sample recordings of whole-cell currents evoked by a 10-ms pressure application of ACh (1 mM) to the various regions of the neuron. Numbers 1–6 represent the position of the ACh-delivery pipette as shown on the left. The distance of the application pipette relative to the center of the cell body is shown in the parentheses. Despite the cable attenuation of the nicotinic currents, the presence of nAChR on the apical dendrite at distances as far as 85 μm from the cell body is clearly demonstrated. Material and methods: neurons dissociated from the hippocampi of 16–18-day-old fetal rats were cultured on polylysine-precoated coverslips following methods described previously (Alkondon and Albuquerque, 1993). The composition of the external solution used to bathe the neurons and to dilute agonists and test compounds was (in mM): NaCl 165, KCl 5, $CaCl_2$ 2, HEPES 5 and dextrose 10 (pH = 7.3; osmolarity = 340 mOsM). The composition of the internal solution was (in mM): CsCl 60, CsF 60, EGTA 10, HEPES 10 (pH = 7.3; osmolarity = 330 mOsM) and to prevent to a great extent the rundown of the nicotinic currents, phosphocreatine (20 mM), creatine phosphokinase (50 units/ml) and ATP (5 mM) were added to this solution. Whole-cell, patch-clamp experiments were performed on 21–29-day-old-cultured hippocampal neurons. Whole-cell patches from the soma of the hippocampal neurons were made under visual control using infrared differential interference contrast (IR-DIC) videomicroscopy (Spruston et al., 1995), utilizing an infrared filter (RG-9, λ 800 nm, Melles Griot) and a Newvicon camera (C2400-07; Hamamatsu, Hamamatsu City, Japan). The images were gathered using a frame grabber and further enhanced using an Argus-10 image processor (Hamamatsu City, Japan). The patch pipettes were pulled from borosilicate capillary glass and had tip diameters ranging from 2 to 4 μm. The resistance of the patch pipettes, when filled with internal solution, was between 3 and 5 MΩ. Whole-cell currents were induced by fast application of ACh (1 mM) to the neurons using a glass "U"-shaped tube positioned about 100 μm from the cell and recorded using an LM-EPC-7 patch-clamp system (List Electronics, Darmstadt, Germany). The signals were filtered at 3 kHz and directly sampled by a microcomputer using the pCLAMP program (version 6.0, Axon Instruments, Foster City, CA). A robotic system was used to control the movements of the micromanipulators that held the patch pipette and the drug-delivery pipette. The movements of these micromanipulators ranged from 20 nm to 1 cm. Whole-cell currents evoked by application of ACh to the various regions on the surface of the neuron were recorded using an LM-EPC-7 patch-clamp system (List Electronics, Darmstadt, Germany) and according to standard patch-clamp technique (Hamill et al., 1981). The signals were filtered at 3 kHz and directly sampled by a microcomputer using the pCLAMP program (for further details see Alkondon et al., 1995).

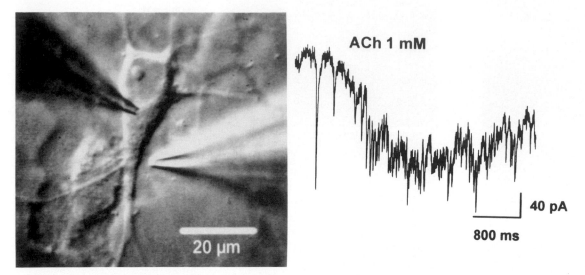

Fig. 2. Infrared image of a pyramidal neuron visualized in the CA1 area of a 200-$\mu$m thick hippocampal slice and samples of the nicotinic currents recorded from the same neuron. On the left: image of the neuron with the recording pipette on the cell soma and the ACh-delivery pipette positioned near the basal dendrite. The image was contrast enhanced using a Hamamatsu Image Processor. On the right: sample recordings of the currents evoked by a 10-ms pressure application of ACh (1 mM). Material and methods: hippocampal slices were prepared according to the procedure described elsewhere (Spruston et al., 1995) and all electrophysiological experiments were made according to the procedures described in Fig. 1. For further details see Alkondon et al. (1995).

dendritic regions surrounding the spines. The activity of neuronal nAChRs located in discrete areas of the dendrites could account for some of the changes in synaptic morphology that are intimately associated with cognitive processes (Desmond and Levy, 1986; Markus and Petit, 1987; Markus et al., 1987).

It has been proposed that increases in intracellular $Ca^{2+}$ levels by $Ca^{2+}$ influx through the NMDA receptor can trigger the activation of the synaptic changes that underlie the processes of learning and memory (Petit, 1988; Bliss and Collingridge, 1993). However, in addition to glutamate, ACh is essential for learning and memory (Winkler et al., 1995), and neuronal nAChRs, particularly the $\alpha$-BGT-sensitive hippocampal nAChRs, have a high relative permeability to $Ca^{2+}$. The $P_{Ca/Na}$ for the $\alpha$-BGT-sensitive neuronal nAChR is about 6, whereas that for the NMDA receptors is approximately 10 (Castro and Albuquerque, 1995). Hence, the activity of neuronal nAChRs could also contribute to control the intracellular levels of $Ca^{2+}$ underlying the changes that account for synaptic plasticity. A number of findings are suggestive of $\alpha$-BGT-sensitive neuronal nAChRs being located

in the dendritic processes of hippocampal neurons: (i) $\alpha$-BGT-sensitive, fast-desensitizing nicotinic currents can be activated by local application of nicotinic agonists to the dendritic processes of pyramidal neurons (Fig. 1); (ii) rhodamine-labeled $\alpha$-BGT can bind not only to the soma of cultured hippocampal neurons but also to the dendrites of these neurons (Alkondon and Albuquerque, 1993); and (iii) [$^{125}$I]$\alpha$-BGT can label postsynaptic dendritic regions of hippocampal neurons (Hunt and Schmidt, 1978). Taking into account that both cholinergic and glutamatergic synapses can be made on a single dendritic spine (Frotscher, 1992), and that dendritic spines may constitute the anatomical substrate for learning and memory, the hypothesis is raised that a fine and well defined integration of signals coming from different chemical synapses at the level of the spines may be a crucial step in the modulation of cell function and brain activity. To understand such an integration of different neurotransmitter systems in the CNS, it is crucial to characterize the properties of the various receptors, including the mechanisms by which their function can be modulated.

# Modulation of the activity of $\alpha$-BGT-sensitive neuronal nAChRs by ions and by allosteric ligands

## *Divalent cations modulate the activation of $\alpha$-BGT-sensitive neuronal nAChRs*

As previously reported, extracellular $Ca^{2+}$ can control the activation of $\alpha$-BGT-insensitive neuronal nAChRs (Mulle et al., 1992). However, it was not until very recently that we have demonstrated that extracellular $Ca^{2+}$ plays a critical role in the regulation of the activation of the $\alpha$-BGT-sensitive neuronal nAChRs in hippocampal neurons. Extracellular $Ca^{2+}$ controls the rectification, and the rates of desensitization and rundown of type IA currents, as well as the affinity of the $\alpha$-BGT-sensitive nAChR for ACh and the cooperativity between the ACh binding sites at this receptor (Castro and Albuquerque, 1995; Bonfante-Cabarcas et al., 1995, 1996).

We have shown that the activation of $\alpha$-BGT-sensitive nAChRs can also be controlled by intracellular $Mg^{2+}$ (Alkondon and Albuquerque, 1993; Alkondon et al., 1994). When a $Mg^{2+}$-containing internal solution is used, type IA currents recorded from hippocampal neurons or from neurons of the olfactory bulb show a very strong inward rectification (Alkondon and Albuquerque, 1993; Alkondon et al., 1994). In fact, the intracellular $Mg^{2+}$-induced inward rectification of currents elicited by activation of $\alpha7$-based nAChRs was recently confirmed in studies carried out in oocytes ectopically expressing these receptors (Forster and Bertrand, 1995). We have now demonstrated that this intracellular $Mg^{2+}$-induced inward rectification of type IA currents is modulated by extracellular $Ca^{2+}$. When nominally $Mg^{2+}$-free, malate-based internal solution was used and the extracellular solution contained 2 mM $Ca^{2+}$, the rectification of type IA currents was confined to a short range of membrane potentials (0–30 mV). Upon raising the intracellular concentration of $Mg^{2+}$ to 10 mM, the inward rectification persisted up to 50 mV, and if concomitantly the $[Ca^{2+}]_o$ was lowered to 0.3 mM or less, the rectification persisted up to 70 mV (Fig. 3). Therefore, in the presence of low levels of

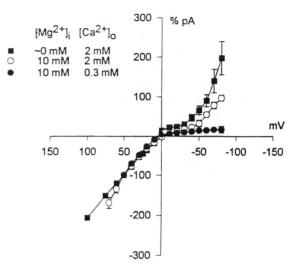

Fig. 3. Effects of extracellular $Ca^{2+}$ on the $Mg^{2+}$-induced inward rectification of type IA currents. After correction for the rundown, the peak amplitude of the ACh (1-s pulse, 1 mM)-evoked currents to hippocampal neurons held at various holding potentials was normalized to that of the currents evoked at −50 mV. Three different experimental conditions were tested: ■, 2 mM $Ca^{2+}$-containing external solution and nominally $Mg^{2+}$-free, malate based internal solution; O, 2 mM $Ca^{2+}$-containing external solution and malate-based internal solution containing 10 mM $Mg^{2+}$; ●, 0.3 mM $Ca^{2+}$-containing extracellular solution and malate-based internal solution containing 10 mM $Mg^{2+}$. The methodology used to record the whole-cell currents was the same as that described in Fig. 1, except that an inverted light microscope was used to visualize the neurons and mechanical micromanipulators were used to hold the patch pipettes and the U-tube. For further details see Bonfante-Cabarcas et al. (1995b).

extracellular $Ca^{2+}$, the activity of the $\alpha$-BGT-sensitive nAChR at the depolarized membrane potentials would be negligible. This intracellular $Mg^{2+}$-induced inward rectification of type IA currents, which is intensified by decreasing $[Ca^{2+}]_o$, may be very relevant to a possible functional integration of cholinergic and glutamatergic inputs onto a dendritic spine.

Similarly to the NMDA receptor, the $\alpha$-BGT-sensitive nAChR is highly permeable to $Ca^{2+}$. Thus, if both the NMDA receptors and the $\alpha$-BGT-sensitive nAChRs were located close to one another on some of the spines known to receive both cholinergic and glutamatergic inputs

(Frotscher, 1992), one would expect that upon simultaneous activation of the cholinergic and glutamatergic terminals, these spines would be flooded with $Ca^{2+}$. Such an overload is unlikely to happen because according to some studies extracellular $Mg^{2+}$ inhibits the activation of the NMDA receptors at negative membrane potentials (Mayer et al., 1984) and potentiates the activation of these receptors at positive potentials (Wang and McDonald, 1995), and as described here intracellular $Mg^{2+}$ inhibits the activation of the $\alpha$-BGT-sensitive nAChRs at positive membrane potentials (Alkondon et al., 1994). Whereas the physiological levels of intracellular $Mg^{2+}$ may be enough to cause the inward rectification of $\alpha$-BGT-sensitive, type IA currents, the increase in the levels of intracellular $Mg^{2+}$ that accompanies the activation of the NMDA receptors (Brocard et al., 1993) would guarantee that the $\alpha$-BGT-sensitive neuronal nAChRs would remain inactive under conditions at which the NMDA receptors are fully operational.

The rundown of type IA currents, i.e. the progressive decrease of the peak current amplitude with the recording time, has been shown to be dependent upon the metabolic state of the neurons. By adding ATP-regenerating compounds to the intracellular solution, this rundown can be prevented to a great extent (Alkondon et al., 1994). Recently, we have provided evidence that if $F^-$ is replaced with malate in the internal solution, and concomitantly the extracellular concentration of $Ca^{2+}$ is decreased to 0.3 $\mu$M or less, the rate of rundown is substantially prolonged, in this way suggesting that the rate of rundown is controlled by the metabolic state of the neurons and to the extracellular levels of $Ca^{2+}$(Bonfante-Cabarcas et al., 1995a,b).

Changes in the $[Ca^{2+}]_o$ also altered the affinity of the $\alpha$-BGT-sensitive neuronal nAChR for ACh, and the cooperative interactions between the agonist binding sites on this receptor. At 10 $\mu$M $[Ca^{2+}]_o$, the values of $EC_{50}$ and Hill coefficient for ACh in eliciting type IA currents were $233 \pm 13\,\mu$M and $2.63 \pm 0.46$, respectively, whereas at 1 mM $[Ca^{2+}]_o$ these values were $174 \pm 2\,\mu$M and $1.39 \pm 0.14$, respectively. Increasing the $[Ca^{2+}]_o$ from 2 to 10 mM altered both the $EC_{50}$ and the Hill

coefficient for ACh, as well as the rate of decay of type IA currents (Bonfante-Cabarcas et al., 1995, 1996; Castro and Albuquerque, 1995). In the presence of 10 mM $[Ca^{2+}]_o$, the $EC_{50}$ and the Hill coefficients for ACh were $291 \pm 4$ and $1.00 \pm 0.12$, respectively, and the decay phase of the currents had a time constant of about 10 ms (in contrast to the decay-time constant of about 20 ms when $[Ca^{2+}]_o$ was 2 mM). The acceleration of the decay phase of type IA currents observed when the $[Ca^{2+}]_o$ was increased from 2 to 10 mM was indicative of extracellular $Ca^{2+}$ playing an important role in the rate of desensitization of the $\alpha$-BGT-sensitive nAChRs.

In addition to $Ca^{2+}$ and $Mg^{2+}$, $Pb^{2+}$ can modulate the activation of both NMDA receptors and $\alpha$-BGT-sensitive neuronal nAChRs (Alkondon et al., 1990; Guilarte and Micele, 1992; Ujihara and Albuquerque, 1992a; Ishihara et al., 1995a,b). These effects of $Pb^{2+}$, some of which are developmentally regulated, may indeed be very important for the toxic actions of this heavy metal (Alkondon et al., 1990; Bressler and Goldstein, 1991; Guilarte and Miceli, 1992). Applying the whole-cell mode of the patch-clamp technique to hippocampal neurons either cultured or acutely dissociated from the brain of 3–30-day-old rats, we have shown that $Pb^{2+}$ acts as a non-competitive inhibitor of the activation of $\alpha$-BGT-sensitive nicotinic currents; the $IC_{50}$ for $Pb^{2+}$ was found to be about 3 $\mu$M (Ishihara et al., 1995a). The blocking action of $Pb^{2+}$ is selective for the $\alpha$-BGT-sensitive nAChR, because only at higher concentrations would $Pb^{2+}$ inhibit the activation of the DH$\beta$E-sensitive nAChR in hippocampal neurons, and practically no inhibition of the activation of the muscle nAChR could be observed in the presence of $Pb^{2+}$ (Atchison and Narahashi, 1984; Ishihara et al., 1995a). Similarly, $Pb^{2+}$ inhibits the activation of the NMDA-type of glutamate receptor without affecting other ionotropic types of glutamate receptors (Alkondon et al., 1990; Ujihara and Albuquerque, 1992a; Ishihara et al., 1995b).

We have reported that hippocampal neurons can respond to co-application of NMDA and glycine with whole-cell currents that have a rapidly and a slowly desensitizing component (Ujihara and Albuquerque, 1992b; Ishihara et al., 1995b). The fast-desensitizing component, which is the major portion of the NMDA-evoked currents recorded from hippocampal neurons cultured for 3–10 days or from neurons acutely dissociated from the hippocampus of 3–10-day-old rats, is very sensitive to the blockade by $Pb^{2+}$. In contrast, the slowly desensitizing component, which is the major portion of the NMDA-evoked currents recorded from hippocampal neurons cultured for more than 15 days or from neurons acutely dissociated from the hippocampus of 15–30-day-old rats, is much less sensitive to the inhibitory action of $Pb^{2+}$ (Ujihara and Albuquerque, 1992a; Ishihara et al., 1995b). Several reports support the concept that each of these responses is subserved by a specific NMDA receptor subtype (Tsumoto et al., 1987; Hestrin, 1992; Kutsuwada et al., 1992; Monyer et al., 1992, 1994; Carmignoto and Vicini, 1993; Petralia et al., 1994), and that the expression of such distinct NMDA receptor subtypes is developmentally regulated in hippocampal neurons (Monyer et al., 1994).

At 10 $\mu$M, $Pb^{2+}$ decreases by about 50% the peak amplitude of the fast-desensitizing NMDA-evoked currents in immature neurons, whereas at the same concentration $Pb^{2+}$ decreases by only 10% the peak amplitude of the slowly desensitizing NMDA responses in mature neurons (Ujihara and Albuquerque, 1992a). The findings that the inhibitory effect of $Pb^{2+}$ on the NMDA receptor can be antagonized competitively by $Ca^{2+}$ (Fig. 4; see also Marchioro et al., 1995) led to the conclusion that $Pb^{2+}$-induced inhibition of the activation of the NMDA receptors is mediated by its binding to an extracellular $Ca^{2+}$ site located at the NMDA receptor. Also, although $Pb^{2+}$ is not a competitive inhibitor of NMDA or glycine at the NMDA receptor, glycine modulates the actions of $Pb^{2+}$ on the NMDA receptor. In the presence of low concentrations of glycine, $Pb^{2+}$ (10 $\mu$M) was unable to decrease the peak amplitude of the NMDA-evoked currents. Instead, in the presence of nanomolar concentrations of glycine, $Pb^{2+}$ increased the total peak current amplitude and the amplitude of the

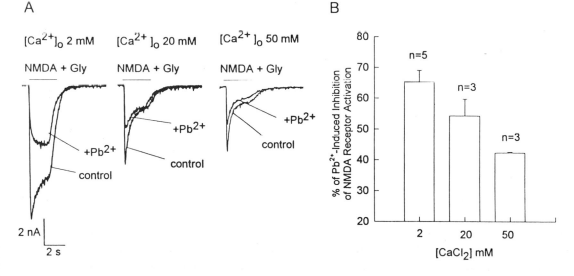

Fig. 4. $Pb^{2+}$-induced inhibition of NMDA-elicited currents is dependent upon $[Ca^{2+}]_o$. (A) Representative traces of whole-cell currents elicited by fast application of NMDA (50 $\mu$M) plus glycine (10 $\mu$M) to neurons bathed in external solution containing different $[Ca^{2+}]_o$ in the presence or in the absence of 30 $\mu$M $Pb^{2+}$ ($Pb^{2+}$ was added to the bathing solution and to the agonist-containing solution). Increasing the $[Ca^{2+}]_o$ reduced the NMDA response and decreased the potency of $Pb^{2+}$. Holding potential, –60 mV. Bars above the traces represent the duration of the agonist application. The magnitude of the blocking effect of $Pb^{2+}$ under different experimental conditions is shown in (B). The methodology used to record the whole-cell currents was the same as that described in Fig. 1, except that an inverted light microscope was used to visualize the neurons and mechanical micromanipulators were used to hold the patch pipettes and the U-tube. For further details see Marchioro et al. (1995).

slowly decaying component of the NMDA-elicited current (Marchioro et al., 1995). In the presence of 50 $\mu$M kynurenic acid, a specific antagonist of glycine at the NMDA receptor, the potentiating effect of $Pb^{2+}$ was no longer observed. Under this condition, $Pb^{2+}$ inhibited the activation of the NMDA receptors. These novel findings support the notion that the potentiation of the NMDA-evoked responses by $Pb^{2+}$ are due to its binding to the glycine site on the NMDA receptors (Marchioro et al., 1995). Given that extracellular $Ca^{2+}$ also controls the function of the $\alpha$-BGT-sensitive nAChRs in hippocampal neurons, it remains to be determined whether the inhibitory action of $Pb^{2+}$ on this nAChR subtype is due to the binding of this cation to the same allosteric sites to which $Ca^{2+}$ binds on the nAChR.

*Ligand-controlled modulation of nAChR activity*

A number of ligands have been shown to modulate allosterically the activation of neuronal nAChRs. For instance, the steroids progesterone and testosterone at low micromolar concentrations can inhibit the activation of the $\alpha 4\beta 2$ neuronal nAChR ectopically expressed in *Xenopus* oocytes (Valera et al., 1992). Also, (i) the tricyclic antidepressants imipramine and desipramine at submicromolar concentrations inhibit the activation of the neuronal nAChRs expressed in the neuroblastoma cell line SH-SY5Y (Rana et al., 1993), (ii) MK-801, a drug believed to be a specific inhibitor of the activation of NMDA receptors, can inhibit the activation of neuronal nAChRs expressed on retinal ganglion cells (Ramoa et al., 1990). The potency of these ligands as negative modulators of the activation of neuronal nAChRs is comparable to their potency as inhibitors of the activation of other receptor types, particularly the NMDA receptors, indicating that simultaneous modulation of the activation of various CNS receptor types by a single ligand should be considered when such a drug is used clinically or in experimental designs.

Electrophysiological studies carried out in our laboratory have shown that amantadine, a drug used to treat patients with Parkinson's disease and previously reported to inhibit the activation of NMDA receptors (Kornhuber et al., 1991) and of the muscle nAChRs (Albuquerque et al., 1978), can also inhibit the activation of neuronal nAChRs (Matsubayashi et al., 1995). Under equilibrium conditions, the $IC_{50}$ for amantadine in inhibiting the activation of these currents is about 10 $\mu$M (Matsubayashi et al., 1995), a value similar to the $K_i$ for amantadine in inhibiting the binding of MK-801 to the NMDA receptors (Kornhuber et al., 1991). The inhibitory effect of amantadine on the $\alpha$-BGT-sensitive nAChR is voltage dependent; the effect of amantadine is more intense at more negative membrane potentials, suggesting that amantadine acts as an open-channel blocker of the $\alpha$-BGT-sensitive nAChR. We have also observed that amantadine can act as an open-channel blocker at the nAChRs that account for type II and type III currents in the hippocampal neurons. In fact, amantadine is apparently much more potent in inhibiting the activation of type III currents than in inhibiting the activation of type II and the IA currents, thus indicating that the $\alpha 3\beta 4$ nAChRs (which presumably subserve type III currents) are much more sensitive than the $\alpha 7$-bearing nAChRs (which account for type IA currents) and the $\alpha 4\beta 2$ nAChRs (which presumably subserve type II currents) to the inhibition by amantadine (Matsubayashi et al., 1995).

Protection of CNS neurons against neurodegeneration may be the key concept to the treatment of Parkinson's disease and of other neurological diseases. It has been suggested that alterations of the glutamatergic system may be implicated in the neurodegeneration that leads to the symptoms observed in Parkinson's disease (Levy and Lipton, 1990; Meldrum and Garthwaite, 1990). However, Parkinson's disease is also characterized by substantial loss of nAChRs, particularly in the pars compacta of the substantia nigra in the midbrain, and such receptor loss appears to be closely related to the primary histopathological changes observed in Parkinson's disease, i.e. loss of dopaminergic cells in the substantia nigra (Whitehouse et al., 1988; James and Nordberg, 1995; Perry et al., 1995). Considering that the activation of the NMDA receptors and of the neuronal nAChRs can be inhibited by low micromolar concentrations of amantadine, and that the concentration of amantadine achieved in the brain of patients with Parkinson's disease 90 min after oral administration of the drug is approximately 100 $\mu$M (Dannysz et al., 1994), it is likely that this drug acts simultaneously on both the glutamatergic neurotransmission mediated by NMDA receptors and on cholinergic neurotransmission mediated by neuronal nAChRs. In light of these findings, it is tempting to speculate that the inhibition of the NMDA receptors by amantadine may contribute to its beneficial effects on patients with Parkinson's disease, whereas the inhibition of the neuronal nAChRs by this drug would explain, at least in part, some of the side effects that develop as a consequence of amantadine therapy.

One of the most provocative new findings in the context of the modulatory control of the nAChR activation is that compounds such as the anticholinesterases physostigmine and galanthamine and the opioid codeine could modulate the activation of neuronal and non-neuronal nAChRs by binding to a newly identified site on these receptors (Kuhlmann et al., 1991; Okonjo et al., 1991; Pereira et al., 1993a,b, 1994; Schrattenholz et al., 1993a,b, 1996; Maelicke et al., 1995a,b). It was initially reported that these compounds could evoke nicotinic single-channel currents in a variety of preparations (Shaw et al., 1985; Pereira et al., 1993a; Schrattenholz et al., 1993a,b; Storch et al., 1995). The agonist effect of these compounds, although insensitive to the inhibition by competitive nicotinic antagonists, is sensitive to the inhibition by the nAChR-specific monoclonal antibody FK1. Using radiolabeled physostigmine, we have shown that the site to which physostigmine and physostigmine-like compounds bind is close to, but distinct from, the ACh-binding sites on the nAChRs. This site is primarily located at the region including and surrounding the residue Lys-125 of the nAChR $\alpha$ subunits, the region of the nAChR $\alpha$ subunits that represents the epitope for FK1 (Schröder et al., 1994). Even though physostigmine, galanthamine, and codeine were seen to activate nicotinic single-channel currents, these compounds were unable to evoke whole-cell currents (Pereira et al., 1993a,b; Storch et al., 1995). It

was then suggested that these compounds were unable to evoke a whole-cell response because (i) the concentrations at which they could activate the nAChR channels overlapped the concentrations at which they could act as non-competitive blockers on the nAChRs, and/or (ii) the probability of the nAChR channels being in the open state was very low when physostigmine and physostigmine-like compounds were used as the agonists. More recently, we demonstrated that galanthamine can potentiate the nicotinic responses evoked by application of nicotinic agonists to PC12 cells (Storch et al., 1995) and to cultured hippocampal neurons (Fig. 5; see also Albuquerque et al., 1995c; Schrattenholz et al., 1996). The finding of an endogenous ligand that can modulate the nAChR activity by binding to this newly discovered "physostigmine-binding site" and of a specific competitive antagonist of this ligand would improve our understanding of the nAChR functions in vivo. Our recent findings that 5-HT and codeine can mimic the effects of physostigmine and of physostigmine-like compounds on nAChR activity (Schrattenholz et al., 1996; Storch et al., 1995) suggested that 5-HT and endogenous opiates could modulate the nAChR activity in vivo by binding to this newly identified site on the nAChRs. The positive modulation of the nAChR activity by physostigmine-like compounds resembles the positive modulation of the GABA receptors by benzodiazepines (Majewska, 1992). It is most likely that multiple modes of modulatory control of receptor activity are related to the higher functions of the mammalian brain, including integrative processes, cognition, and associative memory.

## Concluding remarks

Our present knowledge of brain function is still very much limited not only by the complexity of the neuronal wiring in the CNS but also by the enormous diversity of local and integrated mechanisms of control of the function of the great variety of neurotransmitter receptors. Even though it was long known by the scientists working with neuronal circuitry that nerve terminals containing the same type of neurotransmitter can make mor-

120

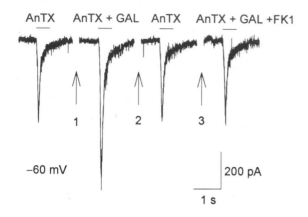

AnTX   AnTX + GAL   AnTX   AnTX + GAL +FK1

1       2       3

−60 mV               200 pA

1 s

Fig. 5. Potentiation by GAL of AnTX-evoked $\alpha$-BGT-sensitive nicotinic currents in cultured hippocampal neurons. Effects of GAL (1 $\mu$M) on $\alpha$-BGT-sensitive currents evoked by a 500-ms pulse application of AnTX (10 $\mu$M) via a 200-$\mu$m aperture on the apex of a U-tube to a cultured hippocampal neuron. The first response of the cells to AnTX (10 $\mu$M) was recorded about 15 min after the patch was obtained, at which time the rate of rundown of the peak current amplitude was negligible. Immediately after the control response was obtained (left), the cell was superfused with GAL (1 $\mu$M)-containing external solution, the latter being applied via a straight tube that was co-assembled with the U-tube (arrow 1). Under these conditions, the potentiating effect of GAL was observed within 1 min after the start of the perfusion (second from left). When the neuron was then superfused for 2 min with GAL-free external solution (arrow 2), the amplitude of the AnTX-evoked current returned to the prior level (third from left). Finally, the neuron was superfused for 14 min with an FK1 (dilution 1:100)-containing external solution, followed by 1-min perfusion with external solution containing GAL (1 $\mu$M) and FK1 (dilution 1:100) (arrow 3), after which time the current evoked by an admixture of AnTX (10 $\mu$M), FK1 (dilution 1:100) and GAL (10 $\mu$M) was recorded from the cell (right). Notice that pre-incubation of the cells with FK1 prevented the potentiating effect of GAL. The methodology used to record the whole-cell currents was the same as that described in Fig. 1, except that an inverted light microscope was used to visualize the neurons and mechanical micromanipulators were used to hold the patch pipettes and the U-tube. For further details see Schrattenholz et al. (1996).

phologically different synapses onto the same postsynaptic neuron, and that a given dendritic spine could receive inputs of various presynaptic terminals each containing different neurotransmit-

ters, it was not until recently that the diversity of many neurotransmitter receptors (e.g. glutamate receptors, GABA receptors, neuronal nAChRs, and 5-HT receptors) could be proven functionally, biochemically, and using molecular biology techniques.

It is apparent that the function of neuronal chemical networks in the CNS would be tightly controlled if the activity of one postsynaptic receptor was modulated by the activity of another receptor or if a given substance could act simultaneously as the primary agonist of a postsynaptic receptor and as an allosteric modulator of the activation of another postsynaptic receptor. As we discussed throughout this paper, several lines of evidence indicate that both modulatory mechanisms take place in the CNS. Therefore, considering the diversity of the neurotransmitter receptors and their binding sites and the diversity of substances that can simultaneously act as a primary agonist of one receptor and an allosteric modulator of a different receptor, an enormous variety of combinatorial possibilities can be achieved in the brain giving rise to very complex neuronal networks.

The characterization of the diversity of many receptors in the CNS, in addition to the very recent findings that (i) functional neuronal nAChRs and NMDA receptors are expressed on the soma and on the dendrites of hippocampal neurons (Alkondon et al., 1995; Spruston et al., 1995), (ii) $Ca^{2+}$, $Mg^{2+}$, $Pb^{2+}$, and drugs such as amantadine, tricyclic antidepressants, and the anticonvulsant MK-801 can allosterically control the activation of both neuronal nAChRs and NMDA receptors (Mulle et al., 1992; Alkondon et al., 1994; Bonfante-Cabarcas et al., 1996; Marchioro et al., 1995; Matsubayashi et al., 1995), (iii) a single substance can act simultaneously as the primary agonist of one receptor type and as an allosteric modulator of another receptor (Johnson and Ascher, 1992; Garcia-Colunga and Miledi, 1995; Schrattenholz et al., 1996), and (iv) the activity of one receptor type can control the activity of a different receptor type (Medina et al., 1994) may provide the basis for a better understanding of cell function, and, consequently, of synaptogenesis, neuronal development, and neuronal plasticity that take place under physiological and pathological conditions.

## Acknowledgements

The authors would like to thank Ms. Mabel Zelle, Mrs. Barbara Marrow, and Mr. Benjamin Cumming for the excellent technical assistance. In the USA, this work was supported by the NIH grants ES05730 and NS25296, the U.S. Army Medical Research Institute of Chemical Defense Contract DAMD17-95-C-5063, a CNPq fellowship from Brazil (M.M.), and a CONICIT fellowship from Venezuela (R.B.-C.). Dr. A. Maelicke is presently on sabbatical leave from the Johannes-Gutenberg University Medical School supported by the Volkswagen Foundation.

## References

Albuquerque, E.X., Eldefrawi, A.T., Eldefrawi, M.E., Mansour, N.A. and Tsai, M.-C. (1978) Amantadine: neuromuscular blockade by suppression of ionic conductance of the acetylcholine receptor. *Science*, 199: 788–790.

Albuquerque, E.X., Castro, N.G., Alkondon, M., Maelicke, A., Reinhardt, S., Schroder, H. and Pereira, E.F.R. (1995a) Nicotinic receptor function in the mammalian central nervous system. *Ann. N. Y. Acad. Sci.*, 757: 48–72.

Albuquerque, E.X., Pereira, E.F.R., Castro, N.G. and Alkondon, M. (1995b) Neuronal nicotinic receptors: function, modulation and structure. *Semin. Neurosci.*, 7: 91–101.

Albuquerque, E.X., Pereira, E.F.R., Schrattenholz, A. and Maelicke, A. (1995c) Galanthamine is a positive modulator of the α-bungarotoxin-sensitive hippocampal nicotinic receptor. *Soc. Neurosci. Abstr.*, 21: 1833.

Alkondon, M. and Albuquerque, E.X. (1993) Diversity of nicotinic acetylcholine receptors in rat hippocampal neurons: I. Pharmacological and functional evidence for distinct structural subtypes. *J. Pharmacol. Exp. Ther.*, 265: 1455–1473

Alkondon, M. and Albuquerque, E.X. (1994) Presence of α-bungarotoxin-sensitive nicotinic acetylcholine receptors in rat olfactory bulb neurons. *Neurosci. Lett.*, 176: 152–156.

Alkondon, M. and Albuquerque, E.X. (1995) Diversity of nicotinic acetylcholine receptors in rat hippocampal neurons: III. Agonist actions of the novel alkaloid epibatidine and analysis of type II currents. *J. Pharmacol. Exp. Ther.*, 274: 771–782.

Alkondon, M., Costa, A.C.S., Radhakrishnan, V., Aronstam, R.S. and Albuquerque, E.X. (1990) Selective blockade of NMDA-activated channel currents may be implicated in learning deficits caused by lead. *FEBS Lett.*, 261: 124–130.

Alkondon, M., Pereira, E.F.R., Wonnacott, S. and Albuquerque, E.X. (1992) Blockade of nicotinic currents in hippocampal neurons defines methyllycaconitine as a potent and specific receptor antagonist. *Mol. Pharmacol.*, 41: 802–808.

Alkondon, M., Reinhardt, S., Lobron, C., Hermsen, B., Maelicke, A. and Albuquerque, E.X. (1994) Diversity of nicotinic acetylcholine receptors in rat hippocampal neurons: II. Rundown and inward rectification of agonist-elicited whole-cell currents and identification of receptor subunits by *in situ* hybridization. *J. Pharmacol. Exp. Ther.*, 271: 494–506.

Alkondon, M., Pereira, E.F.R. and Albuquerque, E.X. (1995) Mapping the location of functional nicotinic acetylcholine receptors (nAChRs) on hippocampal neurons. *Soc. Neurosci. Abstr.*, 21: 1581.

Alkondon, M., Pereira, E.F.R. and Albuquerque, E.X. (1996) Diversity of chemosensitive receptors in rat brain. VI. Mapping the location of acetylcholine and GABA$_A$ functional receptors on hippocampal neurons. *J. Pharmacol. Exp. Ther.*, in press.

Anand, R, Conroy, W.G., Schoepfer, R., Whiting, P. and Lindstrom, J. (1991) Neuronal nicotinic acetylcholine receptors expressed in *Xenopus* oocytes have a pentameric quaternary structure. *J. Biol. Chem.*, 266: 11192–11198.

Atchison, W.D. and Narahashi, T. (1984) Mechanism of action of lead on neuromuscular junctions. *NeuroToxicology*, 5: 267–282.

Barrantes, G.E., Rogers, A.T., Lindstrom, J. and Wonnacott, S. (1995) α-Bungarotoxin binding sites in rat hippocampal and cortical cultures: initial characterisation, colocalisation with α7 subunits and up-regulation by chronic nicotine treatment. *Brain Res.*, 672: 228–236.

Betz, H. (1990) Ligand-gated ion channels in the brain: the amino-acid receptor superfamily. *Neuron*, 5: 383–392.

Bliss, T.V.P. and Collingridge, G.L. (1993) A synaptic model of memory: long-term potentiation in the hippocampus. *Nature*, 361: 31–39.

Bonfante-Cabarcas, R., Swanson, K.L. and Albuquerque, E.X. (1995) Modulation of hippocampal nicotinic receptors by Ca$^{2+}$ and Mg$^{2+}$. *Soc. Neurosci. Abstr.*, 21: 343.

Bonfante-Cabarcas, R., Swanson, K.L., Alkondon, M. and Albuquerque, E.X. (1996) Diversity of nicotinic acetylcholine receptors in rat hippocampal neurons: IV. Modulation of α-bungarotoxin receptor function by external Ca$^{2+}$. *J. Pharmacol. Exp. Ther.*, 277: 432–444.

Bressler, J.P. and Goldstein, G.W. (1991) Mechanisms of lead neurotoxicity. *Biochem. Pharmacol.*, 41: 479–484.

Brocard, J.B., Rajdev, S. and Reynolds, I.J. (1993) Glutamate-induced increases in intracellular free Mg$^{2+}$ in cultured cortical neurons. *Neuron*, 11: 751–757.

Carmignoto, G. and Vicini, S. (1992) Activity-dependent decrease in NMDA receptor responses during development of the visual cortex. *Science*, 258: 1007–1011.

Castro, N.G. and Albuquerque, E.X. (1995) σ-Bungarotoxin-sensitive hippocampal nicotinic receptor channel has a high calcium permeability. *Biophys. J.*, 68: 516–524.

Clarke, P.B.S., Schwarts, R.D., Paul, S.M., Pert, C.B. and Pert, A. (1985) Nicotinic binding in rat brain: autoradiographic comparison of [³H]acetylcholine, [³H]nicotine and [¹²⁵I]alpha-bungarotoxin. *J. Neurosci.*, 5: 1307–1315.

Danysz, W., Gossel, M., Zajaczkowski, W., Dill, D. and Quack, G. (1994) Are NMDA antagonistic properties relevant for antiparkinsonian-like activity in rats? Case of amantadine and memantine. *J. Neural Transm.*, 7: 155–166.

Desmond, N.L. and Levy, W.B. (1986) Changes in the postsynaptic density with long-term potentiation in the dentate gyrus. *J. Comp. Neurol.*, 253: 466–475.

Forster, I. and Bertrand, D. (1995) Inward rectification of neuronal nicotinic acetylcholine receptors investigated by using the homomeric α7 receptor. *Proc. R. Soc. London Ser. B*, 260: 139–148.

Frotscher, M. (1992) Application of the Golgi/Electron Microscopy technique for cell identification in immunocytochemical, retrograde labeling and developmental studies of hippocampal neurons. *Microsc. Res. Technol.*, 23: 306–323.

Garcia-Colunga, J. and Miledi, R. (1995) Effects of serotoninergic agents on neuronal nicotinic acetylcholine receptors. *Proc. Natl. Acad. Sci. USA*, 97: 2919–2923.

Guilarte, T.R. and Miceli, R.C. (1992) Age-dependent effects of lead on [³H]MK-801 binding to the NMDA receptor-gated channel ionophore: in vitro and in vivo studies. *Neurosci. Lett.*, 148: 27–30.

Hamill, O.P., Marty, A., Neher, E., Sakmann, B. and Sigworth, F.J. (1981) Improved patch-clamp techniques for high-resolution current recording from cells and cell-free membrane patches. *Pflügers Arch.*, 391: 85–100.

Heinemann, S., Boulter, J., Connolly, J., Deneris, E., Duvoisin, R., Hartley, M., Hermans-Borgmeyer, I., Hollmann, M., O'Shea-Greenfield, A., Papke, R., Rogers, S. and Patrick, J. (1991) The nicotinic receptor genes. *Clin. Neuropharmacol.*, 14: S45–S61.

Hestrin, S. (1992) Developmental regulation of NMDA receptor-mediated synaptic currents at a central synapse. *Nature*, 357: 686–689.

Hunt, S.P. and Schmidt, J. (1978) The electron microscopic autoradiographic localization of the α-bungarotoxin binding sites within the central nervous system of the rat. *Brain Res.*, 142: 152–159.

Ishihara, K., Alkondon, M., Montes, J.G. and Albuquerque, E.X. (1995a) Nicotinic responses in acutely dissociated rat hippocampal neurons and the selective blockade of fast-desensitizing nicotinic currents by lead. *J. Pharmacol. Exp. Ther.*, 273: 1471–1482.

Ishihara, K., Alkondon, M., Montes, J.G. and Albuquerque, E.X. (1995b) Ontogenically related properties of NMDA receptors in rat hippocampal neurons and the age-specific sensitivity of developing neurons to lead. *J. Pharmacol. Exp. Ther.*, 273: 1459–1470.

James, J.R. and Nordberg, A. (1995) Genetic and environmental aspects of the role of nicotinic receptors in neurodegenerative disorders: emphasis on Alzheimer's disease and Parkinson's disease. *Behav. Genet.*, 25: 149–159.

Johnson, J. and Ascher, P. (1992) Equilibrium and kinetic study of glycine action on the NMDA receptor in cultured mouse brain neurons. *J. Physiol. (London)*, 455: 339–365.

Karlin, A. (1993) Structure of nicotinic acetylcholine receptors. *Curr. Opin. Neurobiol.*, 3: 229–309.

Kornhuber, J., Bormann, J., Hubers, M., Rusche, K. and Riederer, P. (1991) Effects of 1-amino-adamantanes at the MK-801 binding site of the NMDA-receptors-gated ion channel: a human postmortem brain study. *Eur. J. Pharmacol.*, 206: 297–300.

Kuhlmann, J., Okonjo, K.O. and Maelicke, A. (1991) Desensitization is a property of the cholinergic binding region of the nicotinic acetylcholine receptor, not of the receptor-integral ion channel. *FEBS Lett.*, 279: 216–218.

Kutsuwada, T., Kashiwabuchi, N., Mori, H., Sakimura, K., Kushiya, E., Araki, K., Meguro, H., Masaki, H., Kumanishi, T., Arakawa, M. and Mishina, M. (1992) Molecular diversity of the NMDA receptor channel. *Nature*, 358: 36–41.

Levy, D.I. and Lipton, S.A. (1990) Comparison of delayed administration of competitive and uncompetitive antagonists in preventing NMDA receptor-mediated neuronal death. *Neurology*, 40: 852–855.

Lindstrom, J. (1995) Nicotinic acetylcholine receptors. In: Alan North, R. (Ed.), *Handbook of Receptors and Ion Channels*, CRC, Ann Arbor, NY, pp. 153–175.

Lukas, R.J. (1984) Detection of low affinity α-bungarotoxin binding sites in the rat central nervous system. *Biochemistry*, 23: 1160–1164.

Maelicke, A., Schrattenholz, A., Storch, A., Schröder, B., Gutbrod, O., Methfessel, C., Weber, K.-H., Pereira, E.F.R., Alkondon, M. and Albuquerque, E.X. (1995a) Noncompetitive agonist at nicotinic acetylcholine receptors: functional significance for CNS signal transduction. *J. Receptor Signal Transd. Res.*, 15: 333–353.

Maelicke, A., Schrattenholz, A. and Schröder, H. (1995b) Modulatory control by non-competitive agonists of nicotinic cholinergic neurotransmission in the central nervous system. *Semin. Neurosci.*, 7: 103–114.

Majewska, P. (1992) Neurosteroids: endogenous bimodal modulators of GABA$_A$ receptors. Mechanism of action and physiological significance. *Prog. Neurobiol.*, 38: 379–395.

Marchioro, M., Eldefrawi, A.T. and Albuquerque, E.X. (1995) Lead (Pb$^{2+}$) interactions with N-methyl-D-aspartate (NMDA) receptor: modulation by Ca$^{2+}$ and glycine. *Soc. Neurosci. Abstr.*, 21: 351.

Markus, E.J. and Petit, T.L. (1987) Neocortical synaptogenesis, aging and behavior: lifespan development in the motor-sensory system of the rat. *Exp. Neurol.*, 96: 262–278.

Markus, E.J., Petit, T.L. and LeBoutillier, J.C. (1987) Synaptic structural changes during development and aging. *Dev. Brain Res.*, 35: 239–248.

Matsubayashi, H. and Albuquerque, E.X. (1995) Inhibitory

effects of the antiviral agent amantadine on nicotinic acetylcholine responses in rat hippocampal cultured neurons. *Soc. Neurosci. Abstr.*, 21: 1834.

Mayer, M.L., Westbrook, G.L. and Guthrie, P.B. (1984) Voltage-dependent block by $Mg^{2+}$ of NMDA responses in spinal cord neurones. *Nature*, 309: 261–263.

Medina, I., Filippova, N., Barbin, G., Ben-Ari, Y. and Bregestovski, P. (1994) Kainate-induced inactivation of NMDA currents via an elevation of intracellular $Ca^{2+}$ in hippocampal neurons. *J. Neurophysiol.*, 72: 456–465.

Meldrum, B. and Garthwaite, J. (1990) Excitatory aminoacid neurotoxicity and neurodegenerative disease. *Trends Pharmacol. Sci.* ,11: 167–172.

Monyer, H., Sprengel, R., Schoepfer, R., Herb, A., Higuchi, M., Lomeli, H., Burnashev, N., Sakmann, B. and Seeburg, P.H. (1992) Heteromeric NMDA receptors: molecular and functional distinction of subtypes. *Science*, 256: 1217–1221.

Monyer, H., Burnashev, N., Laurie, D.J., Sakmann, B. and Seeburg, P.H. (1994) Developmental and regional expression in the rat brain and functional properties of four NMDA receptors. *Neuron*, 12: 529–540.

Mulle, C., Léna, C. and Changeux, J.-P. (1992) Potentiation of nicotinic receptor response by extracellular calcium in rat central neurons. *Neuron*, 8: 937–945.

Okonjo, K.O., Kuhlmann, J. and Maelicke, A. (1991) A second pathway for the activation of the *Torpedo* acetylcholine receptor. *Eur. J. Biochem.*, 200: 671–677.

Pereira, E.F.R., Alkondon, M., Tano, T., Castro, N.G., Fróes-Ferrão, M.M., Rozental, R., Aronstam, R.S., Schrattenholz, A., Maelicke, A. and Albuquerque, E.X. (1993a) A novel agonist binding site on nicotinic acetylcholine receptors. *J. Receptor Res.*, 13: 413–436.

Pereira, E.F.R., Reinhardt-Maelicke, S., Schrattenholz, A., Maelicke, A. and Albuquerque, E.X. (1993b) Identification and functional characterization of a new agonist site on nicotinic acetylcholine receptors of cultured hippocampal neurons. *J. Pharmacol. Exp. Ther.*, 265: 1474–1491.

Pereira, E.F.R., Alkondon, M., Reinhardt-Maelicke, S., Maelicke, A., Peng, X., Lindstrom, J., Whiting, P. and Albuquerque, E.X. (1994) Physostigmine and galanthamine: probes for a novel binding site on the $\alpha 4 \beta 2$ subtype of neuronal nicotinic acetylcholine receptors stably expressed in fibroblast cells. *J. Pharmacol. Exp. Ther.*, 270: 768–778.

Perry, E.K., Morris, C.M., Court, J.A., Cheng, A., Fairbairn, A.F., McKeith, I.G., Irving, D., Brown, A. and Perry, R.H. (1995) Alteration in nicotine binding sites in Parkinson's disease, Lewy body dementia and Alzheimer's disease: possible index of early neuropathology. *Neuroscience*, 64: 385–395.

Petit, T.L. (1988) The neurobiology of learning and memory: elucidation of the mechanisms of cognitive function. *NeuroToxicology*, 5: 413–428.

Petit, T.L., Markus, E.J. and Milgram, N.W. (1989) Synaptic structural plasticity following repetitive activation in the rat hippocampus. *Exp. Neurol.*, 105: 72–79.

Petralia, R.S., Wang, Y.-X. and Wenthold, R.J. (1994) The NMDA receptor subunits NR2A and NR2B show histological and ultrastructural localization patterns similar to those of NR1. *J. Neurosci.*, 14: 6101–6120.

Ramoa, A.S., Alkondon, M., Aracava, Y., Irons, J., Lunt, G.G., Deshpande, S.S., Wonnacott, S., Aronstam, R.S. and Albuquerque, E.X. (1990) The anticonvulsant MK-801 interacts with peripheral and central nicotinic acetylcholine receptor ion channels. *J. Pharmacol. Exp. Ther.*, 254: 71–82.

Rana, B., McMorn, S.O., Reeve, H.L., Wyatt, C.N., Vaughan, P.F.T. and Peers, C. (1993) Inhibition of neuronal nicotinic acetylcholine receptors by imipramine and desipramine. *Eur. J. Pharmacol.*, 250: 247–251.

Role, L.W. (1992) Diversity in primary structure and function of neuronal nicotinic acetylcholine receptor channels. *Curr. Opin. Neurobiol.*, 2: 254–262.

Romano, C. and Goldstein, A. (1980) Stereospecific nicotine receptors on rat brain membranes. *Science*, 210: 647–649.

Sargent, P.B. (1993) The diversity of neuronal acetylcholine receptors. *Annu. Rev. Neurosci.*, 16: 403–443.

Schrattenholz, A., Coban, T., Schroder, B., Okonjo, O.K., Kuhlmann, J., Pereira, E.F.R., Albuquerque, E.X. and Maelicke, A. (1993a) Biochemical characterization of a novel channel-activating site on nicotinic acetylcholine receptors. *J. Receptor Res.*, 13: 393–412.

Schrattenholz, A., Godovac-Zimmerman, J., Schäfer, H.-J., Albuquerque, E.X. and Maelicke, A. (1993b) Photoaffinity labeling of *Torpedo* acetylcholine receptor by the reversible cholinesterase inhibitor physostigmine. *Eur. J. Biochem.*, 216: 671–677.

Schrattenholz, A., Pereira, E.F.R., Albuquerque, E.X. and Maelicke, A. (1996) Agonist responses of neuronal nicotinic receptors to noncompetitive agonists. *Mol. Pharmacol.*, 49: 1–6.

Schröder, H., Giacobini, E., Struble, R.G., Zilles, K. and Maelicke, A. (1991) Nicotinic cholinoceptive neurons of the frontal cortex are reduced in Alzheimer's disease. *Neurobiol. Ageing*, 12: 259–262.

Schröder, B., Reinhardt-Maelicke, S., Schrattenholz, A., McLane, K.E., Conti-Tronconi, B.M. and Maelicke, A. (1994) Monoclonal antibodies FK1 and WF6 define two neighboring ligand binding sites on *Torpedo* acetylcholine receptor $\alpha$-polypeptide. *J. Biol. Chem.*, 269: 10407–10416.

Séguéla, P., Wadiche, J., Dineley-Miller, K., Dani, J.A. and Patrick, J.W. (1993) Molecular cloning, functional properties and distribution of rat brain $\alpha 7$: a nicotinic cationic channel highly permeable to calcium. *J. Neurosci.*, 13: 596–604.

Shaw, K.-P., Aracava, Y., Akaike, A., Daly, J.W., Rickett, D.L. and Albuquerque, E.X. (1985) The reversible cholinesterase inhibitor physostigmine has channel-blocking and agonist effects on the acetylcholine receptor-ion channel complex. *Mol. Pharmacol.*, 28: 527–538.

Spruston, N., Jonas, P. and Sakmann, B. (1995) Dendritic glutamate receptor channels in rat hippocampal CA3 and

CA1 pyramidal neurons. *J. Physiol. (London)*, 482: 325–352.

Storch, A., Schrattenholz, A., Cooper, J.C., Abdel Ghani, E.M., Gutbrod, O., Weber, K.-H., Reinhardt, S., Lobron, C., Hermsen, B., Šoškiç, V., Pereira, E.F.R., Albuquerque, E.X., Methfessel, C. and Maelicke, A. (1995) Physostigmine, galanthamine and codeine act as "noncompetitive agonists" on clonal rat pheochromocytoma cells. *Eur. J. Pharmacol.*, 290: 207–219.

Tsumoto, T., Hagihara, K., Sato, H. and Hata, Y. (1987) NMDA receptors in the visual cortex of young kittens are more effective than those of adult cats. *Nature*, 327: 513–514.

Ujihara, H. and Albuquerque, E.X. (1992a) Developmental change of the inhibition by lead of NMDA-activated currents in cultured hippocampal neurons. *J. Pharmacol. Exp. Ther.*, 263: 868–875.

Ujihara, H. and Albuquerque, E.X. (1992b) Ontogeny of *N*-methyl-D-aspartate-induced current in cultured hippocampal neurons. *J. Pharmacol. Exp. Ther.*, 263: 859–867.

Valera, S., Ballivet, M. and Bertrand, D. (1992) Progesterone modulates a neuronal nicotinic acetylcholine receptor. *Proc. Natl. Acad. Sci. USA*, 89: 9949–9953.

Vijayaraghavan, S., Pugh, P.C., Zhang, Z.-w., Rathouz, M.M. and Berg, D.K. (1992) Nicotinic receptors that bind $\alpha$-bungarotoxin on neurons raise intracellular free $Ca^{2+}$. *Neuron*, 8: 353–362.

Wang, L.-Y. and McDonald, J.F. (1995) Modulation by magnesium of the affinity of NMDA receptors for glycine in murine hippocampal neurons. *J. Physiol. (London)*, 486: 83–95.

Whitehouse, P.J., Martino, A.M., Wagster, M.V., Price, D.L., Mayeux, R., Atack, J.R. and Kellar, K.J. (1988) Reductions in [$^{3}$H]nicotinic acetylcholine binding in Alzheimer's disease and in Parkinson's disease: an autoradiographic study. *Neurology*, 38: 720–723.

Winkler, J., Suhr, S.T., Gage, F.H., Thal, L.J. and Fisher, L.J. (1995) Essential role of neocortical acetylcholine in spatial memory. *Nature*, 375: 484–487.

J. Klein and K. Löffelholz (Eds.)
*Progress in Brain Research*, Vol. 109
© 1996 Elsevier Science B.V. All rights reserved.

# CHAPTER 10

# Structure and function of neuronal nicotinic acetylcholine receptors

Jon Lindstrom, Rene Anand, Vladimir Gerzanich, Xiao Peng, Fan Wang and Gregg Wells

*Department of Neuroscience, University of Pennsylvania Medical School, Philadelphia, PA 19104-6074 USA*

## Introduction

Neuronal nicotinic acetylcholine receptors (AChRs) are part of a gene family which includes AChRs from skeletal muscle and part of a gene superfamily which includes glycine receptors, $GABA_A$ receptors, and $5HT_3$ receptors. Recent reviews of neuronal AChRs (Chapters 8 and 9) include those by Sargent (1993), Papke (1993), McGehee and Role (1995), and Lindstrom (1995;1996). Probably all of the receptors in this superfamily are formed from five homologous subunits oriented around a central ion channel like barrel staves, and the neuronal AChRs probably have shapes very similar to those of muscle type AChRs which have been determined for AChRs from *Torpedo* electric organ to a resolution of about 9 Å by electron crystallography (Unwin 1993, 1995). The basic homologies in structure of receptors in this superfamily have been illustrated by two elegant experiments by Changeux and his co-workers. Homologies in overall domain relationships were demonstrated by showing that functional mosaics could be made in which the large N-terminal extracellular domain of $\alpha7$ AChRs (which contains the acetylcholine binding site) could be grafted just before the first transmembrane domain to the C-terminal part of $5HT_3$ receptors (which contains the ion channel and large cytoplasmic domain) (Eisile et al., 1993). This resulted in receptors with acetylcholine-gated cation channels having the ion selectivity of $5HT_3$ receptors. Close similarities in the overall struc-

tures of the ion channels were demonstrated by showing that changing only three amino acids in the sequence lining the cation-specific channel of excitatory $\alpha7$ neuronal AChRs to amino acids typical of the anion-specific channels of inhibitory glycine or $GABA_A$ receptors changed the ion selectivity of the mutated $\alpha7$ AChR channels from cations to anions (Galzi et al., 1992).

The AChR gene family can be thought of as having three branches. One branch consists of AChRs from skeletal muscles and fish electric organs which have the subunit composition $(\alpha1)_2\beta1\gamma\delta$ in the fetal form or $(\alpha1)_2\beta1\varepsilon\delta$ in the mature form (reviewed in Changeux, 1990 and Karlin, 1991, 1993). The second branch consists of neuronal AChRs which, unlike muscle AChRs, do not bind the snake venom toxin $\alpha$ bungarotoxin ($\alpha$Bgt). They are formed from combinations of $\alpha2$, $\alpha3$, or $\alpha4$ subunits with $\beta2$ and $\beta4$ subunits, and also sometimes with $\alpha5$ subunits (Sargent, 1993; McGehee and Role, 1995). The predominant form of brain high affinity nicotine binding AChRs is formed from $\alpha4$ and $\beta2$ subunits (Whiting and Lindstrom, 1988). The stoichiometry of these subunits is $(\alpha4)_2 (\beta2)_3$, at least when they are expressed in *Xenopus* oocytes (Anand et al., 1991; Cooper et al., 1991). Ganglionic postsynaptic AChRs contain $\alpha3$ subunits. These are in combination with $\beta4$ and $\alpha5$ subunits, and sometimes also with $\beta2$ subunits (Conroy et al., 1992; Conroy and Berg, 1995). These subunits are in unknown stoichiometries, but probably something like $(\alpha3)_2\beta2\beta4\alpha5$. Many neuronal AChRs appear to be

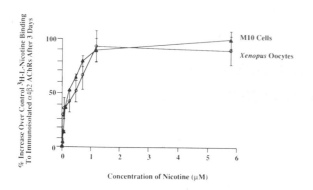

Fig. 1. Dose dependence of nicotine-induced upregulation. M10 cells are mouse fibroblasts permanently transfected with chicken $\alpha4$ and $\beta2$ subunits which exhibit the pharmacological properties of brain $\alpha4\beta2$ AChRs and which function electrophysiologically (Whiting et al., 1991; Periera et al., 1994). *Xenopus* oocytes were injected with cRNA for $\alpha4$ and $\beta2$ RNAs. After 3 days in the indicated concentrations of nicotine, $\alpha4\beta2$ AChRs were solubilized using Triton X-100, immunoisolated on microwells coated with mAb 290 to their $\beta2$ subunits and quantitated using 20 nM [$^3$H]nicotine. Data are reproduced from Peng et al. (1994a).

located presynaptically, where their functional role may be to modulate the release of various transmitters rather than to form a critical postsynaptic link in transmission as they do at neuromuscular junctions (Henley et al., 1986; Swanson et al., 1987; Wonnacott, 1990; Clarke, 1995). Neuronal AChRs are known to be expressed very early in development. Specific developmental changes in subunit composition are likely but have not yet been determined in detail (Hamassaki-Britto et al., 1994; Cimino et al., 1995; Zoli et al., 1995). The third branch of the gene family consists of neuronal AChRs which bind $\alpha$Bgt. These are formed from $\alpha7$, $\alpha8$, or $\alpha9$ subunits (Lindstrom, 1995, 1996; Elgoyhen et al., 1994). All of these can function as homomers when expressed from cRNAs in *Xenopus* oocytes. $\alpha7$ and $\alpha8$ are also found together as heteromers in some native AChRs (Schoepfer et al., 1990; Keyser et al., 1993). It is unknown whether other unknown subunits are associated with these subunits in native AChRs, but they are not co-assembled with other known subunits (Anand et al., 1993a,b). Although $\alpha7$ is expressed as a homomer very efficiently, $\alpha8$ is not (Gerzanich et al., 1994), suggesting that at least $\alpha8$

might normally be associated with other uncharacterized subunits. Preparations of $\alpha$Bgt binding proteins purified from brain have been found to contain several peptide components (Gotti et al., 1994). $\alpha7$, $\alpha8$, and $\alpha9$ AChRs all exhibit high conductance for $Ca^{2+}$ (Seguela et al., 1993; Elgoyhen et al., 1994; Gerzanich et al., 1994) which may allow them to take part in unusual synaptic mechanisms in which $Ca^{2+}$ acts as a second messenger to control other ion channels and other cellular functions (Fuchs and Murrow, 1992; Pugh and Berg, 1994). These AChRs also all exhibit very rapid desensitization, which has inhibited measuring their function in neurons until quite recently (Alkondon et al., 1994). At least some $\alpha7$ AChRs, and perhaps many of these AChRs, have been observed in extrasynaptic or perisynaptic positions (Jacob and Berg, 1983; Sargent and Wilson, 1995), again suggesting that some may participate in unusual synaptic mechanisms where their functional roles might involve sensing leak acetylcholine near a synapse or in a region of synapses, rather than classical postsynaptic function at a single synapse. There are two orphan subunits included in this gene family. $\alpha6$ and $\beta3$ subunit cDNAs have been identified which exhibit extensive homologies with other AChR subunits but which have not yet been reported to be components of native AChRs or to produce functional AChRs when co-expressed with other AChR subunit cDNAs (Sargent, 1993).

In the following two sections some aspects of neuronal AChRs are illustrated in more detail. First, we briefly review some of our recent studies of the mechanism by which chronic exposure to nicotine affects $\alpha4\beta2$ AChRs. Second, we briefly review some of the pharmacological properties of neuronal AChRs, especially the differences between $\alpha7$ and $\alpha8$ AChRs and the usefulness of epibatidine (exo-2-(6-chloro-3-pyridyl)-7-azabicyclo-[2.2.1]-heptane) as a ligand for many types of neuronal AChRs.

## Mechanism of upregulation of $\alpha4\beta2$ AChRs by chronic exposure to nicotine

Tobacco use or chronic exposure to nicotine

TABLE 1

Pharmacological properties of $\alpha4\beta2$ AChRs

| | Upregulation[a] $EC_{50}$ ($\mu$M) | Function $EC_{50}$ ($\mu$M)[a] | | Equilibrium binding[b] $K_I$ ($\mu$M) |
| --- | --- | --- | --- | --- |
| | | Activation | Blocking | |
| *Agonists* | | | | |
| Cytisine | $0.17 \pm 0.02$ | $0.031 \pm 0.003$ | – | $0.00014 \pm 0.00003$ |
| Nicotine | $0.21 \pm 0.04$ | $0.35 \pm 0.02$ | – | $0.0039 \pm 0.00021$ |
| DMPP | $3.0 \pm 0.3$ | $0.073 \pm 0.01$ | – | $0.0094 \pm 0.0002$ |
| Carbamylcholine | $15.0 \pm 4.0$ | $2.5 \pm 0.3$ | – | $0.36 \pm 0.013$ |
| *Competitive antagonist* | | | | |
| Curare | None | – | $2.7 \pm 0.5$ | $25.0 \pm 14.0$ |
| *Noncompetitive antagonist* | | | | |
| Mecamylamine | $65.0 \pm 6.0$ | – | $0.37 \pm 0.5$ | >1000 |

[a]Data from Peng et al. (1994a).
[b]Data from Whiting et al. (1991).

causes an increase in brain $\alpha4\beta2$ AChRs by about twofold (Schwartz and Kellar, 1985; Benwell et al., 1988; Flores et al., 1992). Wonnacott, 1990; Wonnacott et al., 1990) suggested that this increase in AChRs was an adaptive response of neurons to maintain nicotinic transmission despite the accumulation of desensitized AChRs due to chronic exposure to this agonist.

We have found that agonist-induced upregulation of $\alpha4\beta2$ AChRs is an intrinsic property of this protein which is exhibited when the cloned $\alpha4\beta2$ AChRs are expressed in *Xenopus* oocytes or in a permanently transfected fibroblast cell line called M10, as shown in Fig. 1 (Peng et al., 1994a). This has now also been confirmed by others (Zhang et al., 1994).

The $EC_{50}$ for nicotine-induced upregulation of $\alpha4\beta2$ AChRs ($2 \times 10^{-7}$ M, see Fig. 1) is pathologically significant because this concentration is equal to the mean steady state serum concentration of nicotine in smokers (Benowitz et al., 1990). The steady state concentration of nicotine in smokers is interesting because it is near the $EC_{50}$ for activation of $\alpha4\beta2$ AChRs by nicotine when applied acutely ($3.5 \times 10^{-7}$ M), but much greater than the $K_D$ for binding nicotine by desensitized $\alpha4\beta2$

AChRs ($3.9 \times 10^{-9}$ M) (see Table 1 which is adapted from Peng et al., 1994a). Thus, at a steady state nicotine concentration of $2 \times 10^{-7}$ M all $\alpha4\beta2$ AChRs in a tobacco user should be desensitized, and the net behavioral effect should reflect inhibition of these AChRs.

Chronic exposure to other agonists also causes upregulation in the amount of $\alpha4\beta2$ AChRs, as shown in Fig. 2 (Peng et al., 1994a). The competitive antagonist curare prevents nicotine-induced upregulation. The rank order for efficacy in causing upregulation parallels the rank order of equilibrium binding affinity (presumably to the desensitized conformation) but not quite the rank order for activation (Table 1).

The non-competitive antagonist mecamylamine can also causes upregulation and is synergistic with nicotine, as shown in Fig. 3 (Peng et al., 1994a). This parallels the effects of mecamylamine in brain (Collins et al., 1994). Blockage of function by mecamylamine is more effective in the presence of low concentrations of nicotine (Fig. 4). These data are consistent with the idea that mecamylamine is an open channel blocker which is able to bind better when the channel is opened by nicotine (Varanda et al., 1985; Martin et al., 1990).

128

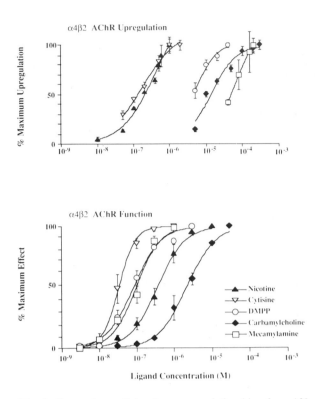

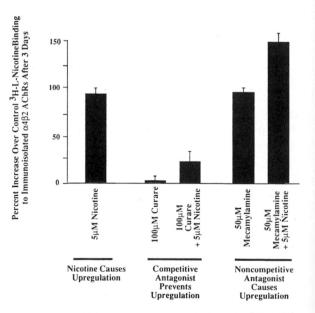

Fig. 2. Comparison of dose/response relationships for $\alpha4\beta2$ upregulation in M10 cells and function in *Xenopus* oocytes. Expression of $\alpha4\beta2$ AChR in M10 cells grown to confluence was induced by 1 $\mu$M dexamethasone for 3 days, then the indicated concentrations of ligands were added for 3 days and finally AChRs were quantitated in solid phase assays using [³H]nicotine to label AChRs immunoisolated through their $\beta2$ subunit as in Fig. 1. Voltage clamp measurements on *Xenopus* oocytes normalized the maximum currents to saturating agonist concentrations. Blockage by mecamylamine was measured at 1 $\mu$M nicotine. Data are reproduced from Peng et al. (1994a).

Ion flow through the cation channel of $\alpha4\beta2$ AChRs does not appear to be necessary to trigger upregulation of these AChRs. This is shown by the observation that a channel blocker like mecamylamine can cause upregulation. It is also shown by the observation that the channel blocker chlorisondamine blocks AChR function but does not prevent nicotine-induced upregulation (El-Bizri and Clarke, 1994). The observation that the rank order of various ligands for equilibrium binding to desensitized $\alpha4\beta2$ AChRs better parallels upregulation than does their order for activation (Table 1) is

also consistent with the idea that cation flow is not necessary to trigger upregulation but that induction of a particular conformation of the AChR may be.

Nicotine-induced upregulation of $\alpha4\beta2$ AChRs is not due to an increase in brain $\alpha4$ or $\beta2$ mRNA (Marks et al., 1992). Similarly, nicotine-induced upregulation of $\alpha4\beta2$ AChRs does not occur at the transcriptional level. Demonstration of AChR upregulation in *Xenopus* oocytes injected with fixed amount of $\alpha4$ and $\beta2$ mRNA is the best evidence that upregulation occurs post transcriptionally (Peng et al., 1994a).

The mechanism of increase in $\alpha4\beta2$ AChRs which is induced by nicotine involves a decrease in the rate of destruction of the AChRs in the surface membrane, as shown in Fig. 5 (Peng et al., 1994a). Nicotine, other agonists, and mecamylamine appear to induce a conformation (a type of desensitized conformation, perhaps) which is turned over more slowly than the resting conformation. This pathological change is an interesting contrast with myasthenia gravis, where chronic

Fig. 3. Effects of antagonists on upregulation of $\alpha4\beta2$ AChRs in M10 cells. M10 cells were grown to confluence, AChR expression was induced with dexamethasone and then treated for 3 days with the indicated ligands before solid phase radioimmunoassay. Data are from Peng et al. (1994a).

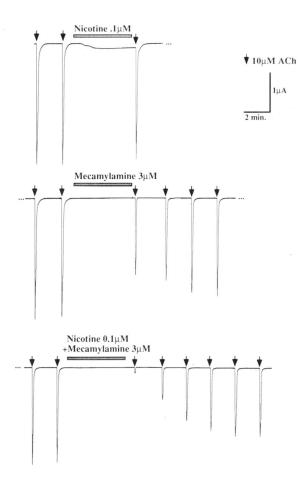

Fig. 4. Mecamylamine blockage of $\alpha 4\beta 2$ AChR function is more effective in the presence of nicotine. $\alpha 4\beta 2$ AChRs were expressed in *Xenopus* oocytes. A continuous series of responses to a saturating concentration of acetylcholine is shown from top to bottom. A low concentration of nicotine causes only a small response and little or no blockage to a subsequent test response to acetylcholine. Mecamylamine caused some blockage to a test response to acetylcholine applied immediately after and some recovery was seen after further washing. Applying nicotine and mecamylamine together results in nearly complete blockage of a test response applied immediately after. Data are from Peng et al. (1994a).

exposure to autoantibodies causes a decrease in the amount of muscle AChRs by crosslinking them and thereby facilitating their endocytosis (Lindstrom et al., 1988).

We found that after exposing oocytes expressing $\alpha 4\beta 2$ AChRs to high concentrations of nicotine for 3 days their net responses to acetylcholine

were greatly reduced, even though the total number of AChRs was doubled, as shown in Fig. 6 (Peng et al., 1994a). Short term desensitization is normally reversed by washing over matters of minutes, but this reduction in response only recovered fully after 3 days, which is about the rate expected for synthesis of new AChRs. Native AChRs also exhibit downregulation of function after chronic exposure to nicotine (Marks et al., 1993). The results in oocytes suggest that long term exposure to high concentrations of nicotine can result in a permanently inactivated $\alpha 4\beta 2$ AChR conformation. It is unknown whether this reflects the same nicotine-induced conformation that is turned over more slowly.

The loss of functional $\alpha 4\beta 2$ AChRs on chronic exposure to nicotine due to both reversible and permanent desensitization provides an explanation for tolerance. Thus, chronic tobacco users are tolerant of much higher nicotine concentrations than are nonusers because, even though the chronic users may have twice as many $\alpha 4\beta 2$ AChRs, most of these excess AChRs can bind nicotine but not function in response to it.

## Some pharmacological properties of neuronal AChRs

$\alpha$Bgt binds virtually irreversibly to muscle type AChRs, but it has lower affinity for $\alpha 7$ AChRs, still lower affinity for $\alpha 8$ AChRs, and substantially lower affinity yet for $\alpha 9$ AChRs, as shown in Table 2 (Keyser et al., 1993; Elgoyhen et al., 1994). The muscarinic antagonists atropine and glycinergic antagonist strychnine are nearly as potent as the nicotinic antagonist curare as antagonists of $\alpha 7$, $\alpha 8$ and $\alpha 9$ AChRs, as shown in Table 2. Despite the similarities in their N-terminal extracellular amino acid sequences, $\alpha 7$ and $\alpha 8$ AChRs differ in their pharmacological properties, with $\alpha 7$ having higher affinity for $\alpha$Bgt but lower affinity for small cholinergic ligands (Anand et al., 1993; Keyser et al., 1993; Gerzanich et al., 1994). This is reflected in the properties of their homomers. Pharmacological properties of $\alpha 8$ are known only in chick and $\alpha 9$ only in rat. $\alpha 7$ from chicks and humans are pharmacologically similar except for

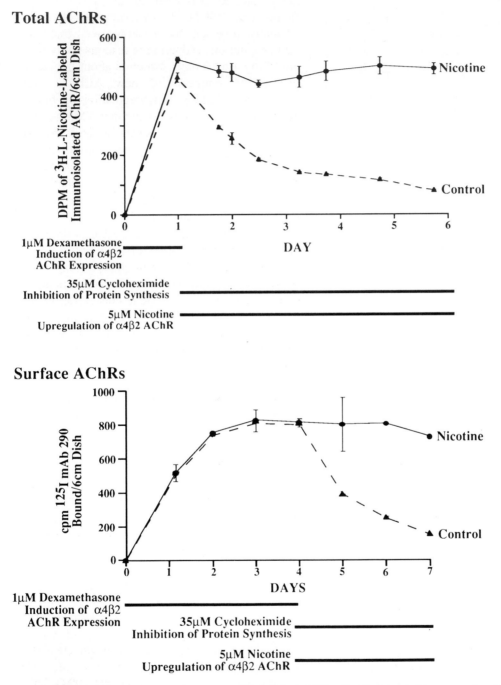

Fig. 5. Nicotine decreases the turnover rate of $\alpha4\beta2$ AChRs in M10 cells. Cycloheximide was used to inhibit the synthesis of new $\alpha4\beta2$ AChRs in dexamethasone-induced M10 cells. In the experiment shown at the top, cycloheximide was added 1 day after dexamethasone induction. Then total $\alpha4\beta2$ AChRs solubilized from the cells was measured by [³H]nicotine binding to immunoisolated AChRs in microwells. In the experiment shown in the lower panel, dexamethasone induction of $\alpha4\beta2$ AChR synthesis was allowed to maximize over 4 days before nicotine was added to one set of cultures. In this case only surface AChRs were measured by binding of ¹²⁵I-labeled mAb290 to $\beta2$ subunits in intact cells. Data are from Peng et al. (1994a).

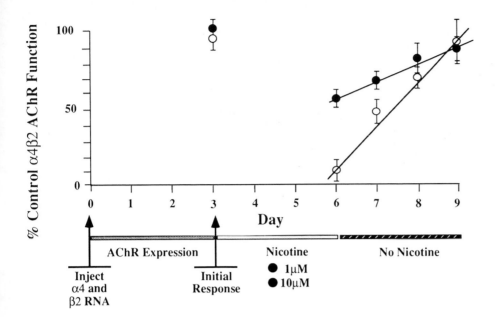

Fig. 6. Time course of recovery of $\alpha4\beta2$ AChR function after chronic treatme t with nicotine of *Xenopus* oocytes. After measuring the initial response to 100 $\mu$M acetylcholine, one group was incubated with 1 $\mu$M nicotine, another with 10 $\mu$M nicotine and a third group was left untreated as a control. After 3 days the oocytes were washed and then responses to 100 $\mu$M acetylcholine were measured at the indicated times over the next 3 days. Data are from Peng et al. (1994a).

TABLE 2

Pharmacological properties of $\alpha7$, $\alpha8$ and $\alpha9$ AChRs

| | IC$_{50}$/EC$_{50}$/$K_D$ ($\mu$M) | | | |
| --- | --- | --- | --- | --- |
| | $\alpha7$ chick | $\alpha7$ human | $\alpha8$ chick | $\alpha9$ rat |
| *Antagonists* | | | | |
| $\alpha$Bgt | 0.0021[e] | 0.001[f] | 0.017[e] | Completely reversed in 10 min wash[b] |
| Curare | 7.3[a] | 20[f] | 50[a] | 0.3[b] |
| Atropine | 120[a] | 1360[f] | 0.58[a] | 1.3[b] |
| Strychnine | 9.9[a] | 9.8[f] | 2.0[a] | 0.02[b] |
| *Agonists* | | | | |
| ACh | 160[a] | 5.8[f] | 0.031[a] | 10[b] |
| Nicotine | 1.3[a] (agonist)[c] | 2.6[f] (agonist)[f] | 0.012[a] (agonist)[c] | 30 (antagonist)[b] |
| DMPP | 83[a] (3.5% partial)[c] | 4.4[f] (full agonist)[f] | 0.39[a] (full agonist)[c] | 5% partial agonist[b] |
| Epibatidine | 2.2[d] agonist | 1.1[d] | 0.0012[d] | ? |

[a]Data from Anand et al. (1993a) for IC$_{50}$ on binding of $^{125}$I$\alpha$Bgt at 2 nM for native $\alpha7$ and 20 nM for native $\alpha8$ AChR.
[b]Data from Elgoyhen et al. (1994) for IC$_{50}$ on function 10 $\mu$M ACh.
[c]Data from Gerzanich et al. (1994) for EC$_{50}$ on $\alpha7$ or $\alpha8$ homomers.
[d]Data from Gerzanich et al. (1995) for EC$_{50}$ of (−) epibatidine on $\alpha7$ or $\alpha8$ homomers on IC$_{50}$ in their function.
[e]Data from Keyser et al. (1993) for $K_D$ of $^{125}$I$\alpha$Bgt.
[f]Data from Peng et al. (1994b) for $K_D$ of $^{125}$I$\alpha$Bgt for IC$_{50}$ at 2 nM $^{125}$I$\alpha$Bgt for native $\alpha7$ AChRs.

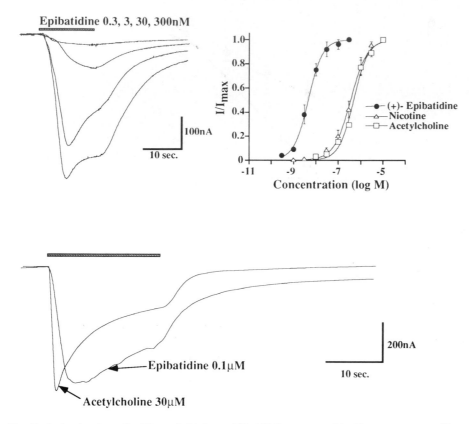

Fig. 7. Activation by epibatidine of chicken $\alpha4\beta2$ AChRs expressed in *Xenopus* oocytes. The top panel shows the responses to different concentrations of epibatidine and compares dose/response curves for epibatidine with those of nicotine and acetylcholine. The bottom panel compares the time courses of responses to saturating concentrations of epibatidine and nicotine. Data are from Gerzanich et al. (1995).

DMPP, which is virtually an antagonist on chick, however, a very potent full agonist on human $\alpha7$ (Gerzanich et al., 1994; Peng et al., 1994b).

It would be extremely useful to have a universal labeling reagent for AChRs. $\alpha$Bgt has long been an excellent ligand for muscle AChRs and $\alpha7$ AChRs (Table 2). Nicotine has been a useful label for $\alpha4\beta2$ AChRs (Table 1). Cholinergic labeling $\alpha3$ AChRs has been difficult because of their relatively low affinity for nicotine, and the failure of neuronal bungarotoxin to bind to detergent solubilized AChRs, so they have long been quantitated using $^{125}$ImAb35 (Smith et al., 1985). This mAb binds to the main immunogenic region on the extracellular surface of $\alpha1$ subunits (Saedi et al., 1990) and to homologous structures on chick $\alpha5$ subunits (Conroy et al., 1992; Conroy and Berg, 1995) and human $\alpha3$ and $\alpha5$ subunits (Wang et al., 1996).

Epibatidine is now proving to be a very useful ligand for several subtypes of neuronal AChRs, including $\alpha3$ AChRs. Epibatidine was initially discovered in the skin of Ecuadoran poison frogs by John Daly (Spande et al., 1992). The frogs derive it from an unknown component of their diet in the wild (Daly, 1995). It has now been synthesized (e.g. Fletcher et al., 1994; reviewed in Broke, 1994). Initially it was discovered to be 200-fold more potent than morphine in preventing nociception. Later it was discovered to be a potent nicotinic agonist (Badio and Daly, 1994) and to elicit a variety of nicotinic effects (Sullivan et al., 1994).

Epibatidine is the most potent neuronal AChR

## Human α3β2 AChRs Expressed in *Xenopus* Oocytes

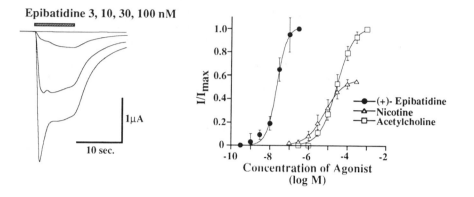

## Human α3β4 AChRs Expressed in *Xenopus* Oocytes

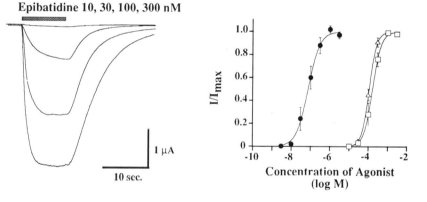

Fig. 8. Activation by epibatidine of human α3β2 and α3β4 AChRs expressed in *Xenopus* oocytes. The top panel shows responses of α3β2 AChRs to various concentrations of epibatidine and compares dose/response curves for epibatidine, nicotine and acetylcholine. The bottom panel makes a similar comparison for α4β4 AChRs. Data are from Gerzanich et al. (1995).

agonist that we have studied on several types of cloned AChRs expressed in *Xenopus* oocytes, and [³H]epibatidine is an excellent labeling reagent which has proven especially useful for α3 AChRs (Gerzanich et al., 1995). Epibatidine is more potent than nicotine as an agonist for α4β2 AChRs, as shown in Fig. 7, or for α3β2 or α3β4 AChRs, as shown in Fig. 8. Note that Fig. 8 also illustrates the important concept that β2 and β4 play an important role in determining the affinity and efficacy of ligands (reviewed in detail by Papke, 1993). This is a consequence of the acetylcholine binding site being formed at the interface between α subunits

and adjacent structural subunits (Blount and Merlie, 1989). The extremely high affinity of [³H]-epibatidine as a labeling reagent is illustrated for human α3β2 AChRs in Fig. 9. Epibatidine is still a more potent agonist than either nicotine or acetylcholine on both α7 and electric organ AChRs, but much less potent than on α4 or α3 AChRs, as shown in Fig. 10.

Thus, despite its potency as an agonist, epibatidine does not quite provide a universal ligand for AChRs. However, the AChRs for which it has relatively low affinity, like α1 and α7 AChRs, can be efficiently labeled with ¹²⁵IαBgt. The affinity of

134

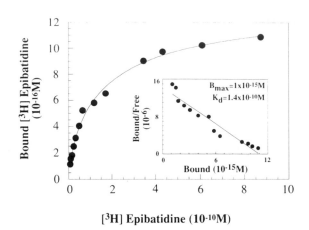

Fig. 9. Binding of racemic [³H]epibatidine to human $\alpha3\beta2$ AChRs expressed in oocytes. Physiological experiments of the types shown in Figs. 7 and 8 with + and − isomers of epibatidine revealed surprisingly small differences in responses to the two isomers. Data are from Gerzanich et al. (1995).

$\alpha4\beta2$ AChRs is highest for epibatidine, making it a better label than [³H]nicotine. But the very high affinity of [³H]epibatidine for $\alpha3$ AChRs provides a really useful breakthrough. Epibatidine is likely to become an important reagent for several types of neuronal AChRs. It is also likely to renew interest in an important role of nicotinic AChRs in nociception. Although epibatidine is not highly selective among neuronal AChR subtypes, which can be a pharmacological drawback, it is a small molecule structure that can provide remarkably high affinity for these AChRs.

## Acknowledgements

The laboratory of Jon Lindstrom is supported by grants from the NINDS (NS11323), The Smokeless Tobacco Research Council, Inc., The Council for Tobacco Research - USA, Inc., and the Muscular Dystrophy Association.

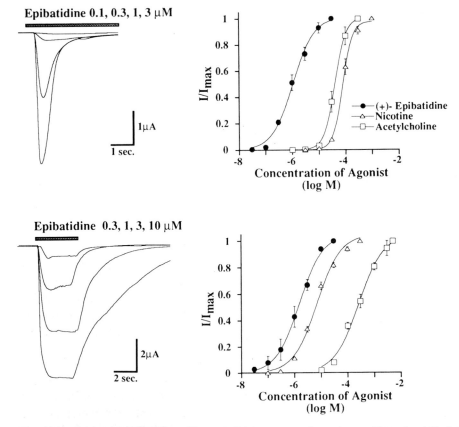

Fig. 10. Activation by epibatidine of human $\alpha7$ homomers and muscle type *Torpedo* $\alpha1\beta1\gamma\delta$ AChRs expressed in oocytes. Data are from Gerzanich et al. (1995).

# References

Alkondon, M., Reinhardt, S., Lobron C., Hermsen, B., Maelicke, A. and Albuquerque, E. (1994) Diversity of nicotinic acetylcholine receptors in rat hippocampal neurons. II. The rundown and inward rectification of agonist-elicited whole cell currents and identification of receptor subunits by in situ hybridization. *J. Pharmacol. Exp. Ther.*, 271: 494–506.

Anand, R., Conroy, W.G., Schoepfer, R., Whiting, P. and Lindstrom, J. (1991) Chicken neuronal nicotinic acetylcholine receptors expressed in *Xenopus* oocytes have a pentameric quaternary structure. *J. Biol. Chem.*, 266: 11192–11198.

Anand, R., Peng, X., Ballesta, J. and Lindstrom, J. (1993a) Pharmacological characterization of $\alpha$ bungarotoxin sensitive AChRs immunoisolated from chick retina: contrasting properties of $\alpha 7$ and $\alpha 8$ subunit-containing subtypes. *Mol. Pharmacol.*, 44: 1046–1050.

Anand, R., Peng, X. and Lindstrom, J. (1993b) Homomeric and native $\alpha 7$ acetylcholine receptors exhibit remarkably similar but nonidentical pharmacological properties suggesting that the native receptors is a heteromeric protein complex. *FEBS Lett.*, 327: 241–246.

Badio, B. and Daly, J. (1994) Epibatidine, a potent analgesic and nicotinic agonist. *Mol. Pharmacol.*, 45: 563–569.

Benowitz, N., Porchet, H. and Jacob, P. (1990) Pharmacokinetics, metabolism and pharmacodynamics of nicotine. In: S. Wonnocott, M. Russell and I. Stolerman (Eds.), *Nicotine Psychopharmacology*, Oxford Science Publications, Oxford, UK, pp. 112–157.

Benwell, M., Balfour, D. and Anderson, J. (1988) Evidence that tobacco smoking increases the density of (–)-[$^3$H] nicotine binding sites in human brain. *J. Neurochem.*, 50: 1243–1247.

Blount, P. and Merlie, J.P. (1989) Molecular basis of the two nonequivalent ligand binding sites of the muscle nicotinic acetylcholine receptor. *Neuron*, 3: 349–357.

Broke, C. (1994) Synthetic approaches to epibatidine. *Med. Chem. Res.*, 4: 440–448.

Changeux, J.P. (1990) Functional architecture and dynamics of the nicotinic acetylcholine receptor: an allosteric ligand-gated ion channel. In: *1988–1989 Fidia Research Foundation: Neuroscience Award Lectures*, Vol. 4, pp. 21–168.

Cimino, M., Marini, P., Colombo, S., Andena, M., Cattabeni, Fornasari, D. and Clemente, F. (1995) Expression of neuronal acetylcholine receptor $\alpha 4$ and $\beta 2$ subunits during postnatal development of the rat brain, *J. Neural Transm.*, 100: 77–92.

Clarke, P. (1995) Nicotinic receptors and cholinergic transmission in the central nervous system, *Ann. N. Y. Acad. Sci.*, 757: 73–83.

Collins, A., Luo, Y., Selvaag, S. and Marks, M. (1994) Sensitivity to nicotine and brain nicotinic receptors are altered by chronic nicotine and mecamylamine infusion, *J. Pharmacol. Exp. Ther.*, 271: 125–133.

Conroy, W. and Berg, D. (1995) Neurons can maintain multiple classes of nicotinic acetylcholine receptors distinguished by different subunit compositions. *J. Biol. Chem.*, 270: 4424–4431.

Conroy, W., Vernallis, A. and Berg, D. (1992) The $\alpha 5$ gene product assembles with multiple acetylcholine receptor subunits to form distinctive receptor subtypes in brain. *Neuron*, 9: 1–20.

Cooper, E., Couturier, S. and Ballivet, M. (1991) Pentameric structure and subunit stoichiometry of a neuronal nicotinic acetylcholine receptor. *Nature*, 350: 235–238.

Daly, J. (1995) The chemistry of poisons in amphibian skin. *Proc. Natl. Acad. Sci. USA*, 92: 9–13.

Eisile, J.L., Bertrand, S., Galzi, J.L., Devillers-Thiery, A., Changeux, J.P. and Bertrand, D. (1993) Chimaeric nicotinic-serotonergic receptor combines distinct ligand binding and channel specificities. *Nature*, 366: 479–409.

El-Bizri, H. and Clarke, P. (1994) Regulation of nicotinic receptors in rat brain following quasi-irreversible nicotinic blockade by chlorisondamine and chronic treatment with nicotine. *Br. J. Pharmacol.*, 113: 917–925.

Elgoyhen, A., Johnson, D., Boulter, J., Vetter, D. and Heinemann, S. (1994) $\alpha 9$: an acetylcholine receptor with novel pharmacological properties expressed in rat cochlear hair cells. *Cell*, 79: 705–715.

Fletcher, S., Baker, R., Chambers, M., Herbert, R., Hobbs, S., Thomas, S., Verrier, H., Watt, A. and Ball, R. (1994) Total synthesis and determination of the absolute configuration of epibatidine. *J. Org. Chem.*, 59: 1771–1778.

Flores, C., Rogers, S., Pabreza, J., Wolfe, B. and Kellar, K. (1992) A subtype of nicotinic cholinergic receptor in rat brain is composed of $\alpha 4$ and $\beta 2$ subunits and is upregulated by chronic nicotine treatment. *Mol. Pharmacol.*, 41: 31–37.

Fuchs, P. and Murrow, B. (1992) A novel cholinergic receptor mediates inhibition of chick cochlear hair cells. *Proc. R. Soc. London Ser B*, 248: 35–40.

Galzi, J.L., Devillers-Thiery, S., Hussy, N., Bertrand, S., Changeux, J.P. and Bertrand, D. (1992) Mutations in the channel domain of a neuronal nicotinic receptor convert ion selectivity from cationic to anionic. *Nature*, 359: 500–505.

Gerzanich, V., Anand, R. and Lindstrom, J. (1994) Homomers of $\alpha 8$ subunits of nicotinic receptors functionally expressed in *Xenopus* oocytes exhibit similar channel but contrasting binding site properties compared to $\alpha 7$ homomers. *Mol. Pharmacol.*, 45: 212–220.

Gerzanich, V., Peng, X., Wang, F., Wells, G., Anand, R., Fletcher, S. and Lindstrom, J. (1995) Comparative pharmacology of epibatidine – a potent agonist for neuronal nicotinic acetylcholine receptors. *Mol. Pharmacol.*, 48: 774–782.

Gotti, C., Hanke, W., Maury, K., Moretti, M., Ballivet, M., Clemente, F. and Bertrand, D. (1994) Pharmacology and biophysical properties of $\alpha 7$ and $\alpha 7$-$\alpha 8$ $\alpha$ bungarotoxin receptor subtypes immunopurified from the chick optic lobe. *Eur. J. Neurosci.*, 6: 1281–1291.

Hamassaki-Britto, D., Gardino, P., Hokoc, J., Keyser, K., Karten, H., Lindstrom, J. and Britto, L. (1994) Differential development of $\alpha$ bungarotoxin-sensitive and $\alpha$ bungarotoxin-insensitive nicotinic acetylcholine receptors in the chick retina. *J. Comp. Neurol.*, 347: 161-170.

Henley, J., Lindstrom, J. and Oswald, R. (1986) Acetylcholine receptor synthesis in retina and transport to the optic tectum in goldfish. *Science*, 232: 1627–1629.

Jacob, M. and Berg, D. (1983) The ultrastructural localization of $\alpha$ bungarotoxin binding sites in relation to synapses on chick ciliary ganglion neurons. *J. Neurosci.*, 3: 260–271.

Karlin, A. (1991) Explorations of the nicotinic acetylcholine receptor. *Harvey Lectures Ser.*, 85: 71–107.

Karlin, A. (1993) Structure of nicotinic acetylcholine receptors. *Curr. Opin. Neurobiol.*, 3: 299-309.

Keyser, K., Britto, L., Schoepfer, R., Whiting, P., Cooper, J., Conroy, W., Brozozowska-Prechtl, A., Karten, H. and Lindstrom, J. (1993) Three subtypes of alpha bungarotoxin-sensitive nicotinic acetylcholine receptors are expressed in chick retina. *J. Neurosci.*, 13: 442–454.

Lindstrom, J. (1995) Nicotinic acetylcholine receptors. In Alan North (Ed.), *CRC Handbook of Receptors*. CRC Press, pp. 153–175.

Lindstrom, J. (1996) Neuronal nicotinic acetylcholine receptors. In: T. Narahashi (Ed.), *Ion Channels*, Vol. IV, Plenum, New York, pp. 377–450.

Lindstrom, J., Shelton, G.D. and Fujii, Y. (1988) Myasthenia gravis. *Adv. Immunol.*, 42: 233-284.

Marks, M., Pauly, J., Gross, D., Deneris, E., Hermans-Borgmeyer, I., Heinemann, S. and Collins, A (1992) Nicotine binding and nicotinic receptor subunit RNA after chronic nicotine treatment. *J. Neurosci.*, 12: 2765–2784.

Marks, M., Grady, S. and Collins, A. (1993) Downregulation of nicotinic receptor function after chronic nicotine infusion. *J. Pharmacol. Exp. Ther.*, 266: 1268–1275.

Martin, T., Suchocki, J., May, E. and Martin, B. (1990) Pharmacological evaluation of the antagonism of nicotines central effects by mecamylamine and pempidine. *J. Pharm. Exp. Ther.*, 254: 45–51.

McGehee, D. and Role, L. (1995) Physiological diversity of nicotinic acetylcholine receptors expressed by vertebrate neurons. *Annu. Rev. Physiol.*, 57: 521–546.

Papke, R. (1993) The kinetic properties of neuronal nicotinic receptor: genetic basis of functional diversity. *Prog. Neurobiol.*, 41: 509–531.

Peng, X., Gerzanich, V., Anand, R., Whiting, P. and Lindstrom, J. (1994a) Nicotine-induced upregulation of neuronal nicotinic receptors results from a decrease in the rate of turnover. *Mol. Pharmacol.*, 46: 523–530.

Peng, X., Katz, M., Gerzanich, V., Anand, R. and Lindstrom, J. (1994b) Human $\alpha$7 acetylcholine receptor: cloning of the $\alpha$7 subunit from the SH-SY5Y cell line and determination of pharmacological properties of native receptors and func-

tional $\alpha$7 homomers expressed in *Xenopus* oocytes. *Mol. Pharmacol.*, 45: 546–554.

Pereira, E., Alkondon, M., Reinhardt-Maelicke, S., Maelicke, A., Peng, X., Lindstrom, J., Whiting, P. and Albuquerque, E. (1994) Physostigmine and galanthamine reveal the presence of the novel binding site on the $\alpha$4 $\beta$2 subtype of neuronal nicotinic acetylcholine receptor stably expressed in fibroblast cells. *J. Pharmacol. Exp. Ther.*, 270: 768–778.

Pugh, P. and Berg, D. (1994) Neuronal acetylcholine receptors that bind $\alpha$ bungarotoxin mediate neurite retraction in a calcium-dependent manner. *J. Neurosci.*, 14: 889–896.

Saedi, M., Anand, R., Conroy, W.G. and Lindstrom, J. (1990) Determination of amino acids critical to the main immunogenic region of intact acetylcholine receptors by in vitro mutagenesis. *FEBS Lett.*, 267: 55–59.

Sargent, P. (1993) The diversity of neuronal nicotinic acetylcholine receptors. *Annu. Rev. Neurosci.*, 16: 403–443.

Sargent, P. and Wilson, H. (1995) Distribution of nicotinic acetylcholine receptor subunit immunoreactivities on the surface of chick ciliary ganglion neurons. In: P. Clarke, M. Quick, F. Adlkofer and K. Thurau (Eds.), *Effects of Nicotine on Biological Systems II*, Birkhauser, Basel.

Schoepfer, R., Conroy, W.G., Whiting, P., Gore, M. and Lindstrom, J. (1990) Brain alpha-bungarotoxin-binding protein cDNAs and mAbs reveal subtypes of this branch of the ligand-gated ion channel superfamily. *Neuron*, 5: 35–48.

Schwartz, R. and Kellar, K. (1985) In vivo regulation of [$^3$H] acetylcholine recognition sites in brain by nicotinic cholinergic drugs. *J. Neurochem.*, 45: 427–433.

Seguela, P., Wadiche, J., Dinelly-Miller, K., Dani, J. and Patrick, J. (1993) Molecular cloning, functional properties and distribution of rat brain $\alpha$7: a nicotinic cation channel highly permeable to calcium. *J. Neurosci.*, 13: 596–604.

Smith, M., Stollberg, J., Lindstrom, J. and Berg, D.K. (1985) Characterization of a component in chick ciliary ganglia that cross-reacts with monoclonal antibodies to muscle and electric organ acetylcholine receptor. *J. Neurosci.*, 5: 2726–2731.

Spande, T., Garraffo, M., Edwards, M., Yeh, H., Pannel, L. and Daly, J. (1992) Epibatidine: a novel (chloropyridyl) azabicyclo-heptane with potent analgesic activity from Ecuadoran poison frog. *J. Am. Chem. Soc.*, 114: 3475–3478.

Sullivan, J., Decker, M., Brioni, J., Donnelly-Roberts, D., Anderson, D., Bannon, A., Kang, C., Adams, P., Piattoni-Kaplan, M., Buckley, M., Gopalakrishnan, M., Williams, M. and Arneric, S. (1994) ($\pm$)-Epibatidine elicits a diversity of in vitro and in vivo effects mediated by nicotinic acetylcholine receptor. *J. Pharmacol. Exp. Ther.*, 271: 624–631.

Swanson, L., Simmons, D., Whiting, P. and Lindstrom, J. (1987) Immunohistochemical localization of neuronal nicotinic receptors in the rodent central nervous system. *J. Neurosci.*, 7: 3334–3342.

Unwin, N. (1993) Nicotinic acetylcholine receptor at 9 Å resolution. *J. Mol. Biol.*, 229: 1101-1124.

Unwin, N. (1995) Acetylcholine receptor channel imaged in the open state. *Nature*, 373: 37–43.

Varanda, W., Aracava, A., Sherby, S., Van Meter, W., Eldefrawi, M. and Albuquerque, E. (1985) The acetylcholine receptor of the neuromuscular junction recognizes mecamylamine as a noncompetitive antagonist. *Mol. Pharmacol.*, 28: 128–137.

Wang, F., Ferzanich, V., Wells, F., Anand, R., Peng, X., Keyser, K. and Lindstrom, J. (1996) Assembly of human neuronal nicotinic receptor $\alpha 5$ subunits with $\alpha 3$, $\beta 2$ and $\beta 4$ subunits. *J. Biol. Chem.* 271: 17656–17665.

Whiting, P.J. and Lindstrom, J.M. (1988) Characterization of bovine and human neuronal nicotinic acetylcholine receptors using monoclonal antibodies. *J. Neurosci.*, 8: 3395–3404.

Whiting, P., Schoepfer, R., Lindstrom, J. and Priestley, T. (1991) Structural and pharmacological characterization of the major brain nicotinic acetylcholine receptor subtype stably expressed in mouse fibroblasts. *Mol. Pharmacol.*, 40: 463–472.

Wonnacott, S. (1990) The paradox of nicotinic acetylcholine receptor upregulation by nicotine. *Trends Pharmacol. Sci.*, 11: 216–219.

Wonnacott, S., Russell, M. and Stolerman, I. (1990) *Nicotine Psychopharmacology: Molecular, Cellular and Behavioral Aspects*, Oxford Science Publications, Oxford, UK.

Zhang, X., Gang, Z.H., Hellstrom-Lindahl, E. and Nordberg, A. (1994) Regulation of $\alpha 4\beta 2$ nicotinic acetylcholine receptors in M10 cells following treatment with nicotinic agents. *NeuroReport*, 6: 313–317.

Zoli, M., LeNovere, N., Hill, J. and Changeux, J.P. (1995) Developmental regulation of nicotinic ACh receptor mRNAs in the central and peripheral nervous systems. *J. Neurosci.*, 15: 1912–1939.

# Section IV


# Muscarinic Receptors: 1. Structural Aspects

J. Klein and K. Löffelholz (Eds.)
*Progress in Brain Research*, Vol. 109
© 1996 Elsevier Science B.V. All rights reserved.

CHAPTER 11

# 3-Heteroarylquinuclidin-2-ene derivatives as muscarinic antagonists: synthesis, structure-activity relationships and molecular modelling

G. Nordvall[1], B.M. Nilsson[1], S. Sundquist[2], G. Johansson[1], G. Glas[2], L. Nilvebrant[2] and U. Hacksell[1]

[1]*Department of Organic Pharmaceutical Chemistry, Uppsala University, 751 23 Uppsala, Sweden and* [2]*Department of Pharmacology, Pharmacia AB, 751 82 Uppsala, Sweden*

## Introduction

Muscarinic receptor antagonists have been used for many years in the treatment of irritable bowel syndrome, obstructive airways disease and urinary urge incontinence. The side effects that frequently occur during treatment with anticholinergics might be reduced by use of selective antagonists. However, most of the available antagonists fail to sufficiently discriminate between the pharmacologically characterized muscarinic receptor subtypes $M_1$, $M_2$, $M_3$ and $M_4$ (Doods et al., 1987). Hence, there is a medical need for subtype-selective muscarinic antagonists. In particular, $M_2$-selective receptor antagonists appear to have potential for the treatment of bradycardic disorders whereas $M_3$-selective receptor antagonists may be useful in the treatment of airway obstruction (Grimm et al., 1994).

We have recently reported (Nilsson et al., 1995) a series of achiral muscarinic antagonists with structures related to previously described chiral quinuclidine derived agonists such as **1** and **2** (Saunders et al., 1990). The most potent antagonist was the 2-benzofuranyl derivative **4**. Replacing the 2-benzofuranyl with a 3-benzofuranyl- (**8**), furanyl- (**3**), thienyl- (**5**), 2- or 3-benzothienyl- (**6** and **9**), or a 2-benzoxazolyl (**7**) moiety gave less potent antagonists (Table 1). In general, the antagonists displayed low selectivity for the investi-

gated muscarinic receptor subtypes ($M_1$, $M_2$ and $M_3$).

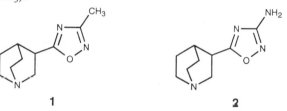

**1**          **2**

An analysis of the structure–affinity relationships among the benzo-fused five-membered heteroaromatic analogues demonstrated that the magnitude of the negative electrostatic potential in the benzene moiety correlated with $M_1$ receptor affinity. In the present work we have examined a variety of C-5 substituted analogues of **4** (**10–18**; Table 1) in an attempt to determine what physicochemical properties influence the $M_1$ receptor affinity. The new compounds were tested in receptor binding and functional assays as previously reported (Nilsson et al., 1995).

## Materials and methods

### Chemistry

The syntheses of the new derivatives were performed by using the Stephens–Castro reaction. This procedure provides an efficient method for substituted benzofuran derivatives from readily

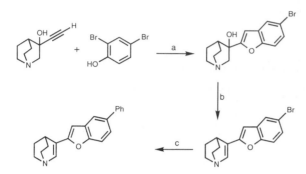

Fig. 1. Synthesis of a benzofuran derivative. Reagents: (a) $Cu_2O$, pyridine; (b) HCOOH, 100°C; (c) phenylboronic acid, $Pd(Ph_3P)_4$, $Na_2CO_3$, DME.

available starting materials. Thus, 3-ethynyl-quinuclidin-3-ol (Clemo and Hoggarth, 1941) and the appropriate *ortho*-halogenated phenol were heated with $Cu_2O$ in pyridine (Fig. 1) (Doad et al., 1989). Dehydration of the resulting alcohols to the corresponding quinuclidin-2-ene derivatives was accomplished by heating in concentrated formic acid (Fig. 1). The substituted 3-(2-benzofuranyl)-quinuclidin-2-ene derivatives were converted into 5-substituted compounds by use of standard functional group transformations.

*Pharmacology*

Receptor binding affinities ($K_i$; Table 1) of the 3-(2-benzofuranyl)quinuclidin-2-ene derivatives for muscarinic receptors in the cerebral cortex ($M_1$), heart ($M_2$), parotid gland ($M_3$) and urinary bladder (data not shown) from guinea pigs were indirectly determined by competition experiments with the non-subtype selective muscarinic radioligand (−)-[$^3$H]QNB (3-quinuclidinyl [*phenyl*-4-$^3$H] benzilate) (Nilvebrant and Sparf, 1982, 1983a,b, 1986). For convenience, we approximate binding to cortical tissue with $M_1$ receptor affinity, binding to heart tissue with $M_2$ receptor affinity and binding to tissue from parotid gland with $M_3$ receptor affinity. Antimuscarinic potencies ($K_B$) were evaluated by functional in vitro studies on isolated guinea pig bladder, using carbachol as the agonist (data not shown). In the presence of antagonist, the concentration response curves to carbachol were

shifted in parallel towards higher concentrations, but the maximal responses remained unaffected. Thus, the inhibition seemed to be competitive since it always could be overcome by an increase in the carbachol concentration. None of the compounds tested in this study exhibited any muscarinic agonist activity in the isolated urinary bladder when tested in concentrations of 10–1000 $\mu$M.

## Results and discussion

As the previously studied compounds (Nilsson et al., 1995), the present series displayed low subtype selectivity for the muscarinic receptors. Hence, structure–affinity comparisons were based on $M_1$-receptor affinities.

Introduction of a fluoro substituent in the 5-position of the benzofuran ring (**10**) increased the affinity 1.5-fold. The 5-bromo- (**11**), 5-cyano- (**16**) and 5-methyl- (**12**) substituted compounds displayed 4-, 6-, 12-fold lower affinity, respectively, compared to **4**. The formyl (**15**) and hydroxymethyl (**13**) derivatives were considerably less potent since they exhibited 38- and 144-fold lower affinity than the unsubstituted compound. The 5-phenyl substituted analogue **18** displayed 30-fold lower affinity than **4**. A 5-nitro (**17**) or 5-amino (**14**) substituent reduced the affinity 14- and 22-fold, respectively, compared to the unsubstituted **4**.

In an attempt to determine what physicochemical properties of the substituted benzofuran derivatives influence the $M_1$-receptor affinity, we performed a QSAR analysis of the 9 derivatives. The properties of the different 5-substituents were described by using $\pi$, MR, $\sigma_m$, $\sigma_p$, and the Sterimol descriptors L, B1 and B5 (Verloop, 1987). The matrix was evaluated using PLS (Wold et al., 1993) and, using two principal components, the resulting model could explain 92% of the variance in affinity.

The main factors that contribute to the $M_1$ muscarinic affinity is the lipophilicity ($\pi$), size and electron-withdrawing effect ($\sigma_m$) of the substituent in the 5-position of the benzofuran ring. The most important size descriptor is B5. The PLS equation for the model generated is: $pK_i$ ($M_1$) =

TABLE 1

Quinuclidine-based muscarinic antagonists: affinities ($K_i$) for muscarinic receptors

| Structure | $K_i$ (nM) | | | Structure | $K_i$ (nM) | | |
|---|---|---|---|---|---|---|---|
| | M$_1$ | M$_2$ | M$_3$ | | M$_1$ | M$_2$ | M$_3$ |
| **3** | 300 | 390 | 1100 | **11** | 39 | 86 | 109 |
| **4** | 9.6 | 31 | 59 | **12** | 116 | 238 | 295 |
| **5** | 290 | 620 | 1200 | **13** | 1400 | 1840 | 10700 |
| **6** | 81 | 270 | 420 | **14** | 212 | 837 | 683 |
| **7** | 100 | 400 | 720 | **15** | 364 | 920 | 1571 |
| **8** | 34 | 99 | 160 | **16** | 60 | 160 | 236 |
| **9** | 37 | 96 | 110 | **17** | 139 | 460 | 515 |
| **10** | 6.3 | 34 | 30 | **18** | 289 | 376 | 410 |

144

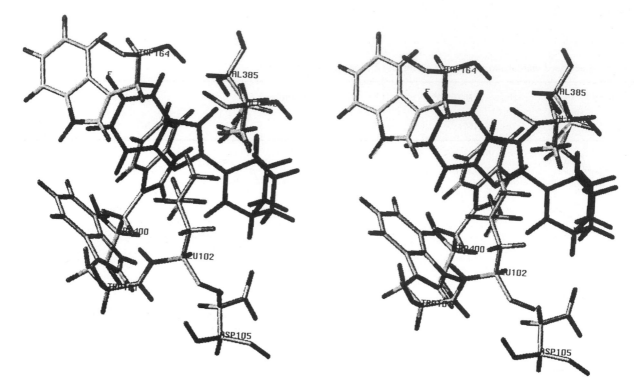

Fig. 2. Stereo representation of the docking of 10 (ligand shown in black) with the muscarinic m1 binding-site model. The 5-fluorobenzofuran derivative 10 forms a reinforced ionic interaction with Asp105 and has an aromatic–aromatic interaction with Trp400.

$9.76 + 0.5443\pi - 0.1308\,MR + 0.2679\sigma_m + 0.1114\sigma_p - 0.1653L - 0.1373B1 - 0.6284B5$. This analysis clearly shows that the substituent in the 5-position should be lipophilic, small and electron-withdrawing for optimal affinity at the muscarinic $M_1$ receptor. This is consistent with the fact that the 5-fluoro derivative **10** displays the highest affinity for the $M_1$ receptor in this series of compounds.

We have previously published a homology based receptor model of the m1 receptor (Nordvall and Hacksell, 1993). The model was used to rationalize the affinity of a number of agonists and antagonists. This model can be used in conjunction with the QSAR data to obtain a more realistic interpretation of the present structure–activity relationship. The muscarinic antagonists bind to the muscarinic receptor model with the protonated nitrogen interacting with Asp105 and the aromatic group located in the vicinity of Trp101 (TM3), Leu102 (TM3), Trp164 (TM4), Val385 (TM7) and

Trp 400 (TM7) (see Fig. 2). Thus, the benzofuran moiety is surrounded by aromatic and lipophilic amino acid side chains. In addition, large substituents on the benzofuran ring should not be tolerated due to steric interactions with the residues in the binding site according to the results from the QSAR model.

## References

Clemo, G.R. and Hoggarth, E. (1941) 81. 5-Ethinylruban-5-ol and related compounds. *J. Chem. Soc.*, 476–477.

Doad, G.J.S., Barltrop, J.A., Petty, C.M. and Owen, T.C. (1989) A versatile and convenient synthesis of benzofurans. *Tetrahedron Lett.*, 30: 1597–1598.

Doods, H.N., Mathy, M.-J., Davidesko, D., van Charldorp, K.J., de Jonge, A. and van Zwieten, P.A. (1987) Selectivity of muscarinic antagonists in radioligand and in vivo experiments for the putative $M_1$, $M_2$ and $M_3$ receptors. *J. Pharmacol. Exp. Ther.*, 242: 257–262.

Grimm, U., Moser, U., Mutschler, E. and Lambrecht, G. (1994) Muscarinic receptors: focus on presynaptic mechanisms and recently developed novel agonists and antagonists. *Pharmazie*, 49: 711–726.

Nilsson, B.M., Sundquist, S., Johansson, G., Nordvall, G., Glas, G., Nilvebrant, L. and Hacksell, U. (1995) 3-Heteroaryl substituted quinuclidin-3-ol and quinuclidin-2-ene derivatives as muscarinic antagonists. synthesis and structure-activity relationships. *J. Med. Chem.*, 38: 473–487.

Nilvebrant, L. and Sparf, B. (1982) Muscarinic receptor binding in the parotid gland. different affinities of some anticholinergic drugs between the parotid gland and ileum. *Scand. J. Gastroenterol.*, 17 (Suppl. 72): 69–77.

Nilvebrant, L. and Sparf, B. (1983a) Muscarinic receptor binding in the guinea pig urinary bladder. *Acta Pharmacol. Toxicol.*, 52: 30–38.

Nilvebrant, L. and Sparf, B. (1983b) Differences between binding affinities of some antimuscarinic drugs in the parotid gland and those in the urinary bladder and ileum. *Acta Pharmacol. Toxicol.*, 53: 304–313.

Nilvebrant, L. and Sparf, B. (1986) Dicyclomine, benzhexol and oxybutynin distinguish between sub-classes of muscarinic binding-sites. *Eur. J. Pharmacol.*, 123: 133–143.

Nordvall, G. and Hacksell, U. (1993) Binding-site modeling of the muscarinic m1 receptor – a combination of homology-based and indirect approaches. *J. Med. Chem.*, 36: 967–976.

Saunders, J., Cassidy, M., Freedman, S.B., Harley, E.A., Iversen, L.L., Kneen, C., MacLeod, A.M., Merchant, K.J., Snow, R.J. and Baker, R. (1990) Novel quinuclidine-based ligands for the muscarinic cholinergic receptor. *J. Med. Chem.*, 33: 1128–1138.

Verloop, A. (1987) *The STERIMOL Approach to Drug Design*. Decker, New York.

Wold, S., Johansson, E. and Cocchi, M. (1993) PLS-partial least-squares projections to latent structures. In: H. Kubinyi (Ed.), *3D QSAR in Drug Design: Theory, Methods and Applications*, ESCOM, Leiden, pp. 523–550.

J. Klein and K. Löffelholz (Eds.)
*Progress in Brain Research*, Vol. 109

CHAPTER 12

# Allosteric regulation of muscarinic receptors

Nigel J.M. Birdsall[1], Sebastian Lazareno[2] and Hideki Matsui[2]

*[1]National Institute for Medical Research, Mill Hill, London, NW7 1AA, UK*
*and [2]MRC Collaborative Centre, Mill Hill, London, NW7 1AD, UK*

## Introduction

Allosteric effects are observed when there are complimentary cross-interactions between two binding processes: the binding of one ligand affects the binding of the second ligand and vice versa. The formation of a ternary complex between the two ligands and, for example, a target protein such as a receptor, is necessary for allosteric interactions to occur. This is illustrated in Scheme 1

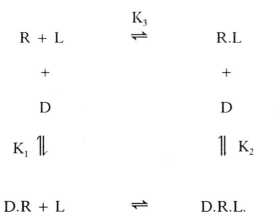

where D and L are the two ligands and R is the protein. If the mutual interactions are favourable, $K_2$ and $K_4$ are greater than $K_1$ and $K_3$ respectively (all $K_i$'s are affinity constants) and there is positive cooperativity. Conversely, disfavoured interactions are observed when $K_2 < K_1$ and $K_4 < K_3$ (negative cooperativity). The magnitude of the cooperativity, $\alpha$, is equal to $K_2/K_1$ (and $K_4/K_3$ because of the

mutual cross-interaction between D and L). Therefore the binding of an allosteric ligand, say D, is characterised by two parameters, $K_1$ and $\alpha$, in contrast to the single affinity constant that describes the binding of a single ligand to form a binary complex. Scheme 1 implies that there are two binding sites on R and that the pharmacology of each site is complex as it is dependent not only on the structure of that site but also on how that structure is perturbed by the binding of ligands to the second site.

In the case of G-protein coupled receptors there is in general a positively cooperative interaction between the binding of agonists and G-protein to the receptor, which favours the formation of the agonist-receptor-G protein complex. This complex has a higher affinity for agonists (and G-proteins) than the receptor alone, and is manifest as the "high affinity state" of the receptor. Disruption of the ternary complex by GTP generates the uncoupled receptor which has a low affinity for agonists. Depending on the agonist, receptor and G-protein, $\alpha$ values can typically be in the range 2–500. Ligands which exhibit a negatively cooperative interaction with G-proteins will tend to disrupt receptor-G protein coupling. These ligands are now commonly called "inverse agonists" and may have pharmacological actions additional to those of classical competitive antagonists if there is a significant level of receptor-G protein coupling (and activation) in the absence of other ligands.

For a small number of G-protein coupled receptors an additional allosteric binding site has been characterised. Most work has been carried out on muscarinic receptors, where an allosteric site on heart ($M_2$) receptors was suggested by the work of Clark and Mitchelson (1976) and confirmed in the binding studies of Stockton et al. (1983) and Dunlap and Brown (1983). Subsequently this allosteric site was shown to be present on all five subtypes of muscarinic receptors (Ellis et al., 1991). The extensive work in this area has been reviewed by Hulme et al. (1990); Lee and El-Fakahany (1991); Birdsall et al. (1995) and Tuček and Proska (1995).

In this paper we describe some of our work on the location of the allosteric site on muscarinic receptors and present recent data on the binding of strychnine to this site.

## The location of the allosteric site

Much of the work on allosteric interactions at muscarinic receptors, including the initial studies, has utilised the neuromuscular blocker gallamine. This is a highly charged ligand, bearing three positive charges, and exhibits negatively cooperative interactions with a range of muscarinic agonists and antagonists (Stockton et al., 1983). Paradoxically, gallamine slows down dramatically the association and dissociation kinetics of quaternised muscarinic antagonists such as $N$-methylscopolamine (NMS) when it binds to muscarinic receptors: in most instances decreases in affinity are associated with faster and not slower kinetics. The decrease in kinetics is so profound that access and egress of NMS of its binding site is prevented when gallamine is bound. In other words there is essentially a compulsory order of binding of these two ligands. One possible interpretation is that gallamine physically prevents NMS dissociation and association (Proska and Tuček, 1994; Matsui et al., 1995). In addition gallamine, an impermeant ligand, rapidly produces its allosteric inhibitory effects on whole cells and tissues. It therefore appears that the allosteric site is located on the extracellular face of the receptor and it may be located extracellular to the antagonist binding site.

The antagonist binding site is thought from chemical labelling and mutagenesis studies to be located within the extracellular one-third of the seven transmembrane domains of muscarinic receptors (reviewed by Wess, 1993; Chapter 13). We therefore considered that the allosteric site could be located on the extracellular loops of the muscarinic receptor or at the boundary of the loops and the transmembrane $\alpha$-helices.

As the allosteric site for gallamine is present on all subtypes of muscarinic receptor (Ellis et al., 1991) it was concluded that certain residues, conserved in all subtypes would be important for the structure of the allosteric site. Accordingly we mutated all the conserved charged, polar, aromatic, glycine and proline residues in the external loops and at the postulated loop-helix interfaces of the human m1 receptor (Matsui et al., 1995). These residues are shown as diamonds in Fig. 1. Initially these residues were mutated to alanine (and also to more homologous amino acids if the alanine mutant exhibited altered binding properties). Most of the mutations did not produce substantial changes in the binding of acetylcholine (ACh), a number of muscarinic antagonists and gallamine. Two tryptophan residues, W101 and W400, appear to be important for gallamine binding (arrowed in Fig. 1). Mutation of these residues to alanine decrease gallamine binding to the unliganded receptor and the NMS-receptor complex over 10-fold. The more conservative mutants W101F and W400F exhibit relatively unchanged and decreased gallamine affinity, respectively, relative to the wild-type receptor. This suggests that, for high affinity gallamine binding, it is important for residue 400 to be tryptophan and for residue 101 to be aromatic. Surprisingly, a third tryptophan, W91, seems to be important for antagonist binding. This residue is in the middle of the postulated first extracellular loop and is topographically separated in the model from the residues in the top third of the transmembrane regions which from published mutation studies are considered to be close to, or part of, the antagonist binding site (reverse mode symbols, Fig. 1). Therefore either the model is wrong and, for example, the middle of the first extracellular loop dips into the transmembrane region or W91 has an important structural role. The latter postulate is in agreement with the view of the fact that this tryptophan is conserved in many G-protein coupled receptors.

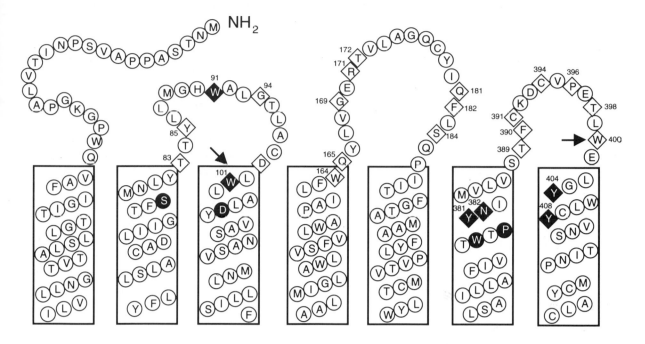

Fig. 1. Model of the external loops and transmembrane $\alpha$-helices (boxed) of the human m1 muscarinic receptor. Diamonds, residues mutated in the study of Matsui et al. (1995) and in our unpublished work. Arrows, residues which when mutated affect gallamine binding. Reverse mode, residues which show altered antagonist binding when mutated. Adapted from Matsui et al. (1995).

Although W101 and W400 are apparently well separated in the linear model in Fig. 1, a helical wheel model suggests that these residues may be close in space and lie just extracellular to the cluster of amino acids which are thought to constitute the antagonist binding site (Matsui et al., 1995). If W101 and W400 constitute part of the allosteric site then it can readily be envisaged how gallamine could sterically interface with the association and disassociation of NMS.

Interestingly, the mutations of W91, 101 and 400 all slowed down the NMS dissociation rate constant despite antagonist affinities being decreased in the case of W101A and W91A and W91F. This paradoxical result can be rationalised if these tryptophans form part of an aromatic lined gorge whose function is to guide and accelerate acetylcholine to its intramembrane binding site by means of interactions of the quaternary ammonium head group with the $\pi$ electron systems of the aromatic rings. This mechanism is analogous to

that postulated as an explanation for the fast kinetics of acetylcholinesterase (Ripoli et al., 1993).

The conserved aspartate residues 71, 99, 105 and 122 and a glutamate in the third intracellular loop, although prime candidates for forming an ion pair with gallamine, do not appear to be part of the allosteric site. Mutation of these residues to asparagine and glutamine do not dramatically affect gallamine binding (Lee et al., 1992; Leppik et al., 1994; Proska and Birdsall, unpublished results).

Chimeras between $M_2$ and $M_5$ and $M_1$ receptors have been generated in order to identify regions of muscarinic receptors responsible for the $M_2$ selectivity of gallamine (Ellis et al., 1993; Leppik et al., 1994). These authors have suggested that residues in the second extracellular loop or within the sixth and seventh transmembrane segments and the interconnecting third extracellular loop could be important. Studies in which specific amino acids from the $M_2$ sequence are introduced into $M_1$ or

150

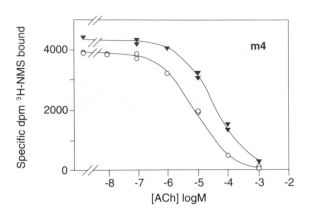

Fig. 2. Inhibition by ACh of [³H]NMS binding to m4 receptors in the presence (▼) and absence (O) of $10^{-4}$ M strychnine. Data from Lazareno and Birdsall (1995).

$M_5$ receptors and result in increases in gallamine affinity would be very informative.

## Strychnine as an allosteric ligand

The early studies showed that gallamine exhibited negative cooperativity with antagonists and agonists (Stockton et al., 1983). This pattern was found for a number of other allosteric ligands (reviewed in Hulme et al., 1990). Some time later, the neuromuscular blocker alcuronium was shown to exhibit positive cooperativity with NMS at $M_2$ receptors (Tuček et al., 1990). This finding provided us with the impetus to look for new allosteric ligands of simpler structure which might exhibit positive cooperativity with muscarinic ligands including acetylcholine. Ligands which enhance acetylcholine binding may have therapeutic use in the alleviation of some of the symptoms of a cholinergic deficit, for example in Alzheimer's disease (Birdsall et al., 1995).

We have discovered that strychnine, in addition to its actions on glycine receptors, acts as an allosteric ligand at m1–m4 muscarinic receptors and have been able to quantitate its binding properties (Lazareno and Birdsall, 1995). At m4 receptors strychnine ($10^{-4}$ M) enhances [³H]NMS binding but inhibits ACh binding (Fig. 2). This concentration of strychnine produces a near maximal change of the NMS and ACh binding properties of the receptor. The relatively small effects are due to the

small positive cooperativity with NMS ($\alpha = 1.7$) and small negative cooperativity with ACh ($\alpha = 0.5$). These values, together with the affinity of strychnine for the unliganded receptor ($1 \times 10^5$ M$^{-1}$) were obtained from the simultaneous analysis of detailed [³H]NMS/ACh competition curves in the presence of varying concentrations of strychnine (Lazareno and Birdsall, 1995). Although strychnine exhibits only small effects on the NMS and ACh binding properties of m4 receptors, it profoundly slows down (like gallamine) the dissociation of [³H]NMS (Fig. 3).

The even smaller effect of strychnine on the binding properties of NMS and ACh at m1 receptors is illustrated in Fig. 4. Strychnine does not affect the equilibrium binding of [³H]NMS but strongly slows down the kinetics. A quantitative analysis of equilibrium and kinetic data at m1 receptors gives $\alpha$ value of 1.0 and 0.4 for NMS and ACh respectively and an affinity of $1 \times 10^5$ M$^{-1}$ at the unliganded receptor (Lazareno and Birdsall, 1995). These data and those for strychnine at m2 and m3 receptors ($\alpha = 2.2$ and 0.7 respectively versus NMS and 0.15 and <1 versus ACh) illustrate the subtle but different ways in which binding properties of individual muscarinic receptor subtypes can be "tuned". The fact that the values of $\alpha$

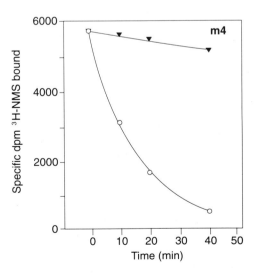

Fig. 3. Dissociation of [³H]NMS from m4 receptors in the presence (▼) and absence (O) of $3 \times 10^{-4}$ M strychnine. Data from Lazareno and Birdsall (1995).

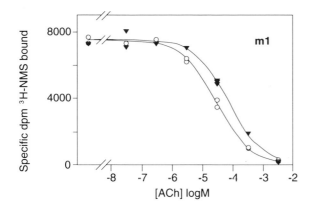

Fig. 4. Inhibition by ACh of [³H]NMS binding to m1 receptors in the presence (▼) and absence (O) of $10^{-4}$ M strychnine. Data from Lazareno and Birdsall (1995).

for strychnine and ACh approach 1 indicates that it should be possible to discover ligands which exhibit positive cooperativity with ACh.

## References

Birdsall, N.J.M., Cohen, F., Lazareno, S. and Matsui, H. (1995) Allosteric regulation of G-protein linked receptors. *Biochem. Soc. Trans.*, 23: 108–111.

Clark, A.L. and Mitchelson, F. (1976) The inhibitory effect of gallamine on muscarinic receptors. *Br. J. Pharmacol.*, 58: 323–331.

Dunlap, J. and Brown, J.H. (1983) Heterogeneity of binding sites on cardiac muscarinic receptors induced by the neuromuscular blocking agents gallamine and pancuronium. *Mol. Pharmacol.*, 24: 15–22.

Ellis, J., Huyler, J. and Brann, M.R. (1991) Allosteric regulation of cloned m1-m5 muscarinic receptor subtypes. *Biochem. Pharmacol.*, 42: 1927–1932.

Ellis, J., Seidenberg, M. and Brann, M.R. (1993) Use of chimeric muscarinic receptors to investigate epitopes involved in allosteric interactions. *Mol. Pharmacol.*, 44: 583–588.

Hulme, E.C., Birdsall, N.J.M. and Buckley N.J. (1990) Muscarinic receptor subtypes. *Annu. Rev. Pharmacol.*, 30: 633–673.

Lazareno, S. and Birdsall, N.J.M. (1995) Detection quantitation and verification of allosteric interactions of agents with labelled and unlabelled ligands at G protein-coupled receptors: interactions of strychnine and acetylcholine at muscarinic receptors. *Mol. Pharmacol.*, 48: 362–378.

Lee, N.H. and El-Fakahany, E.E. (1991) Allosteric antagonists of the muscarinic acetylcholine receptor. *Biochem. Pharmacol.*, 42: 199–205.

Lee, N.H., Jingru, H. and El-Fakahany, E.E. (1992) Modulation by certain conserved aspartate residues of the allosteric interaction of gallamine at the m1 muscarinic receptor. *J. Pharmacol. Exp. Ther.*, 262: 312–316.

Leppik, R.A., Miller, R.C., Eck, M. and Paquet, J.L. (1994) Role of acidic amino acids in the allosteric modulation by gallamine of antagonist binding at the m2 muscarinic acetylcholine receptor. *Mol. Pharmacol.*, 45: 983–990.

Matsui, H., Lazareno, S. and Birdsall, N.J.M. (1995) Probing of the location of the allosteric site on m1 muscarinic receptors by site-directed mutagenesis. *Mol. Pharmacol.*, 47: 88–98.

Proska, J. and Tuček, S. (1994) Mechanisms of steric and cooperative actions of alcuronium on cardiac muscarinic acetylcholine receptors. *Mol. Pharmacol.*, 45: 709–717.

Ripoli, D.R., Faerman, C.H., Axelson, P.H., Silman, I. and Sussman J.L. (1993) An electrostatic mechanism for substrate guidance down the aromatic gorge of acetylcholinesterase. *Proc. Natl. Acad. Sci. USA*, 90: 5128–5132.

Stockton, J.M., Birdsall, N.J.M., Burgen, A.S.V. and Hulme, E.C. (1983) Modification of the binding properties of muscarinic receptors by gallamine. *Mol. Pharmacol.*, 23: 551–557.

Tuček, S., Musilkova, J., Nedoma, J., Proska, J., Shelkovnikov, S. and Vorlicek, J. (1990) Positive cooperativity in the binding of alcuronium and N-methylscopolamine to muscarinic acetylcholine receptors. *Mol. Pharmacol.*, 38: 674–680.

Tuček, S. and Proska, J. (1995) Allosteric modulation of muscarinic acetylcholine receptors. *Trends Pharmacol. Sci.*, 16: 205–212.

Wess, J. (1993) Molecular basis of muscarinic acetylcholine receptor function. *Trends Pharmacol. Sci.*, 14: 308–313.

J. Klein and K. Löffelholz (Eds.)
*Progress in Brain Research*, Vol. 109
© 1996 Elsevier Science B.V. All rights reserved.

CHAPTER 13

# Molecular aspects of muscarinic receptor assembly and function

Jürgen Wess, Nathalie Blin, June Yun, Torsten Schöneberg and Jie Liu

*Laboratory of Bioorganic Chemistry, National Institute of Diabetes and Digestive and Kidney Diseases, Bethesda, MD, USA*

## Introduction

The five mammalian muscarinic acetylcholine receptors (m1–m5) are typical members of the superfamily of G protein-coupled receptors (GPCRs) (Hulme et al., 1990; Wess, 1993). Over the past couple of years, we have used several members of the muscarinic receptor family as model systems to study how GPCRs function at a molecular level. In this chapter, we focus on recent mutagenesis studies that have led to novel insights into the structure and assembly of muscarinic receptors. Moreover, studies aimed at identifying the structural elements governing the selectivity of muscarinic receptor/G protein interactions are also discussed.

## Receptor structure

Like all other GPCRs, the muscarinic receptors are predicted to be composed of seven $\alpha$-helically arranged transmembrane domains (TM I-VII), which are connected by three extracellular (o1–o3) and three intracellular loops (i1–i3). Several lines of evidence suggest that the seven TM helices are arranged in a ring-like fashion, thus forming a tightly packed helical bundle (Baldwin, 1993, 1994; Schwartz, 1994). Such an arrangement is strongly supported by low-resolution electron microscopic images of the photoreceptor, rhodopsin (Schertler et al., 1993), as well as several mutagenesis studies (Baldwin, 1993, 1994; Schwartz, 1994). Residues located on the inner surfaces of different TM helices are known to be involved in

the binding of acetylcholine and other small ligands (Baldwin, 1993, 1994; Wess, 1993; Schwartz, 1994; Strader et al., 1994). Moreover, agonist-induced conformational changes in the TM receptor core are thought to be intimately involved in receptor activation (Dohlman et al., 1991; Wess, 1993; Strader et al., 1994). Detailed structural information about the membrane-embedded portion of GPCRs is therefore essential for understanding how GPCRs function at a molecular level.

In the absence of detailed structural information on any GPCR, we (Pittel and Wess, 1994) and others (Suryanarayana et al., 1992; Rao et al., 1994; Zhou et al., 1994; Elling et al., 1995) have recently employed mutagenesis approaches to identify molecular interactions between individual TM helices. We recently described a series of m2/m5 hybrid muscarinic receptors (C1–C4; Fig. 1) which are unable to bind significant amounts of muscarinic radioligands when transiently expressed in COS-7 cells (Pittel and Wess, 1994). Characteristically, all of these functionally inactive hybrid receptors contained m2 receptor sequence in TM VII and m5 receptor sequence in TM I. Immunocytochemical studies showed that these mutant receptors were properly trafficked to the cell surface (Liu et al., 1995a). The inability of C1–C4 to bind muscarinic ligands thus appears to be due to a specific folding defect. We could demonstrate that replacement of TM I in these misfolded receptors with the corresponding m2 receptor sequence resulted in the appearance of functional muscarinic receptors (Pittel and Wess,

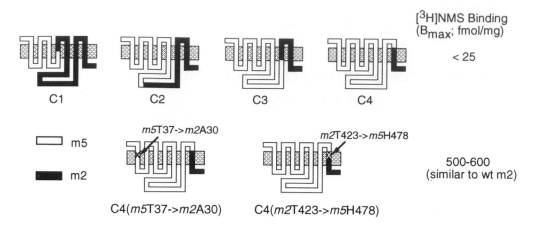

Fig. 1. Structure and [³H]NMS binding properties ($B_{max}$) of hybrid m2/m5 muscarinic receptors (human) expressed in COS-7 cells. Data are taken from Liu et al. (1995a).

1994), indicating that molecular interactions between TM I and TM VII are required for proper receptor folding.

Based on these results, a study was designed to identify specific amino acids on TM I and TM VII which are responsible for the folding defect present in C1–C4. We hypothesized that the identification of such residues should allow predictions as to how TM I and TM VII are oriented relative to each other. It seemed reasonable to assume that the amino acids responsible for the structural defect present in C1–C4 are located at the interface between TM I and TM VII. To identify these residues, the TM I and TM VII sequences (derived from the m5 and the m2 receptor, respectively) of C1–C4 were initially depicted in a helical wheel model assuming an anticlockwise connectivity of the TM helices (as viewed from the extracellular surface of the membrane; Fig. 2A). The positions of the TM helices shown in Fig. 2 is primarily based on the recently published "Baldwin model" (Baldwin 1993, 1994) which is compatible with a low-resolution electron density map of rhodopsin (Schertler et al., 1993).

Strikingly, this model predicts that there are only two residues at the interface between TM I and TM VII in which the m2 and m5 receptors differ. In all inactive hybrid receptors, Thr37 in TM I of the m5 receptor is predicted to face Thr423 in TM VII of the m2 receptor (Fig. 2A). Both threonine residues are thought to be located

about 1–2 helical turns away from the membrane surface. We therefore speculated that the formation of a novel hydrogen bond between these two residues might interfere with proper helix/helix packing, thus leading to misfolded receptor proteins. If this is correct, it should be possible to "pharmacologically rescue" C1–C4 by single point mutations such that m5Thr37 (TM I) faces its "natural partner", m5His478 (TM VII), or, vice versa, that m2Thr423 (TM VII) is located adjacent to m2Ala30 (TM I).

To test this hypothesis experimentally, these two point mutations (m5Thr37 → m2Ala30 and m2Thr423 → m5His478, respectively) were introduced into all pharmacologically inactive hybrid receptors (shown for C4 in Fig. 1), thus creating mutant receptors in which the amino acid configuration at the TM I/TM VII interface mimicked that found in the wild type m2 and m5 muscarinic receptors, respectively (Liu et al., 1995a). We found, consistent with our working hypothesis, that mutant receptors in which either of the two threonine residues (m5Thr37 or m2Thr423) was structurally modified gained the ability to bind significant amounts of muscarinic radioligands (shown for C4 in Fig. 1). Moreover, all resulting mutant receptors were able to bind muscarinic agonists and antagonists with wild type affinities (Liu et al., 1995a).

Characteristically, in each case (C1–C4), replacement of either of the two critical threonine residues had a quantitatively similar effect on the

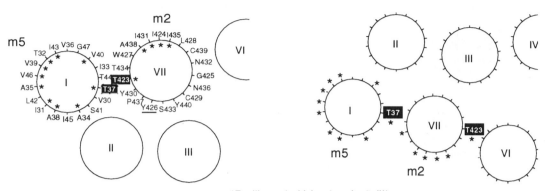

A) Anticlockwise (proposed) connectivity
of TM helices

B) Clockwise (incompatible) connectivity
of TM helices

*Positions at which m2 and m5 differ

Fig. 2. Helical wheel projection of the possible orientation of TM I and VII based on the model proposed by Baldwin (1993, 1994). (A) Anticlockwise, (B) clockwise connectivity of the TM helices (as viewed from the extracellular surface of the membrane). The TM I (m5) and TM VII (m2) sequences of the pharmacologically inactive m2/m5 hybrid receptors, C1–C4 (Fig. 1), are shown. Amino acids which differ between the m2 and m5 muscarinic receptors (human) are marked with asterisks. The two threonine residues (m5Thr37 and m2Thr423) which face each other at the TM I/VII interface in (A) are highlighted. The model shown here slightly differs from the Baldwin projection in that TM VII has been rotated by about 30° (counterclockwise) to allow m2Tyr426 (underlined) to project into the ligand binding cavity formed by TM III–VII. Site-directed mutagenesis studies suggest that this tyrosine residue is critically involved in ligand recognition (Wess et al., 1991; Matsui et al., 1995).

number of "recovered" [$^3$H]NMS binding sites (Liu et al., 1995a). This result would be expected if the virtual lack of ligand binding activity found with C1–C4 is due to a direct (conformationally unfavorable) interaction between these two residues. These data therefore suggest that the model shown in Fig. 2A assuming a counterclockwise connectivity of the TM helices (as viewed from the extracellular membrane surface) is probably correct. In contrast, our results are clearly inconsistent with a receptor model assuming a clockwise arrangement of the TM helices (Fig. 2B). In such a (obviously incorrect) model, m5Thr37 and m2Thr423 would not be located adjacent to each other, making it very difficult to rationalize how substitution of either of these two residues can restore a proper receptor fold.

## Receptor assembly

Little is known about the molecular mechanisms that govern the assembly/folding of GPCRs in the lipid bilayer. Studies with "split" β2-adrenergic (Kobilka et al., 1988) and muscarinic receptors (Maggio et al., 1993a,b) suggested that GPCRs are composed of at least two independent folding units, one containing TM I–V and the other, TM VI and TM VII. This notion is primarily supported by the finding that coexpression of these two receptor domains as two separate polypeptides (in Xenopus oocytes or COS-7 cells) results in the appearance of functional GPCRs.

To test the hypothesis that GPCRs are composed of multiple (rather than only two) building blocks, the rat m3 muscarinic receptor was "split" in all three intracellular (i1–i3) and all three extracellular loops (o1–o3), thus generating six polypeptide pairs (Ni1 + Ci1, No1 + Co1, etc.; Schöneberg et al., 1995). As expected, expression of the individual receptor fragments in COS-7 cells did not result in functional muscarinic receptors. However, coexpression of three of the six polypeptide pairs (Ni2 + Ci2, No2 + Co2, and Ni3 + Ci3, respectively; Fig. 3) led to a significant number of specific [$^3$H]NMS binding sites. Whereas the Ni3 + Ci3 receptor complex displayed wild type-like ligand binding affinities, the Ni2 + Ci2 and No2 + Co2 polypeptide complexes showed 3–20-fold reduced binding affinities for all muscarinic ligands examined (Schöneberg et al.,

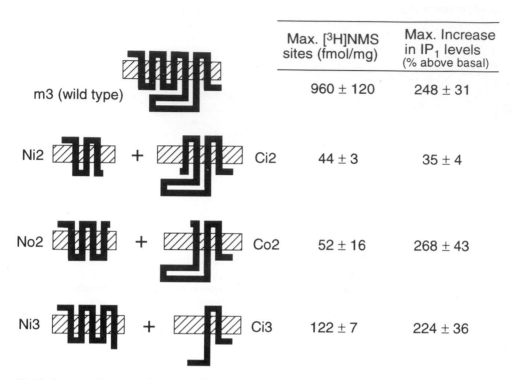

| | Max. [³H]NMS sites (fmol/mg) | Max. Increase in IP₁ levels (% above basal) |
|---|---|---|
| m3 (wild type) | 960 ± 120 | 248 ± 31 |
| Ni2 + Ci2 | 44 ± 3 | 35 ± 4 |
| No2 + Co2 | 52 ± 16 | 268 ± 43 |
| Ni3 + Ci3 | 122 ± 7 | 224 ± 36 |

Fig. 3. Structure of receptor fragments obtained by "splitting" the rat m3 muscarinic receptor in the i2, o2, or i3 loop. The displayed polypeptide pairs were coexpressed in COS-7 cells and studied for their ability to bind muscarinic ligands and to mediate carbachol-induced (1 mM) increases in inositol monophosphate (IP₁) levels (Schöneberg et al., 1995). Data are given as means ± SEM of two or three independent experiments, each performed in duplicate.

1995). This observation suggests that the i2 and o2 loops exert an indirect conformational effect on the proper arrangement of the TM helical bundle where the binding of small ligands such as acetylcholine is thought to occur (Dohlman et al., 1991; Savarese and Fraser, 1992; Wess, 1993; Strader et al., 1994).

To study whether the various polypeptide complexes were still capable of activating G proteins, their ability to mediate agonist-induced PI hydrolysis was examined (Fig. 3). These experiments showed that the No2 + Co2 and Ni3 + Ci3 complexes were able to stimulate the breakdown of phosphoinositide lipids to a similar maximum extent as the wild type m3 receptor. In contrast, splitting the m3 receptor within the i2 loop resulted in a polypeptide complex which showed only residual functional activity, indicating that the structural integrity of the i2 loop is critical for proper receptor/G protein coupling (Schöneberg et al., 1995).

To allow the detection of the individual receptor fragments by immunocytochemical techniques, a 9-amino acid hemagglutinin epitope tag was added to the N-terminus of the N-terminal receptor fragments. In addition, a polyclonal antibody directed against the C-terminus of the m3 muscarinic receptor was used to study the cellular localization of the C-terminal receptor fragments. We found that all N- and C-terminal receptor fragments shown in Fig. 3 were present not only intracellularly (in the ER/Golgi network) but also in the plasma membrane, even when expressed alone (Schöneberg et al., 1995). Thus the various receptor fragments can be stably inserted into lipid bilayers and properly trafficked to the cell surface.

Taken together, these data strongly suggest that muscarinic receptors and, most likely, other GPCRs are composed of multiple structural/functional subunits. These subunits are able to fold independently of each other in a fashion such that

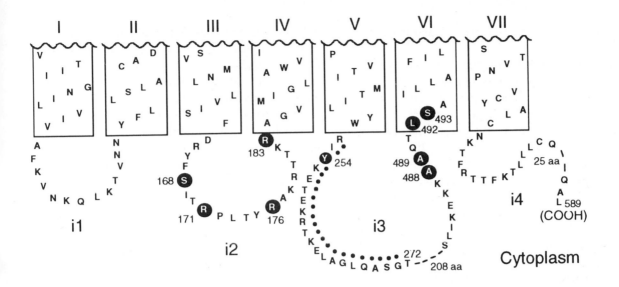

Fig. 4. Residues in the m3 muscarinic receptor required for selective activation of $G_{q/11}$ (modified according to Blin et al., 1995). The intracellular regions (except for the central portion of the i3 loop and a segment of the i4 domain) and the endofacial portions of TM I–VII of the m3 receptor (rat) are shown. The nine amino acids highlighted in black are required for efficient activation of $G_{q/11}$. Moreover, optimum activation of $G_{q/11}$ requires one or more additional residues in the N-terminal portion of the i3 loop (stippled sequence; Blin et al., 1995).

they can interact with each other to form a functional receptor complex. We therefore speculate that the folding of polytopic eucaryotic membrane proteins (containing multiple TM domains) occurs in two consecutive steps, in a fashion similar to the assembly of the bacterial membrane protein, bacteriorhodopsin (Popot and Engelman, 1990). In step I, individually stable folding domains are established across the lipid bilayer, which can then, in step II, interact with each other to form a functional protein complex.

**Molecular basis of receptor/G protein coupling selectivity**

Characteristically, each GPCR can recognize and activate only a limited number of the many structurally closely related G proteins expressed within a cell (Dohlman et al., 1991; Savarese and Fraser, 1992; Hedin et al., 1993). At present, very little is known about which specific amino acids are of particular importance for proper G protein recognition. However, such information is essential to gain deeper insight into the molecular basis of receptor/G protein coupling selectivity.

The muscarinic receptors have served as useful model systems to analyze the structural basis of receptor/G protein interactions (Wess, 1993). Several laboratories have demonstrated that the m1, m3, and m5 receptors preferentially couple to G proteins of the $G_{q/11}$ family (Bonner et al., 1988; Peralta et al., 1988; Berstein et al., 1992; Offermanns et al., 1994), whereas the m2 and m4 receptors selectively activate G proteins of the $G_{i/o}$ class (Peralta et al., 1988; Parker et al., 1991; Offermanns et al., 1994). Studies with hybrid m2/m3 muscarinic receptors have shown that the N-terminal 16–21 amino acids of the i3 loop play an important role in determining the G protein coupling profile of a given muscarinic receptor subtype (Wess et al., 1989, 1990; Lechleiter et al., 1990). Moreover, mutational analysis of the rat m3 muscarinic receptor (Blüml et al., 1994a,b) has shown that Tyr254 (located at the N-terminus of the i3 loop; Fig. 4) is essential for efficient stimulation of m3 receptor-mediated PI hydrolysis, a response known to be mediated by G proteins of the $G_{q/11}$ class (Smrcka et al., 1991; Berstein et al., 1992). This residue is conserved among the m1, m3, and m5 receptors but replaced with a different residue (serine) in the m2 and m4 receptors.

We also found that substitution of Tyr254 into a mutant m3 muscarinic receptor in which the N-terminal segment of the i3 loop was replaced with the corresponding m2 receptor sequence (a modification that eliminated m3 receptor-mediated PI hydrolysis) was able to confer on the resultant mutant receptor the ability to efficiently activate the PI pathway (Blüml et al., 1994b). However, when *m3*Tyr254 was substituted directly into the wild type m2 receptor, the resulting m2 (Ser210 → Tyr) mutant receptor (m2-Y), similar to the wild type m2 receptor, failed to stimulate PI hydrolysis to a significant extent (Blüml et al., 1994b; Table 1). These findings indicated that residues located in other intracellular domains, besides the N-terminal portion of the i3 loop, must also play important roles in proper recognition of G proteins of the $G_{q/11}$ class.

To identify such residues, distinct intracellular loops/segments of the m3 muscarinic receptor were substituted into the wild type m2 receptor as well as into the m2-Y mutant receptor, and the resulting hybrid receptors were studied for their ability to mediate carbachol-induced PI hydrolysis. In a second step, the functional roles of individual amino acids were examined by site-directed mutagenesis (Blin et al., 1995).

We found that a mutant m2 receptor in which the i2 loop was replaced with the corresponding m3 receptor sequence (m2–i2) gained the ability to stimulate PI hydrolysis in a fashion similar to a mutant m2 receptor in which the first 21 amino acids of the i3 loop were replaced with the homologous m3 receptor sequence (m2–Ni3) (Table 1). This result indicated, consistent with a previous study using chimeric adrenergic/muscarinic receptors (Wong et al., 1990), that the i2 loop contributes to proper recognition of $G_{q/11}$ to a similar extent as the N-terminal segment of the i3 loop. Moreover, we could demonstrate that four amino acids in the i2 loop of the m3 receptor (Ser168, Arg171, Arg176, and Arg183) fully account for the ability of this receptor region to support preferential coupling to $G_{q/11}$ (Table 1, Fig. 4). System-

TABLE 1

Carbachol-induced PI hydrolysis mediated by hybrid m2/m3 muscarinic receptors

| Receptor | m2 receptor region containing substitutions | m2 residues replaced with m3 sequence | Max. increase in $IP_1$ above basal (%) | Carbachol $EC_{50}$ ($\mu$M) |
|---|---|---|---|---|
| m3 (wild type) | | | 100 | $0.19 \pm 0.04$ |
| m2 (wild type) | | | $11 \pm 1$ | |
| m2-Y | N-term. of i3 | Ser210 → Tyr254 | $9 \pm 1$ | |
| m2-i2 | i2 | Asp120–Met139 → Asp164–Arg183 | $38 \pm 5$ | $1.59 \pm 0.34$ |
| m2-i2(4aa) | i2 | Cys124 → Ser168, Lys127 → Arg171, Pro132 → Arg176, Met139 → Arg183 | $38 \pm 7$ | $2.4 \pm 0.5$ |
| m2-Ni3 | N-term. of i3 | His208–Glu228 → Arg252-Thr272 | $40 \pm 6$ | $2.9 \pm 0.9$ |
| m2-Ci3 | C-term. of i3 | Val361–Leu390 → Lys464–Ser493 | $3 \pm 2$ | |
| m2-Y-Ci3 | N- and C-term. of i3 | See above | $52 \pm 7$ | $3.1 \pm 1.7$ |
| m2-Ci3(4aa) | C-term. of i3 | Val385 → Ala488, Thr386 → Ala489, Ile389 → Leu492, Leu390 → Ser493 | $4 \pm 1$ | |
| m2-Y-Ci3(4aa) | N- and C-term. of i3 | See above | $49 \pm 11$ | $5.5 \pm 2.2$ |
| m2-i2(4aa)-Ni3-Ci3(4aa) | i2, N- and C-term. of i3 | See above | $160 \pm 16$ | $0.13 \pm 0.02$ |

All wild type and mutant muscarinic receptors were transiently expressed (at similar levels) in COS-7 cells. PI assays were carried out as described by Blin et al. (1995). Carbachol stimulation of the wild type m3 muscarinic receptor (rat) led to a 5–8-fold increase in inositol monophosphate ($IP_1$) levels ($E_{max}$ = 100 %). Data are presented as means ± SEM of two to five independent experiments, each performed in duplicate (data taken from Blin et al., 1995).

atic substitution of these residues into the wild type m2 receptor (or into m2-Y), either individually or in combination, showed that the presence of all four residues is required for optimum efficiency of $G_{q/11}$ recognition (Blin et al., 1995). A mutant m2 receptor (m2–Ci3) in which the C-terminal 30 amino acids of the i3 loop (Val361–Leu390) were replaced with the corresponding m3 receptor sequence (Lys464–Ser493) was unable to stimulate PI hydrolysis to an appreciable extent. However, introduction of the additional *m2*Ser210 → *m3*Tyr254 point mutation (at the N-terminus of the i3 loop) into this hybrid construct yielded a mutant receptor (m2-Y-Ci3) that was capable of activating the PI pathway with the same efficiency as m2-i2 and m2-Ni3 (Table 1). This finding, besides providing additional evidence for the key role of *m3*Tyr254 in $G_{q/11}$ activation (Blüml et al., 1994a,b), is consistent with the notion that specific residues located at the N- and C-terminus of the i3 loop of the m3 receptor form a common recognition site for $G_{q/11}$ proteins.

Systematic mutational analysis showed that four m3 receptor residues (Ala488, Ala489, Leu492, and Ser493) quantitatively account for the contribution of the C-terminal segment of the i3 loop to proper recognition of $G_{q/11}$ proteins (Table 1). Computational methods predict that the region at the i3 loop/TM VI junction is $\alpha$-helically arranged (Strader et al., 1989), suggesting that Ala488, Ala489, Leu492, and Ser493 are located on one side of an $\alpha$-helical receptor domain. Similarly, accumulating evidence suggests that the N-terminal portion of the i3 loop also forms an amphiphilic $\alpha$-helix and that the non-charged (hydrophobic) side of this helical segment is intimately involved in G protein recognition and activation (Cheung et al., 1992; Blüml et al., 1994c). We therefore propose that these two patches of primarily hydrophobic residues, located at the N- and C-terminus of the i3 loop, respectively, form a common binding surface for specific classes of G proteins.

Functional analysis of hybrid m2/m3 muscarinic receptors containing multiple substitutions in various intracellular receptor domains showed that specific residues in the i2 loop and in the membrane-proximal portions of the i3 loop of the m3 receptor act in a concerted fashion to stimulate G proteins of the $G_{q/11}$ family. We found that introduction of Ser168, Arg171, Arg176, and Arg183 (derived from the i2 loop of the m3 receptor) and of Ala488, Ala489, Leu492, and Ser493 (derived from the C-terminus of the i3 loop of the m3 receptor) into m2-Ni3 resulted in a hybrid receptor that was able to stimulate PI hydrolysis with a similarly high carbachol potency as the wild type m3 receptor (Table 1), indicating that these eight amino acids, together with the N-terminal portion of the i3 loop, quantitatively account for the efficiency of m3 receptor-mediated $G_{q/11}$ activation. Surprisingly, this mutant receptor gained the ability to stimulate the production of inositol phosphates to an even greater maximum extent ($E_{max}$ 160% of wild type m3) than the wild type m3 receptor. A possible explanation for this phenomenon is that the central portion of the i3 loop of the m3 receptor contains structural elements (e.g. potential sites of receptor phosphorylation that are not contained in the mutant receptor; Moro et al., 1993; Tobin and Nahorski, 1993) that exert a negative regulatory effect on receptor/G protein coupling.

The nine amino acids highlighted in Fig. 4 are present in all $G_q$-coupled muscarinic receptors (m1, m3 and m5) but absent in the m2 and m4 receptors. Sequence analysis shows that these residues, except for *m3*Ala489, are not well conserved among other classes of receptors which preferentially couple to $G_{q/11}$ proteins. It should therefore be of interest to examine whether the amino acids present at the corresponding positions in other $G_{q/11}$-coupled receptors play similar roles in selective $G_{q/11}$ recognition as described here for the m3 muscarinic receptor. Our data suggest that the G protein coupling preference of a given GPCR is determined by a limited number of amino acids located on various different intracellular receptor regions.

## Identification of a specific receptor/G protein contact site

As discussed above, multiple intracellular receptor

regions appear to be involved in G protein coupling. Similarly, it has been shown that at least three regions on the G protein $\alpha$-subunits can interact with the receptors (Hamm et al., 1988; Conklin and Bourne, 1993; Rasenick et al., 1994). Surprisingly, however, no specific site of contact between a receptor and its cognate G protein(s) has been identified so far. Clearly, such information is required for establishing three-dimensional models of the receptor/G protein interface which should lead to a better understanding of the molecular mechanisms involved in receptor-mediated G protein activation.

Several lines of evidence suggest that the C-terminus of G protein $\alpha$-subunits plays a key role in proper receptor recognition (Conklin and Bourne, 1993; Conklin et al., 1993). Using the m2 muscarinic receptor, a prototypical $G_{i/o}$-coupled receptor as a model system, we have recently initiated a study aimed at identifying the receptor region which can functionally interact with the C-terminus of $G\alpha$-subunits.

Consistent with its G protein coupling preference, the m2 receptor is unable to stimulate PI hydrolysis to a significant extent (Table 1), even when overexpressed with wild type $\alpha_q$ in transfected COS-7 cells (Fig. 5; Liu et al., 1995b). However, consistent with a previous study using other $G_{i/o}$-linked receptors (Conklin et al., 1993), coexpression of the m2 receptor with mutant $\alpha_q$ subunits (qi5, qo5) in which the last five amino acids of $\alpha_q$ were replaced with the corresponding sequences derived from $\alpha_{i2}$ or $\alpha_o$, respectively, led to a pronounced stimulation in inositol phosphate production (Fig. 5; Liu et al., 1995b). The specificity of this interaction could be demonstrated by the relative lack of PI activity observed after coexpression of the m2 receptor with a mutant $\alpha_q$ subunit (qs5) that contained five amino acids of $\alpha_s$ sequence at its C-terminus (Fig. 5). These findings indicated that the m2 muscarinic receptor contains a structural element that can recognize the C-terminal five amino acids of $G\alpha_{i/o}$ with high selectivity and that this interaction is required for G protein activation.

To identify this receptor region, we initially designed a series of loss-of-function experiments in

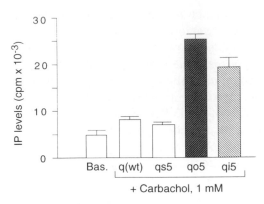

Fig. 5. Stimulation of PI hydrolysis mediated by the wild type m2 muscarinic receptor coexpressed with mutant $G\alpha_q$-subunits. COS-7 cells were cotransfected with expression plasmids coding for the wild type m2 muscarinic receptor and wild type $\alpha_q$ (q(wt)) or mutant $\alpha_q$-subunits in which the C-terminal five amino acids of wild type $\alpha_q$ (*EYNLV*) were replaced with the corresponding sequences from $\alpha_o$ (qo5; *GCGLY*), $\alpha_{i2}$ (qi5; *DCGLF*), or $\alpha_s$ (qs5; *QYELL*). About 48 h after transfections, cells were stimulated with carbachol (1 mM), and increases in intracellular $IP_1$ levels were determined as described (Blin et al., 1995). Basal $IP_1$ levels (Bas.) were determined in cells coexpressing the m2 muscarinic receptor and wild type $\alpha_q$ in the absence of carbachol. These levels were not significantly different from those found in cells cotransfected with the various mutant $\alpha$-subunits. Data are expressed as means ± SD of triplicate determinations in a single experiment; two additional experiments gave similar results (Liu et al., 1995b).

which distinct intracellular segments of the m2 receptor were systematically replaced with the corresponding sequences derived from receptors that couple to G proteins different from $G_{i/o}$ (m3 receptor, $G_q$-coupled; $\beta$2-adrenergic receptor, $G_s$-coupled). The resulting hybrid receptors were then assayed for their ability to functionally interact with the mutant $G\alpha_q$ subunits, qo5 and qi5. We found that the ability of the m2 receptor to productively couple to qo5 (or qi5) was not affected by substitutions involving the i1 and the i2 loops, the N-terminal segment of the i3 domain, or the C-terminal tail (i4) (Liu et al., 1995b). In contrast, replacement of the C-terminal portion of the i3 loop of the m2 receptor (Val361–Leu390) with the homologous m3 or $\beta$2-adrenergic receptor sequence almost completely abolished qo5-mediated PI hydrolysis, suggesting that this region of the m2 receptor is critically involved in the recognition of the C-terminus of $G\alpha_{i/o}$-subunits.

Consistent with these results, gain-of-function experiments showed that substitution of the C-terminal segment of the i3 loop of the m2 receptor (Val361–Leu390) into mutant m3 receptors that are unable to couple to wild type $G_q$ could confer on the resulting hybrid receptors the ability to functionally interact with qo5 (or qi5). More detailed mutagenesis studies showed that only four m2 receptor residues (Val385, Thr386, Ile389, and Leu390), located at the junction between the i3 loop and TM VI, are required for this interaction to occur (Liu et al., 1995b). As discussed earlier in this chapter, this region of GPCRs is likely to be $\alpha$-helically arranged. Based on this notion, the four m2 receptor residues that are predicted to contact the C-terminus of $G\alpha_{i/o}$ are thought to be located on one side of an $\alpha$-helical receptor segment and may thus form a contiguous surface which can bind to the C-terminus of $G\alpha_{i/o}$-subunits. It is conceivable that this receptor site becomes accessible to G proteins only in the agonist-occupied receptor conformation, resulting perhaps from an agonist-induced rotation or movement of TM VI towards the cytoplasm.

In summary, this is the first study describing a specific, functionally relevant site of interaction between a GPCR and its cognate $G\alpha$-subunit(s). Since proper receptor/G protein coupling is thought to involve several intracellular receptor domains (see above) and at least three sites on the $G\alpha$-subunits (including the C-terminus; Hamm et al., 1988; Conklin and Bourne, 1993; Rasenick et al., 1994), it is likely, however, that additional interactions, besides the one described here, modulate the specificity of receptor/G protein coupling and further increase the efficiency of G protein activation. It should be of great interest to investigate whether GPCRs that are linked to G proteins different from $G_{i/o}$ operate through a molecular mechanism similar to that described here.

## References

Baldwin, J.M. (1993) The probable arrangement of the helices in G protein-coupled receptors. *EMBO J.*, 12: 1693–1703.

Baldwin, J.M. (1994) Structure and function of receptors coupled to G proteins. *Curr. Opin. Cell Biol.*, 6: 180–190.

Berstein, G., Blank, J.L., Smrcka, A.V., Higashijima, T., Sternweis, P.C., Exton, J.H. and Ross, E.M. (1992) Reconstitution of agonist-stimulated phosphatidylinositol 4,5-bisphosphate hydrolysis using purified m1 muscarinic receptor, $G_{q/11}$ and phospholipase C-$\beta$1. *J. Biol. Chem.*, 267: 8081–8088.

Blin, N., Yun, J. and Wess, J. (1995) Mapping of single amino acid residues required for selective activation of $G_{q/11}$ by the m3 muscarinic acetylcholine receptor. *J. Biol. Chem.*, 270: 17741–17748.

Blüml, K., Mutschler, E. and Wess, J. (1994a) Identification of an intracellular tyrosine residue critical for muscarinic receptor-mediated stimulation of phosphatidyl inositol hydrolysis. *J. Biol. Chem.*, 269: 402 405.

Blüml, K., Mutschler, E. and Wess, J. (1994b) Functional role of a cytoplasmic aromatic amino acid in muscarinic receptor-mediated activation of phospholipase C. *J. Biol. Chem.*, 269: 11537–11541.

Blüml, K., Mutschler, E. and Wess, J. (1994c) Insertion mutagenesis as a tool to predict the secondary structure of a muscarinic receptor domain determining specificity of G-protein coupling. *Proc. Natl. Acad. Sci. USA*, 91: 7980–7984.

Bonner, T.I., Young, A.C., Brann, M.R. and Buckley, N.J. (1988) Cloning and expression of the human and rat m5 muscarinic acetylcholine receptor genes. *Neuron*, 1: 403–410.

Cheung, A.H., Huang, R.-R.C. and Strader, C.D. (1992) Involvement of specific hydrophobic, but not hydrophilic, amino acids in the third intracellular loop of the $\beta$-adrenergic receptor in the activation of $G_s$. *Mol. Pharmacol.*, 41: 1061–1065.

Conklin, B.R., and Bourne, H.R. (1993) Structural elements of $G\alpha$ subunits that interact with $G\beta\gamma$, receptors, and effectors. *Cell*, 73: 631–641.

Conklin, B.R., Farfel, Z., Lustig, K.D., Julius, D. and Bourne, H.R. (1993) Substitution of three amino acids switches receptor specificity of $G_q\alpha$ to that of $G_i\alpha$. *Nature*, 363: 274–276.

Dohlman, H.G., Thorner, J., Caron, M.G. and Lefkowitz, R.J. (1991) Model systems for the study of seven-transmembrane-segment receptors. *Annu. Rev. Biochem.*, 60: 653–688.

Elling, C.E., Nielsen, S.M. and Schwartz, T.W. (1995) Conversion of antagonist-binding site to metal-ion site in the tachykinin NK-1 receptor. *Nature*, 374: 74–77.

Hamm, H.E., Deretik, D., Arendt, A., Hargrave, P.A., Koenig, B. and Hofmann, K.P. (1988) Site of G protein binding to rhodopsin mapped with synthetic peptides from the $\alpha$ subunit. *Science*, 241 : 832–835.

Hedin, K.E., Duerson, K. and Clapham, D.E. (1993) Specificity of receptor-G protein interactions: searching for the structure behind the signal. *Cell. Sig.*, 5: 505–518.

Hulme, E.C., Birdsall, N.J.M. and Buckley, N.J. (1990) Muscarinic receptor subtypes. *Annu. Rev. Pharmacol. Toxicol.*, 30: 633–673.

Kobilka, B.K., Kobilka, T.S., Daniel, K., Regan, J.W., Caron, M.J. and Lefkowitz, R.J. (1988) Chimeric $\alpha$2-, $\beta$2-adrenergic receptors: delineation of domains involved in effector coupling and ligand binding specificity. *Science*, 240: 1310–1316.

Lechleiter, J., Duerson, K., Ennulat, D., David, N., Clapham, D. and Peralta, E. (1990) Distinct sequence elements control the specificity of G protein activation by muscarinic acetylcholine receptor subtypes. *EMBO J.*, 9: 4381–4390.

Liu, J., Schöneberg, T., van Rhee, M. and Wess, J. (1995a) Mutational analysis of the relative orientation of transmembrane helices I and VII in G protein-coupled receptors. *J. Biol. Chem.*, 270: 19532–19539.

Liu, J., Conklin, B.R., Blin, N., Yun, J. and Wess, J. (1995b) Identification of a receptor/G-protein contact site critical for signaling specificity and G-protein activation. *Proc. Natl. Acad. Sci. USA*, 92: 11642–11646.

Maggio, R., Vogel, Z. and Wess, J. (1993a) Reconstitution of functional muscarinic receptors by coexpression of amino and carboxyl terminal receptor fragments. *FEBS Lett.*, 319: 195–200.

Maggio, R., Vogel, Z. and Wess, J. (1993b) Coexpression studies with mutant muscarinic/adrenergic receptors provide evidence for intermolecular crosstalk between G protein-linked receptors. *Proc. Natl. Acad. Sci. USA*, 90: 3103–107.

Matsui, H., Lazareno, S. and Birdsall, N.J.M. (1995) Probing of the location of the allosteric site on ml muscarinic receptors by site-directed mutagenesis. *Mol. Pharmacol.*, 47: 88–98.

Moro, O., Lameh, J. and Sadee, W. (1993) Serine- and threonine-rich domain regulates internalization of muscarinic cholinergic receptors. *J. Biol. Chem.*, 268: 6862–6865.

Offermanns, S., Wieland, T., Homann, D., Sandmann, J., Bombien, E., Spicher, K., Schultz, G. and Jakobs, K.H. (1994) Transfected muscarinic acetylcholine receptors selectively couple to $G_i$-type G proteins and $G_{q/11}$. *Mol. Pharmacol.*, 45: 890–898.

Parker, E.M., Kameyama, K., Higashijima, T. and Ross, E.M. (1991) Reconstitutively active G protein-coupled receptors purified from baculovirus-infected insect cells. *J. Biol. Chem.*, 266: 519–527.

Peralta, E.G., Ashkenazi, A., Winslow, J.W., Ramachandran, J. and Capon, D.J. (1988) Differential regulation of PI hydrolysis and adenylyl cyclase by muscarinic receptor subtypes. *Nature*, 334: 434–437.

Pittel, Z. and Wess, J. (1994) Intramolecular interactions in muscarinic acetylcholine receptors studied with chimeric m2/m5 receptors. *Mol. Pharmacol.*, 45: 61–64.

Popot, J.-L. and Engelman, D.M. (1990) Membrane protein folding and oligomerization: the two-stage model. *Biochemistry*, 29: 4031–4037.

Rao, V.R., Cohen, G.B. and Oprian, D.D. (1994) Rhodopsin mutation G90D and a molecular mechanism for congenital night blindness. *Nature*, 367: 639–642.

Rasenick, M.M., Watanabe, M., Lazarevic, M.B., Hatta, S.

and Hamm, H.E. (1994) Synthetic peptides as probes for G protein function: carboxyl-terminal $G\alpha_s$ peptides mimic $G_s$ and evoke high affinity agonist binding to $\beta$-adrenergic receptors. *J. Biol. Chem.*, 269: 21519–21525.

Savarese, T.M. and Fraser, C.M. (1992) In vitro mutagenesis and the search for structure-function relationships among G protein-coupled receptors. *Biochem. J.*, 283: 1–19.

Schertler, G.F.X., Villa, C. and Henderson, R. (1993) Projection structure of rhodopsin. *Nature*, 362: 770–772.

Schöneberg, T., Liu, J. and Wess, J. (1995) Plasma membrane localization and functional rescue of truncated forms of a G protein-coupled receptor. *J. Biol. Chem.*, 270: 18000–18006.

Schwartz, T.W. (1994) Locating ligand-binding sites in 7TM receptors by protein engineering. *Curr. Opin. Biotechnol.*, 5: 434–444.

Smrcka, A.V., Hepler, J.R., Brown, K.O. and Sternweis, P.C. (1991) Regulation of polyphosphoinositide-specific phospholipase C activity by purified $G_q$. *Science*, 251: 804–807.

Strader, C.D., Sigal, I.S. and Dixon, R.A.F. (1989) Structural basis of $\beta$-adrenergic receptor function. *FASEB J.*, 3: 1825–1832.

Strader, C.D., Fong, T.M., Tota, M.R., Underwood, D. and Dixon, R.A.F. (1994) Structure and function of G protein-coupled receptors. *Annu. Rev. Biochem.*, 63: 101–132.

Suryanarayana, S., von Zastrow, M. and Kobilka, B.K. (1992) Identification of intramolecular interactions in adrenergic receptors. *J. Biol. Chem.*, 267: 21991–21994.

Tobin, A.B. and Nahorski, S.R. (1993) Rapid agonist-mediated phosphorylation of m3-muscarinic receptors revealed by immunoprecipitation. *J. Biol. Chem.*, 268: 9817–9823.

Wess, J. (1993) Molecular basis of muscarinic acetylcholine receptor function. *Trends Pharmacol. Sci.*, 14: 308–313.

Wess, J., Brann, M.R. and Bonner, T.I. (1989) Identification of a small intracellular region of the muscarinic m3 receptor as a determinant of selective coupling to PI turnover. *FEBS Lett.*, 258: 133–136.

Wess, J., Bonner, T.I., Dörje, F. and Brann, M.R. (1990) Delineation of muscarinic receptor selectivity of coupling to guanine-nucleotide binding proteins and second messengers. *Mol. Pharmacol.*, 38: 517–523.

Wess, J., Gdula, D. and Brann, M.R. (1991) Site-directed mutagenesis of the m3 muscarinic receptor: identification of a series of threonine and tyrosine residues involved in agonist but not antagonist binding. *EMBO J.*, 10: 3729–3734.

Wong, S.K.-F., Parker, E.M. and Ross, E.M. (1990) Chimeric muscarinic cholinergic: $\beta$-adrenergic receptors that activate Gs in response to muscarinic agonists. *J. Biol. Chem.*, 265: 6219–6224.

Zhou, W., Flanagan, C., Ballesteros, J.A., Konvicka, K., Davidson, J.S., Weinstein, H., Millar, R.P., and Sealfon, S.C. (1994) A reciprocal mutation supports helix 2 and helix 7 proximity in the gonadotropin-releasing hormone receptor. *Mol. Pharmacol.*, 45: 165–170.

# Section V

# Muscarinic Receptors: 2. Mechanisms of Desensitization

J. Klein and K. Löffelholz (Eds.)
*Progress in Brain Research*, Vol. 109
© 1996 Elsevier Science B.V. All rights reserved.

CHAPTER 14

# Regulation of muscarinic acetylcholine receptor expression and function

Neil M. Nathanson

*Department of Pharmacology, Box 357750, University of Washington, Seattle, WA 98195-7750, USA*

## Introduction

Muscarinic acetylcholine receptors (mAChR) are members of the superfamily of hormone and neurotransmitter receptors which are presumed to have seven membrane spanning domains and which regulate the activities of both ion channels and enzymes involved in the regulation of intracellular second messenger pathways by interaction with the G-protein family of coupling proteins (Chapter 13). Molecular cloning has demonstrated that there are five subtypes of mAChR in mammals which are products of distinct but homologous genes (Bonner, 1989). In general, the m1, m3, and m5 receptors activate phospholipase C (PLC) using pertussis toxin (IAP)-insensitive G-proteins but do not inhibit adenylyl cyclase, and the m2 and m4 receptors inhibit adenylyl cyclase (using IAP-sensitive G-proteins) but do not stimulate PLC. This specificity is not absolute, however, and m2 and m4 receptors can weakly couple to PLC (again by IAP-sensitive G-proteins) when expressed at high levels in certain cell types (Ashkenazi et al., 1987, 1989a; Tietje et al., 1990; Tietje and Nathanson, 1991; see below). Recent evidence indicates that pertussis toxin-insensitive activation of phospholipase C is mediated by activation of certain isozymes such as PLC$\beta$1 by G$\alpha$q, G$\alpha$11, G$\alpha$14, and G$\alpha$16, while IAP-sensitive activation is mediated by activation of isozymes such as PLC$\beta$2 by $\beta\gamma$ subunits (Katz et al., 1992, and references therein).

Muscarinic receptors play a key role in the functioning of the central nervous system, where they have been implicated in such processes as memory, learning, and control of movement. For example, administration of muscarinic agonists can facilitate the induction of long-term potentiation, (Blitzer et al., 1990; Burgard and Sarvey, 1990), and muscarinic antagonists interfere with learning and memory (Messer et al., 1990, 1991). Alzheimer's disease is characterized by a loss of cholinergic neurons in the nucleus basalis of Meynert. Administration of muscarinic agonists can cause Parkinson's disease-like tremors, and muscarinic antagonists can overcome tremors in Parkinson's disease. Several laboratories have reported altered number of muscarinic receptors in discrete brain regions in humans with Alzheimer's and Parkinson's diseases. In addition, muscarinic agonists can act as mitogens in cultured neural cells, and it has been suggested that muscarinic receptors may play important roles in the development and differentiation of the nervous system (Ashkenazi et al., 1989b; Gutkind et al., 1991).

Muscarinic receptors also modulate synaptic transmission in the ganglia of the autonomic nervous system, and play a major role in regulating the functions of the target organs of the parasympathetic nervous system. For example, stimulation of the parasympathetic ganglia causes the release of ACh which acts on cardiac mAChR. Activation of the mAChR in the heart causes a variety of physiological, biochemical, and electrophysiological responses: decrease in the rate (negative chronotropic effect) and force (negative inotropic effect) of contraction, inhibition of adenylyl cyclase activity, stimulation of phospholipase C and A2 activities, activation of an inward-rectifying potassium channel, inhibition of a calcium channel, and inhibition of the hyperpolarization-activated pacemaker current.

Our laboratory has been using the techniques of molecular and cellular biology to identify the factors which regulate the expression and function of mAChR in order to understand how cells regulate their ability to participate in cholinergic synaptic transmission. This chapter will review some of our findings on the regulation of mAChR expression and function when expressed from endogenous genes in cardiac cells and when expressed from heterologous genes in transfected cells in culture.

## Regulation of mAChR expression and function in chick heart cells

The embryonic chick heart is an attractive system for the study of cardiac mAChR. The development of the heart and its innervation and functional responsiveness have been extensively studied, the embryo is accessible to experimental manipulation, embryonic chick heart cells are particularly easy to grow in cell culture, and there are a number of readily determined functional responses mediated by the mAChR (see above). While the mammalian heart expresses a predominantly single subtype of mAChR (m2), embryonic chick heart has been shown to express the chick m2 (cm2), cm3, and cm4 receptors (Tietje et al., 1990; Tietje and Nathanson, 1991; Gadbut and Galper, 1994). Analysis of both mRNA levels using solution hybridization and protein levels using subtype-selective antibodies demonstrates that the predominate subtype in both chick atria and ventricles during both embryonic development and in posthatched chicks is the cm2 receptor. While the levels of cm2 and cm4 do not change significantly during embryonic development, there is a significant decrease in the expression of the cm3 receptor throughout embryonic development and in posthatched chicks (McKinnon and Nathanson, 1995). This decrease in the expression of cm3 correlates well with the previously described developmental decrease in mAChR-mediated activation of PLC (Orellana and Brown, 1985).

Treatment of chick heart cells in culture with the agonist carbachol leads to significant decreases in the levels of mRNA encoding the cm2 and cm4 receptors, with 6–8 h required to reach a new steady state level of mRNA (Habecker and Nathanson, 1992). These decreases in mRNA were not due to changes in the rate of mRNA degradation, suggesting that receptor activation regulates mRNA levels by affecting receptor gene transcription. The decrease in receptor mRNA levels had an important physiological consequence, as it regulated the rate at which mAChR number returned to control levels following removal of agonist.

Regulation of the activities of both the AC and PLC pathways is required for optimal regulation of mAChR mRNA levels. Treatment of cells with IAP, which blocks carbachol-induced inhibition of AC but does not affect carbachol-mediated activation of PLC, completely blocked the decreases in both cm2 and cm4 mRNAs. Treatment with the agonist pilocarpine, which allows full coupling to AC but does not cause activation of PLC, was much less effective than carbachol in regulation of receptor mRNA. Similarly, simultaneous treatment with heterologous agonists which individually activate only PLC or AC is required for significant regulation of mAChR mRNA levels. These results indicate that there is a complex requirement for multiple second messenger pathways in the regulation of mAChR gene expression (Habecker et al., 1993).

## Specificity of mAChR-G-protein coupling

We have used the regulation of expression of a luciferase reporter gene under the transcriptional control of a cAMP-regulated promoter to study the functional specificity of mAChR in transiently transfected cells (Migeon and Nathanson, 1994). When expressed at high levels, both the m1 and m4 receptors cause agonist-induced activation of the cAMP response element (CRE)-luciferase reporter gene. These increases are not due to increased levels of intracellular calcium or activation of protein kinase C. The increase in CRE-luciferase expression mediated by both the m1 and m4 receptors is not blocked by pertussis toxin pretreatment. Indeed, the m4-mediated increase is potentiated by pertussis toxin treatment, demonstrating that the increases in intracellular cAMP are not due to activation of AC by $\beta\gamma$ subunits released from members of the $G_i$ family of G-proteins. The most likely explanation for these effects is that at high levels of expression, both the m1 and m4 receptors can activate AC by coupling to the stimulatory protein $G_s$. Consistent with this hypothesis, co-expression of the m4 receptor and types I and III AC in stably transfected cells causes G-protein -dependent, IAP-insensitive stimulation of AC activity in membrane homogenates.

When expressed at low levels in transiently transfected JEG-3 cells, the m4 receptor exhibited little if any agonist-dependent inhibition of either basal or forskolin-stimulated CRE-luciferase expression. When cotransfected with either $G_{i\alpha-2}$, $G_{o\alpha A}$, or $G_{o\alpha B}$, the m4 receptor was able to mediate robust inhibition of AC. Cotransfection of the m4 receptor with $G_{i\alpha-1}$ or $G_{i\alpha-3}$ in contrast did not increase the ability of the m4 receptor to mediate inhibition. This indicates that the m4 receptor has a previously unrecognized high level of specificity for coupling to G-protein subtypes to mediate inhibition of AC. In contrast, the m2 receptor is able to couple to all three subtypes of $G_i$ as well as both forms of $G_o$ to mediate inhibition of CRE-luciferase expression (Migeon et al., 1995b). Thus, while the m2 and m4 receptors both inhibit AC activity, they have overlapping but distinct specificities for coupling to G-proteins to mediate this response (Fig. 1).

We have also used the CRE-luciferase reporter gene system to carry out structure-function-analyses of the G-proteins themselves (Migeon et al., 1995a). The validity of this method was demonstrated by the finding that transfection of a mutant $G_{i\alpha-2}$ containing the activating mutation Q205L resulted in inhibition of forskolin-stimulated luciferase expression, while transfection of wildtype $G_{i\alpha-2}$ was without effect. Transfection of a mutant $G_{i\alpha-2}$ containing a G43V mutation also mediated inhibition of luciferase expression. This suggests that this mutation, like the analogous mutation in p21$^{ras}$, confers constitutive activity on a heterotrimeric inhibitory G-protein. We also used

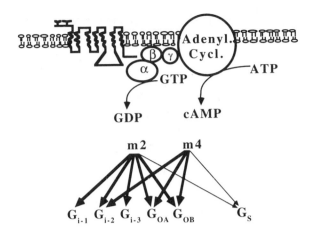

Fig. 1. Specificity of functional coupling of mAChR subtypes to G-proteins. The m2 and m4 receptors exhibit distinct but overlapping specificities for coupling to G-proteins to inhibit forskolin-stimulated CRE-luciferase expression in transfected JEG-3 cells.

the CRE-luciferase activity to demonstrate that the G2A mutation, which eliminates the C-terminal myristoylation site in $G_{i\alpha-2}$ and results in a loss of membrane association, blocked the inhibition of AC by both a constitutively activated Q205L $G_{i\alpha-2}$ and by wildtype $G_{i\alpha-2}$ activated by the m4 receptor.

## Conclusions and future directions

The CRE-luciferase assay has been a valuable tool for determining the specificity of mAChR-G-protein coupling and for structure-function analyses both of the mAChR and the G-proteins. We are currently testing the feasibility of modifying it for use to analyze mechanisms involved in receptor desensitization.

Multiple mechanisms exist for the regulation of mAChR expression and function. The regulation of mAChR gene expression in response to agonist exposure suggests a mechanism for the long-term regulation of postsynaptic responsiveness by varying levels of synaptic activity. We have recently isolated several mAChR genomic regions which contain the putative promoters for these receptor genes. These clones should allow more detailed analysis of the cellular and molecular mechanisms which regulate mAChR gene expression in response to a variety of physiological and developmental signals.

## References

Ashkenazi, A., Winslow, J.W., Peralta, E.G., Peterson, G.L., Schimerlik, M.I., Capon, D.J. and Ramachandran, J. (1987) A single M2 muscarinic receptor subtype coupled to both adenylyl cyclase and phosphoinositide turnover. *Science*, 238: 672–675.

Ashkenazi, A., Peralta, E.G., Winslow, J.W., Ramachandran, J. and Capon, D.J. (1989a) Functionally distinct G-proteins selectively couple different receptors to PI hydrolysis in the same cell. *Cell*, 56: 487–493.

Ashkenazi, A., Ramachandran, J. and Capon, D.J. (1989b) Acetylcholine analogue stimulates DNA synthesis in brain-derived cells via specific muscarinic receptor subtypes. *Nature*, 340: 146–150.

Blitzer, R.D., Gil, O. and Landau, E.M. (1990) Cholinergic stimulation enhances long-term potentiation in the CA1 region of rat hippocampus. *Neurosci. Lett.*, 119: 207–210.

Bonner, T.I. (1989) The molecular basis of muscarinic receptor diversity. *Trends Neurosci.*, 12: 148–151.

Burgard, E.C. and Sarvey, J.M. (1990) Muscarinic receptor activation facilitates the induction of long-term potentiation (LTP) in the rat dentate gyrus. *Neurosci. Lett.*, 116: 34–39.

Gadbut, A.P. and Galper, J.B. (1994) A novel $M_3$ muscarinic acetylcholine receptor is expressed in chick atrium and ventricle. *J. Biol. Chem.*, 269: 25823–25829.

Gutkind, J.S., Novotny, E.A., Brann, M.R., and Robbins, K.C. (1991) Muscarinic acetylcholine receptor subtypes as agonist-dependent oncogenes. *Proc. Natl. Acad. Sci. USA*, 88: 4703–4707.

Habecker, B.A. and Nathanson, N.M. (1992) Regulation of muscarinic acetylcholine receptor mRNA expression by activation of homologous and heterologous receptors. *Proc. Natl. Acad. Sci. USA*, 89: 5035–5038.

Habecker, B.A., Wang, H. and Nathanson, N.M. (1993) Multiple second messenger pathways mediate agonist regula-tion of muscarinic receptor mRNA expression. *Biochemistry*, 32: 4986–4990.

Katz, A., Wu, D. and Simon, M.I. (1992) Subunits $\beta\gamma$ of heterotrimeric G protein activate $\beta2$ isoform of phospholipase C. *Nature*, 360: 686–689.

McKinnon, L.A. and Nathanson, N.M. (1995) Tissue specific regulation of muscarinic acetylcholine receptor expression during embryonic development. *J. Biol. Chem.*, 270: 20636–20642

Messer, Jr., W.S., Bohnett, M. and Stibbe, J. (1990) Evidence for a preferential involvement of $M_1$ muscarinic receptors in representational memory. *Neurosci. Lett.*, 116: 184–189.

Messer, Jr., W.S., Stibbe, J.R. and Bohnett, M. (1991) Involvement of the septohippocampal system in representational memory. *Brain Res.*, 564: 66–72.

Migeon, J.C. and Nathanson, N.M. (1994) Differential regulation of cAMP-mediated gene transcription by m1 and m4 muscarinic acetylcholine receptors. Preferential coupling of m4 receptors to Gi $\alpha$-2. *J. Biol. Chem.*, 269: 9767–9773.

Migeon, J.C., Thomas, S. and Nathanson, N.M. (1995a) Regulation of cAMP-mediated gene expression by wildtype and mutant G-proteins: Inhibition of adenylyl cyclase by muscarinic receptor-activated and constitutively activated $G_{o\alpha}$. *J. Biol. Chem.*, 269: 129146–129152.

Migeon, J.C., Thomas, S. and Nathanson, N.M. (1995b) Differential coupling of m2 and m4 muscarinic receptors to inhibition of adenylyl cyclase by $G_{i\alpha}$ and $G_{o\alpha}$ subtypes. *J. Biol. Chem.*, 270: 16070–16074.

Orellana, S.A. and Brown, J.H. (1985) Stimulation of phosphoinositide hydrolysis and inhibition of cyclic AMP formation by muscarinic agonists in developing chick heart. *Biochem. Pharmacol.*, 34: 1321–1324.

Tietje, K.M. and Nathanson, N.M. (1991) Embryonic chick heart expresses multiple muscarinic acetylcholine receptor subtypes: Isolation and characterization of a gene encoding a novel m2 muscarinic acetylcholine receptor with a high affinity for pirenzipine. *J. Biol. Chem.*, 266: 17382–17387.

Tietje, K.M., Goldman, P.S. and Nathanson, N.M. (1990) Cloning and functional analysis of a gene encoding a novel muscarinic acetylcholine receptor expressed in chick heart and brain. *J. Biol. Chem.*, 265: 2828–2834.

J. Klein and K. Löffelholz (Eds.)
*Progress in Brain Research*, Vol. 109
© 1996 Elsevier Science B.V. All rights reserved.

CHAPTER 15

# The role of G-protein coupled receptor kinases in the regulation of muscarinic cholinergic receptors

M. Marlene Hosey[1], Shubhik K. DebBurman[1], Robin Pals-Rylaarsdam[1],
Ricardo M. Richardson[1] and Jeffrey L. Benovic[2]

[1]*Department of Molecular Pharmacology and Biological Chemistry, Northwestern University Medical School,
303 E. Chicago Avenue-S215, Chicago, IL 60611, USA and* [2]*Department of Pharmacology,
Jefferson Cancer Institute, Thomas Jefferson University, Philadelphia, PA 19107, USA*

## Introduction

The activation of G protein-coupled receptors (GPRs) by ligands results in multiple signalling events. While the events involved in the activation of the many types of GPRs and their effectors have been extensively investigated, the mechanisms that serve to terminate signalling through the GPRs are less well understood. In addition to termination events that include metabolism or removal of the ligands, evidence suggests that multiple intracellular events may occur to turn off signalling through GPRs (Liggett and Lefkowitz, 1994). These termination signals may come into play each time a GPR is activated, although they are more often thought of in terms of playing a role in the "desensitization" of receptors to pharmacological stimuli. However, it seems likely, at least in some systems, that what is often referred to as desensitization is an exaggeration of a process that occurs physiologically. For example, in the visual and olfactory systems, the rapid turn off of the GPRs that allows us to see and smell involves machinery that is very similar to what has been proposed to play a role in the "desensitization" of receptors for biogenic amines (Boekhoff et al., 1994; Liggett and Lefkowitz, 1994). In each of these systems, it appears that a family of unique protein kinases (Inglese et al., 1993; Premont et al., 1995) is involved in terminating the activated states of the receptors and returning them to the resting state so

that they may undergo another round of stimulation. In this chapter we concentrate on the events involved in termination of signalling through the muscarinic cholinergic receptors (mAChRs). Most of the studies have been performed with the m2 mAChR subtype, but recent advances with the m3 mAChR subtype suggest that similar pathways may also regulate other mAChR subtypes.

The term "desensitization", when applied to the GPRs, actually refers to a collection of events that may be independently regulated, and not necessarily sequential (Fig. 1). The hallmarks of desensitization of the GPRs include uncoupling of the receptors from G-proteins and/or loss of high affinity agonist binding. These events occur rapidly and appear to involve phosphorylation of the GPRs by specific and/or generic protein kinases. This phosphorylation induced uncoupling has been proposed to play a physiological role in terminating signalling in the sensory receptor systems (Dawson et al., 1993; Schleicher et al., 1993; Hargrave and Hamm, 1994). A second process that also appears to play a role in signal termination is sequestration or internalization of the receptors, in which the receptors enter an altered membrane environment (Fig. 1). This usually occurs on a slightly slower time course than uncoupling, but is reversible upon removal of the ligand. A third event is downregulation, or net loss of receptors from the cell, which occurs upon prolonged stimulation of the receptors and requires protein synthesis for recov-

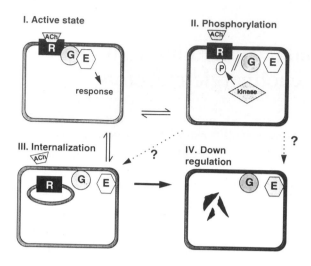

Fig. 1. Events associated with desensitization of G-protein coupled receptors. Desensitization of GPRs may involve several events. Whether these events occur independently, and/or in a sequential manner is not entirely resolved. (I) Active state. Activation of the receptors (R) by ligand is followed by activation of G-proteins (G) and effectors (E), leading to the biological response. (II) Phosphorylation and uncoupling. Following agonist activation of the receptors, GPRs may become phosphorylated (P) by one or more specific and/or generic serine/threonine protein kinases. This phosphorylation is believed to lead to uncoupling of the receptors from G-proteins and loss of high affinity agonist binding. These events appear to be responsible for the acute desensitization and loss of signalling that occurs subsequent to receptor activation (see text for details). These events occur rapidly and are readily reversible. (III) Internalization. The removal of receptors from their normal membrane environment, a process referred to as internalization or sequestration, also may play a role in signal termination. This process has a slower time course than uncoupling and is reversed upon removal of the ligand. Evidence presented in this chapter suggests that internalization may occur independently of receptor phosphorylation and the acute desensitization caused by the uncoupling of receptors from G-proteins and effectors. The role of receptor internalization in the function of the m2 mAChR remains unclear. (IV) Down regulation. The most extreme event associated with desensitization is the down-regulation, or net loss of receptors from the cell, which occurs upon prolonged stimulation of the receptors and requires protein synthesis for recovery to occur. It is not known whether phosphorylation or internalization is a necessary prerequisite for down regulation to occur.

ery to occur (Fig. 1). Desensitization can be homologous, whereby only the activated receptors are desensitized, or heterologous, whereby activation of one receptor system may produce signals that feedback on a heterologous receptor. This chapter will focus exclusively on molecular events underlying homologous desensitization of mAChR by a unique family of protein kinases that appear to specifically recognize GPRs. Other protein kinases, such as protein kinase C, may also be involved in mAChR desensitization (Orellana et al., 1985; Liles et al., 1986; El-Fakahany, 1988; Richardson and Hosey, 1990; Richardson et al., 1992). The role of these kinases is not discussed here and the reader is referred to previous reviews that address this issue (El-Fakahany, 1988; Hosey, 1994).

## Agonists induce rapid phosphorylation and desensitization of m2 mAChR in intact cells

Activation of the m2 mAChRs with agonists has been shown to result in the rapid phosphorylation of these receptors in intact cells (Kwatra and Hosey, 1986; Kwatra et al., 1987, 1989a; Richardson and Hosey, 1992; Pals-Rylaarsdam et al., 1995). This has been demonstrated by incubating preparations in media containing $^{32}P$ and isolating the receptors from control or agonist-treated samples by affinity chromatography (Kwatra and Hosey, 1986; Kwatra et al., 1987, 1989a; Richardson and Hosey, 1992) or immunoprecipitation (Pals-Rylaarsdam et al., 1995). In preparations from avian and mammalian heart and in cultured insect and mammalian cells heterologously expressing the human m2 mAChR, agonists induce phosphorylation of the mAChR on serine and threonine residues to a stoichiometry of 3–5 mol phosphate/mol receptor (Kwatra et al., 1987, 1989a; Richardson and Hosey, 1992; Pals-Rylaarsdam et al., 1995). Most or all of the modified residues are likely to reside in the third intracellular loop of the receptors as this is the only location of intracellular serines, on which 60–70% of the phosphorylation occurs (Kwatra et al., 1987, 1989a). In each of the cases tested, the agonist-induced phosphorylation of the receptors was correlated with events linked to "desensitization" of the receptors. In the chick heart, conditions that led to phosphorylation of the receptors resulted in the

**Activated receptor    Desensitized receptor**

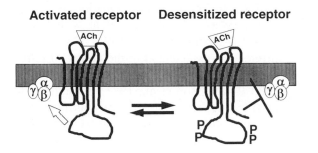

Fig. 2. Two states of m2 mAChRs. Evidence presented in the text suggests that the m2 mAChR can exist in two states in the presence of agonist: an activated state and a desensitized state. Activation is driven by agonist binding and leads to coupling of the receptor to the heterotrimeric G protein ($\alpha\beta\gamma$). The desensitized state occurs when the agonist occupied receptors become phosphorylated on the third intracellular loop of the receptor. This phosphorylation prevents coupling of the receptor to the G proteins and results in loss of high affinity agonist binding.

loss of high affinity agonist binding, and a right-shift in the dose response curve for carbachol to induce a negative inotropic response (Kwatra et al., 1987). In insect cells expressing the human m2 mAChR from recombinant baculovirus, agonist treatments that induced phosphorylation of the receptor also resulted in the loss of the ability of the receptor to activate endogenous insect cell G proteins (Richardson and Hosey, 1992). Thus, these intact cell results strongly suggested that a link might exist between agonist-induced phosphorylation and desensitization of the mAChR. This suggests a model whereby the receptors can exist in at least two states: an activated state where they activate G-proteins and effectors, and a "desensitized" state where phosphorylation of the receptors leads to uncoupling of the receptors from G-proteins and a consequent loss of high affinity binding (Fig. 2).

## Mechanisms of agonist-induced phosphorylation of the mAChR

An important goal is to elucidate the events that are involved in the phosphorylation and desensitization of the mAChRs. Early experiments in intact cells that attempted to identify the molecular events associated with agonist-induced phosphorylation of the mAChRs suggested that agonist occupancy of the receptors was a strict requirement for receptor phosphorylation (Kwatra et al., 1987). In the absence of agonists, no phosphorylation of the receptors occurred when modulators of various signalling pathways were tested, and activation of other GPRs (e.g., adenosine receptors) in the same cells also did not lead to phosphorylation of the mAChRs (Kwatra et al., 1987). That only the agonist activated mAChRs could serve as substrates for phosphorylation recalled the situation that had been described for the light receptor rhodopsin and the $\beta_2$-adrenergic receptors ($\beta_2$ARs), where unique protein kinases are known to phosphorylate the receptors in a manner strictly dependent on agonist occupancy (Palczewski and Benovic, 1991; Premont et al., 1995). Rhodopsin kinase phosphorylates and inactivates light (agonist)-activated rhodopsin, while it does not recognize dark (unactivated) rhodopsin (Shichi and Somers, 1978; Kelleher and Johnson, 1990). Similarly a "$\beta$-adrenergic receptor kinase" ($\beta$ARK) phosphorylates $\beta_2$ARs in an agonist-dependent manner and is inactive towards the non-activated $\beta_2$ARs (Benovic et al., 1986, 1987). Although rhodopsin kinase is expressed primarily in the retina (Lorenz et al., 1991), and is likely to be a true rhodopsin kinase under physiological conditions, $\beta$ARK activity is found in most cells and is likely to have a more wide spread substrate specificity (Palczewski and Benovic, 1991; Premont et al., 1995).

Our initial attempts to identify the kinases that phosphorylate the mAChRs in vivo employed a system whereby we attempted to take apart the system and reconstitute agonist-dependent phosphorylation and desensitization of the receptors in vitro. For this system we purified chick heart or recombinant human m2 mAChRs and reconstituted the receptors into phospholipid vesicles. We tested the receptors as substrates for purified protein kinases with [$\gamma$-$^{32}$P]ATP in the presence or absence of agonists and antagonists. To test the possibility that the m2 mAChRs might be targets of $\beta$ARK or a similar enzyme, the chick heart m2 mAChRs were tested as substrates in vitro for βARK purified from bovine brain. The m2 receptors were found to be substrates for this enzyme preparation (Kwatra et al., 1989b) and the properties of the phosphorylation had striking similarities to what we had observed in vivo: first, the receptors were phosphorylated by the bovine brain βARK strictly in an agonist-dependent manner; second, the stoichiometry of phosphorylation and distribution of phosphates on serine and threonine residues was nearly identical to that observed in vivo. These data provided the first evidence to suggest a potential role of receptor specific kinases in the regulation of the mAChRs.

It is now appreciated that rhodopsin kinase and the original βARK belong to a family of related kinases which are referred to as the G protein-coupled receptor kinases (GRKs) (Palczewski and Benovic, 1991; Inglese et al., 1993; Premont et al., 1995). The family members include: rhodopsin kinase or GRK1; "βARK1" and "βARK2" or GRK2 and GRK3; and GRK4, GRK5 and GRK6. The GRK terminology is preferred because available data concerning substrate specificity and the relatively widespread localization of most of the GRKs make it likely that these kinases will have multiple substrates in the GPR family. In particular the enzymes formerly referred to as βARK1 and βARK2 are now known to recognize multiple GPRs as substrates, at least in vitro (Inglese et al., 1993). The GRKs each have a central catalytic domain, N-terminal domains which may contain substrate recognition sites, and C-terminal domains which are involved in the regulation and potential membrane targeting of several of the isoforms (Premont et al., 1995).

Since there are potentially hundreds of GPRs and only a handful of GRKs, it is important to determine which GRK pairs with which GPRs. The original βARK used in the experiments described above may have contained other GRK isoforms as it was purified from bovine brain. To further probe the role of GRKs in the regulation of mAChR, we have tested the ability of several expressed and purified recombinant GRKs to phosphorylate the m2 and m3 mAChR in vitro. Using the purified and reconstituted human m2 mAChR as a substrate, we observed agonist-dependent phosphory-

lation of the receptors with the kinetics and extent of phosphorylation being equivalent with GRK2 and GRK3 (Richardson et al., 1993). No discernible difference was detected in the ability of either of these enzymes to phosphorylate the m2 mAChRs (Richardson et al., 1993). In addition to recognizing the reconstituted and purified hm2 mAChRs as substrates, we have also recently demonstrated that GRK2 and GRK3 phosphorylate the membrane-bound hm2 mAChR when expressed in Sf9 insect cells (DebBurman et al., 1995a). Phosphorylation occurred in a manner that was dependent on the concentration of agonist used, with the $EC_{50}$ being approximately 10 $\mu$M for carbachol (DebBurman et al., 1995a), which is similar to the $EC_{50}$ for carbachol to induce activation of some responses in intact cells. In contrast, the hm2 mAChRs were poorer substrates for GRK5 (Kunapuli et al., 1994a; DebBurman et al., 1995a), and even poorer substrates for GRK6 (Loudon and Benovic, 1994; DebBurman et al., 1995a).

We recently extended our analyses to include the human m3 mAChRs, which have been shown to undergo agonist-dependent phosphorylation in intact cells (Tobin et al., 1993). The hm3 mAChRs were expressed in insect Sf9 cells from a recombinant baculovirus and the membrane bound receptors were tested as substrates for GRK 2, 3, 5 and 6. Both GRK2 and GRK3 phosphorylated the m3 mAChR in an agonist-dependent manner; the $EC_{50}$ values for carbachol were ~1 $\mu$M for both kinases and the stoichiometry of phosphorylation was approximately 2 mol P/mol protein (DebBurman et al., 1995a). In contrast, neither GRK5 nor GRK6 showed reactivity towards the hm3 mAChR (DebBurman et al., 1995a). Thus, GRK2 and GRK3 appear to be candidates as regulators of both the hm2 and hm3 mAChRs, whereas roles for GRK5 and GRK6 appear less likely. GRK4 has not been tested, and GRK1 has been previously demonstrated not to recognize the hm2 mAChR (Haga and Haga, 1992). Very little is known about the role of phosphorylation in regulating the m1, m4 and m5 mAChR subtypes.

The data discussed above suggest that the GRKs may play a role in the phosphorylation and desensitization of mAChRs. However, the results discussed so far have been obtained in in vitro studies and evidence from intact cell studies has been difficult to obtain. This is in part because there are no specific inhibitors of the GRKs that can be used to implicate a role for these kinases in vivo. To address the role of the GRKs in intact cells we have used a dominant negative approach. A single point mutation was made in GRK2 in which lysine 220 was substituted by an arginine (Kong et al., 1994). This GRK2$^{K220R}$ was demonstrated to be unable to catalyze the phosphotransfer reaction and effectively inhibited wild type GRK2 activity in vitro (Kong et al., 1994). Thus the properties of GRK2$^{K220R}$ are consistent with it being able to act as a dominant negative mutant. We transiently expressed the GRK2$^{K220R}$ mutant in HEK cells and demonstrated that the construct was expressed approximately 20–40-fold higher than the endogenous GRKs (Pals-Rylaarsdam et al., 1995). Cells expressing the dominant negative GRK2$^{K220R}$ and the hm2 mAChR were tested for agonist induced phosphorylation of the hm2 mAChR. The expression of the dominant negative GRK2$^{K220R}$ resulted in a 50% reduction in the ability of carbachol to induce phosphorylation of the receptors (Fig. 3).

With the dominant negative kinase as a tool for in vivo experiments, we were able to directly test for the role of phosphorylation in events associated with desensitization (Pals-Rylaarsdam et al., 1995). In control cells (expressing wild type GRKs) the expressed hm2 mAChRs effectively reduced luteinizing hormone (LH)-receptor stimulated levels of cAMP. In addition, in these control cells, pretreatment of the cells with carbachol, under conditions which we showed led to phosphorylation of the receptors, led to an almost complete desensitization of the receptors (Fig. 3). The carbachol pretreatment abolished the ability of carbachol to attenuate the LH mediated increase in cAMP. In marked contrast, in cells expressing the dominant negative GRK2$^{K220R}$, the receptors did not significantly desensitize in response to a carbachol pretreatment (Fig. 3); in these cells the ability of carbachol to reduce the LH mediated rise in cAMP was identical in naive cells and in cells

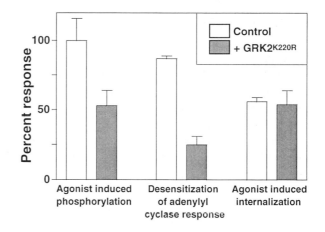

Fig. 3. Effects of the dominant negative allele of GRK2 on phosphorylation, desensitization and internalization of m2 mAChRs expressed in HEK cells. The bar graph summarizes the data discussed in the text concerning the effects of the dominant negative GRK2$^{K220R}$ on m2 mAChRs in transiently transfected HEK cells. Results from cells expressing the GRK2$^{K220R}$ allele are shown in the shaded bars, while those obtained with control cells are in empty bars. Results shown are means ± SEM from 3–4 experiments. In control cells, agonist induced a marked increase (3 fold) in the phosphorylation of the receptors. The maximal response in these cells was defined as 100%. In contrast, in cells expressing GRK2$^{K220R}$, agonist-induced receptor phosphorylation was decreased by ~50% compared to control cells (left bars). The desensitization of the m2 mAChR's ability to attenuate adenylyl cyclase was also markedly decreased by the expression of GRK2$^{K220R}$ (center bars). The percent desensitization was calculated by measuring the ability of carbachol to attenuate the LHR-mediated increase in cAMP in control (naive) cells vs. cells pretreated under desensitizing conditions with carbachol according to the following formula: [(% decrease in cAMP in control cells − % decrease in cAMP in desensitized cells)/% decrease in cAMP in control cells]. The expression of the GRK2$^{K220R}$ allele did not affect the rate or extent of agonist induced internalization, which was measured as an agonist-induced loss of cell surface binding sites for the hydrophilic ligand [$^3$H]$N$-methyl scopolamine. Results shown are % internalization after a 60 min exposure to agonist in either the control cells or cells expressing GRK2$^{K220R}$ (see text for details).

pretreated with carbachol. Thus, the expression of the dominant negative GRK2$^{K220R}$ reduced the agonist dependent phosphorylation of the receptors by ~50% and abolished the short term desensitization of the receptor under the same conditions. We interpret these results to indicate that GRKs play a role in the phosphorylation and desensitization of the receptors in intact cells.

We also tested the ability of the dominant negative GRK2$^{K220R}$ and overexpressed wild type GRK2 to modulate the process of receptor sequestration (Pals-Rylaarsdam et al., 1995). Using a hydrophilic ligand to monitor the disappearance of the receptors from the cell surface, we observed that agonists induced a time dependent sequestration of the receptors. However, the rate and extent of sequestration of the hm2 mAChRs in response to agonist was the same in cells expressing endogenous GRKs and cells overexpressing either wild type GRK2 (data not shown) or the dominant negative GRK2$^{K220R}$ (Fig. 3). Thus GRKs do not appear to play an essential role in the process of receptor sequestration. The consequence of mAChR phosphorylation by GRKs appears to be to acutely desensitize the receptor's ability to attenuate adenylyl cyclase without affecting sequestration. The functional role of receptor sequestration remains to be elucidated. In the $\beta$-adrenergic receptor system, sequestration has been suggested to play a role in the resensitization process (Lohse et al., 1990; Yu et al., 1993; Pippig et al., 1995).

## Regulation of GRK activity by lipids and G-proteins

Originally, the GRKs were considered to be constitutively active cytosolic proteins which would recognize their receptor substrates upon an agonist-induced conformational change in the receptors. However, more recent observations suggest that the GRKs may interact with membranes in multiple ways and that the activity of GRKs may be regulated in a complex manner by a variety of agents. Several GRKs may interact with membranes via lipid modifications on the kinases, including GRK1, which is isoprenylated (Inglese et al., 1992a,b), and GRK6, which is palmitoylated (Stoffel et al., 1994). Other GRKs, GRK2, GRK3 and GRK5 bind membrane phospholipids (Kunapuli et al., 1994b; DebBurman et al., 1995b). This may facilitate the interactions of the kinases

with their substrates. In addition, the lipids may directly modulate the activity of the kinases by complex pathways (see below). Other factors may also regulate the activity of the kinases, including the GPRs themselves (Palczewski et al., 1991; Chen et al., 1993; Haga et al., 1994; Kameyama et al., 1994; Shi et al., 1995), G-proteins (Haga and Haga, 1992; Pitcher et al., 1992; Koch et al., 1993; Richardson et al., 1993), as well as Ca$^{2+}$ binding proteins known as recoverins (Kawamura, 1993; Gorodovikova et al., 1994; Klenchin et al., 1995). Here we limit our comments to how lipids and G-proteins may synergistically regulate the activity of GRK2 and 3.

The $\beta\gamma$ subunits of G proteins were first demonstrated to augment the phosphorylation of GPRs by Haga and Haga (1990). Subsequently a G$_{\beta\gamma}$ binding domain was demonstrated to exist in the carboxyl tail of GRK2 and GRK3 (Pitcher et al., 1992; Koch et al., 1993). In reconstitution assays, we have observed that G$_{\beta\gamma}$ increases the extent of phosphorylation of the hm2 mAChRs by either GRK2 or GRK3 by two-fold (Richardson et al., 1993; DebBurman et al., 1995a,b). More recently, we observed that certain phospholipids increased the phosphorylation of the hm2 mAChRs in a manner that recalled the effects of the G proteins. In addition, other lipids decreased the phosphorylation of the hm2 mAChR. For example, phosphatidylserine (PS) increased and phosphatidylinositol 4,5-bisphosphate (PIP$_2$) decreased the phosphorylation of the receptors under certain conditions (DebBurman et al., 1995b). These effects of the lipids depend on the composition of the phospholipid vesicles used for the reconstitution of the receptors. If a "neutral" (non-activating or non-inhibiting) lipid is used for the reconstitution, PIP$_2$ also stimulates phosphorylation at concentrations lower than those at which it exerts inhibition (DebBurman and Hosey, unpublished observations). Interestingly, the effects of G$_{\beta\gamma}$ and certain lipids can be synergistic. At early time points in the reactions, the effects of PS and PIP$_2$ are synergistic with G$_{\beta\gamma}$, while at later time points the effects are non-additive (DebBurman et al., 1995b; and unpublished observations). The synergistic effects are mainly due to an increase in the rate of phosphorylation of the receptor. We used a GST fusion protein of the C-terminal domain of GRK2 to block the actions of both the phospholipids and $G_{\beta\gamma}$ on kinase activity. These data suggested that the site of interaction of the phospholipids with the GRKs is in the C-terminal domain of the kinase, most likely in a PH (pleckstrin homology) domain, which partly overlaps the $G_{\beta\gamma}$ binding domain (Musacchio et al., 1993; Gibson et al., 1994; Pawson, 1995). While PH domains in general are still poorly understood, the solution structure of pleckstrin suggested that PH domains might be phospholipid binding domains (Harlan et al., 1994;

Yoon et al., 1994). Conceivably the PH domain in GRK2 and GRK3 might play a role in localizing the kinases to membranes and facilitating their interactions with other activators such as the $G_{\beta\gamma}$ subunits and the GPRs.

## Model for the regulation of mAChR activity by GRKs

Taken together the results summarized here allow us to begin to develop a model to explain the regulation and "inactivation" of the mAChRs (Fig. 4). In the resting state the receptors, G proteins and

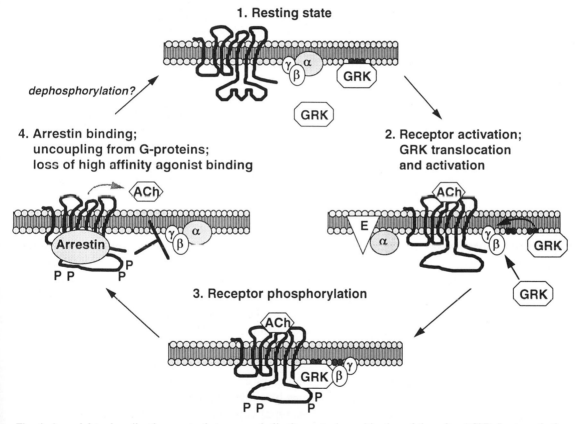

Fig. 4. A model to describe the events that may underlie the acute desensitization of the m2 mAChR. In stage 1, the resting state, the receptor is unliganded, the heterotrimeric G-proteins ($\alpha\beta\gamma$) are inactive (presumably in the GDP liganded state) and the G protein-coupled receptor kinase(s) may be partly membrane-bound through an association with membrane lipids, and/or located in the cytosol. In stage 2, receptor activation by acetylcholine (ACh) leads to dissociation of the G protein $\alpha$ and $\beta\gamma$ subunits and activation of effectors (E). One effector is the GRK, which may translocate to the plasma membrane and become activated in a synergistic manner by $G_{\beta\gamma}$ and phospholipids. In stage 3, agonist dependent receptor phosphorylation by GRK occurs leading to the incorporation of 3–5 mol of phosphate onto sites in the third intracellular domain of the receptor. In stage 4, arrestin-type proteins bind to the phosphorylated receptors and prevent further receptor-G protein coupling and consequently, loss of high affinity agonist binding. A return to the resting state requires dephosphorylation of the receptor and dissociation of arrestin. Details are discussed in the text.

GRKs are in a non-activated state. The GRK may be membrane localized via an interaction with membrane phospholipids or may be cytosolic. Upon agonist activation of the receptor, the receptors are likely to undergo a conformational change that allows for activation and dissociation of G proteins. The GRKs might be recruited to the membranes or locale of the receptors by the $G_{\beta\gamma}$ subunits, and/or lipid localized GRK might be activated by $G_{\beta\gamma}$. The GRK could then phosphorylate the receptor as long as agonist was bound. Once the mAChR is phosphorylated, another protein known as arrestin is likely to bind to the receptor. Arrestin binding to GPRs is thought to prevent interaction of the receptors with G-proteins (Lohse et al., 1990, 1992; Attramadal et al., 1992; Hargrave and Hamm, 1994; Liggett and Lefkowitz, 1994). Thus arrestin binding brings about the ultimate step in the desensitization process, i.e., uncoupling of the receptors and G proteins. A return to the resting state would require dephosphorylation of the receptors and dissociation of the arrestins.

There are two attractive features of this model that allow for specificity of the receptor inactivation mechanisms. One is that the GRKs only recognize the agonist activated receptors; therefore, non-activated receptors cannot be inactivated. This allows for specifically turning off the receptors that are "turned on". Second, arrestins only bind the phosphorylated receptors (Lohse et al., 1990; Attramadal et al., 1992; Gurevich and Benovic, 1992, 1993). Thus arrestins cannot cause uncoupling of non-activated or non-phosphorylated receptors, so uncoupling of receptor/G-protein interaction is not randomly generated.

Despite these attractive aspects of specificity, this model generates many questions and there are various aspects of it that will require further investigation. For example, what really contributes to GRK activation in an intact cell? Do $G_{\beta\gamma}$ subunits and phospholipids synergistically regulate the GRKs in intact cells? This is poorly understood; in particular, it is unclear whether the roles of lipids and $G_{\beta\gamma}$ subunits are simply to recruit the kinases to the membranes, or whether there is an actual activation of the kinases by these modulators.

Much of the GRK2 and GRK3 activity is cytosolic, and agonists have been shown to cause GRK translocation (Strasser et al., 1986; Mayor et al., 1987; Inglese et al., 1992a,b). However, several studies also have demonstrated membrane localized GRK2 or 3 in "resting" cells (Garcia-Higuera et al., 1994; Premont et al., 1995). Another question of importance is: does the agonist activated receptor contribute to activation of the GRKs, and if so, how? The GRKs are very selective in terms of substrates, and substrates such as histones and caseins which are normally used by other kinases, are very poor substrates for GRKs. Is it that the GRKs are very selective in their substrate specificity because the GPRs need to activate the kinases to allow the kinases to phosphorylate the receptors? Data from several studies suggest this may be the case (Palczewski et al., 1991; Chen et al., 1993; Haga et al., 1994; Kameyama et al., 1994; Shi et al., 1995).

Another largely unexplored area concerns the arrestins. Arrestins also are predicted to be cytosolic proteins, and we have shown that arrestins can bind with high affinity to agonist activated and phosphorylated m2 mAChRs in vitro (Gurevich et al., 1993, 1995), but we know very little about the functional effects of these interactions and how and if arrestin binding uncouples the mAChRs from G proteins. Other questions abound. What regulates arrestin activity? What turns off the GRKs? Are proteins like recoverins involved in GRK turn off, and/or are there other proteins, lipids, or other factors that might inhibit the GRKs in the resting state? How is the inactivated receptor reactivated? Does dephosphorylation allow arrestin release or vice-versa? These are the questions and challenges for the future that will ultimately allow refinement of our understanding of the regulation and inactivation of the mAChRs.

## Acknowledgements

The studies discussed in this chapter were supported by grants from the NIH (HL31601 and HL50121 to MMH; GM44944, GM45964 and GM47417 to JLB) and the American Heart Association (MMH). SKDB was the recipient of a predoctoral fellowship from the Pharmaceutical Research and Manufacturers of America Foundation, and RPR was the recipient of a predoctoral fellowship from the Howard Hughes Medical Institute.

# References

Attramadal, H., Arriza, J.L., Aoki, C., Dawson, T.M., Codina, J., Kwatra, M.M., Snyder, S.H., Caron, M.G. and Lefkowitz, R.J. (1992) $\beta$-arrestin2, a novel member of the Arrestin/$\beta$-arrestin gene family. *J. Biol. Chem.*, 267: 17882–17890.

Benovic, J.L., Strasser, R.H., Caron, M.G. and Lefkowitz, R.J. (1986) $\beta$-adrenergic receptor kinase: Identification of a novel protein kinase that phosphorylates the agonist-occupied form of the receptor. *Proc. Natl. Acad. Sci. USA*, 83, 2797–2801.

Benovic, J.L., Mayor, Jr., F., Staniszewski, C., Lefkowitz, R.J. and Caron, M.G. (1987) Purification and characterization of $\beta$-adrenergic receptor kinase. *J. Biol. Chem.*, 262: 9026–9032.

Boekhoff, I., Inglese, J., Scleicher, S., Koch, W.J., Lefkowitz, R.J. and Breer, H. (1994) Olfactory desensitization requires membrane-targetting of receptor kinase mediated by $\beta\gamma$-subunits of heterotrimeric G proteins. *J. Biol. Chem.*, 269: 37–40.

Chen, C.-Y., Dion, S.B., Kim, C.M. and Benovic, J.L. (1993) $\beta$-adrenergic receptor kinase: agonist-dependent receptor binding promotes kinase activation. *J. Biol. Chem.*, 267: 7825–7831.

Dawson, T.M., Arriza, J.L., Jaworsky, D.E., Borisy, F.F., Attramadal, H., Lefkowitz, R.J. and Ronnett, G.V. (1993) $\beta$-adrenergic receptor kinase-2 and $\beta$-arrestin 2 as mediators of odorant induced desensitization. *Science*, 259: 825–829.

DebBurman, S.K., Kunapuli, P., Benovic, J.L. and Hosey, M.M. (1995a) Agonist-dependent phosphorylation of human muscarinic receptors in Spodoptera frugiperda insect cell membranes by G protein-coupled receptor kinases. *Mol. Pharmacol.*, 47: 224–233.

DebBurman, S.K., Ptasienski, J., Boetticher, E., Lomasney, J.L., Benovic, J.L. and Hosey, M.M. (1995b) Lipid-mediated regulation of G protein-coupled receptor kinases 2 and 3. *J. Biol. Chem.*, 270: 5742–5747.

El-Fakahany, E.E., Alger, B.E., Lai, W.S., Pitler, T.A., Worley, P.F. and Baraban, J.M. (1988) Neuronal muscarinic responses: role of protein kinase C. *FASEB J.*, 2: 2575–2583.

Gibson, T.J., Hyvonen, M., Musacchio, A. and Saraste, M. (1994) PH domain: the first anniversary. *Trends Biochem. Sci.*, 19: 349–352.

Gorodovikova, E.N., Gimelbrant, A.A., Senin, I.I. and Philippov, P.P. (1994). Recoverin mediates the calcium effect upon rhodopsin phosphorylation and cGMP hydrolysis in bovine retina rod cells. *FEBS Lett.*, 349: 187–190.

Gurevich, V.V. and Benovic, J.L. (1992) Cell-free expression of visual arrestin: truncation mutagenesis identifies multiple domains involved with rhodopsin interaction. *J. Biol. Chem.*, 267: 21919–21923.

Gurevich, V.V. and Benovic, J.L. (1993) Visual arrestin interaction with rhodopsin: Sequential multisite binding ensures strict selectivity toward light-activated phosphorylated rhodopsin. *J. Biol. Chem.*, 268; 11628–11638.

Gurevich, V.V., Richardson, R.M., Kim, C.M. and Hosey, M.M. (1993) Binding of wild type and chimeric arrestins to the m2 muscarinic cholinergic receptor. *J. Biol. Chem.*, 268: 16879–82.

Gurevich, V.V., Dion, C.M., Onorato, J.J., Ptasienski, J., Kim, C.M., Sterne-Marr, R., Hosey, M.M. and Benovic, J.L. (1995) Arrestin interactions with G protein-coupled receptors: direct binding studies of wild type and mutant arrestins with rhodopsin, $\beta_2$-adrenergic and m2 muscarinic cholinergic receptors. *J. Biol. Chem.*, 270: 720–731.

Haga, K. and Haga, T. (1990) Dual regulation by G proteins of agonist-dependent phosphorylation of muscarinic acetylcholine receptors. *FEBS Lett.*, 268: 43–47.

Haga, K. and Haga, T. (1992) Activation of G protein $\beta\gamma$ subunits of agonist- and light-induced phosphorylation of muscarinic acetylcholine receptors and rhodopsin. *J. Biol. Chem.*, 267: 2222–2227.

Haga, K., Kameyama, K. and Haga, T. (1994) Synergistic activation of a G protein-coupled receptor kinase by G protein $\beta\gamma$ subunits and Mastoparan or related peptides. *J. Biol. Chem.*, 269: 12594–12599.

Hargrave, P. and Hamm, H.E. (1994) Regulation of visual transduction. In: D.R. Sibley and M.D. Houslay (Eds.), *Regulation of Signal Transduction Pathways by Desensitization and Amplification*. Wiley, New York, pp. 25–67.

Harlan, J.E., Hajduk, P.J., Yoon, H.S. and Fesik, S.W. (1994) Pleckstrin homology domains bind to phosphatidylinositol-4,5-bisphosphate. *Nature*, 371: 168–170.

Hosey, M.M. (1994) Desensitization of Muscarinic cholinergic receptors and the role of protein phosphorylation. In: D.R. Sibley and M.D. Houslay (Eds.), *Regulation of Signal Transduction Pathways by Desensitization and Amplification*. Wiley, New York, pp. 113–128.

Inglese, J., Glickman, J.F., Lorenz, W., Caron, M.G. and Lefkowitz, R.J. (1992a) Isoprenylation of a protein kinase: requirement of farnesylation/$\alpha$-carboxyl methylation for full enzymatic activity of rhodopsin kinase. *J. Biol. Chem.* 267: 1422–1425.

Inglese, J., Koch, W.J., Caron, M.G. and Lefkowitz, R.J. (1992b) Isoprenylation in regulation of signal transduction by G protein-coupled receptor kinases. *Nature*, 359: 147–150.

Inglese, J., Freedman, N.J., Koch, W.J. and Lefkowitz, R.J. (1993) Structure and mechanism of the G protein-coupled receptor kinases. *J. Biol. Chem.*, 268: 23735–23738.

Kameyama, K., Haga, K., Haga, T., Moro, O. and Sadee, W. (1994) Activation of GTP binding protein and a GTP-binding-protein-coupled receptor kinase ($\beta$-adrenergic receptor kinase 1) by a muscarinic receptor m2 mutant lacking phosphorylation sites. *Eur. J. Biochem.*, 226: 267–276.

178

Kawamura, S. (1993) Rhodopsin phosphorylation as a mechanism of cyclic GMP phosphodiesterase regulation by S-modulin. *Nature*, 362: 855–857.

Kelleher, D.J. and Johnson, G.J. (1990) Characterization of rhodopsin kinase purified from bovine rod outer segments. *J. Biol. Chem.*, 265: 2632–2639.

Klenchin, V.A., Calvert, P.D. and Bownds, M.D. (1995) Inhibition of rhodopsin kinase by recoverin: Further evidence for a negative feedback system in phototransduction. *J. Biol. Chem.*, 270: 16147–16152.

Koch, W.J., Inglese, J., Stone, W.C. and Lefkowitz, R.J. (1993) The binding site for the $\beta\gamma$ subunits of heterotrimeric G proteins on the $\beta$-adrenergic receptor kinase. *J. Biol. Chem.*, 268: 680–685.

Kong, G., Penn, R. and Benovic, J.L. (1994) A $\beta$-adrenergic receptor kinase dominant negative mutant attenuates desensitization of the $\beta_2$-adrenergic receptor. *J. Biol. Chem.*, 269: 13084–13087.

Kunapuli, P., Onorato, J.J., Hosey, M.M. and Benovic, J.L. (1994a) Expression, purification and characterization of G protein-coupled receptor kinase GRK5. *J. Biol. Chem.*, 269: 1099–1105.

Kunapuli, P., Gurevich, V.V. and Benovic, J.L. (1994b) Phospholipid-stimulated autophosphorylation activates G protein-coupled receptor kinase GRK5. *J. Biol. Chem.*, 269: 10109–10212.

Kwatra, M.M. and Hosey, M.M. (1986) Phosphorylation of the cardiac muscarinic receptor in the intact chick heart and its regulation by a muscarinic agonist. *J. Biol. Chem.*, 261: 12429–12432.

Kwatra, M.M., Leung, E., Maan, A.C., McMahon, K.K., Ptasienski, J., Green, R. and Hosey, M.M. (1987) Correlation of agonist-induced phosphorylation of chick heart muscarinic receptors with receptor desensitization. *J. Biol. Chem.*, 262: 16314–16321.

Kwatra, M.M., Ptasienski, J. and Hosey, M.M.(1989a) The porcine heart m2 muscarinic receptor: agonist-induced phosphorylation and comparison of properties with chick heart receptor. *Mol. Pharmacol.*, 35: 553–558.

Kwatra, M.M., Benovic, J.L., Caron, M.G., Lefkowitz, R.J. and Hosey, M.M. (1989b). Phosphorylation of the chick heart muscarinic receptor with the $\beta$-adrenergic receptor kinase. *Biochemistry*, 28: 4543–4547.

Liggett, S.B. and Lefkowitz, R.J. (1994) Adrenergic receptor-coupled adenylyl cyclase systems: regulation of receptor function by phosphorylation, sequestration and down regulation. In: D.R. Sibley and M.D. Houslay (Eds.), *Regulation of Signal Transduction Pathways by Desensitization and Amplification*. Wiley, New York, pp. 71–98.

Liles, W.C., Hunter, D.D., Meier, K.E. and Nathanson, N.M. (1986) Activation of protein kinase C induces rapid internalization and subsequent degradation of muscarinic acetylcholine receptors in neuroblastoma cells. *J. Biol. Chem.*, 261: 5307–5313.

Lohse, M.J., Benovic, J.L., Caron, M.G. and Lefkowitz, R.J. (1990) Multiple pathways of rapid $\beta_2$-adrenergic receptor desensitization: delineation with specific inhibitors. *J. Biol. Chem.*, 265: 3202–3209.

Lohse, M.J., Andexinger, S., Pitcher, J.A., Trukawinski, S., Codina, J., Faure, J.-P., Caron, M.G. and Lefkowitz, R.J. (1992) Receptor-specific desensitization with purified proteins: kinase dependence and receptor specificity of $\beta$-arrestin and arrestin in the $\beta_2$-adrenergic receptor and rhodopsin systems. *J. Biol. Chem.*, 267: 8558–8564.

Lorenz, W.J., Inglese, J., Palczewski, K., Onorato, J.J., Caron, M.G. and Lefkowitz, R.J. (1991) The receptor kinase family: Primary structure of the rhodopsin kinase family reveals similarities to the $\beta$-adrenergic receptor kinase. *Proc. Natl. Acad. Sci. USA*, 88: 8715–8719.

Loudon, R.P. and Benovic, J.L. (1994) Expression, purification and characterization of G protein-coupled receptor kinase GRK6. *J. Biol. Chem.*, 269: 22691–22697.

Mayor, Jr., F., Benovic, J.L., Caron, M.G. and Lefkowitz, R.J. (1987). Somatostatin induces translocation of the $\beta$-adrenergic receptor kinase and desensitizes somatostatin receptors in S49 lymphoma cells. *J. Biol. Chem.*, 262: 6468–6471.

Musacchio, A., Gibson, T., Rice, P, Thompson, J. and Saraste, M. (1993) The PH domain: a common piece in the structural patchwork of signalling proteins. *Trends Biochem. Sci.*, 18: 343–348.

Orellana, S.A., Solski. P.A. and Brown, J.H. (1985) Phorbol ester inhibits phosphoinositide hydrolysis and calcium mobilization in cultured astrocytoma cells. *J. Biol. Chem.*, 260: 5236–5239.

Palczewski, K. and Benovic, J.L. (1991) G protein-coupled receptor kinases. *Trends Biochem. Sci.*, 16, 387–391.

Palczewski, K., Buczylko, J., Kaplan, M.W., Polans, A.S. and Crabb, J.W. (1991) Mechanism of rhodopsin kinase activation. *J. Biol. Chem.*, 266: 12949–12955.

Pals-Rylaarsdam, R., Xu, Y., Witt-Enderby, P., Benovic, J.L. and Hosey, M.M. (1995) Desensitization and internalization of the m2 muscarinic acetylcholine receptor are directed by independent mechanisms. *J. Biol. Chem.*, 270: 29004–29011.

Pawson, T. (1995) Protein modules and signalling networks. *Nature*, 373: 573–579.

Pippig, S., Andexinger, S. and Lohse, M.J. (1995) Sequestration and recycling of $\beta_2$-adrenergic receptors permit receptor resensitization. *Mol. Pharmacol.*, 47: 666–676.

Pitcher, J.A., Inglese, J., Higgins, J.B., Arriza, J.L., Casey, P.J., Kim, C., Benovic, J.L., Kwatra, M.M., Caron, M.G. and Lefkowitz, R.J. (1992) Role of $\beta\gamma$ subunits of G proteins in targetting $\beta$-adrenergic receptor kinase to membrane-bound receptors. *Science*, 257: 1264–1267.

Premont, R.T., Inglese, J. and Lefkowitz, R.J. (1995) Protein kinases that phosphorylate activated G protein-coupled receptors. *FASEB J.*, 9: 175–182.

Richardson, R.M. and Hosey, M.M. (1990) Agonist-independent phosphorylation of purified cardiac muscarinic cholinergic receptors by protein kinase C. *Biochemistry*, 29: 8555–8561.

Richardson, R.M. and Hosey, M.M. (1992) Agonist-induced phosphorylation and desensitization of human m2 muscarinic acetylcholine receptors in Sf9 insect cells. *J. Biol. Chem.*, 267: 22249–22245.

Richardson, R.M., Ptasienski, J. and Hosey, M.M. (1992) Protein kinase C-mediated phosphorylation of chick heart muscarinic cholinergic receptors. *J. Biol. Chem.*, 267: 10127–10132.

Richardson, R.M., Kim, C., Benovic, J.L. and Hosey, M.M. (1993) Phosphorylation and desensitization of human m2 muscarinic acetylcholine receptors by two isoforms of $\beta$-adrenergic receptor kinase. *J. Biol. Chem.*, 268: 13650–13656.

Schleicher, S., Boekhoff, I., Arriza, J.L., Lefkowitz, R.J. and Breer, H. (1993) A $\beta$-adrenergic receptor kinase-like enzyme is involved in olfactory signal termination. *Proc. Natl. Acad. Sci. USA*, 90: 1420–1424.

Shi, W., Osawa, S., Dickerson, C.D. and Weiss, E.R. (1995) Rhodopsin mutants discriminate sites important for the activation of rhodopsin kinase and $G_t$. *J. Biol. Chem.*, 270: 2112–2119.

Shichi, H. and Somers, R.L. (1978) Light-dependent phosphorylation of rhodopsin: purification and properties of rhodopsin kinase. *J. Biol. Chem.*, 253: 7040–7046.

Stoffel, R.H., Randall, R.R., Premont, R.T., Lefkowitz, R.J. and Inglese, J. (1994) Palmitoylation of G protein-coupled receptor kinase, GRK6: lipid modification diversity in the GRK family. *J. Biol. Chem.*, 45: 27791–27794.

Strasser, R.H., Benovic, J.L., Caron, M.G. and Lefkowitz, R.J. (1986) $\beta$-agonist- and prostaglandin $E_1$-induced translocation of the $\beta$-adrenergic receptor kinase: evidence that the kinase may act on multiple adenylate cyclase-coupled receptors. *Proc. Natl. Acad. Sci. USA*, 83: 6362–6366.

Tobin, A.B., Lambert, D.G. and Nahorski, S.R. (1993) Rapid agonist-induced phosphorylation of m3-muscarinic receptors revealed by immunoprecipitation. *J. Biol. Chem.*, 268: 9817–9823.

Yoon, H.S., Hadjuk, P.J., Petros, A.M., Olejniczak, E.T., Meadows, R.P. and Fesik, S.W. (1994) Solution structure of a pleckstrin-homology domain. *Nature*, 369: 672–679.

Yu, S.S., Lefkowitz, R.J. and Hausdorff, W.P. (1993) $\beta$-adrenergic receptor sequestration: A potential mechanism of receptor resensitization. *J. Biol. Chem.*, 268: 337–341.

J. Klein and K. Löffelholz (Eds.)
*Progress in Brain Research*, Vol. 109
© 1996 Elsevier Science B.V. All rights reserved.

CHAPTER 16

# Activation, cellular redistribution and enhanced degradation of the G proteins $G_q$ and $G_{11}$ by endogenously expressed and transfected phospholipase C-coupled muscarinic m1 acetylcholine receptors

Ian Mullaney[1], Malcolm P. Caulfield[2], Petr Svoboda[3] and Graeme Milligan[1]

[1]*Molecular Pharmacology Group, Division of Biochemistry and Molecular Biology, Institute of Biomedical and Life Sciences, University of Glasgow, Glasgow G12 8QQ, UK,* [2]*Wellcome Laboratory for Molecular, Pharmacology, Department of Pharmacology, University College London, London WC1E 6BT, UK and* [3]*Institute of Physiology, Czech Academy of Sciences, Prague, Czech Republic*

## Introduction

Five distinct mammalian muscarinic acetylcholine receptors have been identified by isolation of corresponding cDNA species. Each of these predicts the protein product to be a single polypeptide containing the seven transmembrane helices that are indicative of guanine nucleotide binding protein (G protein)-linked receptors (Chapter 13). Although some overlap has been noted, usually following high level expression in heterologous cell systems (Offermans et al., 1994), the general consensus is that the predominant cellular signalling cascade regulated in a direct fashion by the muscarinic m1, m3 and m5 acetylcholine receptors is the phospholipase C-mediated hydrolysis of phosphatidylinositol 4,5-bisphosphate ($PIP_2$), whilst agonist activation of the muscarinic m2 and m4 acetylcholine receptors results in inhibition of adenylyl cyclase activity (Chapter 17).

Here we describe experimental evidence to indicate that transfected muscarinic m1 acetylcholine receptors can interact with and regulate both the cellular distribution and total cellular levels of the $\alpha$ subunits of two closely related G proteins, $G_q$ and $G_{11}$ and that muscarinic m1 receptor regulation of the M-type $K^+$ current in superior cervical ganglia is transduced by these G proteins.

## Regulation of phospholipase C activity by G proteins and muscarinic acetylcholine receptors

A number of observations, which included that poorly hydrolysed analogues of GTP could synergize with receptor agonists to promote hydrolysis of inositol containing phospholipids in permeabilised cells and in certain cell free systems (see Litosch, 1990 for details), gave clear indications that such receptors must produce their effects through one or more G proteins. Proof of this, however, lagged substantially behind that for receptors coupled in either stimulatory or inhibitory fashion to adenylyl cyclase. This reflected that neither agents that were able to selectively modulate the activity of the phospholipase C-coupled G protein(s) nor cell systems with a null background for these G proteins were available. It was only a confluence of observations encompassing partial protein sequencing of purified G proteins from fractions containing guanine nucleotide-dependent phospholipase C-stimulating activity (Pang et al., 1990; Taylor et al., 1990; Blank et al., 1991) and the homology cloning of two G protein $\alpha$ subunits termed $G_q$ and $G_{11}$ (Strathmann and Simon, 1990) which resulted in the identification of the sought for pertussis toxin-insensitive phospholipase C-linked G proteins. Formal demonstration of the capacity of these proteins to function in this role was obtained from a combination of reconstitution experiments with purified muscarinic m1 receptor, the G proteins and phospholipase C $\beta$1 (Taylor and Exton 1991; Berstein et al., 1992), transient expression of the two G-protein $\alpha$ subunit cDNA species (Aragay et al., 1992; Wu et al., 1992) and blockade of the ability of phospholipase C-linked receptors to stimulate high affinity GTPase activity (Shenker et al., 1991) and generate inositol phosphates (Gutowski et al., 1991; Aragay et al., 1992) in the presence of antibodies generated against sequences predicted to represent regions close to the C-terminal of these G proteins. Using both immunological means (Mitchell et al., 1991) and molecular probes (Strathmann and Simon, 1990) these two G proteins were shown to be very widely expressed and to be routinely coexpressed (Strathmann and Simon, 1990; Wilkie et al., 1991; Mailleux et al., 1992; Milligan, 1993a; Milligan et al., 1993).

Subsequently, the ability of a range of receptors expressed in haemopoietically derived cell types and cell lines to cause generation of inositol phosphates in a pertussis toxin-sensitive fashion (and thus inherently defined as involving a central role of a $G_i$-like protein) was also shown to involve G proteins. However, in these cases this resulted primarily from the activation of phospholipase C $\beta$2 (Camps et al., 1992a,b) and phospholipase C$\beta$3 (Carozzi et al., 1993) through receptor-mediated release of G protein $\beta\gamma$ complexes (Boyer et al., 1992; Camps et al., 1992b; Katz et al., 1992; Park et al., 1993) from $G_{i2}\alpha$ and $G_{i3}\alpha$ (Gierschik et al., 1989).

### Regulation of G protein distribution and levels in CHO cells transfected to express the human muscarinic m1 acetylcholine receptor

We have used CHO cells transfected to express the human muscarinic m1 acetylcholine receptor to examine agonist-induced redistribution of the receptor and G proteins and how sustained agonist occupancy of the muscarinic acetylcholine receptor might alter the cellular G protein content. Short term exposure of these cells to the cholinergic agonist carbachol resulted in a robust increase in cellular levels of inositol phosphates (in a manner insensitive to pretreatment of the cells with pertussis toxin) if the experiment was performed in the presence of $Li^+$ to inhibit the activity of inositol monophosphate phosphatases. Sustained exposure to carbachol resulted in a downregulation of both the muscarinic m1 receptor, as assessed in ligand binding studies, and some combination of $G_q\alpha$ and $G_{11}\alpha$ as measured in semi-quantitative immunoblotting studies (Mullaney et al., 1993a) (Fig. 1) with an antiserum generated against an epitope shared between these two polypeptides (Mitchell et al., 1991). Reverse transcriptase of RNA isolated from these cells followed by polymerase chain reaction using primer pairs designed to amplify specifically either $G_q\alpha$ or $G_{11}\alpha$ indicated the expression of both G proteins by these cells (Mitchell et al., 1993) and development of SDS-PAGE conditions able to resolve $G_q\alpha$ from $G_{11}\alpha$ confirmed their co-expression and demonstrated that the steady state levels of $G_{11}\alpha$ to be some 2 fold higher than $G_q\alpha$ (Mullaney et al., 1993b). Sustained exposure of the cells to carbachol resulted in downregulation of both of these polypeptides. The degree of loss of each G protein was similar at different time points and with different concentrations of agonist (Mullaney et al., 1993b). We have suggested these results indicate that the muscarinic m1 acetylcholine receptor, at least when expressed in this genetic background, is not able to functionally select between $G_q$ and $G_{11}$ and that it interacts with and activates these two G proteins equally and non-selectively based only on their steady state levels of expression. For this to be true they would necessarily display an equivalence in their cellular distribution. Use of a sucrose density gradient system to fractionate membrane components of homogenised human muscarinic m1 acetylcholine receptor expressing CHO cells demonstrated an identical distribution of $G_q\alpha$ and $G_{11}\alpha$ in unstimulated cells, with the vast bulk of these polypeptides present in the fractions that contained markers for the plasma membrane (Svoboda and Milligan, 1994). Exposure to these cells to carbachol (1 mM) for 30 min did not result in a detectable reduction in levels of $G_q\alpha$ or $G_{11}\alpha$ when crude membrane fractions were analysed.

However, resolution of cellular homogenates on sucrose density gradients indicated a clear redistribution of both the muscarinic m1 receptor and a combination of $G_q\alpha$ and $G_{11}\alpha$ away from the plasma membrane containing fractions to low density membrane fractions (Svoboda and Milligan, 1994). The exact nature of these low density fractions remains to be examined but it is likely that they represent an endosomal pool. Sustained exposure to carbachol (1 mM, 16 h) resulted in the downregulation of both $G_q\alpha/G_{11}\alpha$ and the muscarinic m1 receptor as evidenced by their disappearance from throughout the fractions of the sucrose density gradient (Svoboda and Milligan, 1994). This confirmed the downregulation observed previously using crude membrane preparations (Mullaney et al., 1993b).

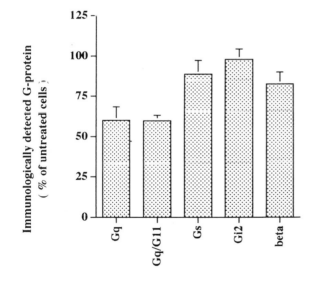

Fig. 1. Sustained exposure of CHO cells expressing the human muscarinic m1 acetylcholine receptor to carbachol results in a marked downregulation of the $\alpha$ subunits of $G_q$ and $G_{11}$. CHO cells expressing the human muscarinic m1 acetylcholine receptor were treated with or without carbachol (1 mM) for 16 h. Membranes prepared from these cells were then immunoblotted to detect the presence of a variety of G protein subunits expressed by these cells. Substantial downregulation of $G_q\alpha$ and $G_{11}\alpha$ was observed without statistically significant alterations in levels of the $\alpha$ subunits of other G proteins. Data are adapted from Mullaney et al. (1993a).

## Mechanisms of $G_q\alpha/G_{11}\alpha$ downregulation in human muscarinic m1 acetylcholine receptor expressing CHO cells exposed to carbachol

A number of possible mechanisms, including alterations in the transcription rate of the $G_q\alpha$ and $G_{11}\alpha$ genes, reductions in stability of their mRNA species and enhanced proteolytic degradation of these polypeptides could be envisaged to contribute to agonist-induced downregulation of $G_q\alpha$ and $G_{11}\alpha$ (Milligan, 1993b). Following pulse [35]S amino acid labelling of human muscarinic m1 receptor expressing CHO cells, immunoprecipitation of $G_q\alpha'/G_{11}\alpha$ was performed with an antiserum able to quantitatively immunoprecipitate these two polypeptides. During a chase in the absence of the [35]S-labelled amino acids such immunoprecipitations demonstrated that the rate of degradation of $G_q\alpha/G_{11}\alpha$ in these cells was consistent with a monoexponential rate of decay with an estimated half-life of some $18 \pm 2$ h (Mitchell et al., 1993). Parallel experiments in which the chase phase was performed in the presence of carbachol (1 mM) clearly showed the degradation of $G_q\alpha/G_{11}\alpha$ to be more rapid than in the absence of ligand. Furthermore, in this situation the degradation of $G_q\alpha/G_{11}\alpha$ displayed a more complex pattern that could not be modelled suitably by a monoexponential function. Computer analysis indicated a proportion of the cellular $G_q\alpha/G_{11}\alpha$ population to be degraded much more rapidly ($t_{0.5} = 3$ h) in the presence of carbachol while the rest was degraded at a rate not different from that in the untreated cells (Mitchell et al., 1993). The simplest interpretations of these results are that the fraction of the cellular $G_q/G_{11}$ population that was rapidly degraded in the presence of carbachol represents that fraction that was activated by the agonist-occupied receptor while the remaining $G_q/G_{11}$ was not activated by the receptor either because it was present in stoichiometric excess over the receptor population or because it was compartmentalised in a pool inaccessible to the receptor. Total cellular levels of other G protein $\alpha$ subunits expressed by these cells were not altered by treatment with carbachol (Mullaney et al., 1993a) and, at least in the case of $G_{i2}\alpha$, direct measurement indicated that carbachol treatment was unable to alter the rate of degradation of this polypeptide (Mitchell et al., 1993). Such selective regulation of cellular levels of $G_q\alpha/G_{11}\alpha$ by enhancement of their rate of degradation is not restricted to muscarinic m1 acetylcholine receptors. Agonist occupation of each of the gonadotrophin releasing hormone receptor that is expressed endogenously by pituitary $\alpha$T3-1 cells (Shah and Milligan, 1994; Shah et al., 1995) and of the $\alpha_{1A/D}$, $\alpha_{1B}$ and $\alpha_{1C}$ adrenoceptor subtypes following their expression in Rat 1 fibroblasts (Wise et al., 1995) has recently been demonstrated to result in a specific downregulation of $G_q$ and $G_{11}$ that reflects agonist-induced increases in the rate of turnover of these G proteins.

In each of the muscarinic m1 acetylcholine receptor expressing CHO cells (Mitchell et al., 1993) and the other systems described above (Shah et al., 1995; Wise et al., 1995), no agonist-induced alterations in either $G_q\alpha$ or $G_{11}\alpha$ mRNA was recorded whether using Northern blots or reverse transcriptase-polymerase chain reaction to detect the presence of the message.

## A role for $G_q\alpha/G_{11}\alpha$ in muscarinic m1 receptor inhibition of M-current

Muscarinic m1 acetylcholine receptors are expressed in neurones of rat superior cervical ganglion (SCG) and therein are able to mediate inhibition of the M-type $K^+$ current (Marrion et al., 1989; Chapter 17). The G-protein(s) involved in this effect were known to be insensitive to pretreatment with pertussis toxin, eliminating a role for either the $\alpha$ subunits of members of the $G_i$ family and probably for the $\beta\gamma$ complex unless the detailed molecular identity of the individual $\beta$ and $\gamma$ gene products in the complex are vital to the effect (Kleuss et al., 1992, 1993). To ascertain a possible role for $G_q\alpha/G_{11}\alpha$ in this effect SCG neurones were microinjected with an antipeptide antiserum directed towards the common C-terminal decapeptide that is shared by these two G proteins (Caulfield et al., 1994). The choice of antiserum was based on an appreciation that the extreme C-terminus of G protein $\alpha$ subunits is a key site for receptor-G protein interactions and because we

(McFadzean et al., 1989) and others (Menon-Johansson et al., 1993; Wilk-Bilaszczak et al., 1994) had previously successfully used antisera of this design to other G proteins to examine the specificity's of interactions between receptors and G proteins using electrophysiological recording. The success of injection of the antiserum into individual neurones was assessed following the electrophysiological recordings by staining the cells with a fluorescein isothiocyanate-labelled secondary antibody. The antiserum to $G_q\alpha/G_{11}\alpha$ reduced, but over a series of cells did not fully eliminate, oxotremorine-M-induced inhibition of M-current (Caulfield et al., 1994). Similar injections with an antiserum to the equivalent regions of $G_o\alpha$ failed to modulate muscarinic regulation of M-current but did reduce $\alpha_2$-adrenoceptor regulation of $Ca^{2+}$ current (Caulfield et al., 1994). Such studies clearly define a role for $G_q\alpha$ and/or $G_{11}\alpha$ in transducing the muscarinic m1 receptor effect. As the reduction of the muscarinic effect was incomplete it has been difficult to state with any degree of certainty whether the entire effect on M-current is produced through activation of $G_q$ and $G_{11}$ or whether other co-expressed pertussis toxin-insensitive G-proteins such as $G_{12}$ and $G_{13}$ (Strathmann and Simon, 1991) might contribute to the effect. While such studies remain to be addressed directly, analysis of individual cells injected with the $G_q\alpha/G_{11}\alpha$ antiserum showed that in a number of them blockade of the effect of oxotremorine-M was virtually complete and could not be overcome by increasing concentrations of the agonist (and thus higher levels of occupancy of the receptor) (Caulfield et al., 1994). By contrast, in a number of the cells the degree of block produced by injection of the antiserum was less impressive and this was further diminished at high agonist concentrations. A possible explanation for these results would reflect the inability to inject sufficient antibody in certain cells to functionally immunoneutralize the bulk of the cellular $G_q\alpha/G_{11}\alpha$ population. If this hypothesis is correct then muscarinic m1 receptor-mediated activation of only $G_q$ and/or $G_{11}$ would be both necessary and sufficient to allow agonist-mediated regulation of M-current. We have observed a very similar pattern of partial blockade of bradykinin $B_2$ receptor-mediated inhibition of M-current by injection of the same anti-$G_q\alpha/G_{11}\alpha$ antiserum into rat SCG neurones and shown that the peptide used to raise the antiserum is able to block the effect of the antibody when they were injected together (Jones et al., 1995). The studies noted above following heterologous expression of the muscarinic m1 acetylcholine receptor in CHO cells indicates that the receptor is unable to functionally select between activation of $G_q$ and $G_{11}$. Both of these are expressed by SCG neurones (Caulfield et al., 1994) and both are capable of regulation of phospholipase C$\beta$1 activity (Taylor and Exton, 1991). It remains undefined, however, whether both are required for and function in muscarinic receptor regulation of M-current. It may be in time that selective antisense elimination of each of these proteins will allow this question to be examined directly.

Although $G_q\alpha$ and $G_{11}\alpha$ stimulate phospholipase C $\beta$1 activity and muscarinic receptor-mediated accumulation of inositol phosphates has been recorded in SCG (Bone et al., 1984) it is far from clear that this is the mechanism for $G_q/G_{11}$-mediated inhibition of M-current. A second messenger is required (Selyanko et al., 1992) but its nature has proved difficult to define (Robbins et al., 1993).

## Conclusions

The muscarinic m1 acetylcholine receptor, whether present endogenously or introduced into cells by transfection, couples predominantly to G proteins of the $G_q$ family. Agonist occupancy of this receptor can result in cellular redistribution of these G proteins and upon sustained exposure to a reduction in cellular levels of these G proteins.

## Acknowledgements

We thank colleagues including David Brown, Noel Buckley, and Fiona Mitchell who have been involved in various aspects of the studies described above. We thank the Wellcome Trust and the Medical Research Council (UK) for support of many of these studies.

## References

Aragay, A.M., Katz, A. and Simon, M.I. (1992) The $G\alpha_q$ and $G\alpha_{11}$ proteins couple the thyrotropin-releasing hormone receptor to phospholipase C in GH3 rat pituitary cells. *J. Biol. Chem.*, 267: 24983–24988.

Berstein, G., Blank, J., Smrcka, A., Higashijima, T., Sternweis, P., Exton, J. and Ross, E. (1992) Reconstitution of agonist-stimulated phosphatidylinositol 4,5-bisphosphate hydrolysis using purified m1 muscarinic receptor, $G_q/G_{11}$ and phospholipase C-$\beta$1. *J. Biol. Chem.*, 267: 8081–8088.

Blank, J.L., Ross, A.H. and Exton, J.H. (1991) Purification and characterisation of two G-proteins that activate the $\beta$1 isozyme of phosphoinositide-specific phospholipase C. Identification as members of the $G_q$ class. *J. Biol. Chem.*, 266, 18206–18216.

Bone, E.A., Fretten, P., Palmer, S., Kirk, C.J. and Michell, R.H. (1984) Rapid accumulation of inositol phosphates in isolated rat superior cervical ganglia exposed to V1 vasopressin and muscarinic cholinergic stimuli. *Biochem. J.*, 22: 803–811.

Boyer, J.L., Waldo, G.L. and Harden, T.K. (1992) $\beta\gamma$-subunit activation of G-protein-regulated phospholipase C. *J. Biol. Chem.*, 267: 25451–25456.

Camps, M., Hou, C., Sidiropoulos, D., Stock, J.B., Jakobs, K.-H. and Gierschik, P. (1992a) Stimulation of phospholipase C by G-protein $\beta\gamma$ subunits. *Eur. J. Biochem.*, 206: 821–831.

Camps, M., Carozzi, A., Schnabel, P., Scheer, A., Parker, P. and Gierschik, P. (1992b) Isozyme-selective stimulation of phospholipase C$\beta$2 by G-protein $\beta\gamma$-subunits. *Nature*, 360: 684–686.

Carozzi, A., Camps, M., Gierschik, P. and Parker, P.J. (1993) Activation of phosphatidyl-inositol lipid-specific phospholipase C-$\beta$3 by G-protein $\beta\gamma$ subunits. *FEBS Lett.*, 315: 340–342.

Caulfield, M.P., Jones, S., Vallis, Y., Buckley, N.J., Kim, G.D., Milligan, G. and Brown, D.A. (1994) Muscarinic M-current inhibition via $G_{\alpha q/11}$ and $\alpha$-adrenoceptor inhibition of $Ca^{2+}$ current via $G_{\alpha o}$ in rat sympathetic neurones. *J. Physiol.*, 477: 415–422.

Gierschik, P., Sidiropoulos, D. and Jakobs, K.-H. (1989) Two distinct $G_i$-proteins mediate formyl peptide receptor signal transduction in human leukaemia (HL-60) cells. *J. Biol. Chem.*, 264: 21470–21473.

Gutowski, S., Smrcka, A., Nowak, L., Wu, D., Simon, M.I. and Sternweis, P.C. (1991) Antibodies to the $\alpha q$ subfamily of guanine nucleotide-binding regulatory protein $\alpha$ subunits attenuate activation of phosphatidylinositol 4, 5 bisphosphate hydrolysis by hormones. *J. Biol. Chem.*, 266: 20519–20524.

Jones, S., Brown, D.A., Milligan, G., Willer, E., Buckley, N.J. and Caulfield, M.P. (1995) Bradykinin excites rat sympathetic neurones by inhibition of M current through a mechanism involving $B_2$ receptors and $G_{\alpha q/11}$. *Neuron*, 14: 399–405.

Katz, A., Wu, D. and Simon, M.I. (1992) Subunits $\beta\gamma$ of heterotrimeric G-protein activate $\beta2$ isoform of phospholipase C. *Nature*, 360: 686–689.

Kleuss, C., Scherubl, H., Hescheler, J., Schultz, G. and Wittig, B. (1992) Different $\beta$ subunits determine G-protein interaction with transmembrane receptors. *Nature*, 358: 424–426.

Kleuss, C., Scherubl, H., Hescheler, J., Schultz, G. and Wittig, B. (1993) Selectivity in signal transduction determined by $\gamma$ subunits of heterotrimeric G-proteins. *Science*, 259: 832–834.

Litosch, I. (1990) Inositol phosphate metabolism and G-proteins. In: M.D. Houslay, and G. Milligan (Eds.), *G-Proteins as Mediators of Cellular Signalling Processes*. Wiley, Chichester, UK, pp. 151–171.

Mailleux, P., Mitchell, F., Vanderhaeghen, J.-J., Milligan, G. and Erneux, C. (1992) Immunohistochemical distribution of neurons containing the G-proteins $G_q\alpha/G_{11}\alpha$ in the adult rat brain. *Neuroscience* 51, 311–316.

Marrion, N.V., Smart, T.G., Marsh, S.J. and Brown, D.A. (1989) Muscarinic suppression of the M-current in the rat sympathetic ganglion is mediated by receptors of the $M_1$-subtype. *Br. J. Pharmacol.*, 98: 557–573.

McFadzean, I., Mullaney, I., Brown, D.A. and Milligan, G. (1989) Antibodies to the GTP binding protein $G_o$ antagonize noradrenaline-induced calcium current inhibition in NG108–15 hybrid cells. *Neuron*, 3: 177–182.

Menon-Johansson, A.S., Berrow, N. and Dolphin, A.C. (1993) $G_o$ transduces $GABA_B$ receptor modulation of N-type calcium channels in cultured dorsal root ganglion neurones. *Pflügers Arch.*, 193: 1–9.

Milligan, G. (1993a) Regional distribution and quantitative measurement of the phosphoinositidase C-linked guanine nucleotide binding proteins $G_{11}\alpha$ and $G_q\alpha$ in rat brain. *J. Neurochem.*, 61: 845–851.

Milligan, G. (1993b) Agonist regulation of cellular G protein levels and distribution: mechanisms and functional implications. *Trends Pharmacol. Sci.*, 14: 413–418.

Milligan, G., Mullaney, I. and McCallum, J.F. (1993) Distribution and relative levels of expression of the phosphoinositidase-C-linked G-proteins $G_q\alpha$ and $G_{11}\alpha$: absence of $G_{11}\alpha$ in human platelets and haemopoietically derived cell lines. *Biochim. Biophys. Acta*, 1179: 208–212.

Mitchell, F.M., Mullaney, I., Godfrey, P.P., Arkinstall, S.J., Wakelam, M.J.O. and Milligan, G. (1991) Widespread distribution of $G_q\alpha/G_{11}\alpha$ detected immunologically by an antipeptide antiserum directed against the predicted C-terminal decapeptide. *FEBS Lett.*, 287: 171–174.

Mitchell, F.M., Buckley, N.J. and Milligan, G. (1993) Enhanced degradation of the phosphoinositidase C-linked guanine-nucleotide binding protein $G_q\alpha/G_{11}\alpha$ following activation of the human M1 muscarinic acetylcholine receptor expressed in CHO cells. *Biochem. J.*, 293: 495–499.

Mullaney, I., Dodd, M.W., Buckley, N.J. and Milligan, G. (1993a) Agonist activation of transfected human M1 muscarinic acetylcholine receptors in CHO cells results in

down-regulation of both the receptor and the $\alpha$ subunit of the G-protein $G_q$. *Biochem. J.*, 289: 125–131.

Mullaney, I., Mitchell. F.M.,. McCallum, J.F., Buckley, N.J. and Milligan, G. (1993b) The human muscarinic M1 acetylcholine receptor, when expressed in CHO cells, activates and downregulates both $G_q\alpha$ and $G_{11}\alpha$ equally and nonselectively. *FEBS Lett.*, 324: 241–245.

Offermans, S., Wieland, T., Homann, D., Sandmann, J., Bombien, E., Spicher, K., Schultz, G. and Jakobs, K.H. (1994) Transfected muscarinic acetylcholine receptors selectively couple to $G_i$-type G proteins and $G_{q/11}$. *Mol. Pharmacol.*, 45, 890–898.

Pang, I.-H. and Sternweis, P.C. (1990) Purification of unique $\alpha$ subunits of GTP-binding regulatory proteins (G proteins) by affinity chromatography with immobilized $\beta\gamma$ subunits. *J. Biol. Chem.*, 265: 18707–18712.

Park, D., Jhon, D.-Y., Lee, C.-W., Lee, K.-H. and Rhee, S.G. (1993) Activation of phospholipase C isozymes by G-protein $\beta\gamma$ subunits. *J. Biol. Chem.*, 268, 4573–4576.

Robbins, J., Marsh, S.J. and Brown, D.A. (1993) On the mechanism of M-current inhibition by muscarinic m1 receptors in DNA-transfected rodent neuroblastoma × glioma cells. *J. Physiol.*, 468: 153–178.

Selyanko, A.A., Stansfeld, C.E. and Brown, D.A. (1993) Closure of potassium M-channels by muscarinic acetylcholine-receptor stimulants requires a diffusible messenger. *Proc. R. Soc. London Ser. B*, 250: 119–125.

Shah, B.H. and Milligan, G. (1994) The gonadotrophin-releasing hormone receptor of $\alpha$T3–1 pituitary cells regulates cellular levels of both of the phosphoinositidase C-linked G proteins, $G_q\alpha$ and $G_{11}\alpha$, equally. *Mol. Pharmacol.*, 46, 1–7.

Shah, B.H., MacEwan, D.J. and Milligan, G. (1995) Gonadotrophin-releasing hormone receptor agonist-mediated downregulation of $G_q\alpha/G_{11}\alpha$ (pertussis toxin-insensitive) G proteins in $\alpha$T3–1 gonadotroph cells reflects increased G protein turnover but not alterations in mRNA levels. *Proc. Natl. Acad. Sci. USA*, 92: 1886–1890.

Shenker, A., Goldsmith, P., Unson, C.G. and Spiegel, A.M. (1991) The G protein coupled to the thromboxane A2 receptor in human platelets is a member of the novel Gq family. *J. Biol. Chem.*, 266: 9309–9313.

Strathmann, M.P and Simon, M.I. (1990) G protein diversity: a distinct class of $\alpha$ subunits is present in vertebrates and invertebrates. *Proc. Natl. Acad. Sci. USA*, 87: 9113–9117.

Strathmann, M.P. and Simon, M.I. (1991) $G\alpha_{12}$ and $G\alpha_{13}$ subunits define a fourth class of G-protein $\alpha$ subunits. *Proc. Natl. Acad. Sci. USA*, 88: 5582–5586.

Svoboda, P. and Milligan, G. (1994) Agonist-induced transfer of the $\alpha$ subunits of the guanine-nucleotide-binding regulatory proteins $G_q$ and $G_{11}$ and of muscarinic m1 acetylcholine receptors from plasma membranes to a light-vesicular membrane fraction. *Eur. J. Biochem.*, 224: 455–462.

Taylor, S.J. and Exton, J.H. (1991) Two $\alpha$ subunits of the $G_q$ class of G proteins stimulate phosphoinositide phospholipase C-$\beta$1 activity. *FEBS Lett.*, 286: 214–216.

Taylor, S.J., Smith, J.A. and Exton, J.H. (1990) Purification from bovine liver membranes of a guanine nucleotide-dependent activator of phosphoinositidase-specific phospholipase C. Immunological identification as a novel G-protein $\alpha$ subunit. *J. Biol. Chem.*, 265: 17150–17156.

Wilk-Blaszczak, M.A., Gutowski, S., Sternweis, P.C. and Belardetti, F. (1994) Bradykinin modulates potassium and calcium currents in neuroblastoma hybrid cells via different pertussis toxin-insensitive pathways. *Neuron*, 12: 119–126.

Wilkie, T.M., Scherle, P.A., Strathmann, M.P., Slepak, V.Z. and Simon, M.I. (1991) Characterization of G-protein $\alpha$ subunits in the $G_q$ class: expression in murine tissues and in stromal and hematopoietic cell lines. *Proc. Natl. Acad. Sci. USA*, 88: 10049–10053.

Wise, A., Lee, T.W., MacEwan, D.J. and Milligan, G. (1995) Degradation of $G_{11}\alpha/G_q\alpha$ is accelerated by agonist occupancy of $\alpha_{1A/D}$, $\alpha_{1B}$ and $\alpha_{1C}$ adrenergic receptors. *J. Biol. Chem.*, 270: 17196–17203.

Wu, D., Katz, A., Lee, C.-H. and Simon, M.I. (1992) Activation of phospholipase C by $\alpha_1$-adrenergic receptors is mediated by the $\alpha$ subunits of Gq family. *J. Biol. Chem.*, 267: 25798–25802.

# Section VI

# Muscarinic Receptors: 3.
# Coupling to Intracellular Signalling Pathways

J. Klein and K. Löffelholz (Eds.)
*Progress in Brain Research*, Vol. 109
© 1996 Elsevier Science B.V. All rights reserved.

CHAPTER 17

# Muscarinic receptors and cell signalling

Konrad Löffelholz

*Department of Pharmacology, University of Mainz, Mainz, Germany*

## Introduction

Cells have developed signal transduction mechanisms in order to communicate with the cell exterior. Acetylcholine as an external signal is recognized by nicotinic (Chapters 8–10) and muscarinic receptors (Chapters 11–16). As shown by Wess et al. (Chapter 13), the muscarinic receptors belong to the superfamily of G protein-coupled receptors consisting of seven transmembrane (TM) helices tightly packed in a ring-like structure and arranged in a counter-clockwise fashion (viewed from outside) (Liu et al., 1995). Agonist binding leads to a conformational change of the receptor thereby activating associated G proteins.

## The specificity of the receptor-G protein-effector cascade

G protein-coupled receptors bind to a certain G protein in a specific way. The amino acid sequence of the receptor molecule that is responsible for the recognition of the particular G protein is currently investigated in several laboratories, frequently by utilizing mutant and hybrid muscarinic receptors (Chapter 13).

Of the five muscarinic receptors (m1–m5; Bonner, 1989), m1, m3 and m5 couple to the $\alpha$-subunit of Gq and G11, whereas m2 and m4 activate Gi and Go ($\alpha$-subunit and/or $\beta\gamma$-dimer) (Chapters 13 and 14; Offermanns et al., 1994). It is remarkable that the murine $G\alpha 15$ and its human counterpart $G\alpha 16$ can be activated by several muscarinic receptor species and even by a large variety of other

G protein-coupled receptors (Offermanns and Simon, 1995). In general, multiple effector systems can be controlled in parallel by more than one type of receptor and heterotrimeric G protein (Hunt et al., 1994).

## The effectors of the muscarinic receptor-G protein activation

Muscarinic stimulation of G proteins leads to (i) activation or inhibition of ion channels such as $K^+$ and $Ca^{2+}$ channels, (ii) activation of enzymes responsible for the formation of second messengers, such as adenylyl cyclase and phospholipases. The relationship between the receptor subtypes and effectors is not exclusive, especially when studied in transfected cells (Caulfield, 1993; Burford et al., 1995).

### *G protein-coupled inwardly rectifying $K^+$ channel ($I_{KACh}$)*

Many receptor-regulated $K^+$ channels are directly gated by G proteins. For example, acetylcholine via the m2-muscarinic receptor through Gi/o activates the atrial $I_{KACh}$ in the heart (Nair et al., 1995) in a PTX-sensitive way. This mechanism is responsible for the parasympathetic slowing of the heart. More specifically, the G protein $\beta\gamma$-dimer binds directly to the C terminus of inwardly rectifying $K^+$ channel (Inanobe et al., 1995). Krapivinsky et al. (1995) have recently shown that this channel is a heteromultimer of two distinct subunits, GIRK1 and a newly cloned member of the family, CIR (cardiac inward rectifier).

## M-current

The M-current ($I_M$) is a voltage-dependent, non-inactivating K$^+$ current that is called "M-current" because of its blockade through muscarinic receptors; $I_M$ operates during depolarization and normally inhibits the cell from firing. Its suppression by acetylcholine or bradykinin can cause a long-lasting facilitation of excitation and, e.g., in sympathetic ganglia, cortical or hippocampal neurons, may provide a mechanism for long-term potentiation (Krnjevic, 1993). The muscarinic receptors couple to inhibit $I_M$ through a direct interaction of G$\alpha$q and G$\alpha$11 with the channel (Chapter 16; Caulfield et al., 1994).

## Ca$^{2+}$- channels

The muscarinic receptor in the heart reduces Ca$^{2+}$ influx by inhibiting the formation of cAMP and the phosphorylation of the L-type Ca$^{2+}$ channel. Muscarinic stimulation of Ca$^{2+}$ influx through "store-operated Ca$^{2+}$ channels" (Mathes and Thompson, 1994) was indirectly caused by the formation of inositol phosphates (see below) and depletion of intracellular Ca$^{2+}$ stores.

It is difficult to prove whether there is direct inhibition or enhancement of Ca$^{2+}$ influx (Clapham, 1994). In smooth muscle cells, muscarinic receptors seem to suppress directly the voltage-gated Ca$^{2+}$ channel through a PTX-insensitive G protein (Unno et al., 1995). In rat striatal neurons, muscarinic agonists reduced Ca$^{2+}$ currents through two distinct signalling pathways: one pathway depended upon PTX-sensitive N- and P-type currents, whereas the other pathway targeted L-type currents and was PTX-insensitive (Howe and Surmeier, 1995). Experiments in GH3 cells have revealed that Go subforms composed of $\alpha$o1/$\beta$3/$\gamma$4 mediate the inhibitory signal from muscarinic m4 to the Ca$^{2+}$ channel (Hescheler and Schultz, 1994). Finally, in non-excitable cells, certain muscarinic receptor types activated directly a "receptor-operated Ca$^{2+}$ channel" in a second messenger-independent manner (Chapter 18).

## Adenylyl cyclase

The muscarinic receptor subtypes m2 and m4 couple preferentially to a different, although overlapping set of Go and Gi $\alpha$-subunits (Chapter 14; Migeon et al., 1995) thereby inhibiting adenylyl cyclase activity and formation of cyclic AMP in a PTX-sensitive manner. Consequently, protein kinase A-dependent phosphorylation of various proteins such as Ca$^{2+}$ channels (see above) and nuclear proteins controlling gene expression is reduced. The receptor subtype-effector relationship is, however, not exclusive (see above).

## Phospholipases

Muscarinic receptors, preferentially m1, m3 and m5, activate phospholipases C and D. The phosphoinositide-specific phospholipase C generates inositol phosphates mobilizing Ca$^{2+}$ from intracellular stores and diacylglycerol activating protein kinase C (Exton, 1994). The activation of G$\alpha$q and G$\alpha$11 leads to a pertussis toxin (PTX)-insensitive activation of phospholipase C$\beta$1, while the $\beta\gamma$-dimer of Gi/o mediates PTX-sensitive activation of phospholipase C$\beta$1–3 (Katz et al., 1992).

Muscarinic activation of phospholipase D (Löffelholz, 1989; Klein et al., 1995; Chapter 19) catalyzes the formation of choline and phosphatidic acid from phosphatidylcholine hydrolysis. In a second step, phosphatidic acid can be dephosphorylated to diacylglycerol. The species of heterotrimeric G protein responsible for phospholipase D activation is still unknown; only in a few cell types, receptor-mediated activation of phospholipase D was sensitive to PTX suggesting an involvement of Gi/o (Liscovitch and Chalifa, 1994). The receptor-mediated activation of phospholipase D appears in a new light since recent observations suggest that the enzyme may not be directly coupled to heterotrimeric G proteins. The complex activation is poorly understood. Tyrosine kinases and/or monomeric G proteins ("small G proteins") are linked into the receptor-activated transduction cascade upstream of phospholipase D (Chapter 20). Most interestingly, evidence for phospholipase D activity as a downstream effector of small

G proteins has been obtained for the ADP-ribosylating factor (ARF; Brown et al., 1993; Cockcroft et al., 1994), for Rho (Bowman et al., 1993; Malcolm et al., 1994; Chapter 20) and for the newly identified Ras/Ral GTPase pathway (Carnero and Lacal, 1995; Jiang et al., 1995). Considering these observations, it is tempting to speculate that the receptor-mediated activation of phospholipase D connects the membrane signal with deeper cellular compartments associated with phenomena such as vesicular trafficking, cytoskeletal functions, mitogenesis (see below), gene expression and growth. Both phosphatidic acid and diacylglycerol, may be involved in these functions. In a certain experimental condition it is difficult to decide whether activation of phospholipase D or C or both are responsible for the generation of these second messengers. The problem is compounded by the fact that activation of, for instance, muscarinic receptors activates in parallel both phospholipases and there exists an intense cross-talk between the two pathways (Chapter 19; Nishizuka, 1992).

*Mitogen-activated protein kinase (MAPK)*

Gi and Gq can activate not only phospholipase C, but also MAPK; G protein $\beta\gamma$-dimer is the primary mediator of Gi-coupled receptor-stimulated MAPK activation through mechanisms utilizing p21$^{ras}$ and p74$^{raf}$ independent of protein kinase C (Koch et al., 1994; Hawes et al., 1995). In contrast, G$\alpha$ mediates Gq-coupled receptor-stimulated MAPK activation using a p21$^{ras}$-independent mechanism that was dependent of protein kinase C and p74$^{raf}$. In cardiac myocytes muscarinic activation of the Gq/phospholipase C-pathway leads to Ras-dependent activation of the atrial natriuretic factor (ANF) gene expression (Ramirez et al., 1995).

## References

Bonner, T.I. (1989) The molecular basis of muscarinic receptor diversity. *Trends Neurosci.*, 12: 148–151.

Bowman, E.P., Uhlinger, D.J. and Lambeth, J.D. (1993) Neutrophil phospholipase D is activated by a membrane-associated Rho family small molecular weight GTP-binding protein. *J. Biol. Chem.*, 268: 21509–21512.

Brown, H.A., Gutowski, S., Moonaw, C.R., Slaughter, C. and Sternweis, P.C. (1993) ADP-ribosylation factor, a small GTP-dependent regulatory protein, stimulates phospholipase D activity. *Cell*, 75: 1137–1144.

Burford, N.T., Tobin, A.B. and Nahorski, S.R. (1995) Differential coupling of m1, m2 and m3 muscarinic receptor subtypes to inositol 1,4,5-trisphosphate and adenosine 3′,5′-cyclic monophosphate accumulation in Chinese hamster ovary cells. *J. Pharmacol. Exp. Ther.*, 274: 134–142.

Carnero, A. and Lacal, J.C. (1995) Activation of intracellular kinases in Xenopus oocytes by p21$^{ras}$ and phospholipases: a comparative study. *Mol. Cell Biol.*, 15: 1094–1101.

Caulfield, M.P. (1993) Muscarinic receptors - Characterization, coupling and function. *Pharmacol. Ther.*, 58: 319–379.

Caulfield, M.P., Jones, S., Vallis, Y., Buckley, N.J., Kim, G.D., Milligan, G. and Brown, D.A. (1994) Muscarinic M-current inhibition via G$\alpha$q/11 and $\alpha$-adrenoceptor inhibition of Ca$^{2+}$ current via G$\alpha$o in rat sympathetic neurons. *J. Physiol. (London)*, 477: 415–422.

Clapham, D.E. (1994) Direct G protein activation of ion channels? *Annu. Rev. Neurosci.*, 17: 441–464.

Cockcroft, S., Thomas, G.M.H., Fensome, A.M., Geny, B., Cunningham, E., Gout, I., Hiles, I., Totty, N.F., Truong, O. and Hsuan, J.J. (1994) Phospholipase D: a downstream effector of ARF in granulocytes. *Science*, 263: 523–526.

Exton, J.H. (1994) Phosphoinositide phospholipases and G proteins in hormone action. *Annu. Rev. Physiol.*, 56: 349–369.

Hawes, B.E., van Biesen, T., Koch, W.J., Luttrell, L.M. and Lefkowitz, R.J. (1995) Distinct pathways of Gi- and Gq-mediated mitogen-activated protein kinase activation. *J. Biol. Chem.*, 270: 17148–17153.

Hescheler, J. and Schultz, G. (1994) Heterotrimeric G proteins involved in the modulation of voltage-dependent calcium channels of neuroendocrine cells. *Ann. N. Y. Acad. Sci.*, 733: 306–312.

Howe, A. R. and Surmeier, D.J. (1995) Muscarinic receptors modulate N-, P- and L-type Ca$^{2+}$ currents in rat striatal neurons through parallel pathways. *J. Neurosci.*, 15: 458–469.

Hunt, T.W., Carroll, R.C. and Peralta, E.G. (1994) Heterotrimeric G proteins containing G$\alpha$ i3 regulate multiple effector enzymes in the same cell. Activation of phospholipases C and A2 and inhibition of adenylyl cyclase. *J. Biol. Chem.*, 269: 29565–29570.

Inanobe, A., Morishige, K., Takahashi, N., Ito, H., Yamada, M., Takumi, T., Nishina, H., Takahashi, K., Kanaho, Y., Katada, T. and Kurachi, Y. (1995) G$_{\beta\gamma}$ directly binds to the carboxyl terminus of the G protein-gated K$^+$ channel, GIRK1. *Biochem. Biophys. Res. Commun.*, 212: 1022–1028.

Jiang, H., Luo, J.Q., Urano, T., Frankel, P., Lu, Z., Foster, D.A. and Feig, L.A. (1995) Involvement of Ral GTPase in v-Src-induced phospholipase D activation. *Nature*, 378: 409–412.

Katz, A., Wu, D. and Simon, M.I. (1992) Subunits $\beta\gamma$ of heterotrimeric G protein activate $\beta2$ isoform of phospholipase C. *Nature,* 360: 686–689.

Klein, J., Chalifa, V., Liscovitch, M. and Löffelholz, K. (1995) Role of phospholipase D activation in nervous system. Physiology and pathophysiology. *J. Neurochem.*, 65: 1445–1455.

Koch, W.J., Hawes, B.E., Allen, L.F. and Lefkowitz, R.J. (1994) Direct evidence that Gi- coupled receptor stimulation of mitogen-activated protein kinase is mediated by G $\beta\gamma$ activation of p21[ras]. *Proc. Natl. Acad. Sci. USA*, 91: 12706–12710.

Krapivinsky, G., Gordon, E.A., Wickman, K., Velimirovic, B., Krapivinsky, L. and Clapham, D.E. (1995) The G protein-gated atrial $K^+$ channel $I_{KACh}$ is a heteromultimer of two inwardly rectifying $K^+$-channel proteins. *Nature,* 374: 135–141.

Krnjevic, K. (1993) Central cholinergic mechanisms and function. In: A.C. Cuello (Ed.), *Cholinergic Function and Dysfunction. Progress in Brain Research,* Vol. 98, Elsevier, Amsterdam, pp. 285–292.

Liscovitch, M. and V. Chalifa (1994) Signal-activated phospholipase D. In: M. Liscovitch (Ed.), *Signal-Activated Phospholipases*, R.G. Landes Company, Austin, TX, pp. 31–63.

Liu, J., Schöneberg, T., van Rhee, M. and Wess, J. (1995) Mutational analysis of the relative orientation of transmembrane helices I and VII in G protein-coupled receptors. *J. Biol. Chem.*, 270: 19532–19539.

Löffelholz, K. (1989) Receptor-regulation of choline phospholipid hydrolysis: a novel source of diacylglycerol and phosphatidic acid. *Biochem. Pharmacol.*, 38: 1543–1549.

Malcolm, K.C., Ross, A.H., Qiu, R.-G., Symons, M. and Exton, J.H. (1994) Activation of rat liver phospholipase D by the small GTP-binding Protein RhoA. *J. Biol. Chem.*, 269: 25951–25954.

Mathes, C. and Thompson, S.H. (1994) Calcium current activated by muscarinic receptors and thapsigargin in neuronal cells. *J. Gen. Physiol.*, 104: 107–121.

Migeon, J.C., Thomas, S.L. and Nathanson, N.M. (1995) Differential coupling of m2 and m4 muscarinic receptors to inhibition of adenylyl cyclase by $G i\alpha$ and $G o\alpha$ subunits. *J. Biol. Chem.*, 270: 16070–16074.

Nair, L.A., Inglese, J., Stoffel, R., Koch, W.J., Lefkowitz, R.J., Kwatra, M.M. and Grant, A.O. (1995) Cardiac muscarinic potassium channel activity is attenuated by inhibitors of G $\beta\gamma$. *Circ. Res.*, 76: 832–838.

Nishizuka, Y. (1992) Intracellular signaling by hydrolysis of phospholipids and activation of protein kinase C. *Science,* 258: 607–614.

Offermanns, S. and Simon, M.I. (1995) $G\alpha15$ and $G\alpha16$ couple a wide variety of receptors to phospholipase C. *J. Biol. Chem.*, 270: 15175–15180.

Offermanns, S., Wieland, T., Homann, D., Sandmann, J., Bombien, E., Spicher, K., Schultz, G. and Jakobs, K.H. (1994) Transfected muscarinic acetylcholine receptors selectively couple to Gi-type G proteins and Gq/11. *Mol. Pharmacol.*, 45: 890–898.

Ramirez, M.T., Post, G.R., Sulakhe, P.V. and Brown, J.H. (1995) M1 muscarinic receptors heterologously expressed in cardiac myocytes mediate Ras-dependent changes in gene expression. *J. Biol. Chem.*, 270: 8446–8451.

Unno, T., Komori, S. and Ohashi, H. (1995) Inhibitory effect of muscarinic receptor activation on $Ca^{2+}$ channel current in smooth muscle cells of guinea-pig ileum. *J. Physiol. (London),* 484: 567–581.

J. Klein and K. Löffelholz (Eds.)
*Progress in Brain Research*, Vol. 109

CHAPTER 18

# Muscarinic receptor activated Ca$^{2+}$ channels in non-excitable cells

Dafna Singer-Lahat[1], Eduardo Rojas[2] and Christian C. Felder[1]

[1]*Laboratory of Cell Biology, National Institute of Mental Health, Bethesda, MD 20892, USA and* [2]*Laboratory of Cell Biology and Genetics, National Institute of Diabetes and Digestive and Kidney Diseases, Bethesda, MD 20892, USA*

## Introduction

The physiological consequences of muscarinic acetylcholine receptor stimulation are determined by the receptor subtype as well as the compliment of signal transduction proteins expressed within a cell. Muscarinic receptors utilize a variety of effector enzymes and ion channels to generate intracellular second messengers leading to a final biological response. For example, the m2 and m4 muscarinic receptor subtypes have been shown to couple to the inhibition of adenylyl cyclase, augmentation of phospholipase A2, activation of inwardly rectifying K$^+$ channels (K$^+_{ir}$), and in some cells a weak coupling to phospholipase C activation. The m1, m3, and m5 receptors stimulate phospholipase A2, phospholipase C, phospholipase D, Ca$^{2+}$ influx, and inhibition of time and voltage gated K$^+$ currents (M-current) (Caulfield, 1993; Felder, 1995). The regulation of effectors by muscarinic receptors can be either dependent or independent of diffusible second messengers released following receptor stimulation. In atrial pacemaker cells, the regulation of K$^+_{ir}$ is independent of second messengers and probably under G protein regulation (Logothetis et al., 1987; Brown and Birnbaumer, 1990). Similarly, G$_q$ mediated activation of phospholipase C-$\beta$ by m3 receptors is second messenger-independent (Hildebrandt and Shuttleworth, 1993; Sawaki et al., 1993). In contrast, stimulation of phospholipases A2, C-, and D by muscarinic receptors in many cells is dependent on influx of extracellular Ca$^{2+}$

(Felder et al., 1995). Therefore, extracellular Ca$^{2+}$ can play an important role as a second messenger in the regulation of a number of signaling effectors, and the mechanism of muscarinic receptor-mediated Ca$^{2+}$ influx is just beginning to be understood.

In excitable cells, such as neurons and muscle cells, Ca$^{2+}$ crosses the membrane predominantly through voltage-sensitive Ca$^{2+}$ channels (Neher, 1992). In nonexcitable cells, such as fibroblasts and epithelial cells, calcium also passes through a family of poorly characterized voltage-insensitive Ca$^{2+}$ channels (VICC) (Fasolato et al., 1994; Felder et al., 1994). VICCs have been preliminarily classified into three general groups; (1) receptor-operated Ca$^{2+}$ channels (ROCCs) which are second messenger independent, (2) second messenger-operated Ca$^{2+}$ channels (SMOCCs), and (3) store-operated Ca$^{2+}$ channels (SOCCs) which open following IP$_3$-mediated depletion of intracellular stores and provide a source of Ca$^{2+}$ for refilling the stores. Through the use of Ca$^{2+}$-sensitive fluorescent dyes, muscarinic receptors have been shown to interact with all known VICC types including ROCCs (Felder et al., 1991, 1993; Lambert et al., 1992), SOCCs (Shuttleworth and Thompson, 1992; Roche et al., 1993) and SMOCCs (Oettling et al., 1992). Recently, electrophysiological studies have provided direct evidence that muscarinic receptor-mediated Ca$^{2+}$ influx occurs through ion channels in non-excitable cells. Direct measurement of m1 receptor-stimulated Ca$^{2+}$ channel activity by whole cell patch clamp in human T cells

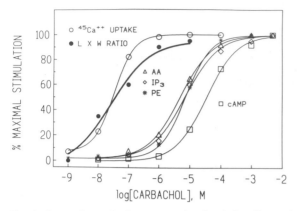

Fig. 1. Comparison of m5 receptor-stimulated signaling pathways with morphology change in CHOm5 cells. CHOm5 cells were stimulated with increasing concentrations of the muscarinic receptor agonist carbachol and $^{45}Ca^{2+}$ uptake, [$^3H$]-arachidonic acid (AA) release, [$^3H$]inositol 1,4,5 trisphosphate (IP3), [$^3H$]phosphatidylethanol (PE), and cAMP accumulation were measured as an index of receptor-operated calcium channels, phospholipase A2, phospholipase C, phospholipase D, and adenylate cyclase activity, respectively. Carbachol-stimulated morphology change was expressed as the length/width ratio. Reproduced with permission from Felder et al. (1993).

indicated that the channel is voltage-independent and resembled the T-type of VOCC in its relative conductance of different cations (McDonald et al., 1993). This T cell channel was shown to be second messenger regulated due to its requirement for IP$_3$. The first direct measurement of muscarinic receptor-stimulated SOCCs in excitable cells was shown by the whole cell nystatin patch clamp technique suggesting that this class of Ca$^{2+}$ channel may be required in most if not all cells for refilling of intracellular Ca$^{2+}$ stores (Mathes and Thompson, 1994).

Changes in intracellular Ca$^{2+}$ following muscarinic receptor stimulation are typically characterized by a rapid and transient spike followed by a plateau phase which persists with agonist stimulation. The transient phase is due to the release of Ca$^{2+}$ from IP$_3$-sensitive intracellular Ca$^{2+}$ stores (Berridge, 1993) and does not appear to play a significant role in regulating phospholipase effector enzymes. For example, phospholipase A2 activation in many cells is independent of the transient release of Ca$^{2+}$ from cytoplasmic stores, but re-

quires extracellular Ca$^{2+}$ for receptor-mediated activation (Brooks et al., 1989; Felder et al., 1990). The plateau phase is maintained by the influx of extracellular Ca$^{2+}$ across the plasma membrane. We have previously demonstrated that Ca$^{2+}$ influx through the plasma membrane plays a central role in the regulation of multiple signaling pathways activated by muscarinic receptors stably expressed in CHO fibroblasts (Fig. 1) (Felder et al., 1993). Muscarinic receptor-stimulated Ca$^{2+}$ influx was shown to be initiated at concentrations of carbachol two orders of magnitude lower than that required for the release of arachidonic acid, phosphatidylethanol, or inositol phosphates and three orders of magnitude lower than for cAMP generation, second messengers measured as an index of phospholipase A2, phospholipase D, phospholipase C, and adenylate cyclase activation respectively. Carbachol-stimulated Ca$^{2+}$ influx correlated with carbachol-mediated suppression of CHO cell tumorigenicity, suggesting a role for Ca$^{2+}$ influx in this process. In this study, the nature of the Ca$^{2+}$ influx regulated by muscarinic receptors was investigated in CHO cells using whole cell patch clamp electrophysiology.

## Methods

### Materials and electrophysiological solutions

As previously described, single channel currents were recorded from cell attached patches (1) with a bath solution of the following composition (in mM): 140 NaCl, 5 KCl, 2.6 CaCl$_2$, 1 MgCl$_2$, 5 glucose, 5 NaHepes at pH 7.4. The electrode solution was composed of (in mM: 25 CaCl$_2$, 100 TEA, 1 NaHepes, pH 7.2). Carbachol was purchased from Sigma (St. Louis, MO).

### Cell culture

Chinese hamster ovary cells were obtained from ATCC (American Type Culture Collection, MD) and transfected with the muscarinic cholinergic receptor m3 clone as described elsewhere (Bonner et al., 1988). A day before the experiment, cells were plated on 35 mm dishes (Corning Glass

Works, NY) at a density of 250 000 cells/ml (2 ml/dish) and cultured at 37°C in an atmosphere of 95% air and 5% $CO_2$.

*Electrophysiology*

Patch clamp experiments were carried out at room temperature (20–25°C). Measurements of membrane potential $(V_m)$ under voltage-clamp conditions were made using an EPC-7 patch-clamp amplifier (List-Electronics, Darmstadt-Eberstadt, Germany). Patch pipettes were pulled from soda glass capillary tubes using a BB-CH-PC puller (Mecanex, Switzerland). The pipettes were coated to the tip with silgard (Corning, NY). Pipettes filled with solutions had tip resistances from 6 to 10 MΩ. $V_m$ and $I_m$ signals at the outputs of the EPC-7 amplifier were made using a 4 channel analog magnetic tape recorder (frequency response 0–25 kHz). The corresponding records were digitized using a digital storage oscilloscope (Nicolet Instrument, Madison, WI).

## Results

*Inward current activated by carbachol in Chinese hamster ovary cells transfected and stably expressing muscarinic type 5 receptors*

Carbachol-activated $Ca^{2+}$ currents were characterized using the cell attached patch clamp technique. Using this technique, single channel events were measured and the second messenger dependency of channel activation was determined (Penner, 1995). Under patch clamp conditions, the pipet was held at a positive potential with high concentrations of $Ca^{2+}$ in the pipet in order to observe carbachol-activated inward currents. When the electrode potential was held at +20 mV, no basal activity could be detected (Fig. 2A). Furthermore, application of carbachol to the bath solution did not activate any channel events (Fig 2B). However, addition of carbachol, to the electrode solution revealed a burst of channel openings (Fig. 2C). Carbachol-activated inward currents were observed only when carbachol was present in the electrode solution, suggesting that activation of the

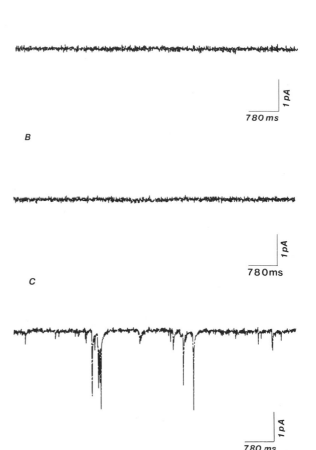

Fig. 2. Carbachol activated inward current in the cell attached configuration. Calcium ion conductance was measured in CHOm5 cells in the cell attached configuration. The holding potential was set at +20 mV. The recorded data were digitized through a low pass 8-pole Bessel filter set at 200 Hz. (A) Basal activity of a typical CHOm5 cell, before application of carbachol. (B) Application of carbachol to the bath solution revealed no activity. (C) Application of carbachol (100 M) into the electrode solution resulted in an inward current. Data shown are a representative record selected from at least three different cells.

channel occurred in a second messengers independent manner.

The carbachol-stimulated $Ca^{2+}$ conductance was further characterized by observing the current–voltage relationship after altering the membrane potential. In the presence of carbachol, the membrane potential was changed manually and the cur-

198

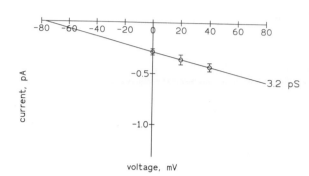

Fig. 3. Current–voltage relationship in the presence of carbachol in cell attached configuration. Current–voltage relationships and single channel conductance for ion channels activated by carbachol (100 mM). Single channel currents were filtered at 200 Hz and measured from records digitized using a digital storage oscilloscope. Changing the voltage was done manually. Each data point is the mean ± standard error of the mean of at least three different cells.

rent across the membrane was measured. Fig. 3 shows the current-voltage relationship of carbachol-activated $Ca^{2+}$ currents. From the slope of the curve we could calculate a single channel conductance of about 3 pS.

## Summary

We have provided preliminary characterization of a single channel $Ca^{2+}$ conductance in CHO cells. We have demonstrated that the channel conducts $Ca^{2+}$, is regulated by m5 receptors, is voltage-independent, has an extremely low conductance, and is second messenger-independent. This channel may be the receptor-operated channel required for downstream activation of several signaling events. It is not known what other cell types express the channel or if it is one of a larger group of related channels. It seems likely that $Ca^{2+}$ influx-dependent signaling pathways, activated by the muscarinic m5 receptor, would utilize a plasma membrane resident $Ca^{2+}$ channel to provide a steady source of $Ca^{2+}$ from outside the cell. The transient nature of IP3-activated increases in intracellular $Ca^{2+}$ make it an unlikely source of the sustained $Ca^{2+}$ rise required for phospholipase regulation. This is especially surprising, since levels of intracellular $Ca^{2+}$ achieved from the release

of intracellular $Ca^{2+}$ stores can be at least one order of magnitude higher than those achieved from extracellular influx (Berridge, 1993). The phospholipase A2 and phospholipase D involved in muscarinic receptor-mediated signaling have not been purified or cloned. It is possible that receptor-activated and $Ca^{2+}$ influx-dependent phospholipases are integral membrane proteins located adjacent to both receptors and channels. The phospholipases may also translocate to the membrane following activation where they would gain access to the continuous $Ca^{2+}$ flow. Purification and cloning of this and other related channels should provide better insight into their role in cell signaling.

## References

Berridge, M.J. (1993) Inositol trisphosphate and calcium signaling. *Nature,* 361: 315–325.

Bonner, T.I., Young, A.C., Brann, M.R. and Buckley, N.J. (1988) Cloning and expression of the human and rat m5 acetylcholine receptor genes. *Neuron,* 1: 403–410.

Brooks, R.C., McCarthy, K.D., Lapetina, E.G. and Morell, P. (1989) Receptor-stimulated phospholipase $A_2$ activation is coupled to influx of external calcium and not to mobilization of intracellular calcium in C62B glioma cells. *J. Biol. Chem.,* 264: 20147–20153.

Brown, A. and Birnbaumer, L. (1990) Ionic channels and their regulation by G protein subunits. *Annu. Rev. Physiol.,* 52: 197–213.

Caulfield, M.P. (1993) Muscarinic receptors- characterization, coupling and function. *Pharmacol. Ther.,* 58: 319–379.

Fasolato, C., Innocenti, B. and Pozzan, T. (1994) Receptor-activated calcium influx: how many mechanisms for how many channels? *Trends Pharmacol. Sci.,* 15: 77–83.

Felder, C.C. (1995) Muscarinic acetylcholine receptors: signal transduction through multiple effectors. *FASEB J.,* 9: 619–625.

Felder, C.C., Dieter, P., Kinsella, J., Tamura, K., Kanterman, R.Y. and Axelrod, J. (1990) A transfected m5 muscarinic acetylcholine receptor stimulates phospholipase A2 by inducing both calcium influx and activation of protein kinase C. *J. Pharmacol. Exp. Ther.,* 255: 1140–1147.

Felder, C.C., Poulter, M. and Wess, J. (1991) Muscarinic receptor-operated $Ca^{2+}$ influx in transfected fibroblast cells is independent of inositol phosphates and release of intracellular calcium. *Proc. Natl. Acad. Sci. USA,* 89: 509–513.

Felder, C.C., MacArthur, L., Ma, A.L., Gusovsky, F. and Kohn, E.C. (1993) Tumor suppressor function of muscarinic acetylcholine receptors is associated with activation of receptor-operated calcium influx. *Proc. Natl. Acad. Sci. USA,* 90: 1706–1710.

Felder, C.C., Singer-Lahat, D. and Mathes, C. (1994) Voltage-insensitive calcium channels: Regulation by receptors and intracellular calcium stores. *Biochem. Pharmacol.*, 48: 1997–2004.

Felder, C.C., Joyce, K.E., Briley, E.M., Mansouri, J., Mackie, K., Blond, O., Lai, Y., Ma, A.L. and Mitchell, R.L. (1995) Comparison of the pharmacology and signal transduction of the human cannabinoid CB1 and CB2 receptors. *Mol. Pharmacol.*, 48: 443–450.

Hildebrandt, J.P. and Shuttleworth, T.J. (1993) A Gq-type G protein couples muscarinic receptors to inositol phosphate and calcium signalling in exocrine cells from the avian salt gland. *J. Membr. Biol.*, 133: 183–190.

Lambert, D.G., Wojcikiewicz, R.J., Safrany, S.T., Whitman, E.M. and Nahorski, S.R. (1992) Muscarinic receptors, phosphoinositide metabolism and intracellular calcium in neuronal cells. *Prog. Neuropsychopharmacol. Biol. Psychiatry*, 16: 253–270.

Logothetis, D., Kurachi, Y., Galper, J., Neer, E. and Clapham, D. (1987) The subunits of GTP-binding proteins activate the muscarinic $K^+$ channel in heart. *Nature*, 325: 321–326.

Mathes, C. and Thompson, S.H. (1994) Calcium current activated by muscarinic receptors and thapsigargin in neuronal cells. *J. Gen. Physiol.*, 104: 107–121.

McDonald, T.V., Premack, B.A. and Gardner, P. (1993) Flash photolysis of caged inositol-1,4,5-trisphosphate activates plasma membrane calcium current in human T cells. *J. Biol. Chem.*, 268: 3889–3896.

Neher, E. (1992) Ion channel for communication between and within cells. *Science,* 256: 498–502.

Oettling, G., Gotz, U. and Drews, U. (1992) Characterization of the calcium influx into embryonic cells after stimulation of the embryonic muscarinic receptor. *J. Dev. Physiol.*, 17: 147–155.

Penner, R. (1995) A partial guide to patch clamping. In: E. Sakmann and E. Neher (Eds.), *Single Channel Recording*, 2nd edn., Plenum Press, New York, pp. 3–30.

Roche, S., Bali, J.P. and Magous, R. (1993) Receptor-operated calcium channels in gastric parietal cells: gastrin and carbachol induce calcium influx in depleting intracellular calcium stores. *Biochem. J.*, 289: 117–124.

Sawaki, K., Hiramatsu, Y., Baum, B.J. and Ambudkar, I.S. (1993) Involvement of G alpha q/11 in m3-muscarinic receptor stimulation of phosphatidylinositol-4,5-bisphosphate-specific phospholipase C in rat parotid gland membranes. *Arch. Biochem. Biophys.*, 305: 546–550.

Shuttleworth, T.J. and Thompson, J.L. (1992) Modulation of inositol(1,4,5)tris-phosphate-sensitive calcium store content during continuous receptor activation and its effects on calcium entry. *Cell Calcium*, 13: 541–551.

J. Klein and K. Löffelholz (Eds.)
*Progress in Brain Research*, Vol. 109

CHAPTER 19

# Muscarinic activation of phosphatidylcholine hydrolysis

Jochen Klein, Ruth Lindmar and Konrad Löffelholz

*Department of Pharmacology, University of Mainz, Mainz, Germany*

## Release of choline, a sensor for receptor-activated phosphatidylcholine hydrolysis

The release of choline from tissues or cells is a sensitive indicator of an enhanced hydrolysis of phosphatidylcholine (PtdCho) and is easily determined by chemiluminescence (Israel and Lesbats, 1982), HPLC (Klein et al., 1993) or radioactive labelling. By following the choline release from the rat cortex in vivo into the "cortical cup" and from the isolated chick heart into the perfusate, Corradetti et al. (1983) were the first to observe the muscarinic activation of PtdCho hydrolysis. The effect in the heart was accompanied by an increase of phosphatidic acid mass, was independent of extracellular calcium and was mimicked by oleate, a well-characterized activator of phospholipase D (PLD; Hattori and Kanfer, 1985), and therefore, a muscarinic activation of PLD was suggested to be responsible for the observed phenomenon (Lindmar et al., 1986, 1988).

In certain cells choline release may reflect the activity of a specific receptor-activated enzyme catalyzing PtdCho hydrolysis; for example, the release of free choline from HL60 cells evoked by the ADP-ribosylating factor (ARF) was correlated to the activation of PLD (Cockcroft et al., 1994; Chapters 17 and 20). Nevertheless, certain precautions have to be taken into account, whenever choline is used as a sensor for receptor-activated PtdCho hydrolysis. Especially other sources of choline or even simple translocation from the cytoplasmic compartment have to be excluded. Exclusion of translocation is relevant for two reasons: first, the intracellular choline concentration is sev-eral-fold higher than the extracellular concentration. Second, choline liberated from PtdCho hydrolysis may at first be released into the cytoplasm. Martinson et al. (1990) found that muscarinic activation of PLD in astrocytoma cells rapidly released choline into the intracellular space and, after a time-lag of several minutes, into the extracellular space. Using hemicholinium as an inhibitor of cellular choline uptake, Lee et al. (1993) also concluded that choline is released from PtdCho into the cytoplasmic space in LA-N-2 cells. It appears likely, therefore, that PLD works on PtdCho pools which are present in the inner leaflet of the plasma membrane or in internal membranes. The problem of translocation can be excluded by measuring net-formation of choline rather than choline efflux which requires a parallel monitoring of changes in the extracellular and tissue/cellular choline concentrations. Another problem, at least in neuronal tissues, is given by the possible contribution of ACh hydrolysis to choline release, a problem that is avoided when the experiments are done in the presence of an acetylcholinesterase inhibitor.

A physiological role of the receptor-mediated release of choline in the brain is given by its role as biosynthetic precursor for ACh and phospholipids. However, the choline concentration normally is not rate-limiting for these syntheses and, in addition, is kept constant within a narrow range due to effective homeostatic mechanisms (Löffelholz et al., 1993). More specifically, the muscarinic mobilization of choline was hypothesized to be a way that couples acetylcholine release to an adequate supply of the precursor, cho-

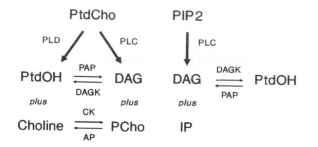

Fig. 1. Products of PtdCho and phosphatidylinositol-4,5-bisphosphate (PIP2) hydrolysis catalyzed by PLD and PI-PLC activity. Further abbreviations: AP, alkaline phosphatase; CK, choline kinase; DAG, diacylglycerol; DAGK, DAG kinase; IP, inositol phosphates; PAP, PtdOH phosphohydrolase; PCho, phosphocholine; PtdOH, phosphatidic acid.

line, under experimental and pathophysiological conditions of enhanced neuronal firing (Löffelholz, 1987; see also Lee et al., 1993).

## Receptor-mediated hydrolysis of PtdCho: PLC or PLD?

When PtdCho hydrolysis is investigated to identify the phospholipase involved, the sole determination of enzymatic products may provide ambiguous results due to rapid metabolic interconversion. Thus, the products of PtdCho-PLC, diacylglycerol and phosphocholine, can be rapidly transformed to phosphatidic acid and choline, which are the products of PLD activation (Fig. 1). The responsible enzymes for these interconversion are ubiquitous and constitutive enzymes which are involved in the synthesis and breakdown of phospholipids. Consequently the release of choline, or the elevation of diacylglycerol per se are not indicative of a certain enzymatic activity. Additional experiments may prove the identity of the phospholipase involved (see above). For example, an early formation of an enzymatic product prior to other metabolites may elucidate the sequence of metabolic steps (e.g., Martinson et al., 1990) or a specific assay of a phospholipase activity identifies the nature of the choline release (see below; Cockcroft et al., 1994).

A specific assay is not available for PtdCho-PLC, but for PLD: the "transphosphatidylation assay". This assay is based on the formation of phosphatidyl-alcohols (e.g., phosphatidylethanol, PEth) in the presence of primary alcohols (e.g., ethanol). To our current knowledge, only PLD is capable of forming phosphatidylalcohols (Pai et al., 1988). Specific inhibitors are lacking for both PtdCho-PLC and PLD. PtdCho-PLC has been partially purified from mammalian tissues (Wolf and Gross, 1985; Sheikhnejad and Srivastava, 1986). Mammalian PLD has been remarkably resistant to purification (Taki and Kanfer, 1979; Chalifa et al., 1990), but Okamura and Yamashita (1994) succeeded in the purification to apparent homogeneity of a PLD species from pig lung. Very recently, the first animal PtdCho-specific PLD cDNA, a human PLD1 (hPLD1), was isolated from a HeLa cDNA library (Hammond et al., 1995; the authors detected a second PLD gene with about 60% identity). The reported sequence belongs to the subspecies activated by ADP-ribosylation factor (ARF) (Massenburg et al., 1994; Chapter 20). At about the same time, a gene encoding PLD activity in yeast has been described which is highly homologous to hPLD1 (Waksman et al., 1996). It is very likely that the identification of these genes will help to advance our knowledge of the physiological functions of PLD in the near future.

## Muscarinic receptor subtypes associated with phospholipid hydrolysis

Hydrolysis of phosphoinositides by phospholipase(s) C (PI-PLC) is mediated preferentially by muscarinic receptors of the m1, m3 and m5 subtype (Chapters 14 and 17) via pertussis toxin-(PTX-)insensitive G proteins of the Gq/G11 family (Caulfield, 1993; Chapter 17). Studies in transfected fibroblasts indicated that these subtypes also mediate PtdCho hydrolysis by PLD while those subtypes that are coupled to Go/Gi and adenylyl cyclase (m2, m4) do not (Pepitoni et al., 1991; Sandmann et al., 1991). However, the type of G protein involved in the muscarinic hydrolysis of PtdCho is as yet unknown. In most studies, the receptor-mediated hydrolysis of PtdCho was PTX-insensitive, but in cells of the hematopoietic system, inhibition by PTX has been observed indicating a role of Go/Gi (Liscovitch and Chalifa, 1994).

## Muscarinic activation of PtdCho hydrolysis in myocardial and other non-neuronal tissues

Most of the work on receptor-mediated hydrolysis of PtdCho has been done in non-neuronal tissues or cells (Löffelholz, 1990; Cockcroft, 1992; Liscovitch and Chalifa, 1994). This is true also for the muscarinic activation of PtdCho breakdown which, for example, has been found in fibroblasts (Diaz-Meco et al., 1989). In m1-transfected fibroblasts, carbachol caused an activation of both PtdCho-PLC and PLD; while PLD activation rapidly desensitized within 2 min, PtdCho-PLC activity was increased for at least 60 min (McKenzie et al., 1992). Rapid desensitization was also observed in m3-transfected fibroblasts (Schmidt et al., 1995). Glandular PLD forms are also activated by muscarinic agonists: In parotid and submandibular glands, a muscarinic receptor-mediated PtdCho hydrolysis was mediated by PLD activation (Guillemain and Rossignol, 1992; Duner-Engström and Fredholm, 1994), and carbachol was found to stimulate PLD activity in lacrimal gland cells (Zoukhri and Dartt, 1995). In contrast, in carbachol-stimulated pancreatic acinar cells, Rubin et al. (1992) observed increased levels of phosphatidic acid and diaclyglycerol but no stimulation of PEth formation.

We have studied the muscarinic activation of PtdCho hydrolysis in myocardial tissues of rats, chicks and guinea-pigs (Löffelholz, 1989). Phorbol ester through activation of protein kinase C enhanced the formations of choline, phosphatidic acid and, in the presence of ethanol, PEth indicating an activation of PLD (Sandmann et al., 1990; Lindmar and Löffelholz, 1992). Aluminium fluoride (AlF), an activator of trimeric G-proteins, also enhanced both PLD and PI-PLC activity (Lindmar and Löffelholz, 1992). It is still unknown whether a G protein is directly coupled to PLD. Recent experiments using non-neuronal cells in culture indicate that the pathway upstream to PLD activation is complex depending on the cell type: tyro-sine kinases, protein kinase C and/or small G proteins (Chapter 20) may be linked into the signal cascade. Thus, in m3-transfected fibroblasts, PLD was activated by muscarinic agonists, and this activation seemed to involve tyrosine kinases (Schmidt et al., 1994).

In heart preparations we found that Ro 31-8220, a specific inhibitor of protein kinase C, blocked the PLD response to phorbol esters, but did not change the response to AlF. Thus, protein kinase C is not linked into the signal cascade coupling trimeric G proteins to PLD, at least in myocardial cells. However, protein kinase C seems to modulate the PLD and PI-PLC pathways: phorbol esters markedly potentiated the AlF-evoked PLD activity, whereas the AlF-evoked PI-PLC activity was inhibited (Lindmar, Klein and Löffelholz, unpublished). As suggested earlier (Löffelholz, 1989), protein kinase C seems to control the G protein-coupled PLD and the PI-PLC pathways in an opposite way, and, therefore, may play a key role in the feedback regulation of phospholipase-mediated signal transduction.

The muscarinic receptor-activation of PLD was previously described for the chick heart preparation (Lindmar et al., 1988). In rat isolated atria, acetylcholine ($10^{-5}$ M) only slightly enhanced the formation of PEth by $65 \pm 24\%$ ($N = 6$, $P < 0.05$) of control, whereas PtdOH was unchanged ($-6.9 \pm 9.2\%$). For carbachol ($10^{-4}$ M), the respective values were $+27 \pm 19\%$ for PEth and $+18 \pm 6\%$ for PtdOH ($N = 4$). These effects are small as compared to the 4-fold increases of PEth caused by either AlF or phorbol ester in the same preparation. Interestingly, the increase of inositol monophosphate ($IP_1$) after carbachol stimulation was also much smaller (2-fold; not documented) than the 5-fold increase caused by AlF (Lindmar and Löffelholz, 1992). The physiological role of a muscarinic activation of PLD as well as of PI-PLC in myocardial function is a matter of speculation. It seems unlikely that the receptor-mediated activation of PLD plays a major role in rapid signal transduction. There is growing evidence for a role in sustained cellular responses and in the modulatory signalling to deeper cellular compartments associated with phenomena such as vesicular trafficking, cytoskeletal functions, mitogenesis, gene expression and growth (Chapter 17; Nishizuka, 1995). This view is corroborated by recent experimental data suggesting that PLD activation is responsible for the mechanical load-induced hypertrophy of myocardial cells (Sadoshima and Izumo, 1993).

It may be added that receptor-mediated hydrolysis of PtdCho could also be catalyzed by a PtdCho-specific PLC type alternatively to the PLD activity. Only few data are available for this pathway. In the heart, a PtdCho-specific PLC has been described by Wolf and Gross (1985) and was found to be activated in cardiac myocytes by anoxic injury (Nachas and Pinson, 1992).

In conclusion, there is evidence for a muscarinic hydrolysis of PtdCho catalyzed by PLD activity in transfected fibroblasts, glandular cells and, to a small extent, mammalian myocardial tissue. To our knowledge, muscarinic activation of PLD has not been observed in smooth muscle cells of, e.g., visceral and bronchial tissue. In contrast to the well-known muscarinic activation of PI-PLC present in these tissues, an enhancement of the PLD activity by carbachol was not found in our laboratory, although basal PLD activity was detectable (Lindmar and Löffelholz, unpublished results).

## Muscarinic activation of PtdCho hydrolysis in cells of neural or glial origin

Several agonists have been shown to elevate PLD activity in nervous tissue and glial cells (Klein et al., 1995). Data shown in Table 1 document the presence of PLD activity in hippocampal slices,

synaptosomes and primary astrocytes prepared from rat brain. Phorbol ester markedly stimulated the PLD activity in all preparations, but there were clearcut differences with respect to receptor-mediated activation.

In rat brain slices, PLD was activated by glutamate (Holler et al., 1993), noradrenaline (Llahi and Fain, 1992), histamine and endothelin in a $Ca^{2+}$-dependent manner (Sarri et al., 1995). The activation by glutamate showed a striking ontogenetic development: glutamate stimulated PLD activity in hippocampal slices taken from young rats, while those of adult animals did not respond (Holler et al., 1993). The data in Table 1 suggest that the glutamate receptors activating PLD may be located at neuronal soma or dendrites as PLD could not be activated by glutamate in either glial cells or presynaptic terminals.

A muscarinic activation of hippocampal PLD could not be observed in mature slices (Llahi and Fain, 1992; Holler et al., 1994; Sarri et al., 1995). However, we found that carbachol in high concentrations (1 mM) stimulated PLD activity in hippocampal slices taken from 8 days old rats (Fig. 2). Similar data were recently published by Costa et al. (1995).

Synaptosomal PLD was investigated in synaptic terminals prepared from adult rat cortex. In this preparation, basal PLD activity was present but did not respond to acetylcholine, glutamate or aluminium fluoride (Seimetz and Klein, unpublished) indicating that PLD activity of nerve terminals is not activated by agonists and not coupled to G proteins.

The PLD activity of primary astrocytes isolated

TABLE 1

Stimulation of phospholipase D activity in different preparations from rat brain

| Stimulus | Hippocampal slice (mature) | Hippocampal slice (immature) | Synaptosomes (cortex; adults) | Astrocytes (cortex) |
|---|---|---|---|---|
| Carbachol | − | + | − | + |
| Noradrenaline | + | + | − | + |
| Glutamate | − | + | − | − |
| Phorbol esters | ++ | ++ | ++ | ++ |

Data taken from Holler et al. (1993, 1994) and unpublished experiments.

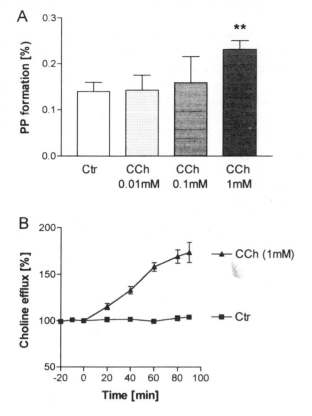

A

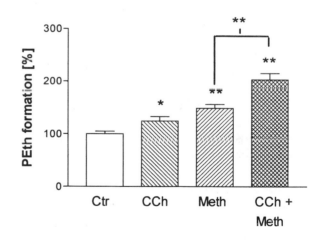

Fig. 2. Effects of the muscarinic agonist carbachol (CCh) on PLD activity (A) and choline efflux (B) from hippocampal slices of 8d old rats. (A) For the determination of phospholipase D activity, the slices were prelabelled by [³H]glycerol. After a washing period, carbachol was added together with 2% propanol. After 30 min, the slices were homogenized, and the phospholipids were separated by two-dimensional thin-layer chromatography. The content of phosphatidylpropanol (PP) which reflects PLD activity was determined by scintillation counting. The results are given as [%] label of total lipid phase and are means ± SEM of 4–6 experiments. For detailed procedures, see Holler et al. (1993). (B) For the determination of choline efflux, the slices were perfused with Krebs–Henseleit buffer, carbachol (CCh; 1 mM) was added at time zero, and choline efflux from the slices was monitored by a chemoluminescence method (for details, see Holler et al., 1994). The results are expressed as % of the basal efflux which was 72 ± 5 pmol/mg protein per min and are the means ± SEM of 4–6 experiments. In (A), **corresponds to $P < 0.01$ versus control (Ctr.). Designations of statistical significance were omitted in (B).

from newborn rats was markedly stimulated by AlF and by noradrenaline through stimulation of

$\alpha$-adrenoceptors. Carbachol ($100\,\mu$M) proved to be a relatively weak stimulator; carbachol and methoxamine, an $\alpha$1-adrenergic agonist, showed an additive activation of the PLD (Fig. 3; Gonzalez, Klein and Löffelholz, unpublished). The noradrenergic activation of PLD in the hippocampal slices (see above) might have been due to the glial cell component of the brain slice.

The muscarinic activation of PtdCho hydrolysis has been investigated in a range of glial and neuronal cell culture systems. A muscarinic PLD activation has been described in HOG oligodendroglioma cells (Dawson et al., 1993) as well as in 1321N1 astrocytoma cells (Martinson et al., 1989). In primary astrocytes from rat brain, Gustavsson et al. (1993) have presented clear-cut data on muscarinic PLD activation while Bruner and Murphy (1990) were unable to detect this effect. The reason for the discrepancies between the data from

Fig. 3. Synergistic effects of alpha-adrenergic and muscarinic stimulation of phospholipase D activity in primary astrocytes. Astrocytes were prepared from the cortices of newborn rats, grown for 2–4 weeks in culture until near confluence, and prelabeled with [³H]glycerol for 24 h. For the determination of PLD activity, the $\alpha$1-adrenergic agonist methoxamine (Meth; $100\,\mu$M) and/or the muscarinic agonist carbachol (CCh; $100\,\mu$M) were added to the cultures together with 2% ethanol. After 10 min, the cells were homogenized, the phospholipids were separated by thin-layer chromatography, and the radioactivity associated with phosphatidylethanol (PEth) was measured and expressed as % of the basal PEth formation in the presence of ethanol but in the absence of agonists. Statistical significance: *$P < 0.05$; **$P < 0.01$.

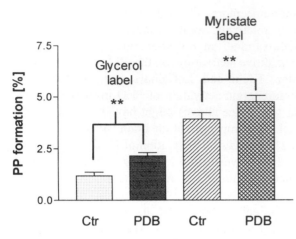

Fig. 4. Influence of the labeling procedure on phospholipase D activity measurements. "Glycerol label": hippocampal slices from adult rats were prelabeled with [³H]glycerol for 2 h resulting in a labeling pattern of phospholipids in which phosphatidylcholine contained 49% of the total label of phospholipids. "Myristate label": hippocampal slices from adult rats were prelabeled with [³H]myristic acid for 2 h resulting in a labeling pattern of phospholipids in which phosphatidylcholine contained 70% of total phospholipids. After superfusion with Krebs–Henseleit buffer containing 0.6 μM phorbol dibutyrate (PDB) and 2% propanol for 90 min, the slices were extracted, and PLD activity was determined as the amount of label associated with phosphatidylpropanol (PP) compared to the label of total phospholipids. **$P < 0.01$ versus control values (Ctr).

different laboratories is unclear. Possibly quantitative differences might have been due to the type of label used for phospholipid analysis (Fig. 4).

In cells of neural origin, muscarinic hydrolysis of PtdCho has been described in human LA-N-2 neuroblastoma cells (Sandmann and Wurtman, 1991) and in SK-N human neuroblastoma cells (Pacini et al., 1993); in the latter case, PtdCho hydrolyses may have been due to a PtdCho-PLC activation. A muscarinic activation of PtdCho hydrolysis has also been described in adrenal chromaffin cells (del Carmen Garcia et al., 1992), but the mechanism of this effect is a matter of controversy (Boarder, 1993).

In conclusion, muscarinic activation of PLD in neuronal tissue is an exception rather than a rule. As this was similarly found in non-neuronal tissues, the general role of muscarinic activation of PLD in cellular activities may have to be reconsidered.

## Phospholipase D: a role in signal transduction and cellular activities?

While the functional role of PLD remains mysterious, recent findings seem to indicate that the activity of PLD is coupled more likely to sustained cellular responses than to rapid responses to individual transmitters or hormones. Constitutive cellular functions which may relate to PLD activity include cytoskeletal arrangement, and growth and mitogenesis (Boarder, 1994). The presence of muscarinic activation of PLD in immature but not in mature brain tissue may be explained by some association of PLD with developmental processes in the central nervous system. In this respect one is reminded to the observation that acetylcholine has been identified as a mitogenic signal for developing astrocytes (Ashkenazi et al., 1989). The recent demonstration that small G-proteins such as ARF and rho interact with PLD (see Chapter 20) has led to speculation that PLD may be involved in the transport and membrane modification of neuronal vesicles (Kahn et al., 1993). In conclusion, muscarinic receptor activation (and possibly receptor activation in general) may not be essential as an excitatory input for PLD activation, at least in many of the neuronal and non-neuronal tissues so far investigated. Instead, PLD may be linked to intrinsic cellular activities which may receive modulatory signals from transmembrane receptors.

## Acknowledgements

The original studies of the authors were supported by the Deutsche Forschungsgemeinschaft (DFG) and the German-Israeli-Foundation (GIF). We are grateful for the excellent technical assistance by Mrs. M. Campanini and Mrs. U. Kreis.

## References

Ashkenazi, A., Ramachandran, J. and Capon, D.J. (1989) Acetylcholine analogue stimulates DNA synthesis in brain-derived cells via specific muscarinic receptor subtypes. *Nature*, 340: 146–150.

Boarder, M.R. (1993) Phospholipase D in chromaffin cells ? *J. Neurochem.*, 60: 1978–1979.

Boarder, M.R. (1994) A role for phospholipase D in control of mitogenesis. *Trends Pharmacol. Sci.*, 15: 57–62.

Bruner, G. and Murphy, S. (1990) Regulation of phospholipase D in astroglial cells by calcium- activated protein kinase C. *Mol. Cell. Neurosci.*, 1: 146–150.

Caulfield, M.P. (1993) Muscarinic receptors - Characterization, coupling and function. *Pharmacol. Ther.*, 58: 319–379.

Chalifa, V., Möhn, H. and Liscovitch, M. (1990) A neutral phospholipase activity from rat brain synaptic plasma membranes. *J. Biol. Chem.*, 265: 17512–17519.

Cockcroft, S. (1992) G-protein-regulated phospholipases C, D and $A_2$-mediated signalling in neutrophils. *Biochim. Biophys. Acta,* 1113: 135–160.

Cockcroft, S., Thomas, G.M.H., Fensome, A., Geny, B., Cunningham, E., Gout, I., Hiles, I., Totty, N.F., Truong, O. and Hsuan, J.J. (1994) Phospholipase D: a downstream effector of ARF in granulocytes. *Science*, 263: 523–526.

Corradetti, R., Lindmar, R. and Löffelholz, K. (1983) Mobilization of cellular choline by stimulation of muscarine receptors in isolated chicken heart and rat cortex in vivo. *J. Pharmacol. Exp. Ther.*, 226: 826–832.

Costa, L.G., Balduini, W. and Reno, F. (1995) Muscarinic receptor stimulation of phospholipase D activity in the developing rat brain. *Neurosci. Res. Commun.*, 17: 169–176.

Dawson, G., Dawson, S.A. and Post, G.R. (1993) Regulation of phospholipase D activity in a human oligodendroglioma cell line (HOG). *J. Neurosci. Res.*, 34: 324–330.

del Carmen Garcia, M., Lopez, M.G., Garcia, A.G. and Crespo, M.S. (1993) Muscarinic acetylcholine receptor enhances phosphatidylcholine hydrolysis via phospholipase D in bovine chromaffin cells in culture. *J. Neurochem.*, 59: 2244–2250.

Diaz-Meco, M.T., Larrodera, P., Lopez-Barahona, M., Cornet, M.E., Barreno, P.G. and Moscat, J. (1989) Phospholipase C-mediated hydrolysis of phosphatidylcholine is activated by muscarinic agonists. *Biochem. J.*, 263: 115–120.

Duner-Engström, M. and Fredholm, B.B. (1994) Carbachol-induced phosphatidylcholine hydrolysis and choline efflux in rat submandibular gland involves phospholipase D activation and is modulated by protein kinase C and calcium. *Acta Physiol. Scand.*, 151: 515–525.

Guillemain, I. and Rossignol, B. (1992) Evidence for receptor-linked activation of phospholipase D in rat parotid glands. *FEBS Lett.*, 314: 489–492.

Gustavsson, L., Lundqvist, C. and Hansson, E. (1993) Receptor-mediated phospholipase D activity in primary astroglial cultures. *Glia*, 8: 249–255.

Hammond, S.M., Altshuller, Y.M., Sung, T.-C., Rudge, S.A., Rose, K., Engebrecht, J., Morris, A.J. and Frohman, M.A. (1995) Human ADP-ribosylation factor-activated phosphatidylcholine-specific phospholipase D defines a new and highly conserved gene family. *J. Biol. Chem.*, 270: 29640–29643.

Hattori, H. and Kanfer, J.N. (1985) Synaptosomal phospholipase D potential role in providing choline for acetylcholine synthesis. *J. Neurochem.*, 45: 1578–1584.

Holler, T., Cappel, E., Klein, J. and Löffelholz, K. (1993) Glutamate activates phospholipase D in hippocampal slices of newborn and adult rats. *J. Neurochem.*, 61: 1569–1572.

Holler, T., Klein, J. and Löffelholz, K. (1994) Phospholipase C and phospholipase D are independently activated in rat hippocampal slices. *Biochem. Pharmacol.*, 47: 411–414.

Israel, M. and Lesbats, B. (1982) Application to mammalian tissues of the chemiluminescent method for detecting acetylcholine. *J. Neurochem.*, 39: 248–250.

Kahn, R.A., Yucel, J.K. and Malhotra, V. (1993) ARF signalling: a potential role for phospholipase D in membrane traffic. *Cell,* 75: 1045–1048.

Klein, J., Gonzalez, R., Köppen, A. and Löffelholz, K. (1993) Free choline and choline metabolites in rat brain and body fluids: sensitive determination and implications for choline supply to the brain. *Neurochem. Int.*, 22: 293–300.

Klein, J., Chalifa, V., Liscovitch, M. and Löffelholz, K. (1995) Role of phospholipase D activation in nervous system physiology and pathophysiology. *J. Neurochem.*, 65: 1445- 1455.

Lee, H.-C., Fellenz-Malloney, M.-P., Liscovitch, M. and Blusztajn, J.K. (1993) Phospholipase D-catalyzed hydrolysis of phosphatidylcholine provides the choline precursor for acetylcholine synthesis in a human neuronal cell line. *Proc. Natl. Acad. Sci. USA*, 90: 10086–10090.

Lindmar, R. and Löffelholz, K. (1992) Phospholipase D in heart: basal activity and stimulation by phorbol esters and aluminum fluoride. *Naunyn Schmiedberg's Arch. Pharmacol..*, 346: 607–613.

Lindmar, R., Löffelholz, K. and Sandmann, J. (1986) Characterization of choline efflux from the perfused heart at rest and after muscarinic receptor activation. *Naunyn-Schmiedeberg's Arch. Pharmacol.*, 332: 224–229.

Lindmar, R., Löffelholz, K. and Sandmann, J. (1988) On the mechanism of muscarinic hydrolysis of choline phospholipids in the heart. *Biochem. Pharmacol.*, 37: 4689–4695.

Liscovitch, M. and Chalifa, V. (1994) Signal-activated phospholipase D. In: M. Liscovitch (Ed.), *Signal-activated Phospholipases*, R.G. Landes, Austin, TX, pp. 31–63.

Llahi, S. and Fain, J.N. (1992) alpha1-adrenergic receptor-mediated activation of phospholipase D in rat cerebral cortex. *J. Biol. Chem.*, 267: 3679–3685.

Löffelholz, K. (1987) Nutrients affecting brain composition and behaviour (commentary). *Integr. Psych.*, 5: 242–244.

Löffelholz, K. (1989) Receptor regulation of choline phospholipid hydrolysis: a novel source of diacylglycerol and phosphatidic acid. *Biochem. Pharmacol.*, 38: 1543–1549.

Löffelholz, K. (1990) Receptors linked to hydrolysis of choline phospholipids: the role of phospholipase D in a putative mechanism of signal transduction. In: N.N. Osborne (Ed.), *Current Aspects of the Neurosciences*, Vol. 2, Macmillan, New York, pp. 49–76.

Löffelholz, K., Klein, J. and Köppen, A. (1993) Choline, a

precursor of acetylcholine and phospholipids in the brain. *Prog. Brain Res.*, 98: 199–202.

Martinson, E.A., Goldstein, D. and Brown, J.H. (1989) Muscarinic receptor activation of phosphatidylcholine hydrolysis: relationship to phosphoinositide hydrolysis and diacylglycerol metabolism. *J. Biol. Chem.*, 264: 14748–14754.

Martinson, E.A., Trilivas, I. and Brown, J.H. (1990) Rapid protein kinase C-dependent activation of phospholipase D leads to delayed 1,2-diglyceride accumulation. *J. Biol. Chem.*, 265: 22282–22287.

Massenburg, D., Han, J.-S., Liyanage, M., Patton, W.A., Rhee, S.G., Moss, J. and Vaughan, M. (1994) Activation of rat brain phospholipase D by ADP-ribosylation factors 1, 5 and 6: separation of ADP-ribosylation factor-dependent and oleate-dependent enzymes. *Proc. Natl. Acad. Sci. USA*, 91: 11718–11722.

McKenzie, F.R., Seuwen, K. and Pouysségur, J. (1992) Stimulation of phosphatidylcholine breakdown by thrombin and carbachol but not by tyrosine kinase receptor ligands in cells transfected with M1 muscarinic receptors. *J. Biol. Chem.*, 267: 22759–22769.

Nachas, N. and Pinson, A. (1992) Anoxic injury accelerates phosphatidylcholine degradation in cultured cardiac myocytes by phospholipase C. *FEBS Lett.*, 298: 301–305.

Nishizuka, Y. (1995) Protein kinase C and lipid signaling for sustained cellular responses. *FASEB J.*, 9: 484–496.

Okamura, S.-i. and Yamashita, S. (1994) Purification and characterization of phosphatidylcholine phospholipase D from pig lung. *J. Biol. Chem.*, 269: 31207–31213.

Pacini, L., Limatola, C., Frati, L., Luly, P. and Spinedi, A. (1993) Muscarinic stimulation of SK-N-BE(2) human neuroblastoma cells elicits phosphoinositide and phosphatidylcholine hydrolysis: relationship to diacylglycerol and phosphatidic acid accumulation. *Biochem. J.*, 289: 269–275.

Pai, J.K., Siegel, M.I., Egan, R.W. and Billah, M.M. (1988) Phospholipase D catalyzes phospholipid metabolism in chemotactic-peptide stimulated HL-60 granulocytes. *J. Biol. Chem.*, 263, 12472–12477.

Pepitoni, S., Mallon, R.G., Pai, J.-K., Borkowski, J.A., Buck, M.A. and McQuade, R.D. (1991) Phospholipase D activity and phosphatidylethanol formation in stimulated HeLa cells expressing the human m1 muscarinic acetylcholine receptor gene. *Biochem. Biophys. Res. Commun.*, 176: 453–458.

Rubin, R.P., Hundley, T.R. and Adolf, M.A. (1992) Regulation of diacylglycerol levels in carbachol-stimulated pancreatic acinar cells: relationship to the breakdown of phosphatidylcholine and metabolism to phosphatidic acid. *Biochim. Biophys. Acta*, 1133: 127–132.

Sadoshima, J. and Izumo, S. (1993) Mechanical stretch rap-

idly activates multiple signal transduction pathways in cardiac myocytes: potential involvement of an autocrine/paracrine mechanism. *EMBO J.*, 12: 1681–1692.

Sandmann, J. and Wurtman, R.J. (1991) Stimulation of phospholipase D activity in human neuroblastoma (LA-N-2) cells by activation of muscarinic acetylcholine receptors or by phorbol esters: relationship to phosphoinositide turnover. *J. Neurochem.*, 56: 1312–1319.

Sandmann, J., Leissner, J., Lindmar, R. and Löffelholz, K. (1990) The effects of phorbol esters on choline phospholipid hydrolysis in heart and brain. *Eur. J. Pharmacol.*, 188: 89–95.

Sandmann, J., Peralta, E.G. and Wurtman, R.J. (1991) Coupling of transfected muscarinic acetylcholine receptor subtypes to phospholipase D. *J. Biol. Chem.*, 266: 6031–6034.

Sarri, E., Picatoste, F. and Claro, E. (1995) Histamine H1 and endothelin ETB receptors mediate phospholipase D stimulation in rat brain hippocampal slices. *J. Neurochem.* 65, 837–841.

Schmidt, M., Hüwe, S.M., Fasselt, B., Homann, D., Rümenapp, U., Sandmann J. and Jakobs, K.H. (1994) Mechanisms of phospholipase D stimulation by m3 muscarinic acetylcholine receptors. *Eur. J. Biochem.*, 225: 667–675.

Schmidt, M., Fasselt, B., Rümenapp, U., Bienek, C., Wieland, T., van Koppen, C.J. and Jakobs, K.H. (1995) Rapid and persistent desensitization of m3 muscarinic acetylcholine receptor-stimulated phospholipase D. *J. Biol. Chem.*, 270: 19949–19956.

Sheikhnejad, R.G. and Srivastava, P.N. (1986) Isolation and properties of a phosphatidylcholine-specific phospholipase C from bull seminal plasma. *J. Biol. Chem.*, 261: 7544–7549.

Taki, T. and Kanfer, J.N. (1979) Partial purification and properties of a rat brain phospholipase D. *J. Biol. Chem.*, 254: 9761–9765.

Waksman, M., Eli, Y., Liscovitch, M. and Gerst, J.E. (1996) Identification and characterization of a gene encoding phospholipase D activity in yeast. *J. Biol. Chem.*, 271: 2361–2364.

Wolf, R.A. and Gross, R.W. (1985) Identification of neutral active phospholipase C which hydrolyzes choline glycerophospholipids and plasmalogen selective phospholipase $A_2$ in canine myocardium. *J. Biol. Chem.*, 260: 7295–7303.

Zoukhri, D. and Dartt, D.A. (1995) Cholinergic activation of phospholipase D in lacrimal gland acini is independent of protein kinase C and calcium. *Am. J. Physiol.*, 268: C713–720.

J. Klein and K. Löffelholz (Eds.)
*Progress in Brain Research*, Vol. 109
© 1996 Elsevier Science B.V. All rights reserved.

# CHAPTER 20

# Participation of small GTP-binding proteins in m3 muscarinic acetylcholine receptor signalling to phospholipase D and C

Ulrich Rümenapp, Martina Schmidt, Miklós Geiszt and Karl H. Jakobs

*Institut für Pharmakologie, Universität GH Essen, Hufelandstrasse 55, D-45122 Essen, Germany*

## Introduction

Cleavage of the membrane phospholipid, phosphatidylinositol 4,5-bisphosphate ($PtdIns(4,5)P_2$), by phospholipase C (PLC) is one of the major cellular signalling pathways of cell surface receptors. The hydrolysis of $PtdIns(4,5)P_2$ by PLC generates the two second messengers, inositol 1,4,5-trisphosphate and diacylglycerol, causing release of calcium ions from internal stores and activation of protein kinase C enzymes, respectively (Berridge, 1993; Nishizuka, 1995). The receptor coupling mechanisms to PLC isoenzymes have been elucidated in great detail. $PLC\gamma$ isoenzymes are apparently activated by tyrosine phosphorylation and mediate the phosphoinositide turnover evoked by receptors either possessing tyrosine kinase activity or coupled to tyrosine kinases. On the other hand, $PLC\beta$ isoenzymes are activated by seven-transmembrane helix receptors coupled to heterotrimeric guanine nucleotide-binding proteins (G proteins), either by activated $\alpha$ subunits of pertussis toxin (PTX)-insensitive G proteins of the $G_q$ family or by free $\beta\gamma$ dimers of PTX-sensitive G proteins of the $G_i$ family (Cockcroft and Thomas, 1992; Lee and Rhee, 1995). In intact cells, however, receptor signalling to PLCs is apparently more complex. Particularly, an "on-going" synthesis of phosphoinositides involving various distinct proteins has been shown to be essential for $PtdIns(4,5)P_2$-dependent signalling pathways. Specifically, CDP-diacylglycerol synthase (Wu et al., 1995) and the phosphatidylinositol transfer

protein (Thomas et al., 1993; Cunningham et al., 1995; Kauffmann-Zeh et al., 1995) have recently been reported to be required for receptor and G protein-mediated activation of PLC. It can be assumed that the lipid kinases which catalyze the final steps of $PtdIns(4,5)P_2$ synthesis, i.e. phosphatidylinositol 4-kinase and phosphatidylinositol 4-phosphate 5-kinase, which are under regulatory control by tyrosine phosphorylation and small molecular weight G proteins (Carpenter and Cantley, 1990; Stephens et al., 1993; Chong et al., 1994), are similarly essential for receptor signalling to PLC in intact cells.

Stimulation of phospholipase D (PLD) has been reported in a wide range of cell types in response to many hormones, neurotransmitters and growth factors, including those that act on receptors that possess intrinsic tyrosine kinase activity as well as those acting on receptors coupled to heterotrimeric G proteins (Billah, 1993; Exton, 1994). The hydrolysis of the major membrane phospholipid, phosphatidylcholine, by PLD leads to the formation of phosphatidic acid and choline. Phosphatidic acid may act by itself as a second messenger (Fukami and Takenawa, 1992). Moreover, it can be converted to the extra- and/or intracellular signalling molecules, lysophosphatidic acid (Van Corven et al., 1992) and diacylglycerol, which if derived from phosphatidylcholine apparently activates specific protein kinase C isoenzymes (Ha and Exton, 1993). The receptor coupling mechanisms to PLD are only poorly understood. Multiple signalling pathway components, including heterotrimeric and small molecular weight G proteins, tyrosine kinases, protein kinase C, $Ca^{2+}$, and PtdIns(4,5)P$_2$, have been implicated in the regulation of cellular PLD activity. The involvement of heterotrimeric G proteins in the activation of PLD by G protein-coupled receptors can be hypothesized, however, direct interaction of PLD with heterotrimeric G proteins or G protein subunits has as yet not been demonstrated. On the other hand, recent reports provided evidence that several small molecular weight G proteins, namely ADP-ribosylation factors (ARFs) (Brown et al., 1993; Cockcroft et al., 1994), Rho proteins (Bowman et al., 1993; Malcolm et al., 1994; Singer et al., 1995) and Ras (Jiang et al., 1995), can stimulate PLD activity in cell-free preparations.

To study the coupling of muscarinic acetylcholine receptor (mAChR) to PLC and PLD and the role of the small GTP-binding proteins, ARF and Rho, in these receptor signalling pathways, we used human embryonic kidney (HEK) cells, transfected with and stably expressing the human m3 mAChR subtype. Marked stimulation of both phospholipases, PLC and PLD, can be observed upon mAChR activation in these cells (Peralta et al., 1988; Sandmann et al., 1991). The m3 mAChR-mediated stimulation of either phospholipase is apparently mediated by PTX-insensitive G proteins (Offermanns et al., 1994; Schmidt et al., 1994). Furthermore, although PLD activity can be stimulated in these cells by various mechanisms (protein kinase C, $Ca^{2+}$ and tyrosine phosphorylation), the coupling of the m3 mAChR to PLD is largely independent of the concomitant activation of PLC, including increase in cytosolic $Ca^{2+}$ and protein kinase C, but rather involves a tyrosine kinase-dependent mechanism (Schmidt et al., 1994).

## Involvement of ARF proteins in mAChR signalling to PLD

In the m3 mAChR-expressing HEK cells, agonist (carbachol) stimulation of PLD can be mimicked in a non-additive manner by direct activation of heterotrimeric G proteins with AlF$_4^-$, strongly implicating involvement of heterotrimeric G pro-

teins in mAChR signalling to PLD. Stimulation of PLD activity by the stable GTP analogue GTPγS in permeabilized HEK cells, however, is apparently mediated, at least in part, by the small molecular weight G protein ARF. HEK cells permeabilized for several minutes with digitonin or streptolysin O lose cytosolic proteins. In such pre-permeabilized cells, stimulation of PLD activity by GTPγS was substantially reduced, to about 30% of that seen without prior cytosol depletion, indicating loss of a cytosolic factor necessary for efficient PLD stimulation. A similar dependence of PLD stimulation on a cytosolic factor has been described for HL-60 cells, and there it has been identified as ARF (Brown et al., 1993; Cockcroft et al., 1994). In unstimulated HEK cells, two electrophoretically distinct ARF species, most likely ARF I and ARF II (Moss and Vaughan, 1993), were found by immunoblot analysis, with ARF II proteins being predominantly located in the cytosol. As exemplified in Fig. 1, ARF proteins are released from digitonin-permeabilized HEK cells. To study whether this loss of cytosolic ARF proteins could explain the decrease in PLD stimulation by GTPγS in pre-permeabilized HEK cells, ARF proteins were reconstituted. Upon addition of purified recombinant ARF 1, cytosol-depleted pre-permeabilized cells exhibited increased stimulation of PLD by GTPγS (Rümenapp et al., 1995).

To study the role of ARF proteins in m3 mAChR signalling to PLD, two approaches were used, a biochemical and a pharmacological approach. First, we studied the subcellular localization of ARF proteins upon m3 mAChR activation (Rümenapp et al., 1995). ARF proteins in the inactive GDP-bound state are located in the cytosol or only loosely attached to membranes, and therefore are released from permeabilized cells or are found in the cytosolic fraction upon cell fractionation. In contrast, the active GTP-bound ARF proteins are found tightly associated with membranes (Regazzi et al., 1991; Serafini et al., 1991). Thus, this distinct subcellular localization of inactive and active ARF proteins can be used to monitor the activity state of ARF proteins. Second, the influence of brefeldin A (BFA) on receptor-stimulated PLD was studied in m3 mAChR-expressing HEK

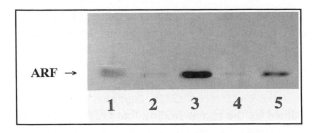

Fig. 1. Translocation of ARF proteins by GTP$\gamma$S and carbachol. HEK cells ($2 \times 10^6$) pretreated for 10 min with or without 1 mM carbachol were permeabilized with 8 $\mu$M digitonin in the absence or presence of 100 $\mu$M GTP$\gamma$S, and proteins released into the supernatant were analyzed for ARF immunoreactivity with an anti-recombinant ARF 1 antibody after SDS-PAGE and blotting to a nitrocellulose membrane. An autoluminogram of the immunoblot is shown. Lane 1, 1 $\mu$g purified recombinant ARF 1; lane 2, non-permeabilized cells; lane 3, digitonin-permeabilized cells; lane 4, cells permeabilized in the presence of GTP$\gamma$S; lane 5, carbachol-pretreated permeabilized cells.

cells (Rümenapp et al., 1995). Like other small G proteins of the Ras superfamily, ARF proteins do not readily release bound GDP in exchange for GTP. Instead an accessory guanine nucleotide exchange factor is required, whose activity has been described in Golgi membranes and which was partially purified from bovine brain (Tsai et al., 1994). This nucleotide exchange activity is inhibited by the fungal macrolide antibiotic BFA, thereby preventing ARF activation and vesicle transport (Donaldson et al., 1992; Helms and Rothman, 1992; Randazzo et al., 1993). Recent purification studies suggest that BFA does not directly act at the catalytic domain of the nucleotide exchange enzyme (Tsai et al., 1994).

*mAChR activation of ARF proteins*

To study the activity state of endogenous ARF proteins upon mAChR activation, the membrane-cytosol localization of ARF proteins was monitored. For this, HEK cells were treated with and without carbachol prior to permeabilization performed in the absence or presence of GTP$\gamma$S. Proteins released into the supernatant or remaining cell-attached were subsequently analyzed for their ARF content by SDS-PAGE and immunoblotting.

During permeabilization without stimulus addition, substantial release of immunoreactive ARF proteins from HEK cells was observed (compare lanes 2 and 3 of Fig. 1), concomitantly with a decrease in cell-attached, mainly cytosolically located ARF II proteins. When the permeabilization was performed in the presence of GTP$\gamma$S, release of ARF proteins was virtually completely prevented (compare lanes 3 and 4 of Fig. 1), and the cell-attached ARF II content was identical to that in non-permeabilized cells. These data, thus, indicate that GTP$\gamma$S can activate endogenous ARF proteins in HEK cells, and that these GTP$\gamma$S-liganded ARF proteins activate PLD activity.

When the m3 mAChR-expressing HEK cells were pretreated with carbachol before permeabilization, release of immunoreactive ARF proteins was reduced by about 60% (compare lanes 3 and 5 of Fig. 1). Activation of heterotrimeric G proteins by AlF$_4^-$ in intact cells mimicked the effect of the receptor agonist. The extent of ARF redistribution induced by carbachol was consistently not complete, thus smaller than in response to GTP$\gamma$S. This is most likely due to the fact that GTP$\gamma$S-bound ARFs are persistently activated and thus translocated to membranes, while upon mAChR activation in intact cells ARF proteins are activated by the natural, but hydrolyzable GTP. Under this condition, only a fraction of ARF proteins will be in the GTP-bound state and thus translocated to membranes. Alternatively, it is also conceivable that the number of ARF proteins activated and thus translocated to membranes upon addition of GTP$\gamma$S to permeabilized cells is larger than the number of ARF proteins available for activation by agonist-activated mAChR in intact cells.

*Inhibition of mAChR-stimulated PLD activity by BFA*

The activation of ARF proteins, measured as translocation to membranes, by agonist-activated mAChR suggested that ARF proteins are involved in the signalling pathway leading from m3 receptor activation to PLD stimulation. To corroborate this hypothesis, we studied the effect of BFA, which is known to specifically inhibit ARF activation by its

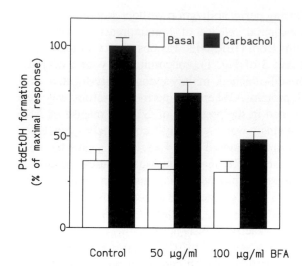

Fig. 2. Inhibition of m3 mAChR-stimulated PLD by BFA. Basal and carbachol (1 mM)-stimulated formation of [³H]phosphatidylethanol ([³H]PtdEtOH) was studied in [³H]oleic acid-prelabeled m3 mAChR-expressing HEK cells pretreated for 15 min with BFA at the indicated concentrations, as described before (Schmidt et al., 1994). Maximal carbachol-stimulated [³H]PtdEtOH formation was 0.66 ± 0.03% of total labeled phospholipids.

guanine nucleotide exchange factor(s), on mAChR-stimulated PLD. Pretreatment of the m3 mAChR-expressing HEK cells with BFA caused an inhibition of carbachol-stimulated PLD activity (Fig. 2). The BFA concentrations required to observe this inhibition are higher than those reported to inhibit ARF binding to Golgi membranes (Donaldson et al., 1992; Helms and Rothman, 1992). However, rather similar BFA concentrations were needed to inhibit guanine nucleotide binding to ARF proteins stimulated by a soluble exchange protein (Tsai et al., 1994). Control experiments indicated that BFA does not unspecifically interfere with mAChR signalling to PLD. First, under a condition causing half-maximal inhibition of mAChR-stimulated PLD, BFA treatment of HEK cells also reduced the inhibitory effect of carbachol on ARF release from permeabilized cells. Second, BFA treatment affected neither cell surface mAChR number nor receptor coupling to heterotrimeric G proteins. Third, BFA is not a direct inhibitor of PLD itself, since basal PLD ac-

tivities in both intact and permeabilized HEK cells were not affected by BFA treatment. Finally, BFA treatment did not affect inhibition of ARF release and stimulation of PLD activity by GTPγS in permeabilized cells (Rümenapp et al., 1995). The lack of a BFA effect on GTPγS actions may be explained by the recent finding that GTPγS can bind to endogenous, myristoylated ARF proteins at physiological Mg²⁺ levels in the presence of phospholipids even without the requirement of a guanine nucleotide exchange factor (Franco et al., 1995).

In conclusion, ARF proteins can apparently regulate PLD activity in HEK cells as described before in other cell types. Most important, by using a biochemical (ARF translocation) as well as a pharmacological approach (BFA) strong evidence is presented that ARF proteins are involved in m3 mAChR signalling to PLD.

## Involvement of Rho proteins in mAChR signalling to PLD and PLC

To study the role of Rho proteins in mAChR signalling to phospholipases, we used the cytotoxin B of *Clostridium (C.) difficile* and the C3 exoenzyme of *C. botulinum* (Schmidt et al., 1996a,b). Toxin B, which enters cells by receptor-mediated endocytosis (Ciesielski-Treska et al., 1989), has recently been shown to inactivate members of the Rho protein family (Rho A, Rac1 and Cdc42), by causing monoglucosylation of these proteins (Just et al., 1994, 1995). *C. botulinum* C3 exoenzyme, which does not or only very poorly enters intact cells, also inactivates Rho proteins (Rho A, B and C), but by catalyzing incorporation of ADP-ribose into these proteins (Aktories and Just, 1993).

### Inhibition of mAChR-stimulated PLD activity by toxin B and C3 exoenzyme

Treatment of the m3 mAChR-expressing HEK cells with *C. difficile* toxin B caused a concentration- and time-dependent decrease in receptor-stimulated PLD activity. As exemplified in Fig. 3, treatment of the cells with toxin B (30 pg/ml, 24 h) reduced the carbachol-induced PLD stimulation by

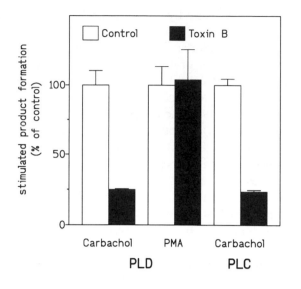

Fig. 3. Inhibition of mAChR-stimulated PLD and PLC by *C. difficile* toxin B. Carbachol (1 mM)- and PMA (0.1 μM)-stimulated formation of [³H]PtdEtOH (PLD) and [³H]inositol phosphates (PLC) was studied in [³H]oleic acid- and *myo*-[³H]inositol-prelabeled m3 mAChR-expressing HEK cells, respectively, pretreated or not for 24 h with 30 pg/ml toxin B, as described before (Schmidt et al., 1994). Stimulated product formation is given as % of controls in untreated cells.

about 75%. Toxin B treatment also inhibited PLD stimulation by direct activation of heterotrimeric G proteins with AlF₄⁻. In contrast, stimulation of PLD activity by direct activation of protein kinase C with the phorbol ester PMA as well as basal PLD activity were not affected by this toxin B treatment, indicating that toxin B did not modify PLD by itself. Furthermore, toxin B did also not inactivate, e.g. down-regulate, the mAChR or interfere with receptor coupling to heterotrimeric G proteins. Finally, treatment of HEK cells with cytochalasin B or *C. botulinum* C2 toxin, which cause similar morphological alterations of HEK cells as toxin B, did not affect mAChR-stimulated PLD (Schmidt et al., 1996a).

To confirm the hypothesis that small molecular weight G proteins of the Rho family are the intracellular target of toxin B, we studied the effects of toxin B and *C. botulinum* C3 exoenzyme, which both inactivate Rho proteins but by distinct mechanisms, on the GTPγS-induced PLD stimulation in digitonin-permeabilized HEK cells. Pretreatment of intact HEK cells with toxin B (30 pg/ml, 24 h) decreased GTPγS-stimulated PLD activity by about 75% (Fig. 4A). Similarly, when digitonin-permeabilized cells were treated

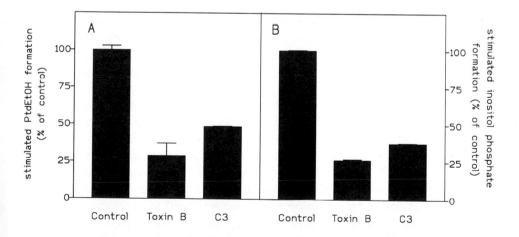

Fig. 4. Inhibition of GTPγS-stimulated PLD and PLC activities by toxin B and C3 exoenzyme. In (A), GTPγS (100 μM)-stimulated [³H]PtdEtOH formation was measured in digitonin-permeabilized [³H]oleic acid-prelabeled HEK cells pretreated or not with toxin B (30 pg/ml, 24 h) and in permeabilized cells additionally treated with C3 exoenzyme (12 μg/ml) and 50 μM NAD, as described before (Schmidt et al., 1994). In (B), GTPγS (100 μM)-stimulated [³H]inositol phosphate formation was measured in permeabilized *myo*-[³H]inositol-prelabeled HEK cells, treated or not with toxin B or C3 exoenzyme as in A. GTPγS-stimulated formation of [³H]PtdEtOH and [³H]inositol phosphates are given as% of controls in untreated cells.

with C3 exoenzyme in the presence of NAD, stimulation of PLD activity by GTPγS was reduced by about 50%. Neither toxin B nor C3 exoenzyme treatment had an effect on basal PLD activity (Schmidt et al., 1996a). Thus, Rho proteins are apparently involved in mAChR signalling to PLD in HEK cells.

*Inhibition of mAChR-stimulated PLC activity by toxin B and C3 exoenzyme*

Since in the m3 mAChR-expressing HEK cells, the agonist-activated mAChR, in addition to activating PLD, also mediates a large PLC response, we wondered whether Rho proteins may also be involved in mAChR signalling to PLC in this cell type. Therefore, similar experiments as performed on regulation of PLD activity were carried out to study the effects of toxin B and C3 exoenzyme on regulation of PLC activity. As exemplified in Fig. 3, toxin B treatment (30 pg/ml, 24 h) not only inhibited mAChR-stimulated PLD activity but also reduced PLC stimulation by the agonist-activated mAChR by about 80%. In addition, $AlF_4^-$ stimulation of inositol phosphate formation was also inhibited by toxin B treatment, whereas basal PLC activity was not affected (Schmidt et al., 1996b). Furthermore, pretreatment of intact HEK cells with toxin B (30 pg/ml, 24 h) reduced GTPγS-stimulated formation of inositol phosphates from endogenous phosphoinositides by about 70% (Fig. 4B). Finally, this inhibitory effect of intact cell treatment with toxin B was mimicked by treating permeabilized HEK cells with C3 exoenzyme, reducing GTPγS-stimulated inositol phosphate formation by about 60%. Again, basal PLC activity was not reduced by either treatment (Schmidt et al., 1996b).

Toxin B and C3 exoenzyme modify and thereby inactivate members of the Rho protein family by highly specific covalent modifications. Toxin B monoglucosylates Rho A, Rac1 and Cdc42, while C3 exoenzyme ADP-ribosylates Rho A, B and C. The combined analysis with both tools working on distinct members of the Rho protein family allows us to conclude that toxin B and C3 exoenzyme evoke their effects on mAChR-stimulated PLD and PLC most likely by an action on Rho A proteins. Furthermore, the data suggest that Rho proteins and their cellular targets are involved in receptor stimulation of both phospholipases. Rho proteins have recently been reported to stimulate phosphatidylinositol 4-phosphate 5-kinase activity (Chong et al., 1994), the enzyme finally responsible for $PtdIns(4,5)P_2$ synthesis. Since $PtdIns(4,5)P_2$ is not only the PLC substrate but also stimulates guanine nucleotide exchange on ARF and acts as a cofactor for PLD activity (Liscovitch and Cantley, 1995), it may be speculated that Rho proteins, being substrates for both toxin B and C3 exoenzyme, regulate both signalling pathways by controlling the cellular level of $PtdIns(4,5)P_2$. Preliminary results indeed support this hypothesis (Schmidt et al., 1996b).

**Conclusions**

As briefly summarized herein, the signalling pathway from the m3 mAChR to PLD apparently does not only involve heterotrimeric PTX-insensitive G proteins but also small molecular weight GTP-binding proteins of the ARF family. Future experiments will have to analyze the possible link between heterotrimeric G proteins and the ARF guanine nucleotide exchange factor as well as the role and position of tyrosine kinase-based mechanisms in this PLD signalling pathway. Furthermore, the data obtained with *C. difficile* toxin B and *C. botulinum* C3 exoenzyme strongly suggest that a member of the Rho protein family, most likely Rho A, is involved in mAChR signalling to both PLD and PLC. These data, additionally, support the notion that receptor stimulation of PLC in intact cells is apparently far more complex than formerly realized. The common or even distinct Rho targets finally being responsible for the observed toxin actions need to be elucidated. However, whatever the final mechanisms are, it is feasible to assume that *C. difficile* toxin B will be a very powerful tool to analyze not only the cellular role of Rho proteins but also those of PLC and PLD and the signalling molecules formed by these phospholipases in cellular actions of hormones and neurotransmitters.

## Acknowledgements

The studies reported here were supported by the Deutsche Forschungsgemeinschaft (SFB 354) and a DAAD fellowship to M.G.

## References

Aktories, K. and Just, I. (1993) GTPases and actin as targets for bacterial toxins. In: B.F. Dickey and L. Birnbaumer (Eds.), GTPases in Biology I, Springer-Verlag, Berlin, pp. 87–112.

Berridge, M.J. (1993) Inositol trisphosphate and calcium signalling. Nature, 361: 315–325.

Billah, M.M. (1993) Phospholipase D and cell signaling. Curr. Opin. Immunol., 5: 114–123.

Bowman, E.P., Uhlinger, D.J. and Lambeth, J.D. (1993) Neutrophil phospholipase D is activated by a membrane-associated Rho family small molecular weight GTP-binding protein. J. Biol. Chem., 268: 21509–21512.

Brown, H.A., Gutowski, S., Moonaw, C.R., Slaughter, C. and Sternweis, P.C. (1993) ADP-ribosylation factor, a small GTP-dependent regulatory protein, stimulates phospholipase D activity. Cell, 75: 1137–1144.

Carpenter, C.L. and Cantley, L.C. (1990) Phosphoinositide kinases. Biochemistry, 29: 11147–11156.

Chong, L.D., Traynor-Kaplan, A., Bokoch, G.M. and Schwartz, M.A. (1994) The small GTP-binding protein Rho regulates a phosphatidylinositol 4-phosphate 5-kinase in mammalian cells. Cell, 79: 507–513.

Ciesielski-Treska, J., Ulrich, G., Rihn, B. and Aunis, D. (1989) Mechanism of action of Clostridium difficile toxin B: role of external medium and cytoskeletal organization in intoxicated cells. Eur. J. Cell Biol., 48: 191–202.

Cockcroft, S. and Thomas, G.M.H. (1992) Inositol-lipid-specific phospholipase C isoenzymes and their differential regulation by receptors. Biochem. J., 288: 1–14.

Cockroft, S., Thomas, G.M.H., Fensome, A.M., Geny, B., Cunningham, E., Gout, I., Hiles, I., Totty, N.F., Truong, O. and Hsuan, J.J. (1994) Phospholipase D: A downstream effector of ARF in granulocytes. Science, 263: 523–526.

Cunningham, E., Thomas, G.M.H., Ball, A., Hiles, I. and Cockroft, S. (1995) Phosphatidylinositol transfer protein dictates the rate of inositol trisphosphate production by promoting the synthesis of PIP$_2$. Curr. Biol., 5: 775–783.

Donaldson, J.G., Finazzi, D. and Klausner, R.D. (1992) Brefeldin A inhibits Golgi membrane-catalysed exchange of guanine nucleotide onto ARF protein. Nature, 360: 350–352.

Exton, J.H. (1994) Phosphatidylcholine breakdown and signal transduction. Biochim. Biophys. Acta, 1212: 26–42.

Franco, M., Chardin, P., Chabre, M. and Sonia, P. (1995) Myristoylation of ADP-ribosylation factor 1 facilitates nucleotide exchange at physiological Mg$^{2+}$ levels. J. Biol. Chem., 270: 1337–1341.

Fukami, K. and Takenawa, T. (1992) Phosphatidic acid that accumulates in platelet-derived growth factor-stimulated Balb/c 3T3 cells is a potential mitogenic signal. J. Biol. Chem., 267: 10988–10993.

Ha, K.-S. and Exton, J.H. (1993) Differential translocation of protein kinase C isoenzymes by thrombin and platelet-derived growth factor. A possible function for phosphatidylcholine-derived diacylglycerol. J. Biol. Chem., 268: 10534–10539.

Helms, J.B. and Rothman, J.E. (1992) Inhibition by brefeldin A of a Golgi membrane enzyme that catalyses exchange of guanine nucleotide bound to ARF. Nature, 360: 352–354.

Jiang, H., Lu, Z., Luo, J.-Q., Wolfman, A. and Foster, D.A. (1995) Ras mediates the activation of phospholipase D by v-Src. J. Biol. Chem., 270: 6006–6009.

Just, I., Fritz, G., Aktories, K., Giry, M., Popoff, M.R., Boquet, P., Hegenbarth, S. and von Eichel-Streiber, C. (1994) Clostridium difficile toxin B acts on the GTP-binding protein Rho. J. Biol. Chem., 269: 10706–10712.

Just, I., Selzer, J., Wilm, M., von Eichel-Streiber, C. and Aktories, K. (1995) Glucosylation of Rho proteins by Clostridium difficile toxin B. Nature, 375: 500–503.

Kauffmann-Zeh, A., Thomas, G.M.H., Ball, A., Prosser, S., Cunningham, E., Cockcroft, S. and Hsuan, J.J. (1995) Requirement for phosphatidylinositol transfer protein in epidermal growth factor signaling. Science, 268: 1188–1190.

Lee, S.B. and Rhee, S.G. (1995) Significance of PIP$_2$ hydrolysis and regulation of phospholipase C isozymes. Curr. Opin. Cell Biol., 7: 183–189.

Liscovitch, M. and Cantley, L.C. (1995) Signal transduction and membrane traffic: the PITP/phosphoinositide connection. Cell, 81: 659–662.

Malcolm, K.C., Ross, A.H., Qui, R.-G., Symons, M. and Exton, J.H. (1994) Activation of rat liver phospholipase D by the small GTP-binding protein Rho A. J. Biol. Chem., 269: 25951–25954.

Moss, J. and Vaughan, M. (1993) ADP-ribosylation factors, 20,000 M$_r$ guanine nucleotide-binding protein activators of cholera toxin and components of intracellular vesicular transport systems. Cell. Signal., 5: 367–379.

Nishizuka, Y. (1995) Protein kinase C and lipid signaling for sustained cellular response. FASEB J., 9: 484–496.

Offermanns, S., Wieland, T., Homann, D., Sandmann, J., Bombien, E., Spicher, K., Schultz, G. and Jakobs, K.H. (1994) Transfected muscarinic acetylcholine receptors selectively couple to G$_i$-type G proteins and G$_{q/11}$. Mol. Pharmacol., 45: 890–898.

Peralta, E.G., Ashkenazi, A., Winslow, J.W., Ramachandran, J. and Capon, D.J. (1988) Differential regulation of PI hydrolysis and adenylyl cyclase by muscarinic receptor subtypes. Nature, 334: 434–437.

Randazzo, P.A., Yang, Y.C., Rulka, C. and Kahn, R.A. (1993) Activation of ADP-ribosylation factor by Golgi membranes. Evidence for a brefeldin A- and protease-sensitive activating factor on Golgi membranes. J. Biol. Chem., 268: 9555–9563.

216

Regazzi, R., Ullrich, S., Khan, R.A. and Wollheim, C.B. (1991) Redistribution of ADP-ribosylation factor during stimulation of permeabilized cells with GTP analogues. *Biochem. J.*, 275: 639–644.

Rümenapp, U., Geiszt, M., Wahn, F., Schmidt, M. and Jakobs, K.H. (1995) Evidence for ADP-ribosylation-factor-mediated activation of phospholipase D by m3 muscarinic acetylcholine receptor. *Eur. J. Biochem.*, 234: 240–244.

Sandmann, J., Peralta, E.G. and Wurtman, R.J. (1991) Coupling of transfected muscarinic acetylcholine receptor subtypes to phospholipase D. *J. Biol. Chem.*, 266: 6031–6033.

Schmidt, M., Hüwe, S.M., Fasselt, B., Homann, D., Rümenapp, U., Sandmann, J. and Jakobs, K.H. (1994) Mechanisms of phospholipase D stimulation by m3 muscarinic acetycholine receptors. Evidence for involvement of tyrosine phosphorylation. *Eur. J. Biochem.*, 225: 667–675.

Schmidt, M., Rümenapp, U., Bienek, C., Keller, J., von Eichel-Streiber, C. and Jacobs, K.H. (1996a) Inhibition of receptor signaling to phospholipase D by *Clostridium difficile* toxin B. Role of Rho proteins. *J. Biol. Chem.*, 271: 2422–2426.

Schmidt, M., Bienek, C., Rümenapp, U., Zhang, C., Lümmen, G., Jakobs, K.H., Just, I., Aktories, K., Moos, M. and von Eichel-Streiber, C. (1996b) A role for Rho in receptor- and G protein stimulated phospholipase C. Reduction in phosphatidylinositol 4,5-bisphosphate by *Clostridium difficile* toxin B. *Naunyn-Schmiedeberg's Arch. Pharmacol.*, 354: 87–94.

Serafini, T., Orci, L., Amherdt, M., Brunner, M., Kahn, R.A. and Rothman, J.E. (1991) ADP-ribosylation factor is a subunit of Golgi-derived COP-coated vesicles: A novel role for a GTP-binding protein. *Cell*, 67: 239–253.

Singer, W.D., Brown, H.A., Bokoch, G.M. and Sternweis, P.C. (1995) Resolved phospholipase D activity is modulated by cytosolic factors other than Arf. *J. Biol. Chem.*, 270: 14944–14950.

Stephens, L., Jackson, T.R. and Hawkins, P.T. (1993) Activation of phosphatidylinositol 4,5-bisphosphate supply by agonists and non-hydrolysable GTP analogues. *Biochem. J.*, 296: 481–488.

Thomas, G.M.H., Cunningham, E., Fensome, A., Ball, A., Totty, N.F., Truong, O., Hsuan, J.J. and Cockcroft, S. (1993) An essential role for phosphatidylinositol transfer protein in phospholipase C-mediated inositol lipid signaling. *Cell*, 74: 919–928.

Tsai, S.-C., Adamik, R., Moss, J. and Vaughan, M. (1994) Identification of a brefeldin A-insensitive guanine nucleotide-exchange protein for ADP-ribosylation factor in bovine brain. *Proc. Natl. Acad. Sci. USA*, 91: 3063–3066.

Van Corven, E.J., Van Rijswijk, A., Jalink, K., Van Der Bend, R.L., Blitterswijk, W.J. and Moolenaar, W.H. (1992) Mitogenic action of lysophosphatidic acid and phosphatidic acid on fibroblasts. Dependence on acyl-chain length and inhibition by suramin. *Biochem. J.*, 281: 163–169.

Wu, L., Niemeyer, B., Colley, N., Socolich, M. and Zuker, C.S. (1995) Regulation of PLC-mediated signalling *in vivo* by CDP-diacylglycerol synthase. *Nature*, 373: 216–222.

# Section VII

# Presynaptic Modulation of Acetylcholine Release

J. Klein and K. Löffelholz (Eds.)
*Progress in Brain Research*, Vol. 109
© 1996 Elsevier Science B.V. All rights reserved.

CHAPTER 21

# Modulation of acetylcholine release by nitric oxide

H. Kilbinger

*Department of Pharmacology, University of Mainz, Mainz, Germany*

## Introduction

Chemical neurotransmission in the central and peripheral nervous system can be controlled by presynaptic auto- and heteroreceptors (for reviews see Starke et al., 1989; Fuder and Muscholl, 1995). Stimulation of these receptors by neurotransmitters influences the release process leading to facilitation or, more general, inhibition of transmitter release. During the past few years evidence has accumulated that, in addition to the receptor-mediated regulatory mechanisms, release of transmitter can also be modulated by nitric oxide (NO). NO is a novel unconventional neuromodulator or neurotransmitter proper which is not stored in vesicles and diffuses from its site of synthesis to influence indiscriminately pre- and postsynaptic targets (for recent reviews see Schuman and Madison, 1994; Garthwaite and Boulton, 1995).

Neuronal NO is synthesized by the isoform I of the enzyme NO synthase (Förstermann et al., 1994). NO synthase is co-localized with choline acetyltransferase in certain cholinergic neurons, e.g. in basal forebrain, in the septohippocampal pathway and in the myenteric plexus (Snyder and Bredt, 1991; Vincent, 1994) and this suggests that NO may influence cholinergic synaptic transmission by modulating the release of acetylcholine. The present chapter reviews studies in which the effects of endogenous and exogenous NO on the release of acetylcholine were investigated. As tools for the involvement of endogenous NO the effects of NO synthase inhibitors on release of acetylcholine were studied.

## Central nervous system

### Effect of NO synthase inhibitors

Evidence for the involvement of endogenous NO in the regulation of acetylcholine release was provided in push-pull experiments with NO synthase inhibitors (Table 1). Superfusion of basal forebrain and striatum of conscious, freely moving rats with $N^G$-nitro-L-arginine (L-NA) and $N^G$-nitro-L-arginine methyl ester (L-NAME) caused a decrease in basal acetylcholine release (Prast and Philippu, 1992; Prast et al., 1995). The authors therefore concluded that the cholinergic neurons are tonically stimulated by endogenous NO. On the other hand, NO synthase inhibitors failed to change either basal or electrically evoked acetylcholine release from slices of rat hippocampus and striatum (Suzuki et al., 1993; Sandor et al., 1995). This suggests that in the isolated brain slices endogenous NO is not permanently secreted to affect acetylcholine release.

### Effect of NO donors

A variety of NO donors increased basal release of acetylcholine measured in vivo and in vitro in different parts of the central nervous system (Table 2). The increase was attenuated by hemoglobin which suggests that the effect is mediated via NO liberation from the NO donors (Lonart et al., 1992; Guevara-Guzman et al., 1994; Ohkuma et al., 1995). Furthermore, the facilitatory effect of the NO donors was calcium-dependent (Guevara-

TABLE 1

Modulation of acetylcholine release by NO synthase inhibitors

| Species | Tissue | Inhibitor | Effect on release | Reference |
|---|---|---|---|---|
| Rat | Basal forebrain, in vivo | L-NA | Inhibition of basal release | Prast and Philippu (1992) |
| Rat | Ventral striatum, in vivo | L-NA, L-NAME | Inhibition of basal release | Prast et al. (1995) |
| Rat | Hippocampus | L-NA | None | Suzuki et al. (1993) |
| Rat | Striatum | L-NAME | None | Sandor et al. (1995) |
| Rat | Trachea | L-NMMA | Increase of evoked release | Sekizawa et al. (1993) |
| Guinea pig | Trachea | L-NA | None | Brave et al. (1991) |
| Guinea pig | Ileum, MPLM | L-NA, L-NMMA | Increase of evoked release | Kilbinger and Wolf (1994), Present paper |
| Guinea pig | Ileum, circular muscle | L-NA | Increase of evoked release | Present paper |
| Dog | Ileum, circular muscle | L-NAME, L-NA | Increase of evoked release | Hryhorenko et al. (1994) |
| Man | Trachea | L-NAME | None | Ward et al. (1993) |

L-NA, $N^G$-nitro-L-arginine; L-NAME, $N^G$-nitro-L-arginine methyl ester; L-NMMA, $N^G$-monomethyl-L-arginine; MPLM, myenteric plexus longitudinal muscle preparation.

Guzman et al., 1994) indicating that NO enhances exocytotic release. Controversial results were obtained on rat hippocampal slices. Lonart et al. (1992) who measured release of [$^{14}$C]acetylcholine observed an increase by hydroxylamine, whereas Suzuki et al. (1993) found no effect of sodium nitroprusside on release of endogenous acetylcholine determined by radioimmunoassay. It is not clear whether there are differences in the properties of the NO donors or whether the methods utilized are not comparable.

NO donors enhance also the release of various other neurotransmitters such as glutamate, GABA and noradrenaline (Lonart et al., 1992; Guevara-Guzman et al., 1994). It is therefore possible that NO does not directly affect acetylcholine release, but instead acts indirectly, e.g. via stimulation of GABAergic or glutamatergic neurons.

TABLE 2

Modulation of acetylcholine release by NO donors

| Species | Tissue | NO donor | Effect on release | Reference |
|---|---|---|---|---|
| Mouse | Cerebral cortex, primary culture | SNAP, SNP | Increase of basal release | Ohkuma et al. (1995) |
| Rat | Basal forebrain, in vivo | SIN-1 | Increase of basal release | Prast and Philippu (1992) |
| Rat | Medial striatum, in vivo | SNAP, SNOG, NO gas | Increase of basal release | Guevara-Guzman et al. (1994) |
| Rat | Ventral striatum, in vivo | SIN-1, SNAP | Increase of basal release | Prast et al. (1995) |
| Rat | Hippocampus | Hydroxylamine | Increase of basal release | Lonart et al. (1992) |
| Rat | Hippocampus | SNP | None | Suzuki et al. (1993) |
| Guinea pig | Ileum, MPLM | NaNO$_2$ in HCl | Inhibition of evoked release | Wiklund et al. (1993) |
| Guinea pig | Ileum, MPLM | SIN-1 | Increase of basal release; inhibition of evoked release | Present paper |

SNAP, S-nitroso-N-acetylpenicillamine; SNP, sodium nitroprusside; SIN-1, 3-morpholino-sydnonimine; SNOG, S-nitroso-glutathione.

## Mechanisms

Membrane-permeable analogues of $3',5'$-cyclic guanosine monophosphate (cGMP) such as 8-Br-cGMP and dibutyryl-cGMP increased the release of acetylcholine from striatum (Guevara-Guzman et al., 1994). It is therefore likely that exogenous and endogenous NO enhance acetylcholine release in the central nervous system by stimulating guanylyl cyclase and the formation of cGMP. The striatum contains the highest guanylyl cyclase activity in the brain; however, the cGMP system appears to be absent from the cholinergic interneurones of the striatum (Vincent, 1994). This favors the hypothesis that the NO-mediated increase in acetylcholine release is secondary to the modulation of other neurotransmitters.

A recent study suggests that peroxynitrite is involved in the stimulatory effect of NO on acetylcholine release (Ohkuma et al., 1995). NO can react with superoxide to yield peroxynitrite which may alter membrane functions. In support of this hypothesis exogenous peroxynitrite enhanced the release of acetylcholine and superoxide dismutase diminished the stimulatory effects of NO donors.

## Enteric nervous system

NO is considered as a major non-adrenergic, non-cholinergic neurotransmitter causing inhibition of gastrointestinal motility (Sanders and Ward, 1992) and NO synthase has been found throughout the gastrointestinal tract in neuronal cell bodies and processes of the myenteric plexus (Bredt et al., 1990). More detailed studies have revealed that about one-quarter of all myenteric neurons in the guinea pig small intestine contain constitutive NO synthase immunoreactivity (Costa et al., 1992). NO synthase immunoreactive axons were found to make synapses with other myenteric neurons (Llewellyn-Smith et al., 1992) which suggests that NO acts as a transmitter within myenteric ganglia. In addition, the circular muscle of the intestine is innervated by NO synthase immunoreactive nerves which were found closer than 100 nm to the muscle cells (Llewellyn-Smith et al., 1992). NO has therefore been suggested to be also a neurotrans-

mitter of inhibitory motor neurons innervating the circular muscle.

We have studied the effects of NO synthase inhibitors and NO donors on acetylcholine release on two isolated preparations of the guinea pig small intestine, i.e. the circular muscle preparation and the myenteric plexus longitudinal muscle preparation. In all experiments the stimulation frequency was 10 Hz since only high frequency stimulation (8–32 Hz) causes a release of endogenous NO from guinea pig intestine (Wiklund et al., 1993a).

## Effect of NO synthase inhibitors

Release of [$^3$H]acetylcholine from the circular muscle was determined as described recently (Dietrich and Kilbinger, 1995). The strips were loaded with [$^3$H]choline, and subsequently stimulated twice (S1, S2) (10 Hz, trains of 30 pulses every 30 s during 10 min). The release of [$^3$H]-acetylcholine evoked during S1 was $2.74 \pm 0.82\%$ ($n = 6$) of the tritium tissue content. L-NA ($100\,\mu$M) added to the tissue 30 min before S2 significantly increased the electrically evoked release ($145 \pm 16\%$ of control value, $n = 5$). Basal outflow was not affected by L-NA. Similar findings have been obtained on the circular muscle of the canine ileum where inhibition of NO synthase enhanced the field stimulated release of [$^3$H]-acetylcholine (Hryhorenko et al., 1994).

L-NA also increased the outflow of [$^3$H]-acetylcholine from myenteric plexus longitudinal muscle preparations that had been preincubated with [$^3$H]choline. The strips were stimulated twice (S1, S2) with monophasic pulses (1 ms) delivered for 1 min periods from Grass S6 stimulators. When the strips were stimulated with supramaximal current strength (400 mA) the release caused by S1 amounted to $6.86 \pm 1.08\%$ ($n = 9$) of the tissue tritium content. L-NA ($100\,\mu$M, added 30 min before S2) did not significantly change the electrically evoked release ($96 \pm 4\%$ of the control value; $n = 4$). If the current strength was reduced to 200 mA (submaximal stimulation) a release (S1) of $1.79 \pm 0.08\%$ ($n = 37$) of the tissue tritium content was evoked. Under these conditions L-NA increased in a concentration-dependent fashion the release of acetylcholine (Fig. 1). Basal outflow was not affected by L-NA. The facilitatory effect of 10 $\mu$M L-NA was stereospecifically prevented by 1 mM L-arginine but not by 1 mM D-arginine. L- and D-arginine alone did not change the electrically evoked release of acetylcholine. Similar results have previously been obtained (Kilbinger and Wolf, 1994). L-NA and $N^G$-monomethyl-L-arginine (L-NMMA) facilitated the release of acetylcholine from guinea pig myenteric plexus stimulated by biphasic pulses (10 Hz, 1 min, supramaximal currents). In these experiments S1 caused a release of 1.22 $\pm$ 0.10 % ($n = 12$) which is not different from the above mentioned release by monophasic pulses and a current of 200 mA.

Taken together, NO synthase inhibitors facilitate the electrically evoked release of acetylcholine from circular and longitudinal muscle preparations of guinea pig and dog intestine. This suggests that endogenous NO which is released during field stimulation of the intestine (Wiklund et al., 1993a) exerts a physiological inhibitory effect on the evoked acetylcholine release.

NO synthase inhibitors have failed to affect acetylcholine release from the isolated trachea of man and guinea-pig, but increased the evoked release from the rat trachea (see Table 1). It remains unclear whether this variability is due to the stimulation conditions when release is studied, or whether this reflects tissue and species differences.

*Effect of NO donors*

We have tested the effects of SIN-1 (3-morpholino-sydnonimine) on acetylcholine release from the myenteric plexus longitudinal muscle preparation. SIN-1 (300 $\mu$M) caused a large transient increase in basal release of [³H]acetylcholine (2.70 $\pm$ 0.24 % of the tritium tissue content; $n = 6$). Omission of calcium from the medium or tetrodotoxin (300 nM) abolished the release-enhancing effect of SIN-1. Indirect evidence for an acetylcholine liberating action of NO has previously been obtained in functional experiments on the guinea pig intestine: exogenous NO elicited a contraction that was abolished by tetrodotoxin or at-

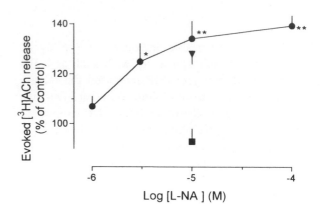

Fig. 1. Increase by L-NA of electrically evoked release of [³H]acetylcholine. Myenteric plexus longitudinal muscle preparations from the guinea pig small intestine preincubated with [³H]choline were stimulated twice (10 Hz, 1 min, 200 mA) 45 min apart (S1, S2). L-NA ($\bullet$; $n = 5$–8) was added 30 min before S2. In interaction experiments 1 mM D-arginine ($\blacktriangledown$; $n = 6$) or 1 mM L-arginine ($\blacksquare$; $n = 4$) was in addition present from 15 min before S1 onwards. Effects of drugs on the evoked release were calculated from the ratio S2/S1 and expressed as percentage of the equivalent ratio obtained in control experiments. Means $\pm$ SEM shown by vertical lines. Significance of increase in [³H]acetylcholine release: *$P < 0.05$; **$P < 0.01$.

ropine (Barthó and Lefebvre, 1994). In addition, SIN-1 (300 $\mu$M, added 30 min before S2) significantly reduced the electrically induced release to 79 $\pm$ 5% of the control value. Likewise, exogenous NO caused a slight but significant inhibition of the evoked acetylcholine release from guinea pig small intestine (Wiklund et al., 1993b).

To summarize, NO is able of either increasing basal or inhibiting the depolarization-evoked acetylcholine release from enteric neurons. Such dual effects of a drug on basal and evoked acetylcholine release from the small intestine have been seen in several instances (e.g. with serotonin, substance P, GABA, muscarine; for review see Fuder and Muscholl, 1995). The increase in release is blocked by tetrodotoxin and therefore involves action potential propagation. NO is thus likely to stimulate cholinergic cell bodies within myenteric ganglia. The inhibitory effect of NO has functional consequences, since the electrically evoked cholinergic contractions of intestinal preparations were inhibited by exogenous NO and enhanced by

NO synthase inhibitors (Knudsen and Tottrup, 1992; Wiklund et al., 1993b). A presynaptic inhibitory action of exogenous NO at slow excitatory synapses has also been observed in electrophysiological experiments on guinea pig myenteric plexus (Tamura et al., 1993).

*Mechanisms*

As in the central nervous system, the excitatory effect of exogenous NO may be due to the activation of guanylyl cyclase. Guanylyl cyclase has been identified in a subpopulation of neurons in myenteric ganglia, and exogenous NO or electrical field stimulation caused an accumulation of cGMP-like immunoreactivity in these neurons (Shuttleworth et al., 1993). Interestingly, none of the myenteric neurons which produced cGMP-like immunoreactivity were stained for NO synthase by the NADPH-diaphorase technique. This suggests the possibility that NO released from nitrergic neurons stimulates the guanylyl cyclase activity of nearby neurons that do not contain NO synthase. The NO mediated increase in cGMP levels of enteric neurons in turn leads to the enhancement in basal acetylcholine release. In support of this possibility, 8-Br-cGMP increased the basal release of acetylcholine from the guinea pig myenteric plexus (Matusak et al., 1991).

The inhibition of the electrically evoked acetylcholine release by NO is unlikely to be mediated via cGMP since 8-Br-cGMP did not modify the evoked acetylcholine release from the intestine (Alberts and Stjärne, 1982). NO is known to activate calcium-dependent $K^+$ channels which could cause hyperpolarization of vascular smooth muscle (Bolotina et al., 1994). A similar hyperpolarization of neuronal varicosities might lead to the inhibition of acetylcholine release.

## Conclusions

Endogenous and exogenous NO increase basal acetylcholine release from central and peripheral cholinergic neurons. The effect is tetrodotoxin-sensitive and calcium-dependent and thus reflects stimulation of exocytotic release. The facilitation of acetylcholine release is likely to be mediated via the cGMP system.

In addition, NO has been shown to decrease the action potential-evoked acetylcholine release from peripheral neurons. The mechanism of inhibition is not known. Contraction experiments on various intestinal preparations suggest that the presynaptic nitrergic inhibition of acetylcholine release has functional consequences.

## Acknowledgements

This study was supported by a grant from the Deutsche Forschungsgemeinschaft (Ki 210/8-1).

## References

Alberts, P. and Stjärne, L. (1982) Secretion of [³H]-acetylcholine from guinea-pig ileum myenteric plexus is enhanced by 8-Br adenosine 3′,5′-cyclic monophosphate but not changed by 8-Br guanosine 3′,5′-cyclic monophosphate. *Acta Physiol. Scand.*, 115: 269–272.

Barthó, L. and Lefebvre, R. (1994) Nitric oxide induces acetylcholine-mediated contractions in the guinea-pig small intestine. *Naunyn Schmiedeberg´s Arch. Pharmacol.*, 350: 582–584.

Bolotina, V.M., Najibi, S., Palacino, J.J., Pagano, P.J. and Cohen, R.A. (1994) Nitric oxide directly activates calcium-dependent potassium channels in vascular smooth muscle. *Nature,* 368: 850–853.

Bravc, S.R., Hobbs, A.J., Gibson, A. and Tucker, J.F. (1991) The influence of L-N(G)-nitro-arginine on field stimulation induced contractions and acetylcholine release in guinea pig isolated tracheal smooth muscle. *Biochem. Biophys. Res. Commun.*, 179: 1017–1022.

Bredt, D.S., Hwang, P.M. and Snyder, S.H. (1990) Localization of nitric oxide synthase indicating a neural role for nitric oxide. *Nature,* 347: 768–770.

Costa, M., Furness, J.B., Pompolo, S., Brookes, S.J.H., Bornstein, J.C., Bredt, D.S. and Snyder, S.H. (1992) Projections and chemical coding of neurons with immunoreactivity for nitric oxide synthase in the guinea-pig small intestine. *Neurosci. Lett.,* 148: 121–125.

Dietrich, C. and Kilbinger, H. (1995) Prejunctional M1 and postjunctional M3 muscarinic receptors in the circular muscle of the guinea pig ileum. *Naunyn-Schmiedeberg´s Arch. Pharmacol.,* 351: 237–243.

Förstermann, U., Closs, E.I., Pollock, J.S., Nakane, M., Schwarz, P., Gath, I. and Kleinert, H. (1994) Nitric oxide synthase isozymes – characterization, purification, molecular cloning, and functions. *Hypertension,* 23: 1121–1131.

224

Fuder, H. and Muscholl, E. (1995) Heteroreceptor-mediated modulation of noradrenaline and acetylcholine release from peripheral nerves. *Rev. Physiol. Biochem. Pharmacol.*, 126: 265–412.

Garthwaite, J. and Boulton, C.L. (1995) Nitric oxide signaling in the central nervous system. *Annu. Rev. Physiol.*, 57: 683–706.

Guevara-Guzman, R., Emson, P.C. and Kendrick, K.M. (1994) Modulation of in vivo striatal transmitter release by nitric oxide and cyclic GMP. *J. Neurochem.*, 62: 807–810.

Hryhorenko, L.M., Woskowska, Z. and Fox-Threlkeld, J.E.T. (1994) Nitric oxide (NO) inhibits release of acetylcholine from nerves of isolated circular muscle of the canine ileum: relationship to motility and release of nitric oxide. *J. Pharmacol. Exp. Ther.*, 271: 918–926.

Kilbinger, H. and Wolf, D. (1994) Increase by NO synthase inhibitors of acetylcholine release from guinea-pig myenteric plexus. *Naunyn-Schmiedeberg's Arch. Pharmacol.*, 349: 543–545.

Knudsen, M.A. and Tottrup, A. (1992) A possible role of the L-arginine-nitric oxide pathway in the modulation of cholinergic transmission in the guinea-pig taenia coli. *Br. J. Pharmacol.*, 107: 837–841.

Llewellyn-Smith, I.J., Song, Z.M., Costa, M., Bredt, D.S. and Snyder, S.H. (1992) Ultrastructural localization of nitric oxide synthase immunoreactivity in guinea-pig enteric neurons. *Brain Res.*, 577: 337–342.

Lonart, G., Wang, J. and Johnson, K.M. (1992) Nitric oxide induces neurotransmitter release from hippocampal slices. *Eur. J. Pharmacol.*, 220: 271–272.

Matusak, O., Kuchel, O., Hamet, P. (1991) Effect of atrial natriuretic factor and 8-bromo cyclic guanosine 3'-5'-monophosphate on [$^3$H]acetylcholine outflow from myenteric-plexus longitudinal muscle of the guinea pig. *J. Pharmacol. Exp. Ther.*, 257: 107–113.

Ohkuma, S., Katsura, M., Guo, J.L., Hasegawa, T. and Kuriyama, K. (1995) Participation of peroxynitrite in acetylcholine release induced by nitric oxide generators. *Neurosci. Lett.*, 183: 151–154.

Prast, H. and Philippu, A. (1992) Nitric oxide releases acetylcholine in the basal forebrain. *Eur. J. Pharmacol.*, 216: 139–140.

Prast, H., Fischer, H.,Werner, E., Werner-Felmayer, G. and Philippu, A. (1995) Nitric oxide modulates the release of acetylcholine in the ventral striatum of the freely moving rat. *Naunyn-Schmiedeberg's Arch. Pharmacol.*, 352: 67–73.

Sanders, K.M. and Ward, S.M. (1992) Nitric oxide as a media-tor of nonadrenergic noncholinergic neurotransmission. *Am. J. Physiol.*, 262: G379–G392.

Sandor, N.T., Brassai, A., Puskas, A. and Lendvai, B. (1995) Role of nitric oxide in modulating neurotransmitter release from rat striatum. *Brain Res. Bull.*, 36: 483–486.

Schuman, E.M. and Madison, D.V. (1994) Nitric oxide and synaptic function. *Annu. Rev. Neurosci.*, 17: 153–183.

Sekizawa, K., Fukushima, T., Ikarashi, Y., Maruyama, Y. and Sasaki, H. (1993) The role of nitric oxide in cholinergic neurotransmission in rat trachea. *Br. J. Pharmacol.*, 110: 816–820.

Shuttleworth, C.W., Xue, C., Ward, S.M., de Vente, J. and Sanders, K.M. (1993) Immunohistochemical localization of 3',5'-cyclic guanosine monophosphate in the canine proximal colon – responses to nitric oxide and electrical stimulation of enteric inhibitory neurons. *Neuroscience*, 56: 513–522.

Snyder, S.H. and Bredt, D.S. (1991) Nitric oxide as a neuronal messenger. *Trends Pharmacol. Sci.*, 12: 125–127.

Starke, K.,Göthert, M. and Kilbinger, H. (1989) Modulation of neurotransmitter release by presynaptic autoreceptors. *Physiol. Rev.*, 69: 864–989.

Suzuki, T., Nonaka, H., Fujimoto, K. and Kawashima, K. (1993) Effects of physostigmine and some nitric oxide-cyclic GMP-related compounds on muscarinic receptor-mediated autoinhibition of hippocampal acetylcholine release. *J. Neurochem.*, 60: 2285–2289.

Tamura, K., Schemann, M. and Wood, J.D. (1993) Actions of nitric oxide-generating sodium nitroprusside in myenteric plexus of guinea pig small intestine. *Am. J. Physiol.*, 265: G887-G893

Vincent, S.R. (1994) Nitric oxide – a radical neurotransmitter in the central nervous system. *Prog. Neurobiol.*, 42: 129–160.

Ward, J.K., Belvisi, M.G., Fox, A.J., Miura, M., Tadjkarimi, S., Yacoub, M.H and Barnes, P.J. (1993) Modulation of cholinergic neural bronchoconstriction by endogenous nitric oxide and vasoactive intestinal peptide in human airways in vitro. *J. Clin. Invest.*, 92: 736–742.

Wiklund, N.P., Leone, A.M., Gustafsson, L.E. and Moncada, S. (1993a) Release of nitric oxide evoked by nerve stimulation in guinea-pig intestine. *Neuroscience*, 53: 607–611.

Wiklund, C.U, Olgart, C., Wiklund, N.P. and Gustafsson, L.E. (1993b) Modulation of cholinergic and substance P-like neurotransmission by nitric oxide in the guinea-pig ileum. *Br. J. Pharmacol.*, 110: 833–839.

J. Klein and K. Löffelholz (Eds.)
*Progress in Brain Research*, Vol. 109
© 1996 Elsevier Science B.V. All rights reserved.

# CHAPTER 22

# Presynaptic interactions between acetylcholine and glycine in the human brain

Mario Marchi[1], Gian Carlo Andrioli[2], Paolo Cavazzani[2], Sabrina Marchese[1] and Maurizio Raiteri[1]

[1]*Istituto di Farmacologia e Farmacognosia, Università degli Studi di Genova, Viale Cembrano 4, 16148 Genova and* [2] *Divisione di Neurochirurgia, Ospedali Galliera, Via A. Volta 8, 16128 Genova, Italy*

## Introduction

Several neurotransmitters are probably involved in cognitive processes. There is strong evidence that a decrease in the function of acetylcholine (ACh) may be in part responsible for the decrease of cognitive performances in humans (Perry et al., 1981; Bartus et al., 1982). Also the glutamatergic system is likely to play an important role in the processes of learning and memory and, in this context, glycine (Gly), as a co-agonist of glutamate at the excitatory $N$-methyl-D-aspartate (NMDA) receptors may be of particular interest. It has been proposed that activation of NMDA receptors either by an increased availability of the endogenous amino acid or by administering exogenous glycinomimetics, may be of help in cases of impaired cognitive function (Monahan et al., 1989; Flood et al., 1992; Thompson et al., 1992; Pittaluga et al., 1993).

The above considerations prompted us to investigate the interaction between cholinergic and glycinergic systems in the human cerebral cortex. We found that ACh can potentiate Gly release, a finding which suggests that the cholinergic system may enhance indirectly, through a release of Gly, the glutamatergic transmission in the human brain.

## Depolarization-evoked release of [³H]Gly from human cortex synaptosomes

Glycine has long been known as an inhibitory transmitter in the lower part of the neuraxis, i.e. spinal cord and brain stem, but recent evidence supports a function for glycine as a transmitter/modulator also in telencephalic areas.

The widespread distribution throughout the central nervous system of excitatory NMDA receptors, which seem to require glycine for their activation, implies the presence of glycine-releasing structures in the vicinity of NMDA receptors. Synaptosomal preparations from rat cortex (Levi et al., 1982) and hippocampus (Raiteri et al., 1990a) have been shown to release previously captured [³H]Gly in a $Ca^{2+}$-dependent manner. Moreover, a high-affinity uptake system for Gly exists also in supraspinal regions of the CNS (Valdes et al., 1977; Debler and Lajtha, 1987; Fedele and Foster, 1992). Thus, the existence in telencephalic regions of structures able to release and to recapture Gly together with that of Gly receptors support the putative role for the amino acid as a transmitter also in these areas. In the present work we have studied the release of [³H]Gly from human cortex synaptosomes. The human specimens used had to be removed by the neurosurgeon to reach deeply located tumors. The samples represented parts of frontal (4), temporal (10), and parietal (4) lobes either of female (10) and male (8) patients. After premedication with atropine and meperidine, anaesthesia was induced with pentothal and maintained with 70% nitrous oxide in 30% oxygen and 0.5–1% isoflurane. Pancuronium was employed to obtain muscular relaxation. Immediately after re-

226

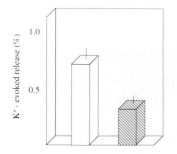

Fig. 1. Depolarization-evoked release of tritium from human synaptosomes prelabeled with [³H]Gly. Empty column, controls; hatched column, Ca²⁺-free, Mg² (10 μM)-enriched medium.

moval the tissues were placed in a physiological salt solution kept at 2–4°C and synaptosomal fractions were obtained within 60 min. Crude synaptosomes were prepared and superfused according to Raiteri et al. (1984).

Human cerebrocortex nerve endings prelabeled with [³H]Gly can release tritium radioactivity when moderately depolarized with 15 mM KCl. (Fig. 1). Part of this release (about 60%) was blocked by Mg²⁺ in keeping with an exocytotic release of [³H]Gly requiring influx of Ca²⁺ through voltage-dependent, Mg²⁺-sensitive channels (Fig. 1) (Russo et al., 1993).

## Effect of ACh on [³H]Gly release from human cortex

The K⁺-evoked overflow of [³H]Gly was augmented by ACh (5–100 μM) or by oxotremorine (10 μM; Fig. 2; Russo et al., 1993). Considering the characteristics of the experimental set-up employed to study release (a thin layer of synaptosomes up-down superfused in which indirect effects are minimized) ACh and oxotremorine are likely to act directly on the particles releasing [³H]Gly. The basal release of Gly was unaffected. The concentration of ACh causing half-maximal effect (EC₅₀) amounted to 7 μM.

The findings that oxotremorine, a muscarinic receptor agonist, mimicked ACh and that the effect of ACh was insensitive to the nicotinic receptor antagonist mecamylamine (100 μM), whereas it

was blocked by the muscarinic antagonist atropine (0.1 μM), indicate that the depolarization-evoked release of [³H]Gly was potentiated through the activation of muscarinic receptors (Fig. 2). These receptors are possibly located on glycinergic nerve terminals, although the involvement of non-neuronal structures (glia, for instance) cannot be ruled out at present.

## Antagonism of the ACh-stimulated [³H]Gly overflow

On the basis of the innumerable reports published during the last decade, multiple subtypes of the muscarinic receptor appear to exist. Functional pharmacological studies have permitted the identification of at least four subtypes, designated M₁, M₂, M₃ and M₄ (Hulme et al., 1990; Raiteri et al., 1990b; Caulfield and Brown, 1991). Interestingly, five muscarinic receptors (m1–m5) have been cloned and shown to be expressed also in the human brain (Peralta et al., 1987; Bonner et al., 1988; Maeda et al., 1988; Buckley et al., 1989; Weiner et al., 1990). Pharmacological characterization of the five cloned receptors was carried out by antagonist binding (Buckley et al., 1989; Dörje et al., 1991). Results from these binding studies as well as from experiments of functional pharmacology indicate that only a limited number of selective drugs can be of help when a muscarinic receptor involved in a given function needs to be classified.

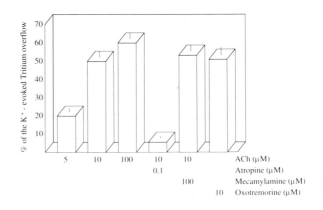

Fig. 2. Antagonism of ACh (10 μM)-stimulated [³H]Gly overflow.

For the pharmacological characterization of the muscarinic receptor presently found to mediate enhancement of Gly release in the human neocortex, three receptor antagonists were selected (pirenzepine, himbacine and AF-DX 116) on the basis of the data shown in the paper by Dörje et al. (1991) reporting the antagonist binding profiles of five cloned human muscarinic receptor subtypes stably expressed in CHO-K1 cells. The calculated $pA_2$ values for the three antagonists were as follows: 7.27 for pirenzepine, 6.65 for AF-DX 116 and 8.34 for himbacine (Russo et al., 1993).

When the rank of potencies of the three receptor antagonists at the muscarinic receptor potentiating [$^3$H]Gly release is compared with that at each of the cloned receptors as reported by Dörje et al. (1991), this muscarinic receptor appears to resemble both the m3 and the m4 subtypes. On the other hand, the absolute values of the affinities for the m3 receptor are lower than those for the m4 subtype and, in turn, the values for the m4 are much closer to those for the muscarinic receptor potentiating [$^3$H]Gly release. Thus the comparison between the m1–m5 human cloned receptors and this receptor indicates that the muscarinic receptor involved in the enhancement of Gly release in human cerebral cortex resembles the m4 human cloned receptor.

The relatively high $pA_2$ value obtained with pirenzepine may at a first glance appear surprising. Pirenzepine has long been known as a selective $M_1$ receptor antagonist (see Hulme et al., 1990). The pattern of the affinities of the drug for the different cloned human receptors confirms this view. However the drug also displays moderate to high affinity (7.43) for the m4 subtype. This value is almost identical to the $pA_2$ of the antagonist as calculated in the case of receptor regulating Gly release.

As to himbacine, the drug was originally classified as an $M_2$ selective antagonist (Gilani and Cobbin, 1986; Lazareno and Roberts, 1989). However, more recent functional pharmacological studies (Caulfield and Brown, 1991) and the work by Dörje et al. (1991) with human cloned receptors showed that the drug possesses high affinities to both m2 and m4 receptors. Thus the drug may be considered a valuable tool to discriminate between

$M_2/M_4$ receptors versus all other subtypes (Caulfield and Brown, 1991; Waelbroeck et al., 1992). Our data with himbacine tend therefore to exclude an involvement of $M_1$, $M_3$ or $M_5$ receptors.

An involvement of $M_1$, $M_2$ or $M_3$ receptors is also unlikely from the results obtained with oxotremorine. It has been reported that the odd-numbered receptors are positively linked to the phosphoinositide metabolism (Bonner et al., 1987; Peralta et al., 1987). Oxotremorine behaves as a partial agonist in the phosphoinositide cycle with very low intrinsic activity (Fisher et al., 1984; Marchi et al., 1988). The finding that also oxotremorine enhanced Gly release with a potency similar to that of ACh tends to exclude the involvement of a receptor ($M_1$, $M_3$ or $M_5$) linked to phosphoinositide hydrolysis.

## Concluding remarks

The results of the present work are compatible with the existence in human cerebral cortex of a direct functional interaction between cholinergic and glycinergic systems possibly mediated by $M_4$ muscarinic receptors situated on Gly releasing structures. Such an interaction has not been observed in animals in which, however, the reverse effect, i.e. a modulation by Gly of ACh release, was reported (Beani et al., 1983; Taylor et al., 1988). Interestingly, it has been suggested that ACh acting at the $M_4$ receptor subtype could have a more general significance by modulating the release of neurotransmitters (Caulfield and Brown, 1991; McKinney et al., 1993; Van Der Zee and Luiten, 1993).

It seems important to note that the autoreceptors situated on the axon terminals releasing ACh in human brain were found to be insensitive to pirenzepine (Marchi et al., 1990). The presumptive pharmacological difference between muscarinic autoreceptors and receptors modulating Gly release may allow the ACh–Gly interaction to be modified by using drugs selective for the different receptor subtypes.

The possibility for ACh to activate Gly release may have important implications. Both cholinergic and glutamatergic transmissions have been deeply implicated in the processes of learning and memory. The well known amnesia caused by scopolamine, besides other evidence, is in favor of an involvement of muscarinic receptors. On the glutamatergic side, the importance of NMDA receptors in cognitive processes is well recognized. Functional evidence for the presence of NMDA receptors in human cerebral cortex has recently been provided (Fink et al., 1992). Our results propose a link between cholinergic and glutamatergic systems occurring through the mediation of Gly released by ACh onto NMDA receptors. As previously mentioned, Gly appears to be obligatory for NMDA receptor activation. Very recent behavioral and neurochemical studies carried out in aged animals support the idea that the Gly site on the NMDA receptor is an important target, activation of which may lead to improved learning (Flood et al., 1992; Thompson et al., 1992; Pittaluga et al., 1993). The ACh–Gly interaction found to exist in the human brain may thus represent a biological substrate for the rational development of both selective muscarinic receptor agonists and glycinomimetic drugs as potential cognitive enhancers.

Finally a recent report has provided a clear description of the distribution of the muscarinic receptor subtypes in human brain and their regulation in Alzheimer's disease (Flynn et al., 1995). Interestingly the m4 receptor subtype was upregulated significantly in Alzheimer patients and this subtype can be therefore a target for cholinergic drugs useful in Alzheimer's disease.

## Acknowledgements

This work was supported by grants from the Italian M.U.R.S.T. and from the Italian C.N.R. The authors thank Mrs. Maura Agate for her secretarial assistance.

## References

Bartus, R.T., Dean, III, R.L., Beer, B. and Lippa, A.S. (1982) The cholinergic hypothesis of geriatric memory dysfunction. *Science*, 217: 408–417.

Beani, L., Bianchi, C., Siniscalchi, A. and Tanganelli, S. (1983) Glycine-induced changes in acetylcholine release from guinea-pig brain slices. *Br. J. Pharmacol.*, 79: 623–628.

Bonner, T.I., Buckley, N.J., Young, A.C. and Brann, M.R. (1987) Identification of a family of muscarinic acetylcholine receptor genes. *Science*, 237: 527–532.

Bonner, T.I., Young, A.C., Brann, M.R. and Buckley, N.J. (1988) Cloning and expression of the human and rat m5 muscarinic acetylcholine receptor genes. *Neuron*, 1: 403–410.

Buckley, N.J., Bonner, T.I., Buckley, C.M. and Brann, M.R. (1989) Antagonist binding properties of five cloned muscarinic receptors expressed in CHO-K1. *Cell Mol. Pharmacol.*, 35: 469–476.

Caulfield, M.P. and Brown, D.A. (1991) Pharmacology of the putative $M_4$ muscarinic receptor mediating Ca-current inhibition in neuroblastomaxglioma hybrid (NG 108-15) cells. *Br. J. Pharmacol.*, 104: 39–44.

Debler, E.A. and Lajtha, A. (1987) High affinity of gamma-aminobutyric acid, glycine, taurine, L-aspartic acid, and L-glutamic acid in synaptosomal ($P_2$) tissue: a kinetic and substrate specificity analysis. *J. Neurochem.*, 48: 1851–1856.

Dörje, F., Wess, J., Lambrecht, G., Tacke, R., Mutschler, E. and Brann, M.R. (1991) Antagonist binding profiles of five cloned human muscarinic receptor subtypes. *J. Pharmacol. Exp. Ther.*, 256: 727–733.

Fedele, E. and Foster, A.C. (1992) [$^3$H]Glycine uptake in rat hippocampus: kinetic analysis and autoradiographic localization. *Brain Res.*, 572: 154–163.

Fink, K., Schultheiss, R. and Göthert, M. (1992) Stimulation of noradrenaline release in human cerebral cortex mediated by N-methyl-D-aspartate (NMDA) and non-NMDA receptors. *Br. J. Pharmacol.*, 106: 67–72.

Fisher, S.K., Figueiredo, J.C. and Bartus, R.T. (1984) Differential stimulation of inositol phospholipid turnover in brain by analogs of oxotremorine. *J. Neurochem.*, 43: 1171–1179.

Flood, J.F., Morley, J.E. and Lanthorn, T.H. (1992) Effect on memory processing by D-cycloserine, an agonist of the NMDA/glycine receptor. *Eur. J. Pharmacol.*, 221: 249–254.

Flynn, D.D., Ferrari-DiLeo, G., Mash, D.C. and Levey, A.I. (1995) Differential regulation of molecular subtypes of muscarinic receptors in Alzheimer's disease. *J. Neurochem.*, 64: 1888–1891.

Gilani, S.A.H. and Cobbin, L.B. (1986) The cardioselectivity of himbacine: a muscarine receptor antagonist. *Naunyn-Schmiedeberg's Arch. Pharmacol.*, 332: 16–20.

Hulme, E.C., Birdsall, N.J.M. and Buckley, N.J. (1990) Muscarinic receptor subtypes. *Annu. Rev. Pharmacol. Toxicol.*, 30: 633–673.

Lazareno, S. and Roberts, F.F. (1989) Functional and binding studies with muscarinic $M_2$-subtype selective antagonists. *Br. J. Pharmacol.*, 98: 309–317.

Levi, G., Bernardi, G., Cherubini, E., Gallo, V., Marciani, M.G. and Stanzione, P. (1982) Evidence in favor of a neurotransmitter role of glycine in the rat cerebral cortex. *Brain Res.*, 236: 121–131.

Maeda, A., Kubo, T., Mishina, M. and Numa, S. (1988) Tissue distribution of mRNAs encoding muscarinic acetylcholine receptor subtype. *FEBS Lett.*, 239: 339–342.

Marchi, M., Fontana, G., Paudice, P. and Raiteri, M. (1988) The activation of phosphatidylinositol turnover is not directly involved in the modulation of neurotransmitter release mediated by presynaptic muscarinic receptors. *Neurochem. Res.*, 13: 903–907.

Marchi, M., Ruelle, A., Andrioli, G.C. and Raiteri, M. (1990) Pirenzepine-insensitive muscarinic autoreceptors regulate acetylcholine release in human neocortex. *Brain Res.*, 520: 347–350.

McKinney, M., Miller, J.H. and Aagard, P.J. (1993) Pharmacological characterization of the rat hippocampal muscarinic autoreceptor. *J. Pharmacol. Exp. Ther.*, 264: 74–78.

Monahan, J.B., Handelmann, G.E., Hood, W.F. and Cordi, A.A. (1989) D-cycloserine, a positive modulator of the *N*-methyl-D-aspartate receptor, enhances performance of learning tasks in rats. *Pharmacol. Biochem. Behav.*, 34: 649–653.

Peralta, E.G., Ashkenazi, A., Winslow, J.W., Smith, D.H., Ramachandran, J. and Capon, D.J. (1987) Distinct primary structures, ligand-binding properties and tissue-specific expression of four human muscarinic acetylcholine receptors. *EMBO J.*, 6: 3923–3929.

Perry, E.K., Blessed, G. and Tomlinson, B.E. (1981) Neurochemical activities in human temporal lobe related to aging and Alzheimer-type changes. *Neurobiol. Aging*, 2: 251–256.

Pittaluga, A., Fedele, E., Risiglione, C. and Raiteri, M. (1993) Age-related decrease of the *N*-methyl-D-aspartate receptor-mediated noradrenaline release in rat hippocampus and partial restoration by D-cycloserine. *Eur. J. Pharmacol.*, 231: 129–134.

Raiteri, M., Bonanno, G., Marchi, M. and Maura, G. (1984) Is there a functional linkage between neurotransmitter uptake mechanisms and presynaptic receptors? *J. Pharmacol. Exp. Ther.*, 231: 671–677.

Raiteri, M., Fontana, G. and Fedele, E. (1990a) Glycine stimulates [$^3$H]noradrenaline release by activating a strychnine-sensitive receptor present in rat hippocampus. *Eur. J. Pharmacol.*, 184: 239–250.

Raiteri, M., Marchi, M. and Paudice, P. (1990b) Presynaptic muscarinic receptors in the CNS. *Ann. N. Y. Acad. Sci.*, 604: 113–129.

Russo, C., Marchi, M., Andrioli, G.C., Cavazzani, P. and Raiteri, M. (1993) Enhancement of glycine release from human brain cortex synaptosomes by acetylcholine acting at $M_4$ muscarinic receptors. *J. Pharmacol. Exp. Ther.*, 266: 142–146.

Taylor, C.A., Tsai, C. and Lehmann, J. (1988) Glycine-evoked release of [$^3$H]acetylcholine from rat striatal slices is independent of the NMDA receptor. *Naunyn-Schmiedeberg's Arch. Pharmacol.*, 337: 552–555.

Thompson, L.T., Moskal, J.R. and Disterhoft, J.F. (1992) Hippocampus-dependent learning facilitated by a monoclonal antibody or D-cycloserine. *Nature*, 359: 638–641.

Valdes, F., Munoz, C., Feria-Velasco, A. and Orrego, F. (1977) Subcellular distribution of rat brain cortex high-affinity, sodium dependent, glycine transport sites. *Brain Res.*, 122: 95–112.

Van Der Zee, E.A. and Luiten, P.G.M. (1993) GABAergic neurons of the rat dorsal hippocampus express muscarinic acetylcholine receptors. *Brain Res. Bull.*, 32: 601–609.

Waelbroeck, M., Camus, J., Tastenoy, M. and Christophe, J. (1992) Binding properties of nine 4-diphenyl-acetoxy-*N*-methyl-piperidine (4-DAMP) analogues to $M_1$, $M_2$, $M_3$ and putative $M_4$ muscarinic receptor subtypes. *Br. J. Pharmacol.*, 105: 97–102.

Weiner, D., Levey, A. and Brann, M.R. (1990) Expression of muscarinic acetylcholine and dopamine receptor mRNAs in rat basal ganglia. *Proc. Natl. Acad. Sci. USA*, 87: 7050–7054.

J. Klein and K. Löffelholz (Eds.)
*Progress in Brain Research*, Vol. 109
© 1996 Elsevier Science B.V. All rights reserved.

# CHAPTER 23

# Purinergic regulation of acetylcholine release

J. Alexandre Ribeiro[1], Rodrigo A. Cunha[1], Paulo Correia-de-Sá[2] and Ana M. Sebastião[1]

[1]*Laboratory of Pharmacology, Gulbenkian Institute of Science, 2781-Oeiras, Portugal
and* [2]*Laboratory of Pharmacology, ICBAS, University of Porto, Portugal*

## Storage, release and extracellular metabolism of ATP. Formation of adenosine

There is now a consensus that in the cholinergic motor nerve terminals, adenosine triphosphate (ATP) is stored together with acetylcholine (ACh) in cholinergic synaptic vesicles (Dowdall et al., 1974), as well as in vesicles containing ATP but not ACh (Fariñas et al., 1990). It has also been demonstrated that ATP is co-released with ACh from motor nerve terminals (Silinsky, 1975) and also released independently of ACh secretion (Marsal et al., 1987). When ATP is released at the neuromuscular junction it should be catabolized in the same manner as when it is exogenously applied. In these conditions, ATP is degraded into ADP, AMP, IMP, adenosine, inosine and hypoxanthine (Fig. 1). Evidence that different metabolites (IMP, adenosine, inosine and hypoxanthine) are formed from AMP degradation is provided by interfering with the ectoenzymes that catalyze formation of these metabolites; thus, formation of adenosine is prevented through inhibition of the ecto-5′-nucleotidase using the ADP analogue, $\alpha,\beta$-methylene ADP (AOPCP). AOPCP also inhibits inosine formation from IMP. The formation of IMP from AMP can be inhibited by the AMP deaminase inhibitor, coformycin (Fig. 1) (Cunha and Sebastião, 1991).

The total release of adenine nucleotides evoked upon electrical stimulation of innervated skeletal muscle preparations in the presence of a supramaximal concentration of tubocurarine represents approximately half of the total release evoked in the absence of tubocurarine, suggesting that one half of the released adenine nucleotides originates from the cholinergic nerve terminals and the other half from the skeletal muscle fibers (Cunha and Sebastião, 1993).

Adenosine as such also accumulates at the neuromuscular junction (Cunha and Sebastião, 1993). During nerve stimulation, in the presence of the ecto-5′-nucleotidase inhibitor, AOPCP, and providing that a supramaximal concentration of tubocurarine is used to block the nicotinic receptors, the amount of adenosine detected is approximately half of that detected in the absence of tubocurarine. This suggests that, as happens with the adenine nucleotides, half of the adenosine being released as such comes from the nerves, and the other half originates from the muscle fibers. In the absence of external calcium, but in the presence of tubocurarine, release of adenosine was not detected, supporting its neural origin. Removal of ecto-5′-nucleotidase inhibition reveals that the catabolism of the adenine nucleotides released during stimulation, at the neuromuscular junction, contributes to about 50% of the amount of endogenous adenosine that accumulates extracellularly (Cunha and Sebastião, 1993).

The amount of adenine nucleotides released by an innervated frog sartorius muscle during 30 min stimulation at 0.2 Hz (360 stimuli) in the presence of tubocurarine (i.e. evoked release from the nerve endings) and in the presence of AOPCP to prevent AMP catabolism, is approximately 0.15 pmol per stimulus (Cunha and Sebastião, 1993). Assuming (Katz and Miledi, 1977), that a frog sartorius muscle has about 1000 endplates, $1.5 \times 10^{-16}$ mol of adenine nucleotides are released per stimulus per endplate. If each cleft in the frog sartorius neuromuscular junction has approximately 5980 $\mu m^3$, the concentration of adenine nucleotides in the cleft might increase transiently by 25 $\mu M$. This amount is about 2-fold higher than the previously estimated transient increase of adenosine levels at the frog sartorius endplates upon stimulation of the nerve terminal (Ribeiro and Sebastião, 1987), and it is also 2–3-fold higher than that estimated for the rat phrenic-diaphragm without inhibition of adenine nucleotide catabolism (Silinsky, 1975). A much higher value was estimated by Smith (1991) at the rat neuromuscular junction, because the amount of adenine nucleotides released from the contracting muscle fibers, partially being outside the endplate area, was included in the calculations.

In the central nervous system there is also the machinery to metabolize extracellular ATP into AMP (Cunha et al., 1992). Exogenously added AMP is catabolized into adenosine and then into inosine in the rat hippocampal cholinergic nerve

terminals and in hippocampal slices, as well as in cortical slices. In contrast with the neuromuscular junction, IMP formation from extracellular AMP was not detected in the central nervous system. In the immunopurified cholinergic nerve terminals obtained from the rat brain cortex, AMP catabolism was not detected, whereas in immunopurified cholinergic nerve terminals of the hippocampus the activity of ecto-5′-nucleotidase is about 50 times higher than in a crude synaptosomal fraction (Cunha et al., 1992). The explanation for this finding could reside upon the fact that the cholinergic nerve terminals in the hippocampus are enriched in ecto-5′-nucleotidase, whereas in the cerebral cortex ecto-5′-nucleotidase activity seems to be located preferentially outside the cholinergic nerve terminals. Since immunopurified cholinergic nerve terminals have negligible contamination

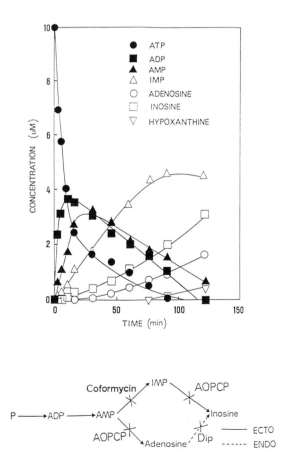

Fig. 1. Pathways involved in the extracellular metabolism of adenine nucleotides at the innervated sartorius muscle of the frog. The upper panel shows the time course of ATP degradation and appearance of its products of metabolism. ATP (10 $\mu M$) was incubated at zero time with an innervated frog sartorius muscle in a 2 ml bath, in the presence of the adenosine uptake inhibitor, dipyridamole (0.5 $\mu M$). Samples (75 $\mu l$) were collected at the times indicated in the abscissa and analyzed by RP-HPLC. ATP was catabolized into ADP, AMP, IMP, adenosine, inosine and hypoxanthine. Sequential addition of each of the ATP metabolites as initial substrates (Cunha and Sebastião, 1991) allowed us to establish the steps by which adenine nucleotides are extracellularly metabolised at the frog innervated sartorius muscle, which is shown in the lower panel. All steps are mediated by ecto-enzymes (ECTO) except that indicated by the broken line (ENDO). ATP is sequentially converted into ADP and AMP. AMP can be either deaminated into IMP by an ecto-AMP deaminase, which is inhibited by coformycin, or dephosphorylated into adenosine by an ecto-5′-nucleotidase, which is inhibited by $\alpha,\beta$-methylene ADP (AOPCP). Adenosine has to be taken up intracellularly to be metabolised into inosine, a step inhibited by the adenosine uptake inhibitor, dipyridamole (Dip), while IMP is extracellularly metabolised by an AOPCP-sensitive ecto-5′-nucleotidase also yielding inosine (modified from Cunha and Sebastião, 1991).

with glial cells, these findings also suggest that ecto-5'-nucleotidase is located on neuronal structures in the hippocampus.

## Effects of purines on ACh release

### Inhibitory effects

Adenosine (Ginsborg and Hirst, 1972) and ATP (Ribeiro and Walker, 1973) inhibit both evoked (decrease in the amplitude of evoked endplate potentials) and spontaneous (decrease in the frequency of miniature endplate potentials (MEPPs)) release of ACh from motor nerve terminals. These purines did not significantly affect the average amplitude of MEPPs, suggesting that, under the conditions used, they do not have post-junctional effects. In the absence of calcium in the external medium, and in the presence of the calcium chelating agent, ethylenodiaminotetraacetic acid (EDTA), adenosine also decreased the frequency of MEPPs (Ribeiro and Dominguez, 1978), which suggests that the effect of adenosine on spontaneous release of ACh is not related to its ability to inhibit the entry of calcium into the nerve terminal. Comparing the efficiency of adenosine to inhibit evoked release with its efficiency to inhibit spontaneous release of ACh from the motor nerve terminals, in both the rat and frog, it was found that adenosine is consistently more efficient to decrease evoked rather than spontaneous release. Furthermore, there is a poor correlation between the effects of adenosine on evoked and spontaneous release in the rat, and no correlation was detected between these two forms of ACh release in the frog (Ribeiro and Sebastião, 1986). These results suggest that two different mechanisms could be involved in these effects of adenosine: one on evoked and the other on spontaneous release of ACh. This possibility does not preclude that common aspects (e.g. intracellular calcium) participate in both forms of ACh release. Whether this different efficiency reflects activation of two different entities (receptors and/or transducing systems?) by adenosine, or different effects in the synaptic vesicle proteins that regulate acetylcholine release from the motor nerve terminals, is presently un-

known. It has been suggested that activation of inhibitory adenosine receptors at motor nerve endings decreases the affinity of intracellular calcium for the release process (Silinsky, 1984). Since adenosine also reduces calcium uptake by nerve endings (Ribeiro et al., 1979), these two mechanisms might simultaneously contribute to the inhibitory effect of adenosine on the evoked release, whereas the spontaneous (asynchronous) release might only be affected by modifications of intracellular calcium efficiency.

To find out whether adenosine and ATP, which are equipotent to inhibit ACh release (Ribeiro and Walker, 1975), operate through a similar mechanism, a supramaximal concentration of ATP was applied in the presence of a supramaximal concentration of adenosine. Under these conditions, ATP had no further inhibitory effect on the amplitude of evoked endplate potentials, suggesting that the inhibitory effects of both purines on ACh release are mediated by the same mechanism. Knowing this, the obvious question was: does ATP as such causes inhibition of ACh release or does the effect of ATP depend upon its hydrolysis to adenosine? Experiments using the ecto-5'-nucleotidase inhibitor AOPCP together with ATP showed that this nucleotide has no inhibitory effect on ACh release (Ribeiro and Sebastião, 1987). The conclusion that the ATP intact molecule is non-effective to inhibit ACh release is supported by the observation that the stable ATP analogue $\alpha,\beta$-methylene ATP which cannot be hydrolyzed into adenosine by ecto-5'-nucleotidase (Cascalheira and Sebastião, 1992), is inactive.

There are, however, situations in which ATP as such inhibited ACh release. This was seen by measuring [$^3$H]ACh release from the rat brain cortex, where the ATP analogue, ATPγS, inhibited ACh release, in spite of not being metabolized into adenosine (Cunha et al., 1994a). Since the cholinergic terminals of the rat brain cortex do not possess ecto-5'-nucleotidase (see above), these results stress the importance of ecto-5'-nucleotidase for adenosine to act as a neuromodulator of ACh release. It appears that ATP can subserve a role usually played by adenosine only when ecto-5'-nucleotidase is absent. Surprisingly, the effect of

ATP itself on ACh release in the cortex is antagonized by the $A_1$ selective antagonist, 1,3 dipropyl-8-cyclopentylxanthine (DPCPX), but not by the $P_2$ antagonist, suramin, or by substances that interfere with adenosine accumulation (Cunha et al., 1994a). This suggests that the effect of ATP was mediated either by a nucleotide-sensitive adenosine $A_1$ receptor or by a xanthine-sensitive adenine nucleotide receptor. The precise nature of this receptor needs further investigation.

### Excitatory effects

If one increases the frequency of stimulation of the nerve terminals in the innervated rat diaphragm and measures [$^3$H]ACh release, it is possible to observe enhancement in [$^3$H]ACh release by applying exogenous adenosine. This is not the case when the release of ACh is determined as quantal content of evoked endplate potentials, because low frequency of stimulation has to be used in this experimental approach. Up to 30 $\mu$M, the nucleoside is inhibitory but in higher concentrations (e.g. 100–500 $\mu$M) it is excitatory (Fig. 2). An excitatory effect was observed also with the use of low nanomolar concentrations of the selective $A_{2a}$ agonist, CGS 21680. This compound in both the electrophysiological (low frequency nerve stimulation) experiments and in the [$^3$H]ACh release (high frequency nerve stimulation) experiments, facilitates ACh release (Correia-de-Sá et al., 1991). The comparison of the excitatory effect of CGS 21680 with the inhibitory effect of the $A_1$ agonist, R-$N^6$-phenylisopropyladenosine (R-PIA), in the same endplate, provided evidence for the presence of both inhibitory ($A_1$) and excitatory ($A_{2a}$) adenosine receptors in the same nerve terminal.

### Endogenous adenosine

Evidence that endogenous adenosine causes both inhibition and excitation was obtained with experiments in which the adenosine uptake blocker nitrobenzylthioinosine (NBTI) or the adenosine deaminase inhibitor erythro-9-(2-hydroxy-3-nonyl)adenine (EHNA) were used. These compounds, both mimicked (depending upon the concentration)

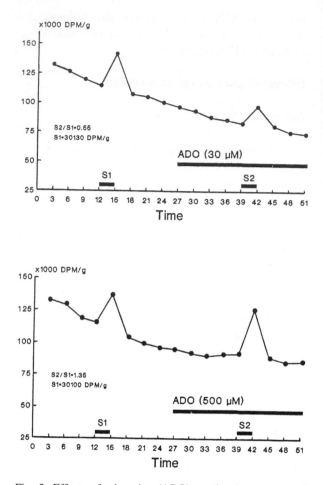

Fig. 2. Effects of adenosine (ADO) on the time course of tritium outflow (dpm per g of tissue) from the rat phrenic nerve terminals. After the labelling and washout periods (see e.g. Correia-de-Sá et al., 1991), [$^3$H]ACh release was evoked twice by electrical stimulation (5 Hz, 40 $\mu$s, for 3 min) of the phrenic nerve, at the indicated times ($S_1$ and $S_2$) Tritium outflow was measured in samples collected every 3 min. ADO was applied 15 min before $S_2$ and remained in the bath until the end of the experiments, as represented by the horizontal bars. The $S_2/S_1$ ratio in the absence of any drugs (i.e. in control conditions) was 0.81. Note that this ratio was decreased by 30 $\mu$M adenosine (upper panel) but increased by 500 $\mu$M adenosine (lower panel).

the inhibitory and the excitatory effects of exogenously applied adenosine on [$^3$H]ACh release. The inhibition caused by low concentrations of NBTI was mediated by $A_1$ receptors, since it was antagonized by the $A_1$ selective antagonist, DPCPX, but not by the $A_2$ antagonist, PD 115,199. In contrast, the excitatory effect of high concentrations of NBTI is an $A_2$-mediated effect, since the $A_1$ antagonist DPCPX did not antagonize the effect, but even enhanced it. The enhancement observed with this $A_1$ antagonist could result from $A_1$ to $A_2$ interactions, so that $A_1$ receptor activation counteracts $A_2$ receptor activation and vice versa, as shown to occur in the CA3 area of the hippocampus (Cunha et al., 1994b).

The results obtained with NBTI and EHNA provide two types of information: (1) endogenous adenosine activates both inhibitory and excitatory receptors, the concentrations of adenosine needed to activate excitatory receptors being probably higher than those needed to activate inhibitory receptors, and (2) the inactivation system of the adenosine effects at the rat neuromuscular junction is via adenosine uptake and deamination (see also Sebastião and Ribeiro, 1988). It has been shown that in the frog neuromuscular junction the inhibition caused by adenosine on quantal release of ACh is inactivated only through uptake (Ribeiro and Sebastião, 1987; Cunha and Sebastião, 1991).

In experiments where adenosine deaminase was exogenously applied to remove the tonic action of adenosine on ACh release at low frequency stimulation (0.5 Hz) and recording EPPs, adenosine deaminase increased EPPs amplitude (Fig. 3). In the experiments, where a higher frequency was used and [³H]ACh release was measured, the excitatory effect of adenosine deaminase was progressively reduced when the number of stimuli applied to the phrenic nerve was decreased. This is consistent with previous observations showing that adenosine is released during stimulation (Cunha and Sebastião, 1993), and that the preponderant effect of endogenous adenosine on acetylcholine release from motor nerve endings is inhibitory (Ribeiro and Sebastião, 1987; Sebastião and Ribeiro, 1988). However, by increasing pulse duration as well as the number of pulses applied to the phrenic nerve, an inhibition of [³H]ACh release by adenosine deaminase was observed (Correia-de-Sá and Ribeiro, 1996). Again, this suggests that higher concentrations of endogenous adenosine are needed to activate excitatory receptors than those needed to activate inhibitory receptors.

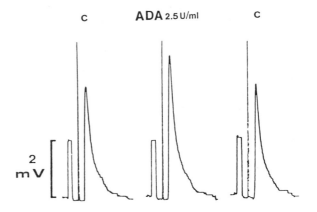

Fig. 3. Effect of adenosine deaminase (ADA) on the averaged amplitude of evoked endplate potentials (EPPs) recorded from an innervated frog sartorius muscle fiber. c: pre-, and post-control. The effect shown represents the average of EPPs recorded 8–10 min after starting perfusion of adenosine deaminase (2.5 U/ml). Membrane resting potential was −85 mV. Solutions contained 10 mM $Mg^{2+}$ to prevent muscle action potentials. Each evoked response is preceded by a 2 mV calibration pulse of 2 ms duration and is the computed average of 64 successive EPPs. Supramaximal stimulation with 10 μs electrical pulses delivered to the nerve at a frequency of 0.5 Hz, was used.

## Effects of adenosine on ACh release in the central nervous system

In the central nervous system there are also both inhibitory (see e.g. Ribeiro, 1995, for a review) and excitatory effects of adenosine on ACh release (see Sebastião and Ribeiro, 1996). The non-selective adenosine $A_2$ receptor agonist, 5′-N-ethylcarboxamidoadenosine (NECA) increases ACh release from cortical slices (Spignoli et al., 1984). The selective adenosine $A_{2a}$ receptor agonist, CGS21680 (a NECA analogue) facilitates [³H]ACh release from the rat hippocampal synaptosomes. This effect is mediated by an $A_{2a}$ receptor, since it is antagonized by 8-(3-chloro-styryl)caffeine (CSC) and by 3,7-dimethyl-1-propargylxanthine (DMPX) but not by DPCPX (Cunha et al., 1995). This excitatory effect of CGS21680 on [³H]ACh release was observed in CA3 and dentate gyrus regions of the hippocampus, but not in the CA1 region (Cunha et al., 1994b). It is worth-noting that the electrically

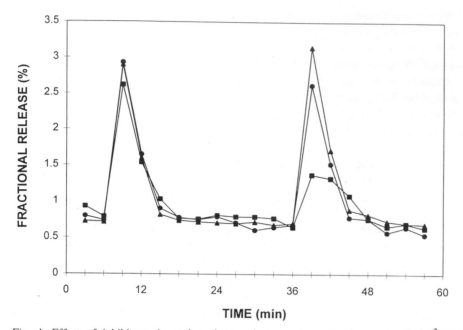

Fig. 4. Effect of inhibitory $A_1$ and excitatory $A_{2a}$ receptor activation on evoked [³H]ACh release from rat hippocampal $CA_3$ subslices. [³H]ACh release was measured as tritium outflow and expressed as the percentage of the tritium retained in the preparations at the time of sample collection (fractional release). Two periods ($S_1$ and $S_2$) of field-electrical stimulation (40 V, 1 ms, 2 Hz for 2 min) were delivered 6 ($S_1$) and 36 min ($S_2$) after starting sample collection. In the absence of any drugs added to the bath, i.e. under control conditions (●), the $S_2/S_1$ ratio was 0.92. The effect of drugs was calculated from comparison of the $S_2/S_1$ ratio obtained in the presence of drugs (added 21 min before $S_2$) with that obtained in control conditions. The mixed $A_1/A_2$ agonist, 2-chloroadenosine (CADO, 5 $\mu$M) decreased the $S_2/S_1$ ratio to 0.45 (■), and the $A_{2a}$ agonist, CGS 21680 (30 nM) increased the $S_2/S_1$ ratio to 1.32 (▲). The amount of tissue in each chamber ranged from 17.4 to 20.6 mg.

evoked release of [³H]ACh in the three areas of the rat hippocampus can be differentially modulated by adenosine. In the CA1 area, only $A_1$ inhibitory receptors activated by agonists or by endogenous adenosine modulate ACh release, whereas in the CA3 area, both $A_{2a}$ excitatory and $A_1$ inhibitory adenosine receptors respond either to adenosine agonists (Fig. 4) or to endogenous adenosine. In the dentate gyrus, both $A_1$ inhibitory and $A_{2a}$ excitatory receptors are present, but endogenous adenosine did not activate them. This might result from rapid inactivation and/or poor release not allowing enough adenosine accumulation to activate the receptors (Cunha et al., 1994b).

## Adenosine receptors and transducing systems

Four subtypes of adenosine receptors have been proposed (Fredholm et al., 1994) based on functional studies and on cloning experiments: $A_1$ and $A_3$ are inhibitory, $A_{2a}$ and $A_{2b}$ are excitatory. In the rat central and peripheral nervous systems, adenosine $A_1$ receptors were mainly detected via the use of the selective $A_1$ antagonist, DPCPX, and adenosine $A_{2a}$ receptors via the use of the selective $A_{2a}$ agonist, CGS21680.

Comparing the $K_i$ values for DPCPX at the rat and frog neuromuscular junctions, a typical $A_1$ receptor was found in the rat, with a $K_i$ value of 0.54 nM (Sebastião et al., 1990), but not in the frog, where the $K_i$ value for DPCPX was 35 nM (Sebastião and Ribeiro, 1989). These different values supported the notion that the adenosine receptor inhibiting ACh release in the frog could be different from that of the rat (Ribeiro and Sebastião, 1994; see also Linden, 1994).

The signal transducing systems operated by adenosine receptors were also investigated. Adenosine $A_{2a}$ receptors are, in general, positively coupled to the adenylate cyclase/cyclic AMP system, whereas the $A_1$ receptor might be negatively coupled to the adenylate cyclase/cyclic AMP. Suggestions that adenosine $A_1$ receptors are coupled to other transducing systems (e.g. phospholipase C, calcium channels, potassium channels) have also been put forward (Fredholm and Dunwiddie, 1988).

To investigate whether the adenosine $A_{2a}$ receptors present in the rat motor nerve terminals also activate the adenylate cyclase/cyclic AMP transducing system, a pharmacological approach was employed. Supramaximal activation of the adenylate cyclase by forskolin was expected to prevent the excitatory effect of the $A_{2a}$ receptor agonists on [$^3$H]ACh release. This was, in fact, observed. Similar results were obtained by using a supramaximal concentration of the phosphodiesterase inhibitor rolipram. Also the compound that mimics cyclic AMP itself, 8-bromo-cyclic AMP, prevented the excitatory effect of the $A_{2a}$ agonist, CGS21680 on [$^3$H]ACh release. This suggests the involvement of protein kinase A in this process. The effect of forskolin was selective, since the adenylate cyclase inactive analogue of forskolin, dideoxyforskolin, was without effect (Correia-de-Sá and Ribeiro, 1994b). As expected, when submaximal concentrations of rolipram were used, the effect of this $A_{2a}$ agonist was potentiated (Correia-de-Sá and Ribeiro, 1994b). The conclusion from these results was that the adenosine excitatory $A_{2a}$ receptors of the cholinergic nerve terminals are positively coupled to the adenylate cyclase/cyclic AMP system.

In the case of the inhibitory adenosine receptor of the frog motor nerve terminals, the following results were obtained suggesting involvement of phospholipase C: the action of adenosine receptor agonists, but not the affinity for antagonists, was inhibited by the active phorbol ester, $\beta$-phorboldiacetate, but not by the inactive phorbol ester, $\alpha$-phorboldidecanoate (Sebastião and Ribeiro, 1990). Inhibitors of protein kinase C, such as polymyxin B, which prevented the excitatory effect of active phorbol esters on transmitter release, did not modify the inhibitory action of adenosine receptor agonists, suggesting that inhibition of protein kinase C was not the unique pathway involved in

the inhibitory action of adenosine on neurotransmitter release. In fact, using an agent that interferes with phosphoinositides turnover, lithium ions in low millimolar concentrations, it was observed that adenosine receptor agonists inhibit the excitatory effect of lithium on neurotransmitter release. The enhancement of ACh release caused by lithium is probably due to its ability to prevent dephosphorylation of inositol phosphates, with the consequent accumulation of inositol triphosphate. Moreover, since the accumulation of inositol triphosphate in the presence of lithium should be greater with increasing activity of phospholipase C, the observation that adenosine receptor agonists markedly reduce the excitatory effect of lithium prompted us to suggest that adenosine receptor activation might inhibit phospholipase C, at least in the frog motor nerve terminals. The intracellular $Ca^{2+}$ available for ACh release might have been decreased through this mechanism (Sebastião and Ribeiro, 1990).

## Role of the presynaptic adenosine $A_{2a}$-receptors

Examples for the role of adenosine $A_{2a}$ receptors have been obtained at the neuromuscular junction, where activation of these receptors is essential for the regulatory function of some endogenous substances on ACh release.

### Calcitonin gene related peptide (CGRP)

The neuropeptide, calcitonin gene related peptide (CGRP) is localized in the motor nerve terminals and its exogenous application to the rat neuromuscular junction facilitated ACh release (Correia-de-Sá and Ribeiro, 1994a). This effect depends upon the presence of endogenous adenosine, since the hydrolysis of this nucleoside with adenosine deaminase prevented the facilitatory effect of CGRP on ACh release. The effect of endogenous adenosine is mediated through an adenosine $A_{2a}$ receptor, since application of the $A_{2a}$ agonist, CGS21680 in nanomolar concentrations, in the presence of adenosine deaminase, restored the facilitation of [$^3$H]ACh release induced by CGRP. This effect of CGRP was antagonized by the adenosine $A_2$ receptor antagonist, PD115,199, but not by the $A_1$ selective antagonist, DPCPX, which further indicates that the interaction between CGRP and adenosine is mediated by adenosine $A_2$ but not by $A_1$ receptors. The CGRP effect was even enhanced in the presence of DPCPX, indicating that there is some counteraction of the $A_{2a}$ mediated effect by the adenosine $A_1$ inhibitory action on [$^3$H]ACh release. This supports the idea that adenosine $A_1$ and $A_{2a}$ receptors are interacting to regulate ACh release from the rat phrenic motor nerve terminals. The CGRP facilitation of ACh release is not a consequence of presynaptic nicotinic facilitation, since tubocurarine did not significantly modify the facilitation caused by CGS21680 on [$^3$H]ACh release.

### Dimethylphenylpiperazinium (DMPP)

Another example of adenosine $A_{2a}$ receptor involvement in ACh release was detected by the use of the presynaptic nicotinic agonist, dimethylphenylpiperazinium (DMPP), which causes facilitation of ACh release when used in low concentrations, and inhibition when tested in high concentrations or applied for long periods. This change of facilitation into inhibition is due to desensitization of the autofacilitatory nicotinic receptors (Correia-de-Sá and Ribeiro, 1994c). The DMPP facilitation is increased if one increases $A_1$ receptor activation by blocking the $A_{2a}$ influence with the adenosine $A_2$ antagonist, DMPX. In the case of the $A_1$ antagonist, DPCPX, the effect of DMPP was reduced, suggesting that $A_1$ receptors enhance facilitation of ACh release by DMPP, probably by reducing desensitization of presynaptic nicotinic autofacilitatory receptors. In contrast, the nicotinic desensitization was prevented, when adenosine was inactivated, e.g. by adenosine deaminase. This phenomenon was an $A_{2a}$-mediated effect, since CGS21680 enhanced the desensitization, and the $A_2$ antagonist, DMPX antagonized this effect.

### Forskolin

Adenosine $A_{2a}$ receptor activation revealed to be crucial also for the enhancement of [$^3$H]ACh release caused by the adenylate cyclase activator, forskolin. This effect was specific, since the forskolin analogue dideoxyforskolin which is inactive on adenylate cyclase, did not facilitate [$^3$H]ACh release. As the adenosine $A_{2a}$ receptor is positively coupled to the adenylate cyclase/cyclic AMP system, it is likely that interactions involving adenosine $A_2$ receptor activation are operated via cyclic AMP (Correia-de-Sá and Ribeiro, 1993; Correia-de-Sá and Ribeiro, 1994b).

### Origin of adenosine that activates adenosine $A_{2a}$ receptors

As shown above, endogenous adenosine activated both $A_1$ and $A_2$ receptors, and these two subtypes of adenosine receptors can co-exist in the same motor nerve terminal. Hence the question is: how does adenosine "choose" to activate $A_1$ or $A_2$ receptors and which condition favors activation of one or the other? We observed that AMP which is hydrolyzed by ecto-5′-nucleotidase to form adenosine, enhanced the evoked release of [$^3$H]ACh from motor nerve endings. When we added to the bath the enzyme that causes hydrolysis of adenosine, adenosine deaminase, the excitatory effect of AMP was prevented. The same response occurred when AMP was added in the presence of the ecto-5′-nucleotidase inhibitor, AOPCP. Thus, our interpretation of these results was that the excitatory effect of AMP is a consequence of its metabolite, adenosine, which activates excitatory receptors. An indication that this excitatory effect is mediated by $A_2$ receptors was provided by the fact that the $A_2$ receptor antagonist DMPX antagonized the enhancement caused by AMP on [$^3$H]ACh release from motor nerve endings, whereas the $A_1$ antagonist was devoid of a similar effect (Cunha et al., 1996).

An indication that adenosine originated from released adenine nucleotides preferentially activates $A_2$ receptors was obtained by investigations on the effect of the ecto-5′-nucleotidase inhibitor AOPCP. This compound reduced [$^3$H]ACh release from the rat motor nerve endings stimulated at a high frequency (5 Hz). As occurred with the excitatory effect of AMP, the inhibitory effect of

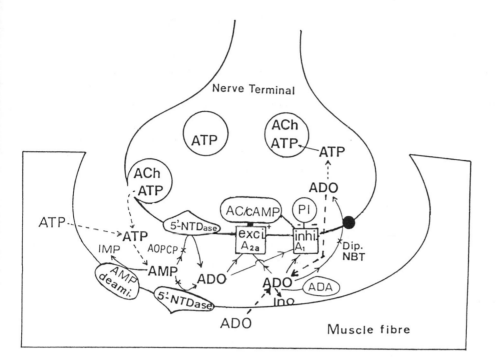

Fig. 5. Schematic representation of a cholinergic nerve terminal innervating a skeletal muscle fiber. ATP is released into the synaptic cleft and it is hydrolyzed to adenosine (ADO), which is the active substance, and preferentially activates excitatory adenosine $A_{2a}$ receptors positively coupled to adenylate cyclase (AC). Adenosine formation from AMP is inhibited by the ecto-5'-nucleotidase (5'-NTDase) inhibitor, $\alpha,\beta$-methylene ADP (AOPCP). Adenosine released as such preferentially activates inhibitory adenosine $A_1$ receptors, which might be negatively coupled to adenylate cyclase or to phosphoinositide (PI) turnover. Adenosine is inactivated through uptake, a process inhibited by dipyridamole (Dip.) or nitrobenzylthioinosine (NBT). In the rat neuromuscular junction, but not in the frog, adenosine deaminase (ADA) also contributes to extracellular adenosine inactivation into inosine (INO). AMP deaminase (AMP deami.) forms inosine from AMP and thus might reduce the levels of adenosine formed from released adenine nucleotides.

AOPCP was prevented by adenosine deaminase and was antagonized by $A_2$, but not by $A_1$ antagonists. This suggests that the inhibitory effect of AOPCP on [3H]ACh release is a result of its ability to prevent formation of adenosine, which enhances [3H]ACh release by acting on $A_2$ receptors.

The AMP-excitatory and the AOPCP-inhibitory effects taken together with the observation that adenosine deaminase under similar experimental conditions enhanced [3H]ACh release from the rat motor nerve endings, tempted us to propose (Cunha et al., 1996) a. novel view of the role of endogenous adenosine available at the synaptic cleft: adenosine activates $A_1$ receptors independently of being released as such or formed from adenine nucleotides, whereas adenosine formed from adenine nucleotides acts preferentially on the

$A_{2a}$ receptors. Adenosine released as such seems to prefer $A_1$ receptors. Whether this phenomenon is a consequence of a differential localization of $A_1$ and $A_2$ receptors via adenosine release sites and ecto-5'-nucleotidase localization or is a consequence of a burst-like formation of adenosine from released adenine nucleotides (see James and Richardson, 1993) leading to suprathreshold concentrations with respect to activation of $A_2$ receptors, awaits further investigation. An indication that a burst-like adenosine formation from released adenine nucleotides might be, at least in part, responsible for its preferential action on $A_2$ receptors is the finding that when the motor nerve endings are stimulated at low frequencies, i.e., in conditions where smaller amounts of adenine nucleotides are released, AOPCP enhanced (Ribeiro and Sebastião, 1987; Redman and Silinsky, 1994) rather than inhibited (Cunha et al., 1996) ACh release.

## Summary and conclusions

At the neuromuscular junction and possibly also at the synaptic level in the brain, the main sequence of events (see Fig. 5) that involves purines in modulation of ACh release includes the following observations: (1) storage of ATP and its release either together with, or independently of acetylcholine. ATP is also released from the postjunctional component. Adenosine as such is released either from the motor nerve terminals or from the post-junctional component. (2) There is extracellular hydrolysis of ATP to adenosine, which is the active substance to modulate transmitter release. The key enzyme in the conversion of AMP into adenosine is the ecto 5′-nucleotidase. When ecto-5′-nucleotidase is not available (e.g. in cholinergic nerve terminals of the cerebral cortex) ATP as such exerts the neuromodulatory role normally fulfilled by adenosine. (3) Both the inhibition and the excitation induced by adenosine on ACh release in the rat is inactivated through uptake and deamination. (4) Adenosine-induced inhibition of ACh release is mediated via $A_1$ receptors and the excitation via $A_{2a}$ receptors. The $A_{2a}$ receptors are positively coupled to the adenylate cyclase/cyclic AMP system, whereas the presynaptic $A_1$ receptors (a) may be negatively linked to adenylate cyclase and (b) to phospholipase C, and, upon stimulation, (c) increase potassium conductance and (d) decrease calcium conductance. (5) Activation of $A_{2a}$ receptors is essential for substances that facilitate ACh release (e.g. CGRP, forskolin) to exert their effects, as well as for induction of nicotinic autofacilitatory receptor desensitization. (6) There are interactions between $A_1$ and $A_{2a}$ receptors. Thus, the net adenosine neuromodulatory response is the resultant, at each moment, of the relative degree of activation of each one of these receptors. This relative activation depends upon the intensity (frequency, pulse duration) of stimulation of the motor nerve terminals. (7) Adenosine released as such seems to preferentially activate $A_1$ receptors, whereas the adenosine

formed from metabolism of adenine nucleotides prefers to activate the $A_{2a}$ receptors.

In conclusion, to find out precisely what occurs with ACh in transmitting its message at the synaptic level, one has to consider the subtle ways used by purines to modulate the ACh response. It therefore appears of interest that pharmacological and therapeutic strategies use this knowledge to approach cholinergic transmission deficiencies based upon reduction of ACh release.

## Acknowledgements

This work has been supported by grants from the Gulbenkian Foundation, European Union, and Junta Nacional de Investigação Científica e Tecnológica (Portugal).

## References

Cascalheira, J.F. and Sebastião, A.M. (1992) Adenine nucleotide analogues, including γ-phosphate-substituted analogues, are metabolised extracellularly in innervated frog sartorius muscle. *Eur. J. Pharmacol.*, 222: 49–59.

Correia-de Sá, P. and Ribeiro, J.A. (1993) Facilitation of [³H]-ACh release by forskolin depends on $A_2$-adenosine receptor activation. *Neurosci. Lett.*, 151: 21–24.

Correia-de-Sá, P. and Ribeiro, J.A. (1994a) Potentiation by tonic $A_{2a}$-adenosine receptor activation of CGRP-facilitated [³H]-ACh release from rat motor nerve endings. *Br. J. Pharmacol.*, 111: 582–588.

Correia-de-Sá, P. and Ribeiro, J.A. (1994b) Evidence that the presynaptic $A_{2a}$-adenosine receptor of the rat motor nerve endings is positively coupled to adenylate cyclase. *Naunyn-Schmiedeberg's Arch. Pharmacol.*, 350: 514–522.

Correia-de-Sá, P. and Ribeiro, J.A. (1994c) Tonic adenosine $A_{2a}$ receptor activation modulates nicotinic autoreceptor function at the rat neuromuscular junction. *Eur. J. Pharmacol.*, 271: 349–355.

Correia-de-sá, P. and Ribeiro, J.A. (1996) Adenosine uptake and deamination regulate A2$_a$-receptor facilitation of evoked [³H]-ACh release from the rat motor nerve terminals. *Neuroscience*, 73: 85–92.

Correia-de-Sá, P., Sebastião, A.M. and Ribeiro, J.A. (1991) Inhibitory and excitatory effects of adenosine receptor agonists on evoked transmitter release from phrenic nerve endings of the rat. *Br. J. Pharmacol.*, 103: 1614–1620.

Cunha, R.A. and Sebastião, A.M. (1991) Extracellular metabolism of adenine nucleotides and adenosine in the innervated skeletal muscle of the frog. *Eur. J. Pharmacol.*, 197: 83–92.

Cunha, R.A. and Sebastião, A.M. (1993) Adenosine and adenine nucleotides are independently released from both the nerve terminals and the muscle fibres upon electrical stimulation of the innervated skeletal muscle of the frog. *Pflügers Arch.,* 424: 503–510.

Cunha, R.A., Sebastião, A.M. and Ribeiro, J.A. (1992) Ecto-5'-nucleotidase is associated with cholinergic nerve terminals in the hippocampus but not in the cerebral cortex of the rat. *J. Neurochem.,* 59: 657–666.

Cunha, R.A., Ribeiro, J.A. and Sebastião, A.M. (1994a) Purinergic modulation of the evoked [3H]-acetylcholine release from the hippocampus and cerebral cortex of the rat: role of the ectonucleotidases. *Eur. J. Neurosci.,* 6: 33–42.

Cunha, R.A., Milusheva, E., Vizi, E.S., Ribeiro, J.A. and Sebastião, A.M. (1994b) Excitatory and inhibitory effects of $A_1$ and $A_{2a}$ adenosine receptor activation on the electrically evoked [3H]-acetylcholine release from different areas of the rat hippocampus. *J. Neurochem.,* 63: 207–214.

Cunha, R.A., Johansson, B., Fredholm, B.B., Ribeiro, J.A. and Sebastião, A.M. (1995) Adenosine $A_{2a}$ receptors stimulate acetylcholine release from nerve terminals of the rat hippocampus. *Neurosci. Lett.,* 196: 41–44.

Cunha, R.A., Correia-de-sá, P., Sebastião, A.M. and Ribeiro, J.A. (1996) Preferential activation of excitatory adenosine receptors at rat hippocampal and neuromuscular synapses by adenosine formed from released adenine nucleotides. *Br. J. Pharmacol.,* in press.

Dowdall, M.J., Boyne, A.F. and Whittaker, V.P. (1974) Adenosine triphosphate. A constituent of cholinergic synaptic vesicles. *Biochem. J.,* 140: 1–12.

Fariñas, I., Solsona, C. and Marsal., J. (1992) Omega-conotoxin differentially blocks acetylcholine and adenosine triphosphate releases from *Torpedo* synaptosomes. *Neuroscience,* 47: 641–648.

Fredholm, B.B. and Dunwiddie, T.V. (1988) How does adenosine inhibit transmitter release? *Trends Pharmacol. Sci.,* 9: 130–134.

Fredholm, B.B, Abbracchio, M.P., Burnstock, G., Daly, J.W., Harden, T.K., Jacobson, K.A., Leff, P. and Williams, M. (1994) Nomenclature and classification of purinoceptors. *Pharmacol. Rev.,* 46: 143–156.

Ginsborg, B.L. and Hirst, G.D.S. (1972) The effect of adenosine on the release of the transmitter from the phrenic nerve of the rat. *J. Physiol. (London),* 224: 629–645.

James, S. and Richardson, P.J. (1993) Production of adenosine from extracellular ATP at striatal cholinergic synapse. *J. Neurochem.,* 60: 219–227.

Katz, B. and Miledi, R. (1977) Transmitter linkage from motor nerve endings. *Proc. R.. Soc. London B.,* 196: 59–72.

Linden, J. (1994) Cloned adenosine A3 receptors: pharmacological properties, species differences and receptor functions. *Trends Pharmacol. Sci.,* 15: 298–306.

Marsal, J., Solsona, C., Rabasseda, X., Blasi, J. and Casanova, A. (1987) Depolarization-induced release of ATP from cholinergic synaptosomes is not blocked by botulinium toxin type A. *Neurochem. Int.,* 10: 295–302.

Redman, R.S. and Silinsky, E.M. (1994) ATP released together with acetylcholine as the mediator of neuromuscular depression at frog motor nerve endings. *J. Physiol. (London),* 477: 117–127.

Ribeiro, J.A. (1995) Purinergic inhibition of neurotransmitter release in the central nervous system. *Pharmacol. Toxicol.,* 77: 299–305.

Ribeiro J.A. and Dominguez, M.L. (1978) Mechanisms of depression of neuromuscular transmission by ATP and adenosine. *J. Physiol. (Paris),* 74: 491–496.

Ribeiro, J.A. and Sebastião, A.M. (1986) Adenosine receptors and calcium: basis for proposing a third (A3) adenosine receptor. *Prog. Neurobiol.,* 26: 179–209.

Ribeiro, J.A. and Sebastião, A.M. (1987) On the role, inactivation and origin of endogenous adenosine at the frog neuromuscular junction. *J. Physiol. (London),* 384: 571–585..

Ribeiro, J.A. and Sebastião, A.M. (1994) Further evidence for adenosine A3 receptors. *Trends Pharmacol. Sci.,* 15: 13.

Ribeiro, J.A. and Walker, J. (1973) Action of adenosine triphosphate on endplate potentials recorded from muscle fibres of the rat-diaphragm and frog sartorius. *Br. J. Pharmacol.,* 49: 724–725.

Ribeiro, J.A. and Walker, J. (1975) The effects of adenosine triphosphate and adenosine diphosphate on transmission at the rat and frog neuromuscular junctions. *Br. J. Pharmacol.,* 54: 213–218.

Ribeiro, J.A., Sá-Almeida, A.M. and Namorado, J.M. (1979) Adenosine and adenosine triphosphate decreases $^{45}$Ca uptake by synaptosomes stimulated by potassium. *Biochem. Pharmacol.,* 28: 1297–1300.

Sebastião, A.M. and Ribeiro, J.A. (1988) On the adenosine receptor and adenosine inactivation at the rat diaphragm neuromuscular junction, *Br. J. Pharmacol.,* 94: 109–120.

Sebastião, A.M. and Ribeiro, J.A. (1989) 1,3,8- and 1,3,7-substituted xanthines: relative potency as adenosine receptor antagonists at the frog neuromuscular junction. *Br. J. Pharmacol.,* 96: 211–219.

Sebastião, A.M. and Ribeiro, J.A. (1990) Interactions between adenosine and phorbol esters or lithium at the frog neuromuscular junction. *Br. J. Pharmacol.,* 100: 55–62.

Sebastião, A.M. and Ribeiro, J.A. (1996) $A_2$ receptor mediated excitatory actions of adenosine in the nervous system. *Prog. Neurobiol.;* 48, 167–189.

Sebastião, A.M., Stone, T.W. and Ribeiro, J.A. (1990) The inhibitory adenosine receptor at the neuromuscular junction and hippocampus of the rat:: antagonism by 1,3,8-substituted xanthines. *Br. J. Pharmacol.,* 101: 453–459.

Silinsky, E.M. (1975) On the association between transmitter secretion and the release of adenine nucleotides from mammalian motor nerve terminals. *J. Physiol. (London),* 247: 145–162.

Silinsky, E.M. (1984) On the mechanism by which adenosine receptor activation inhibits the release of acetylcholine from motor nerve terminals. *J. Physiol. (London),* 346: 243–256.

Smith, D.O. (1991) Sources of adenosine released during neuromuscular transmission in the rat. *J. Physiol. (London),* 432: 343–354.

Spignoli, G., Pedata, F. and Pepeu, G. (1984) $A_1$ and $A_2$ adenosine receptors modulate acetylcholine release from brain slices. *Eur. J. Pharmacol.,* 97: 341–342.

J. Klein and K. Löffelholz (Eds.)
*Progress in Brain Research*, Vol. 109
© 1996 Elsevier Science B.V. All rights reserved.

CHAPTER 24

# Activity-related modulation of cholinergic transmission

B. Collier

*Department of Pharmacology and Therapeutics, McGill University, 3655 Drummond Street, Montreal, Quebec H3G 1Y6, Canada*

## Introduction

This contribution surveys some aspects of long-term potentiation (LTP) of cholinergic synaptic transmission, as revealed by studies upon mammalian sympathetic ganglia. The phenomena to be described are examples of the modulation of cholinergic transmission that are the consequence of changed presynaptic activity. This form of modulation differs mechanistically and functionally from the other mechanisms of modulation presented in the preceding chapters of this section. The short-term regulation of transmitter release effected by autoreceptors and heteroreceptors, and the longer-term modulation of synaptic strength caused by altered synaptic input, afford complementary mechanisms by which cholinergic transmission can be modified either physiologically or pharmacologically.

The term LTP is used to describe various phenomena characterized by an increase of the efficacy of synaptic transmission that lasts an hour or more following altered afferent activity. Much attention has focused on LTP of hippocampal glutaminergic transmission since its initial detailed description by Bliss and his colleagues (Bliss and Gardner-Medwin, 1973; Bliss and Lomo, 1973). This attention is well directed as an important support for the ideas expressed by Hebb (1949) that included the notion that the cellular basis of memory rests upon an altered synaptic efficacy in response to repetitive activity.

Nevertheless, as for many items related to synaptic transmission, the origin of the evidence that altered presynaptic activity can provoke relatively long-lasting changes of synaptic functions is to be found in the literature that reports on cholinergic transmission. The first appears to be the studies of Rosenblueth and colleagues upon what they called the five stages of neuromuscular transmission (see Rosenblueth et al., 1939). They described their observations of skeletal muscle tension during high-frequency motor nerve stimulation as an initial rise (1st stage), a subsequent fade (2nd stage), then another increase (3rd stage), another decrease (4th stage) and a late augmentation (5th stage). They interpreted their stages 1 4 as reflecting an adjustment between the rate of formation of transmitter and its rate of release, transition from stage 4 to 5 was attributed to the ability of nerves to produce and release more acetylcholine (ACh) than they normally did, and they supported that conclusion by measures of ACh content.

A clearer example of a relatively long-lasting potentiation of neuromuscular transmission following tetanic stimulation of motor nerves is that of Feng (1941). He measured end-plate potentials to show an appreciable synaptic enhancement following tetanic stimulation of the nerves that far outlasted the fleeting post-tetanic potentiation (PTP) already known from Feng's earlier work.

Lastly, the report of Dunant and Dolivo (1968) also preceded the discovery of hippocampal LTP. They described a long-lasting potentiation of synaptic transmission at the rat superior cervical ganglion that resulted from a brief preganglionic tetanus.

Thus, the essential description of LTP-like modulation at cholinergic synapses are of mature age; this contribution describes some possible mechanisms that might be involved in initiating LTP phenomena at sympathetic ganglia. The literature holds several examples of ganglionic LTP (reviewed in Kuba and Kumamoto, 1990); two examples are featured here: a potentiation induced as the consequence of a short tetanus and a potentiation that follows a long period of tetanic stimulation.

## Potentiation following a brief tetanus

As mentioned above, LTP of transmission at sympathetic ganglia following a few seconds of high-frequency stimulation of preganglionic nerves was described clearly by Dunant and Dolivo (1968). The experiment used an ex-vivo preparation of the rat superior cervical ganglion. This LTP as shown by this preparation has been analyzed with some detail by McAfee and his colleagues (Brown and McAfee, 1982; Briggs et al., 1985a,b; Briggs and McAfee, 1988; Wu et al., 1991), experiments that clearly identify presynaptic mechanisms as responsible for the potentiation. The tetanus resulted in a prolonged enhancement of ACh release that was of magnitude and duration sufficient to account for the potentiation of synaptic transmission assessed by measures of postganglionic compound action potentials. There appeared to be no important change in the sensitivity of postsynaptic cells to nicotinic agonists under the conditions of these experiments.

In contrast, analysis of LTP induced at sympathetic ganglia of the cat, studied in situ, yields quite the opposite conclusion to that mentioned above: potentiation appears to result from a postsynaptic change in sensitivity to ACh. The LTP of synaptic transmission at stellate ganglia (Bachoo and Polosa, 1991) and at superior cervical ganglia (Bachoo et al., 1992) can be induced by a 5 s 40 Hz stimulation of their preganglionic input; potentiation lasting 1–2 h was demonstrated to either homosynaptic or heterosynaptic test stimuli. The cat's superior cervical ganglion also shows LTP when perfused through its normal vasculature with an artificial medium, and this allowed measure of ACh release to be made in parallel to a measure of synaptic transmission (Morales et al.,

1994). Preganglionic tetanic stimulation induced a LTP of synaptic transmission but provoked only a small and transient enhancement of ACh release. Furthermore, the neurogenic LTP was associated with an increased response to exogenously applied ACh or other nicotinic agonist. Both of these results are indicative of a postsynaptic change in the sensitivity to ACh as being causative to the LTP, rather than the change in synaptic transmission being the consequence of increased transmitter release.

Thus, although there is a consistency in the evidence that supports the idea that a brief preganglionic tetanus can induce a LTP of synaptic transmission at sympathetic ganglia, there is no consistency in the results of analyses that test its pre- or post-synaptic site of change. The difference might rest with the species or it might reflect both pre- and post-synaptic changes, the magnitude of each being affected by the synaptic milieu as determined by experimental conditions. The superior cervical ganglion as used ex vivo is not perfused through its normal vasculature as is the in vivo preparation. This difference is appreciable with respect to the rate of exchange of active agents from medium to synapse (see Birks and Isacoff, 1988) and presumably also in the reverse direction, allowing for considerable differences in the potential for local modulation of nerve terminal and target cell properties. The author's opinions, for what they are worth, are that both sets of experiments are sound and provide reliable results that both sites of LTP maintenance are possible, that some aspect of synaptic micro-chemistry can direct a particular mechanism to dominate, and that a preparation perfused through its normal vasculature is closer to physiological than one that is not. It might be noted that the question about pre- or post-synaptic changes in hippocampal LTP remains not clearly answered (see Collingridge and Bliss, 1995; Liao et al., 1995).

The initiation of the LTP of synaptic transmission in the cat superior cervical ganglion appears to be due to the release and action of a non-cholinergic mediator. This is evident from the frequency requirement: ACh release from preganglionic nerve terminals is optimal at low frequency but the induction of LTP requires high frequency stimulation. Furthermore, the induction of the LTP appears not affected by nicotinic receptor antagonists, it is evident when muscarinic receptors are blocked, and its magnitude and duration appears not altered significantly by AChE inhibition. Potential non-cholinergic mediators of this LTP include catecholamines released from sif cells, but this mechanism appears not to account for the potentiation, which survives the application of adrenoceptor or dopamine receptor antagonists; also, LTP is apparent in the presence of atropine, which blocks the release of amines from sif cells in response to preganglionic stimulation.

The requirement for high frequency impulses to initiate LTP is reminiscent of the requirement for high frequency stimulation to release some neuropeptides (e.g., Agoston and Lisziewicz, 1989; Lundberg et al., 1989). The sympathetic ganglia contains a variety of such neuropeptides (reviewed by Elfvin et al., 1993). The nerve terminal peptides appear to be stored in dense cored vesicles (e.g., Fried et al., 1985; Morales et al., 1993) that are larger than the clear vesicles that contain ACh.

The dense cored vesicles of preganglionic sympathetic nerve terminals appear to be released by a calcium-dependent mechanism (Weldon et al., 1993) and the characteristics of this release seem to parallel the conditions necessary to induce the LTP phenomenon at superior cervical ganglia. Thus, the range of stimulation frequencies effective in releasing the dense-core vesicles (20–40 Hz) was similar to that able to induce LTP. The continuous preganglionic stimulation at 40 Hz for 10–20 min released most of the dense core vesicles; when the ability of a 5 s, 40 Hz train to induce LTP was tested with ganglia that had previously experienced 10–20 min high frequency stimulation, little LTP was evident, a phenomenon called use-dependent fade by Bachoo et al. (1992). The recovery of the ability of ganglia to initiate LTP in response to test conditioning stimuli is slow (90% recovery in some 5 days), just as is the reappearance of dense-core vesicles after their exhaustion (incomplete recovery at 4 days, complete recovery at 9 days). The slow return of the nerve terminal dense core vesicles after their depletion

by prolonged high frequency stimulation was attributed to the need for axonal transport to deliver them to the synaptic varicosities, a notion supported by the ability of axonal colchicine to prevent this recovery (Weldon et al., 1993). This blockade of axonal transport also prevented the recovery of the LTP phenomenon (Bachoo et al., 1992).

These parallels between the analysis of dense core vesicle dynamics (Weldon et al., 1993) and of the characteristics of LTP (Bachoo et al., 1992) in sympathetic ganglia might be fortuitous, but it is more attractive to consider that they indicate a relationship. If the relationship is accepted, the component of the dense-core vesicles that, upon release, initiates the LTP remains to be determined. The mimicry of the neurogenic potentiation with respect to magnitude and duration by selective neuropeptides has not yet been achieved by agents tested.

The second unanswered questions are those related to the mechanisms that sustain the LTP phenomenon. Under the experimental conditions that point to a post-synaptic change related to the expression of LTP, the mechanism could be particular to the nicotinic receptor, its ion channel or a reflection of a less specific change of membrane biophysics. There exists an older literature to suggest that a preganglionic train can potentiate ganglion cell responses to diverse stimuli (see review by Volle, 1969), but the LTP phenomenon discussed here seems more particular to nicotinic agonists (Morales et al., 1994).

## Potentiation following a prolonged tetanus

When preganglionic axons are subject to tetanic stimulation for rather longer than necessary to induce the LTP phenomenon discussed above, a second form of potentiation can be induced. This is the feature mentioned in the introduction as discovered by Rosenblueth et al. (1939). This phenomenon was rediscovered by Friesen and Khatter (1971) and explored by Bourdois et al. (1975). The essential feature is an increase of the ACh content of sympathetic ganglia induced as the result of 15–60 min of preganglionic stimulation at 15–60 Hz, and manifest primarily after the conditioning stimulation has ceased. This latter characteristic has suggested the term "rebound ACh" to describe the increase.

This adaptive increase of transmitter content following a conditioning stimulation is relatively long-lasting with a half-life estimated as about 2 h (Bourdois et al., 1975). The increased ACh results from increased ACh synthesis associated with enhanced choline uptake activity, rather than a change of cholineacetyltransferase activity (O'Regan and Collier, 1981; Collier et al., 1983).

The important feature of the post-conditioning increase of the ACh store is its releasability. Collier et al. (1983), labelled the extra ACh by exposing conditioned ganglia to radiolabeled choline during the period of rest following the tetanic stimulation, to show its ready incorporation into a releasable transmitter store, but they did not pursue the time-course of this release. The measure of ACh release from ganglia conditioned by 15 Hz stimulation for 45 min has shown a persistent enhancement of release (Tandon and Collier, 1994). In this paradigm, the conditioning promoted, during a subsequent rest, an increase of some 45% for ganglionic ACh content and the subsequent release of ACh was increased over non-conditioned controls by about the same proportion. Moreover, this increase in release was maintained for longer than was the increased tissue content, as if the adaptive changes induced by the conditioning stimulation are multiple, including not only the formation of "rebound ACh" but also a re-setting of the relationship between ACh synthesis and release.

As with the LTP of ganglionic transmission, the initiating factor for the post-conditioning presynaptic adaptive changes appears to be non-cholinergic. First, the frequency-dependence of the conditioning stimulation does not match the requirement for ACh release; the latter is effected by low and high frequency stimulation, but "rebound ACh" is induced only by high impulse traffic. Second, blockade of ACh responses, nicotinic or muscarinic, does not consistently prevent the phenomenon (Bourdois et al., 1975; Collier et al., 1983). Nevertheless, if the conditioning stimulation is presented in the absence of calcium, the adaptive increase of ACh content is not apparent, as if the calcium-dependent release of an initiating factor by preganglionic impulses is required to effect the change.

This increase of ganglionic ACh content following a 45 min 15 Hz conditioning stimulation appears to be prevented by some, but not all, nucleoside transport inhibitors (Tandon and Collier, 1994). The most effective agent was dipyridamole and the inhibitory effect of this drug appeared to be particular to this "rebound" phenomenon; it had little effect on ACh turnover of ganglia at rest or those stimulated to release ACh at stimulation frequencies that do not induce the "rebound". The pharmacology of nucleoside transport inhibitors as inhibitors of the initiation of this post-conditioning increase of ACh content was somewhat similar, although certainly not identical, to that of the same drugs when tested as inhibitors of an adenosine-induced increase of ACh content of resting ganglia (Tandon and Collier, 1993).

Thus, one possible factor involved in the genesis of the "rebound ACh" is an adenosine-like agent. The adenosine is best known for its inhibitory effects on transmitter release, an effect manifest when release is activated; there are also reports of adenosine-induced increases in transmitter release and both aspects are well reviewed in Chapter 23 of this volume by Ribeiro. The effect of adenosine to somehow increase ACh synthesis and storage under certain conditions is one more example of the diverse effects of this compound. The mechanism by which adenosine might cause an increased ACh synthesis by preganglionic nerve terminals is unclear. One possibility is that it does so as the result of altering intraterminal phosphatase activity, an effect reported for adenosine (Mateo et al., 1995); we have some preliminary evidence that choline transport can be regulated by changes in phosphatase activity.

If adenosine or an adenosine-like agent is involved in the generation of the post-conditioning change in ACh content of ganglia, its action appears not to be manifest during the presynaptic tetanic stimulation because the dipyridamole, if present only during that time, had little inhibitory effect to the subsequent formation of "rebound

ACh". Thus, our present hypothesis is that some other non-cholinergic mediator is released during the conditioning stimulation and the action of this mediator provokes the efflux of adenosine or like agent. The origin of the latter was postulated to be ganglion cells, rather than presynaptic nerve terminals and the postulate is supported by tests of antidromic conditioning.

The stimulation of postganglionic fibers can provoke a change in the ACh content of sympathetic ganglia (Tandon and Collier, 1995). Considering that ChAT and ACh synthesis is a feature of cholinergic nerve terminals not ganglionic cells, it seems likely that antidromic impulses in this experiment releases some mediator that induces preganglionic nerve terminals to increase their ACh synthesis and storage capacity. The extra ACh contained in ganglia activated antidromically, like that induced by orthodromic stimulation, appeared to be synthesized mainly following the period of tetanic invasion, rather than during it. Like the "rebound ACh" produced in response to orthodromic activity, that induced following the antidromic stimulation increases the subsequent release of ACh to test impulses applied to preganglionic nerves. Thus, it seems that ganglionic cells, when activated, can be caused to release some mediator that diffuses to presynaptic structures where it has an effect upon neurotransmitter turnover. The nature of this retrograde messenger is not fully known, but the phenomenon of antidromic stimulation-induced increase of ACh content is sensitive to inhibition by dipyridamole, suggesting that the retrograde messenger might be an adenosine-like compound. Whether this effect of dipyridamole is to inhibit an influx or an efflux of an adenosine-like mediator is not clear; the nucleoside transport system is complex and can operate bidirectionally (see e.g. Gu et al., 1995). Also, it remains possible that the dipyridamole acts on some process unrelated to a nucleoside transporter to account for its apparent effect on both the antidromic- and the orthodromic-stimulation induced change of ACh synthesis.

With respect to the "rebound ACh" of ganglionic transmitter induced by preganglionic tetanic activity, the above evidence suggests the involvement of two non-cholinergic mechanisms in this example of synaptic modulation. The first is postulated as being released during the stimulation, with release characteristics that are particular to high frequency not low frequency impulses. The second is postulated to be released, primarily following the stimulation from target cells as the consequence of the action of the initial mediator; this second mediator acts as a retrograde message to alter ACh synthesis in the nerve terminals.

The idea of retrograde modulation of presynaptic function appears not to have been applied to potentiating phenomena at cholinergic synapses, but it has been invoked in LTP of non-cholinergic transmission (see review by Hawkins et. al., 1993) and it has been suggested as a mechanism involved in long-term depression at cholinergic junctions (Chapter 34). Indeed, the phenomenon of retrograde signalling in the nervous system is likely to have multiple roles both during development and in adult life (see, e.g., Jessell and Kandel, 1993; Davis and Murphy, 1994).

There are many remaining questions about the presynaptic modulation of cholinergic transmission illustrated by this "rebound ACh" despite its history now lasting some 56 years. The initial activator presumed to be released by a calcium-dependent mechanism by nerve impulse invasion of presynaptic varicosities, is yet to be identified. The nature of the action of this initiating mediator and the signal transduction mechanisms involved are not known. The origin and identity of the proposed retrograde mediator as well as its mechanism of release is uncertain, at best. And the mode of action of the retrograde signal to alter ACh dynamics in the presynaptic nerve terminals is unknown.

## Conclusions

The above summary presents two examples of potentiation of cholinergic transmission that can be analyzed as resulting from activity-related changes. As such, they add to the diversity of mechanisms by which cholinergic synaptic transmission can be modulated and they represent rather different, yet complementary, mechanisms to others discussed in this section of this volume.

The two examples result from studies of sympathetic ganglionic transmission and it seems likely that they are mechanism that represent functional processes of adaptation of autonomic control. There is evidence that activity-dependent changes of ganglionic transmission can occur with physiological stimuli (see Birks, 1977).

It is now apparent that LTP phenomena can display two distinct phases; an early one that does not require protein synthesis and a more persistent phase that does require new proteins to be produced (see Huang et al., 1994). The characteristics of ganglionic LTP and "rebound ACh" formation appear closer to the former than the latter, but this has yet to be directly tested.

If peripheral cholinergic synapses can manifest adaptive change in response to variation of neuronal traffic, it appears probable that central cholinergic neurones might also do so. There seems to be little evidence in the literature to indicate that this has been explored, although the apparent adaptation of hippocampal ACh turnover to a partial lesion of the fimbria-fornix (Lapchak et al., 1991) might be suggestive if that partial lesion altered activity in residual cholinergic neurones. It is evident that the normal discharge rate of the septal-hippocampal cholinergic neurones is relatively low (see Dutar et al., 1995), suggestive of the potential for an imposed period of high-frequency activity to modify their terminal dynamics.

Certainly, the postulated importance of the LTP mechanism to learning and memory combined with the evidence that implicates the central cholinergic systems as of importance to cognitive function, suggests that activity-related changes of cholinergic synaptic transmission might be worthy of study in the central nervous system.

## References

Agoston, D.V. and Lisziewicz, J. (1989) Calcium uptake and protein phosphorylation in myenteric neurones, like the release of vasoactive intestinal polypeptide and acetylcholine, are frequency dependent. *J. Neurochem.*, 52: 1637–1640.

Bachoo, M. and Polosa, C. (1991) Long-term potentiation of nicotinic transmission by a heterosynaptic mechanism in the stellate ganglion of the cat. *J. Neurophysiol.*, 65: 639–647.

Bachoo, M., Morales, M.A. and Polosa, C. (1992) Use-dependent fade and slow recovery of long-term potentiation in the superior cervical ganglion of the cat. *J. Neurophysiol.*, 67: 470–476.

Birks, R.I. (1977) A long-lasting potentiation of transmitter release related to an increase in transmitter stores in a sympathetic ganglion. *J. Physiol.*, 271: 847–862.

Birks, R.I. and Isacoff, E.Y. (1988) Burst-patterned stimulation promotes nicotinic transmission in isolated perfused rat sympathetic ganglia. *J. Physiol.*, 402: 515–532.

Bliss, T.V.P. and Gardner-Medwin, A.R. (1973) Long-lasting potentiation of synaptic transmission in the dentate area of the unanesthetized rabbit following stimulation of the perforant path. *J. Physiol.*, 232: 357–374.

Bliss, T.V.P. and Lomo, T. (1973) Long-lasting potentiation of synaptic transmission in the dentate area of anaesthetized rabbit following stimulation of the perforant path. *J. Physiol.*, 232: 331–356.

Bourdois, P.S., McCandless, D.L. and MacIntosh, F.C. (1975) A prolonged after-effect of intense synaptic activity on acetylcholine in a sympathetic ganglion. *Can. J. Physiol. Pharmacol.*, 53: 155–165.

Briggs, C.A. and McAfee, D.A. (1988) Long-term potentiation at nicotinic synapses in the rat superior cervical ganglion. *J. Physiol.*, 404: 129–144.

Briggs, C.A., Brown, T.H. and McAfee, D.A. (1985a) Neurophysiology and pharmacology of long-term potentiation in the rat sympathetic ganglion. *J. Physiol.*, 359: 503–521.

Briggs, C.A., McAfee, D.A. and McCaman, R.E. (1985b) Long-term potentiation of synaptic acetylcholine release in the superior cervical ganglion of the rat. *J. Physiol.*, 363: 181–190.

Brown, T.H. and McAfee, D.A. (1982) Long-term potentiation in the superior cervical ganglion. *Science*, 215: 1411–1413.

Collier, B., Kwok, Y.N. and Welner, S.A. (1983) Increased acetylcholine synthesis and release following presynaptic activity in a sympathetic ganglion. *J. Neurochem.*, 40: 91–98.

Collingridge, G.L. and Bliss, T.V.P. (1995) Memories of NMDA receptors and LTP. *Trends Neurosci.*, 18: 54–56.

Davis, G.W. and Murphy, R.K. (1994) Long-term regulation of short term transmitter release properties: retrograde signalling and synaptic development. *Trends Neurosci.*, 17: 9–13.

Dunant, Y. and Dolivo, M. (1968) Plasticity of synaptic functions in the excised sympathetic ganglion of the rat. *Brain Res.*, 10: 271–273.

Dutar, P., Bassant, M.-H., Senut, M.-C. and Lamour, Y. (1995) The septohippocampal pathway: structure and function of a central cholinergic system. *Physiol. Rev.*, 75: 393–427.

Elfvin, L.-G., Lindh, B. and Hokfelt, T. (1993) The chemical neuroanatomy of sympathetic ganglia. *Annu. Rev. Neurosci.*, 16: 471–507.

Feng, T.P. (1941) Studies on the neuromuscular junction XXVI. The changes of the end-plate potential during and after prolonged stimulation. *Clin. J. Physiol.,* 16: 341–371.

Fried, G., Lundberg, J.M. and Theodorsson-Nordheim, E. (1985) Subcellular storage and axonal transport of neuropeptide Y in relation to adenosine transporters in the release of L-[$^3$H]adenosine from rat brain synaptosomal preparations. *J. Neurochem.,* 64: 2105–2110.

Friesen, A.J.D. and Khatter, J.C. (1971) The effect of pregangionic stimulation on the acetylcholine and choline content of a sympathetic ganglion. *Can. J. Physiol. Pharmaol.,* 49: 375–381.

Gu, J.G., Foga, I.O., Parkinson, F.E. and Geiger, J.D. (1995) Involvement of bidirectional adenosine transporters in the release of L-[$^3$H]adenosine from rat brain synaptosomal prepartations. *J. Neurochem.,* 64: 2105–2110.

Hawkins, R.D., Kandel, E.R. and Siegelbaum, A. (1993) Learning to modulate transmitter release: themes and variations in synaptic plasticity. *Annu. Rev. Neurosci.,* 16: 625–665.

Hebb, D.O. (1949) *The Organization of Behavior: A Neuropsychological Theory.* Wiley, New York.

Huang, X.-Y., Li, X.-C. and Kandel, E.R. (1994) cAMP contributes to mossy fiber LTP by initiating both a covalently mediated early phase and macromolecular synthesis-dependent late phase. *Cell,* 79: 69–79.

Jessel, T.M. and Kandel, E.R. (1993) Synaptic transmission: a bidirectional and self-modifiable form of cell-cell communication. *Neuron,* 10: 1–30.

Kuba, K. and Kumamoto, E. (1990) Long-term potentiations in vertebrate synapses: a variety of cascades with common subprocesses. *Prog. Neurobiol.,* 34: 197–269.

Lapchak, P.A., Jenden, D.J. and Hefti, F. (1991) Compensatory elevation of acetylcholine synthesis in vivo by cholinergic neurones surviving partial lesions of the septohippocampal pathway. *J. Neurosci.,* 11: 2821–2828.

Liao, D., Hessler, N.A. and Malinow, R. (1995) Activation of postsynaptically silent synapses during pairing-induced LTP in CAI region of hippocampal slice. *Nature,* 375: 400–404.

Lundberg, J.M., Rudehill, A., Sollevi, A., Fried, G. and Wallin, G. (1989) Co-release of neuropeptide Y and noradrenaline from pig spleen in vivo: importance of subcellular storage, nerve impulse frequency and pattern, feedback regulation and resupply by axonal transport. *Neuroscience,* 28: 475–486.

Mateo, J., Castro, E., Zwiller, J., Aunis, D. and Miras-Portugal, M.T. (1995) 5′-(Ethylcarboxamido) adenosine inhibits Ca$^{2+}$ influx and activates a protein phosphatase in bovine adrenal chromaffin cells. *J. Neurochem.,* 64: 77–84.

Morales, M.A., Bachoo, M., Beaudet, A., Collier, B. and Polosa, C. (1993) Ultrastructural localization of neurotensin immunoreactivity in the stellate ganglion of the cat. *J. Neurocytol.,* 22: 1017–1021.

Morales, M.A., Bachoo, M., Collier, B. and Polosa, C. (1994) Pre- and post-synaptic components of nicotinic long-term potentiation in the superior cervical ganglion of the cat. *J. Neurophysiol.,* 72: 819–824.

O'Regan, S. and Collier, B. (1981) Factors affecting choline transport by the cat superior cervical ganglion during and following stimulation. *Neuroscience,* 6: 511–520.

Rosenblueth, A., Lissak, K. and Lanari, A. (1939) An explanation of the five stages of neuromuscular and ganglionic synaptic transmission. *Am. J. Physiol.,* 128: 31–44.

Tandon, A. and Collier, B. (1993) Increased acetylcholine content induced by adenosine in a sympathetic ganglion and its subsequent mobilization by electrical stimulation. *J. Neurochem.,* 60: 2124–2133.

Tandon, A. and Collier, B. (1994) The role of endogenous adenosine in a poststimulation increase in the acetylcholine content of a sympathetic ganglion. *J. Neurosci.,* 14: 4927–4936.

Tandon, A. and Collier, B. (1995) Increased acetylcholine content induced by antidromic stimulation of a sympathetic ganglion: a possible retrograde action of adenosine. *J. Neurochem.,* 65: 2116–2123.

Volle, R.L. (1969) Ganglionic transmission. *Annu. Rev. Pharmacol.,* 9: 135–146.

Weldon, P., Bachoo, M., Morales, M.A., Collier, B. and Polosa, C. (1993) Dynamics of large dense-cored vesicles in synaptic boutons of the cat superior cervical ganglion. *Neuroscience,* 55: 1045–1054.

Wu, R., McKenna, D. and McAfee, D.A. (1991) Age-related changes in the synaptic plasticity of rat superior cervical ganglia. *Brain Res.,* 542: 324–329.

# Section VIII

# Cholinergic Dysfunction: 1. Animal Models

J. Klein and K. Löffelholz (Eds.)
*Progress in Brain Research*, Vol. 109

CHAPTER 25

# Immunolesion by 192IgG-saporin of rat basal forebrain cholinergic system: a useful tool to produce cortical cholinergic dysfunction

Reinhard Schliebs, Steffen Roßner and Volker Bigl

*Paul Flechsig Institute for Brain Research, Medical Faculty, University of Leipzig, D-04109 Leipzig, Germany*

## Introduction

The basal forebrain cholinergic system is known to play an important role in cortical arousal and normal cognitive function. Cortical cholinergic dysfunction has been implicated in cognitive deficits that occur in Alzheimer's disease, and the cholinergic projection from the nucleus basalis of Meynert to areas of the cerebral cortex is the pathway that is earliest and severely affected in brains from Alzheimer patients (for review, see e.g. Bigl et al., 1989). In the rat this nucleus is not yet developed into a delineated nuclear structure but corresponds to a more heterogeneous region of cholinergic neurons which often is referred to as nucleus basalis magnocellularis (Nbm) providing the main source of cholinergic terminals to the cerebral cortex (Wenk et al., 1980; Bigl et al., 1982). The changes in markers of the neocortical cholinergic system found in brains of Alzheimer patients are complemented by alterations in other cortical transmitter systems like glutamate, GABA, noradrenaline or serotonin receptors (for reviews, see Nordberg, 1992; Carlson et al., 1993; Greenamyre and Maragos, 1993), suggesting (i) an important influence of the cholinergic basal forebrain system on cortical neurotransmission, and (ii) a cholinergic role in cortical reorganization and in adaptive processes following injury. This is consistent with the hypothesis that the basal forebrain cholinergic system is not directly involved in the formation of learning and memory but acts as a modulatory system to control cortical information processing.

Characterization of the mechanisms underlying this adaptive response might be of particular importance (i) to elucidate the cascade of events initiated by decreased cortical cholinergic activity and (ii) to further derive rationales to pharmacologically intervene in this process with respect e.g. to find a therapeutic strategy to treat Alzheimer's disease or to characterize the role of the cholinergic system in cortical information processing, learning and memory and in cognitive behaviour in greater detail. Such investigations on the functions of the central cholinergic system require adequate animal models to produce specific cholinergic deficits in vivo. This would allow for a detailed evaluation of the neurochemical, neuropathological, and behavioural sequela as well as functional implications of plastic repair mechanisms following cholinergic hypofunction, and provide information that cannot or only partially be obtained in humans. At present there is no adequate animal model available which could mimic all the biochemical, behavioural, and histopathological abnormalities as observed in patients with Alzheimer's disease. However, partial success can be achieved with so called "isomorphic models" (Fisher and Hanin, 1986) representing partial parallelism between model and some human conditions. The value of such models is to delineate mechanisms underlying the pathological processes as well as to test for new potential therapeutic strategies. In the last decades a number of different paradigms have been introduced to produce cortical cholinergic dysfunction which are briefly discussed in the following chapter.

## Paradigms to produce cholinergic lesions in the basal forebrain

### Mechanical lesion of the Nbm

Mechanical lesion of the Nbm (e.g. by radiofrequency or electrolysis) results in damage to all neural tissue at the lesion site including cell populations residing in the basal forebrain as well as passing (e.g. noradrenergic and dopaminergic) fibre bundles. Thus, besides a loss of cortical cholinergic input a considerable reduction of dopaminergic and noradrenergic innervation of the cortex as well as degeneration of non-cholinergic neurons which are present in varying proportions depending on the lesion site must be taken into account. Furthermore, if the size of the lesion is not kept small enough, it encroaches on the neighbouring nuclei and might damage, e.g. the globus pallidus, the nucleus caudate-putamen or hypothalamic nuclei and even the capsula interna which might also considerably affect the results obtained.

Although the lesion is relatively non-specific for the cholinergic system destroying all the neural tissue at the lesion site including all passing fibres, the destruction of cholinergic cells is relatively massive as revealed by considerable decrease in cortical choline acetyltransferase (ChAT) and acetylcholinesterase (AChE).

### Lesion of basal forebrain nuclei by excitotoxins

Excitotoxins are conformationally restricted analogues of the excitatory amino acid neurotransmitter glutamate (e.g. ibotenic acid, quisqualic acid, kainic acid, $N$-methyl-D-aspartic acid (NMDA), $\alpha$-amino-hydroxy-5-methyl-4-isoxazole propionic acid (AMPA)). These compounds act as glutamate receptor agonists and exert their toxic action by prolonged activation of the receptors resulting in increased influx of chloride and calcium ions, excess water entry and osmotic lysis of

the cell. The excitotoxins cited have differential affinities to distinct glutamate receptor subtypes (e.g. quisqualate acts as agonist of the AMPA-type receptor, whereas ibotenic acid preferentially binds to NMDA receptors), thus the cytotoxicity of each glutamate analogue is dependent on the presence of a particular glutamate receptor subtype on the neuron. This might partly explain the differential cytotoxic effects of the known excitotoxins in different regions of the brain. Quinolinic, ibotenic, and quisqualic acid destroy cholinergic cells in the ventral pallidum and substantia innominata-complex, whereas quinolinic and quisqualic acid do not degenerate cholinergic neurons in the medial septum (see e.g. Wenk et al., 1992). Ibotenate and quisqualate induce loss of neuronal cells throughout the Nbm complex but they produce different behavioural impairments in a variety of tasks indicating the presence of heterogeneous cell populations with differential sensitivity to a certain excitotoxin. This might be partly due to the fact that the various excitotoxins differentially affect cholinergic neurons in the basal forebrain. Quisqualate has been seen to produce a greater destruction of Nbm-cholinergic neurons than ibotenic acid. Cortical ChAT depletion by ibotenic acid infusion into the Nbm ranges between 27 and 46%, whereas quisqualate lesions of the Nbm result in cortical ChAT depletions by 41–74%. AMPA is even more effective in destroying cholinergic neurons of the Nbm (greater than 70%), but sparing dorsal pallidum and other non-cholinergic neurons in the basal forebrain (Boegman et al. 1992; for overview and references, see Dunnett et al., 1991).

### Lesion of basal forebrain by ethylcholine aziridinium ion (AF64A)

Ethylcholine aziridinium ion (AF64A) is a neurotoxic analog of choline and exerts its toxic action by disrupting the high affinity choline transport system that regulates the rate and extent of acetylcholine (ACh) synthesis. At higher concentrations AF64A also inhibits AChE activity in vitro. Several authors have suggested that AF64A completely lacks selectivity for cholinergic markers (see e.g. McGurk et al., 1987). However, in further studies it was demonstrated that the specificity of AF64A depends on both the dosage applied and the site of injection. Local administration of AF64A at concentrations higher than 0.02 nmol into various brain regions was shown to produce considerable non-specific tissue destruction at the site of injection (McGurk et al., 1987). However, intraventricular injection of AF64A at low concentrations (less than 5 nmol) produces a relatively specific loss of cholinergic neurons restricted to the medial septal nucleus and the vertical limb of the diagonal band, but sparing cholinergic neurons in the Nbm, and without inducing histological damage to overlying cortex, fimbria fornix, or adjacent structures (Chrobak et al., 1988, 1989; Johnson et al., 1988; Gower et al., 1989; Potter et al., 1989; Lorens et al., 1991; Hörtnagl et al., 1992). These AF64A-induced degenerations are accompanied by decreased choline uptake, ChAT, and ACh synthesis in the hippocampus (Lorens et al., 1991). Therefore low doses of AF64A allow for a selective lesion of medial septal cholinergic neurons, which might be useful when separately studying the role of the septo-hippocampal cholinergic system.

*Lesion of basal forebrain cholinergic system by 192IgG-saporin, a novel cholinergic immunotoxin*

Cholinergic neurons of the basal forebrain possess nerve growth factor (NGF) receptors (Chapter 33) whereas other neurons in this region including the cholinergic cells in the nearby striatum do not express detectable levels of NGF receptors (Gage et al., 1989; Yan and Johnson, 1989). It was demonstrated that a well-characterized monoclonal antibody to the low-affinity NGF receptor, 192IgG, accumulates bilaterally exclusively in cholinergic neurons of the basal forebrain following intracerebroventricular administration (see e.g. Thomas et al., 1991). Employing these properties of 192IgG, a cholinergic immunotoxin was developed by chemical linking of 192IgG via a disulfide bond to the ribosome inactivating protein saporin (192IgG-saporin; see Wiley et al., 1991; Wiley, 1992; for details of preparation, see Wiley and Lappi, 1993). The immunotoxin can be applied both systemically

and by intraventricular as well as parenchymal injection. The most promising results, however, have been obtained by intracerebroventricular applications. Intracerebroventricular administration of $4 \mu g$ of 192IgG-saporin conjugate (at concentrations of 0.3–0.4 mg/ml) results in substantial reductions in ChAT activity in widespread areas of the cortex and hippocampus and in a nearly complete disappearance of ChAT-positive, NGF receptor immunoreactive neurons in the medial septum, in both the vertical and horizontal limbs of the nucleus of the diagonal band of Broca and in the Nbm, whereas cholinergic interneurons in the striatum are not affected (Book et al., 1992; Berger-Sweeney et al., 1994; Heckers et al., 1994; Roßner et al., 1995b). Seven days following injection of the immunotoxin there was a dramatic loss of AChE staining in frontal, parietal, piriform, temporal and occipital cortices, hippocampus and olfactory bulb, but not in the striatum and cerebellum (Heckers et al., 1994; Roßner et al., 1994a, 1995a). Non-cholinergic septal neurons containing parvalbumin and non-cholinergic substantia innominata neurons containing calbindin-D$_{28K}$ or NADPH-diaphorase were not affected by 192IgG-saporin (Heckers et al., 1994). The number of parvalbumin-containing GABAergic projection neurons in the septum-diagonal band of Broca complex and Nbm was not reduced following intraventricular 192IgG-saporin application (Lee et al., 1994; Leanza et al., 1995; Roßner et al., 1995b). Moreover, 192IgG-saporin did not destroy neurotensin, galanin, somatostatin, or neuropeptide neurons within the Nbm (Wenk et al., 1994). Corresponding to the topographic location of cholinergic neurons in the basal forebrain a dramatic increase in microglia has been demonstrated (Roßner et al., 1995b), suggesting that the immunotoxin is lethal to cholinergic cells in the Nbm rather than suppressing the expression of cholinergic markers (e.g. ChAT) in these cells (Book et al., 1994).

It was found that 192IgG-saporin affects two neuronal groups outside of the basal forebrain which express p75NGF receptors: NGF-reactive cerebellar Purkinje cells after intraventricular injection and cholinergic striatal interneurons after injections into the substantia innominata (Heckers et al., 1994). There are ChAT-positive, but NGF-receptor negative neurons in the rat Nbm-substantia innominata complex innervating the amygdala and parts of the rhinal paralimbic areas (see e.g. Woolf et al., 1989; Bickel and Kewitz, 1990) which are spared or only partially affected by the immunotoxin (Heckers et al., 1994). Similarly, cholinergic neurons in the ventral pallidum and sublenticular substantia innominata not expressing p75NGF receptors are not affected by the immunotoxin.

Complete cholinergic lesion by 192IgG-saporin did not produce any deficit in the Morris water maze task (Torres et al., 1994). Despite the high depletion in cortical ChAT activity by 192IgG-saporin acquisition, performance of the delayed alternation or passive avoidance tasks were not impaired by the lesions suggesting that selective loss of cholinergic cells is not sufficient to produce functional impairments (Wenk et al., 1994). In contrast, other authors reported that intracerebral administration of 192IgG-saporin induced dose-dependent (ranging between 1 and 10 $\mu$g) impairments in the water maze task and passive avoidance retention, but only weak effects on locomotor activity (Leanza et al., 1995; Waite et al., 1995). Intracerebroventricular injections of 192IgG-saporin severely affected spatial and cued navigation (Nilsson et al., 1992; Berger-Sweeney et al., 1994). However, an almost 90% reduction in ChAT activity is needed to produce substantial behavioural deficits (Waite et al., 1995).

## Effect of different cholinergic lesion procedures on cortical cholinergic markers

As outlined in the previous chapter the procedures to lesion cholinergic nuclei in rat basal forebrain differ in selectivity and specificity to degenerate cholinergic cells. A comparison of the effects of the various lesion procedures on cholinergic markers in cholinoceptive cortical target regions should therefore allow a further valuation of the different lesion techniques. Therefore, the same experimental design was applied to both electrolytic, ibotenic acid and immunolesion of basal forebrain cholinergic nuclei. Seven days after lesion receptor au-

toradiography and in situ hybridization were performed in adjacent coronal brain sections at six selected distances from the bregma ranging from +2.7 to −5.3 mm according to the atlas of Zilles (1985). The levels of cryocutting were selected to include for data analysis all cortical areas which receive a prominent cholinergic innervation from the basal forebrain. To prove the efficiency of the lesion histochemistry for AChE and ChAT were performed in adjacent brain sections (Roßner et al., 1994a,b, 1995a; Schliebs et al., 1994). The data obtained are summarized in Table 1.

Seven days following unilateral electrolytic Nbm lesion we found a small reduction in $M_2$-muscarinic ACh receptor (mAChR) binding restricted to frontal and parietal cortices, but no change in $M_1$-receptor binding sites in any of the cortical regions studied as compared to the unlesioned brain side. These alterations in cortical $M_2$-mAChR binding are complemented by corresponding changes in the $m_2$-and $m_4$-mRNA transcripts (Schliebs et al., 1994; Table 1).

Ibotenic acid lesion resulted in a striking loss of AChE-staining in the lesioned Nbm which is associated with a 60% decrease in AChE staining and a 30% reduction in [$^3$H]hemicholinium-3 binding in frontal and parietal cortical regions as well as fore/hindlimb areas ipsilateral to the lesion, being more prominent in the more rostral cortical regions. $M_1$-mAChR binding was not changed in any of the cortical regions studied 1 week after lesion. $M_2$-mAChR binding levels are slightly increased in the parietal cortex only. The lesion-induced increase in parietal cortical $M_2$-mAChR binding is complemented by an increase in the hybridization signal for the corresponding $m_4$-mRNA transcript (Roßner et al., 1994b; Table 1).

Seven days following an intracerebroventricular injection of the cholinergic immunotoxin 192IgG-saporin hemicholinium-3 binding to high-affinity choline uptake sites was considerably decreased by up to 45% in all cortical regions and in the hippocampus as compared to the corresponding control values. In contrast, $M_1$-mAChR sites were increased over the corresponding control values in the anterior parts of cingulate, frontal, and piriform cortex by about 20%, in the hindlimb/forelimb

## TABLE I

Pattern of changes in cholinergic markers in selected rat brain regions 1 week after electrolytic and ibotenic acid lesion of the nucleus basalis magnocellularis as well as immunolesion of rat basal forebrain cholinergic system by 192IgG-saporin

| Region | Cholinergic lesion procedure (significant relative changes over control in %) | | |
|---|---|---|---|
| | Electrolytic | Ibotenic acid | 192IgG-saporin |
| *Acetylcholinesterase* | | | |
| Cg | -12 | – | -19 |
| Fr | -40 | -50 | -80 |
| Par | -18 | -50 | -80 |
| Pir | – | -25 | -46 |
| Occ | – | – | -85 |
| Temp | – | – | -79 |
| *High-affinity choline uptake sites* | | | |
| Cg | n.d. | – | -40 |
| Fr | n.d. | -22 | -40 |
| Par | n.d. | -20 | -38 |
| Pir | n.d. | – | -30 |
| Occ | n.d. | – | -40 |
| Temp | n.d. | – | -38 |
| *Nicotinic acetylcholine receptor* | | | |
| Cg | n.d. | n.d. | – |
| Fr | n.d. | n.d. | – |
| Par | n.d. | n.d. | – |
| Pir | n.d. | n.d. | – |
| Occ | n.d. | n.d. | – |
| Temp | n.d. | n.d. | – |
| *$M_1$-muscarinic acetylcholine receptor* | | | |
| Cg | – | – | -18 |
| Fr | – | – | -18 |
| Par | – | – | -35 |
| Pir | – | – | -18 |
| Occ | – | – | -17 |
| Temp | – | – | -25 |
| *$M_2$-muscarinic acetylcholine receptor* | | | |
| Cg | – | – | – |
| Fr | -10 | – | – |
| Par | -8 | +15 | +22 |
| Pir | – | – | – |
| Occ | – | – | +20 |
| Temp | – | – | – |
| *m1-mAChR* | | | |
| Cg | – | – | – |
| Fr | – | – | – |
| Par | – | – | – |
| Occ | – | – | +20 |
| Pir | – | – | – |
| Temp | – | – | +20 |
| *m2-mAChR* | | | |
| Cg | – | – | – |
| Fr | -10 | – | – |
| Par | – | – | – |
| Pir | – | – | – |
| Occ | – | – | – |
| Temp | – | – | – |
| *m3-mAChR* | | | |
| Cg | – | – | – |
| Fr | – | – | – |
| Par | – | – | – |
| Pir | – | – | – |
| Occ | – | – | +20 |
| Temp | – | – | +20 |
| *m4-mAChR* | | | |
| Cg | – | – | – |
| Fr | -10 | – | – |
| Par | – | +15 | – |
| Pir | -20 | – | – |
| Occ | – | – | +25 |
| Temp | – | – | +25 |

The alterations in various neurotransmitter receptors and AChE staining in selected brain regions 1 week after lesion are summarized and given as relative changes over the corresponding control value ($P < 0.05$ or higher, two-tailed Student's $t$-test) obtained from vehicle-injected control animals (immunolesion) or from the unlesioned brain side (electrolytic and ibotenic acid lesion). Unilateral electrolytic and ibotenic acid lesion of the nucleus basalis magnocellularis were performed as previously described (Schliebs et al., 1994; Roßner et al., 1994b). Immunolesion of rat basal forebrain cholinergic system by intracerebroventricular injection of 192IgG-saporin was carried out as described by Roßner et al. (1994a). AChE was measured by histochemical staing of adjacent brain sections and quantified by image analysis; high-affinity choline uptake sites, $M_1$- and $M_2$-mAChR were assayed by receptor autoradiography and quantitative image analysis (Schliebs et al., 1994; Roßner et al., 1994a,b; Roßner et al., 1995a). m1–m4-mAChR subtypes were determined by in situ hybridization using $^{35}S$-labeled oligonucleotide probes (Roßner et al., 1993, 1994c). –, No significant change over control; n.d., not determined; Cg, cingulate cortex; Fr, frontal cortex; Par, parietal cortex; Pir: piriform cortex; Occ, occipital cortex; Temp, temporal cortex.

areas (18%), in the parietal cortex (35%), in the occipital cortex (17%) as well as in the temporal cortex (25%) following immunolesion. $M_2$-mAChR levels were found to be significantly enhanced in the posterior part of the parietal cortex (by about 22%) and in the occipital cortex area (20%) only (Roßner et al., 1995a; Table 1).

The increase in $M_1$-mAChR binding in the temporal and occipital cortex as a consequence of immunolesion is complemented by an increase in the amount of m1 and m3 mAChR mRNA by about 20% in these regions. The elevated levels of $M_2$-mAChR sites in the occipital and temporal cortex following immunolesion are accompanied by an increase in the m4 (by 25%) but not m2 mAChR mRNA. There was no effect of immunolesion on the m1–m4 mAChR mRNA in frontal cortical regions. In the basal forebrain, however, immunolesioning resulted in a considerable decrease in the level of m2 mAChR mRNA in the medial and lateral septum as well as in the vertical and horizontal limb of the diagonal band by about 40%, whereas $M_1$-and $M_2$-mAChR binding and the levels of m1, m3, and m4 mAChR mRNA were not affected by immunolesion in any of the basal forebrain nuclei studied. Seven days following a single dosage of the 192IgG-saporin no change in the level of cortical nicotinic acetylcholine receptor sites in any of the regions studied was observed as compared to the corresponding control values (Roßner et al., 1995a).

One week after lesion a reduced high-affinity uptake of [$^3$H]choline into cholinergic nerve terminals in the cerebral cortex and hippocampus was observed, which was accompanied by a decreased K$^+$-stimulated release of [$^3$H]ACh from cortical and hippocampal slices of immunolesioned rats (Roßner et al., 1995c). Cholinergic immunolesion led to enhanced cortical $M_1$-mAChR numbers, but did not alter mAChR sensitivity as measured by carbachol-stimulated inositol phosphate production or phorbol ester binding to membrane-bound protein kinase C (Roßner et al., 1995c). In the hippocampal formation differential enhancements in binding levels of $M_1$-mAChR sites in the CA1 region and in the dentate gyrus were observed, whereas the nicotinic and $M_2$-mAChR subtype are seemingly not affected by the immunotoxin in either of the subfields studied. Cholinergic immunolesioning did not result in any alterations in the hybridization signals for m1–m4 mAChR mRNA in any region or layer of the hippocampus (Roßner et al., 1995c).

In summary, electrolytic Nbm lesions which destroy both cells, nerve terminals and passing fibres, did not change $M_1$-mAChR but resulted in reduced $M_2$-mAChR in frontal and parietal cortices 1 week after lesion. Nbm ibotenic acid lesion which likely affects both cholinergic and GABAergic cells but spares crossing fibres, did not alter $M_1$-mAChR in any cortical region but resulted in enhanced $M_2$-mAChR binding in the parietal cortex only. When applying the cholinergic immunotoxin 192IgG-saporin which specifically and selectively destroys basal forebrain cholinergic cells only, both $M_1$-and $M_2$-mAChR binding sites were increased in a number of cortical areas 1 week after lesion. From this comparison it can be suggested that the various lesion procedures differentially affect populations of mAChR localized on distinct cortical cholinoceptive cell populations. In particular, the different effects of the specific and selective cholinergic immunolesion and the less specific ibotenic acid lesion on $M_1$-mAChR suggest that other transmitter systems, probably GABAergic projection neurons, contribute to the different cortical effects.

## Effect of cholinergic immunolesion by 192IgG-saporin on cortical glutamate and GABA neurotransmission

Glutamate is used as an excitatory transmitter in corticofugal as well as cortico-cortical systems and plays an important role in realizing cortico-cortical information transfer. GABA represents the major inhibitory transmitter in the cerebral cortex. It is generally accepted that the precise interaction of excitatory and inhibitory signals seems to be a major step in efficient processing of cortical information transfer. In a current study alterations in dendritic morphology of cortical neurons after basal forebrain lesions have been described (Wellman and Sengelaub, 1995) suggesting that the cholinergic input plays an important modulatory role in cortical function and plasticity. To study the impact of reduced cortical cholinergic activity on glutamatergic und GABAergic transmission in the cerebral cortex and to elucidate possible adaptive responses, glutamate and GABA receptor subtypes were assayed by quantitative receptor autoradiography 1 week after a single intracerebroventricular injection of 4 μg of 192IgG-saporin. Receptor autoradiography and AChE staining were performed in adjacent brain sections, which allows simultanous detection of the consequences of lesions on various parameters in a distinct cortical area, and thus provides an appropriate tool to reveal correlations between cortical cholinergic hypoactivity and lesion-induced adaptive response in distinct cholinoceptive target regions. One week after cholinergic lesion by 192IgG-saporin, NMDA receptor binding was markedly reduced in cortical regions displaying a reduced activity of AChE and high-affinity choline uptake sites as a consequence of cholinergic lesion, whereas AMPA and kainate binding sites were significantly increased in these regions (Roßner et al., 1995d; Table 2). Muscimol binding to GABA$_A$ receptors was increased in the caudal portions of frontal and parietal cortices as well as occipital and temporal cortex as compared to the corresponding brain regions from vehicle-injected control rats (Table 2). Binding levels of benzodiazepine receptors were not affected by the lesion in any of the cortical regions studied (Roßner et al., 1995d). Equivalent changes in cortical glutamate and GABA receptor subtype levels have been observed 7 days after electrolytic (Schliebs et al., 1994) or ibotenic acid lesion (Roßner et al., 1994b) of the Nbm applying the same experimental design. In Table 2 the pattern of changes in selected cortical rat brain regions due to various cholinergic lesion procedures are summarized. Despite some differences in the specificity of the various lesion procedures applied, the data support the view that the alterations in cortical glutamate and GABA receptor subtypes following immunolesion are mainly due to the loss of cortical cholinergic input originating preferentially in the Nbm. To study whether the lesion-induced alterations in glutamate and GABA transmission are consequences of reduced cortical cholinergic input, the parietal cortex with its wide rostral-caudal extension was used as an appropriate cortical model region displaying a gradient in the lesion-induced decreases in AChE activity and choline uptake sites from rostral to the caudal extension. The loss of cholinergic input as assayed by choline uptake sites is significantly correlated with the changes in binding levels of glutamate subtype and GABA$_A$ receptors (Roßner et al., 1995d) suggesting that the receptor changes might be the consequence of the imbalance between cortical cholinergic innervation and intracortical glutamatergic and GABAergic neurotransmission. This is supported by a recent report demonstrating that primate cortical M$_1$-and M$_2$-mAChR are associated with asymmetric synapses thus providing morphological evidence for cholinergic modulation of excitatory transmission via M$_1$-and M$_2$-receptors (Mrzljak et al., 1993). In the rat, cholinergic fibres from the basal forebrain terminate preferentially in cerebral cortical layers I and V, and M$_1$-mAChR are mainly concentrated in layers II/III and VI (Eckenstein et al., 1988; Schliebs and Roßner, 1995). Glutamate-containing neurons are concentrated in cortical layer V and the deep part of layer VI, whereas glutamate-containing axon terminals show the highest density in layers I–IV (Zilles et al., 1990). Each glutamate receptor subtype exhibits a distinct cortical laminar pattern (Kumar et al., 1993) suggesting that glutamate exerts a different influence in each particular cortical layer by inducing different cellular responses through distinct receptor subtypes. The decrease in NMDA receptor binding following immunolesion could be explained when assuming that at least some of the cortical NMDA receptors are located on cholinergic terminals originating in the basal forebrain. However, there is no evidence that cortical NMDA receptors may exist on presynaptic terminals of cholinergic neurons originating in the Nbm (Maragos et al., 1991). But the basal forebrain magnocellular complex receives among others also a strong glutamatergic innervation from the cortex (Martin et al., 1993) suggesting that the cholinergic immunolesion of the Nbm should have

TABLE 2

Pattern of changes in glutamatergic and GABAergic markers in selected rat brain regions one week after electrolytic and ibotenic acid lesion of the nucleus basalis magnocellularis as well as immunolesion of rat basal forebrain cholinergic system by 192IgG-saporin

| Region | Cholinergic lesion procedure (significant relative changes as compared to controls in %) | | |
|---|---|---|---|
| | Electrolytic | Ibotenic acid | 192IgG-saporin |
| *NMDA receptor* | | | |
| Cg | −8 | − | − |
| Fr | −15 | −18 | −15 |
| Par | −20 | −18 | −20 |
| Pir | − | − | − |
| Occ | − | − | −20 |
| Temp | − | − | −15 |
| *AMPA receptor* | | | |
| Cg | − | − | − |
| Fr | +25 | +20 | +12 |
| Par | − | +18 | +15 |
| Pir | − | − | − |
| Occ | − | − | − |
| Temp | − | − | +12 |
| *Kainate receptor* | | | |
| Cg | − | − | − |
| Fr | +30 | +18 | +20 |
| Par | +40 | +20 | +25 |
| Pir | +12 | − | − |
| Occ | +18 | − | − |
| Temp | − | − | +20 |
| *GABA$_A$ receptor* | | | |
| Cg | n.d. | − | − |
| Fr | n.d. | +16 | +18 |
| Par | n.d. | +16 | +20 |
| Pir | n.d. | − | − |
| Occ | n.d. | − | +18 |
| Temp | n.d. | − | +18 |
| *Benzodiazepine receptor* | | | |
| Cg | n.d. | − | − |
| Fr | n.d. | − | − |
| Par | n.d. | − | − |
| Pir | n.d. | − | − |
| Occ | n.d. | − | − |
| Temp | n.d. | − | − |

The alterations in various neurotransmitter receptors in selected brain regions 1 week after lesion are summarized and given as relative changes over the corresponding control value ($P < 0.05$ or higher, two-tailed Student's $t$-test) obtained from vehicle-injected control animals (immunolesion) or from the unlesioned brain side (electrolytic and ibotenic acid lesion). Unilateral electrolytic and ibotenic acid lesion of the nucleus basalis magnocellularis were performed as previously described (Schliebs et al., 1994; Roßner et al., 1994b). Immunolesion of rat basal forebrain cholinergic system by intracerebroventricular injection of 192IgG-saporin was carried out as described by Roßner et al. (1994a). Glutamate receptor subtypes like NMDA, AMPA, and kainate as well as GABA and benzodiazepine receptors were assayed by receptor autoradiography and quantitative image analysis (Schliebs et al., 1994; Roßner et al., 1994a,b, 1995c). −, No significant change over control; n.d., not determined; Cg, cingulate cortex; Fr, frontal cortex; Par, parietal cortex; Pir, piriform cortex; Occ, occipital cortex; Temp, temporal cortex.

functional consequences also on cortical glutamatergic neurons. Therefore, the changes in the number of cortical NMDA receptors following lesion could be considered as a loss and/or down-regulation of NMDA receptor sites. In contrast, the increased kainate and AMPA binding following lesion should be considered as up-regulation of receptor sites. Up-regulation of postsynaptic glutamate receptors is assumed to compensate for reduced presynaptic input. This would suggest that cholinergic terminals directly affect glutamate transmission on presynaptic glutamatergic elements. However, MK-801 is assumed to bind to a site within the NMDA receptor ion-channel and thus can also be considered as a marker of the agonist-bound, open state of the channel (Seeburg, 1993). Therefore, the immunotoxin-induced decline in MK-801 binding also indicates a lower amount of glutamate bound to the NMDA receptor channel. This supports the suggestion that cholinergic hypofunction reduces cortical glutamatergic activity by less release of glutamate from presynaptic elements presumably due to enhanced inhibition by GABA.

Seven days after immunolesion we found significantly increased GABA$_A$ but not benzodiazepine binding sites in the frontal and parietal cortices (Table 2) suggesting an up-regulation of postsynaptically localized GABA$_A$ receptors as an adaptive response to the reduced GABAergic input as measured by Gomeza et al. (1992). In a recent study it was suggested that cholinergic excitation of GABAergic interneurons is mediated via $M_2$-mAChR (Mrzljak et al., 1993). But whether the immunolesion-induced increase in $GABA_A$ receptor binding is a direct consequence of the enhanced $M_2$-mAChR level observed in these regions 1 week after immunolesion (Roßner et al., 1995a) cannot be concluded from these data and must await further analysis. It is interesting to note that $GABA_A$ receptors can up-regulate while benzodiazepine receptors remain unchanged 1 week after immunolesion, although both receptors should exist within the same protein receptor complex. However, from in vitro studies it is well known that a functional coupling exists between $GABA_A$ and benzodiazepine receptors: benzodiazepines can affect GABA binding and vice-versa by altering binding affinities (see e.g. Bureau and Olsen, 1993). Thus immunolesion-induced changes in GABA release could affect the binding states for benzodiazepines by altering the binding affinity and this could cover some changes in benzodiazepine binding.

However, regardless of possible interpretations the immunotoxin-induced differential changes in glutamate and GABA receptor subtypes in cortical regions displaying reduced cholinergic activity clearly demonstrate that cortical glutamatergic and GABAergic markers are partially driven by cholinergic activity. Moreover, it is interesting to note that the same sort of alterations in glutamate and GABA receptor subtypes observed in rat cortex following basal forebrain cholinergic immunolesion have been detected in cortical brain areas from patients with Alzheimer's disease (Nordberg, 1992). This supports the suggestion that the receptor changes observed might indicate compensatory mechanisms due to presumably cholinergic degenerative events. However, these data further support a glutamatergic strategy which might be therapeutically potential in treating Alzheimer's disease (Advokat and Pelligrini, 1992; Carlson et al., 1993; Burney, 1994). Moreover, they suggest that cholinergic immunolesion by 192IgG-saporin exhibits a valuable tool to produce specific cholinergic deficits in rats, which can be used as a model to study the effect of treatment with various drugs.

## Summary

Cholinergic lesion paradigms have been used to study the role of the cholinergic system in cortical arousal and cognitive function, and its implication in cognitive deficits that occur in Alzheimer's disease. In the last few years an increasing number of studies have applied neurotoxins including excitotoxins or cholinotoxins (e.g. AF64A) by stereotaxic injection into the Nbm to produce reductions in cortical cholinergic activity. One of the most serious limitations of these lesion paradigms is the fact that basal forebrain cholinergic neurons are always intermingled with populations of non-cholinergic cells and that the cytotoxins used are far from being selective to cholinergic cells. Excitotoxins when infused directly into the Nbm destroy non-specifically cell bodies but spare axons passing the injection site, whereas the specificity of AF64A to destroy cholinergic neurons depends on both the dosage applied and the site of injection.

Recently, a monoclonal antibody to the low-affinity nerve growth factor (NGF) receptor, 192IgG, coupled to a cytotoxin, saporin, has been described as an efficient and selective immunotoxin for the NGF-receptor bearing cholinergic neurons in rat basal forebrain. Intraventricular administration of the 192IgG-saporin conjugate appears to induce a nearly complete and specific lesion of neocortical and hippocampal cholinergic afferents. Other neuronal systems in the basal forebrain are spared by the immunotoxin.

Electrolytic, ibotenic acid, and cholinergic immunotoxic lesions of cholinergic basal forebrain nuclei resulted in slightly different effects on cortical cholinergic markers: Electrolytic lesion of the Nbm did not change $M_1$-mAChR but resulted in reduced $M_2$-mAChR in frontal and parietal cortices 1 week after lesion. Ibotenic acid lesion of the nucleus basalis did not alter $M_1$-mAChR in any cortical region but led to enhanced $M_2$-mAChR binding in the parietal cortex only. When applying the cholinergic immunotoxin 192IgG-saporin, both $M_1$- and $M_2$-mAChR binding sites were increased in a number of cortical areas 1 week after lesion. This comparison suggests that possibly the destruction of non-cholinergic basal forebrain cells by ibotenic acid and electrolytic lesion, might partly contribute to these different cortical effects.

NMDA receptor binding was markedly reduced and AMPA, kainate, and $GABA_A$ receptor binding has been significantly increased in cortical regions displaying a reduced activity of AChE and decreased levels of high-affinity choline uptake sites due to immunolesion of the basal forebrain cholinergic system. Equivalent changes in cortical glutamate and GABA receptor subtype levels have been observed 7 days after electrolytic or ibotenic acid lesion of the Nbm.

The data suggest that cholinergic immunolesion by 192IgG-saporin exhibits a valuable tool to produce specific cholinergic deficits in rats, which can be used as a model to study the effect of treatment with various drugs for compensating the impaired cortical cholinergic input.

## Acknowledgements

This work was partly supported by a grant of the Bundesministerium für Forschung und Technik to R.S., no. FKZ 01 ZZ 9103/2.8.

## References

Advokat, C. and Pelligrini, A.I. (1992) Excitatory amino acids and memory, evidence from research on Alzheimer's disease and behavioral pharmacology. *Neurosci. Behav. Rev.*, 16: 13–24.

Berger-Sweeney, J., Heckers, S., Mesulam, M.-M., Wiley, R.G., Lappi, D.A. and Sharma, M. (1994) Differential effects on spatial navigation of immunotoxin-induced cholinergic lesions of the medial septal area and nucleus basalis magnocellularis. *J. Neurosci.*, 14: 4507–4519.

Bickel, U. and Kewitz, H. (1990) Colocalization of choline acetyltransferase and nerve growth factor receptor in the rat basal forebrain. *Dementia*, 1: 146–150.

Bigl, V., Woolf, N.J. and Butcher, L.L. (1982) Cholinergic projections from the basal forebrain to frontal, parietal, temporal, occipital, and cingulate cortices: a combined fluorescent tracer and acetylcholinesterase analysis. *Brain Res. Bull.*, 8: 727–749.

Bigl, V., Arendt, T. and Biesold, D. (1989) The nucleus basalis Meynert during aging and in dementing disorders. In: M. Steriade and D. Biesold (Eds.), *Cholinergic Systems of the Basal Forebrain*, Oxford University Press, Oxford, pp. 364–386.

Boegman, R.J., Cockhill, J., Jhamandas, K. and Beninger, R.J. (1992) Excitoxic lesions of rat basal forebrain: differential effects on choline acetyltransferase in the cortex and amygdala. *Neuroscience*, 51: 129–135.

Book, A.A., Wiley, R.G. and Schweitzer, J.B. (1992) Specificity of 192 IgG-saporin for NGF receptor-positive cholinergic basal forebrain neurons in the rat. *Brain Res.*, 590: 350–355.

Book, A.A., Wiley, R.G. and Schweitzer, J.B. (1994) 192 IgG-saporin: I. Specific lethality for cholinergic neurons in the basal forebrain of the rat. *J. Neuropathol. Exp. Neurol.*, 53: 95–102.

Bureau, M.H. and Olsen, R.W. (1993) $GABA_A$ receptor subtypes: Ligand binding heterogeneity demonstrated by photoaffinity labeling and autoradiography. *J. Neurochem.*, 61: 1479–1491.

Burney, R.N. (1994) Therapeutic potential of NMDA antagonists in neurodegenerative diseases. *Neurobiol. Aging*, 15: 271–273.

Carlson, M.D., Penney, Jr., J.B. and Young, A.B. (1993) NMDA, AMPA, and benzodiazepine binding site changes in Alzheimer's disease visual cortex. *Neurobiol. Aging*, 14: 343–352.

Chrobak, J.J., Hanin, I., Schmechel, D.E. and Walsh, T.J. (1988) AF64A-induced working memory impairment: Behavioural, neurochemical and histological correlates. *Brain Res.*, 463: 107–117.

Chrobak, J.J., Spates, M.J., Stackman, R.W. and Walsh, T.J. (1989) Hemicholinium-3 prevents the working memory impairments and the cholinergic hypofunction induced by ethylcholine aziridinium ion (AF64A). *Brain Res.*, 504: 269–275.

Dunnett, S.B., Everitt, B.J. and Robbins, T.W. (1991) The basal forebrain-cortical cholinergic system: interpreting the functional consequences of excitotoxic lesions. *Trends Neurosci.*, 14: 494–501.

Eckenstein, F.P., Baughman, R.W. and Quinn, J. (1988) An anatomical study of cholinergic innervation in rat cerebral cortex. *Neuroscience*, 25: 457–474.

Fisher, A. and Hanin, I. (1986) Potential animal models for senile dementia of Alzheimer's type, with emphasis on AF64A-induced cholinotoxicity. *Annu. Rev. Pharmacol. Toxicol.*, 26: 161–181.

Gage, F.H., Batchelor, P., Chen, K.S., Chin, D., Deputy, S., Rosenberg, M.B., Higgins, G.A., Koh, S., Fischer, W. and Björklund, A. (1989) NGF-receptor expression and NGF-mediated cholinergic neuronal hypertrophy in the damaged adult neostriatum. *Neuron*, 2: 1177–1184.

Gomeza, J., Aragón, C. and Giménez, C. (1992) High-affinity transport of choline and amino acid neurotransmitters in synaptosomes from brain regions after lesioning the nucleus basalis magnocellularis of young and aged rats. *Neurochem. Res.*, 17: 345–350.

Gower, A.J., Rousseau, D., Jamsin, P., Gobert, J., Hanin, I.

and Wulfert, E. (1989) Behavioral and histological effects of low concentrations of intraventricular AF64A. *Eur. J. Pharmacol.*, 166: 271–281.

Greenamyre, J.T. and Maragos, W.F. (1993) Neurotransmitter receptors in Alzheimer disease. *Cerebrovasc. Brain Met. Rev.*, 5: 61–94.

Heckers, S., Ohtake, T., Wiley, R.G., Lappi, D.A., Geula, C. and Mesulam, M.M. (1994) Complete and selective denervation of rat neocortex and hippocampus but not amygdala by an immunotoxin against the p75 NGF receptor. *J. Neurosci.*, 14: 1271–1289.

Hörtnagl, H., Sperk, G., Sobal, G. and Maas, D. (1990) Cholinergic deficit induced by ethylcholine aziridinium (AF64A) transiently affects somatostatin and neuropeptide Y levels in rat brain. *J. Neurochem.*, 54: 1608–1613.

Johnson, G.V.W., Simonato, M. and Jope, R.S. (1988) Dose- and time dependent hippocampal cholinergic lesions induced by ethylcholine mustard aziridinium ion: effects of nerve growth factor, GM1 ganglioside, and vitamin E. *Neurochem. Res.*, 13: 685–692.

Kumar, A., Schliebs, R. and Bigl, V. (1993) Development of NMDA, AMPA, and kainate receptors in individual layers of rat visual cortex and the effect of monocular deprivation. *Int. J. Dev. Neurosci.*, 12: 31–41.

Leanza, G., Nilsson, O.G., Wiley, R.G. and Björklund, A. (1995) Selective lesioning of the basal forebrain cholinergic system by intraventricular 192 IgG-saporin: behavioural, biochemical and stereological studies in the rat. *Eur. J. Neurosci.*, 7: 329–343.

Lee, M.G., Chrobak, J.J., Sik, A., Wiley, R.G. and Buzsáki, G. (1994) Hippocampal theta activity following selective lesion of the septal cholinergic system. *Neuroscience*, 62: 1033–1047.

Lorens, S.K., Kindel, G., Dong, X.W., Lee, J.M. and Hanin, I. (1991) Septal choline acetyltransferase immunoreactive neurons: dose dependent effects of AF64A. *Brain Res. Bull.*, 26: 965–971.

Maragos, W.F., Greenamyre, J.T., Chu, D.C.M., Penney, J.B. and Young, A.B. (1991) A study of cortical and hippocampal NMDA and PCP receptors following selective cortical and subcortical lesions. *Brain Res.*, 538: 36–45.

Martin, L.J., Blackstone, C.D., Levey, A.I., Huganir, R.L. and Price, D.L. (1987) Cellular localization of AMPA glutamate receptors within the forebrain magnocellular complex of rat and monkey. *J. Neurosci.*, 13: 2249–2263.

McGurk, S.R., Hartgraves, S.L., Kelly, P.H., Gordon, M.N. and Butcher, L.L. (1987) Is ethylcholine mustard aziridinium ion a specific cholinergic neurotoxin? *Neuroscience*, 22: 215–224.

Mrzljak, L., Levey, A.I. and Goldman-Rakic, P.S. (1993) Association of m1 and m2 muscarinic receptor proteins with asymmetric synapses in the primate cerebral cortex:morphological evidence for cholinergic modulation of excitatory neurotransmission. *Proc. Natl. Acad. Sci. USA*, 90: 5194–5198.

Nilsson, O.G., Leanza, G., Rosenblad, C., Lappi, D.A., Wiley,

R.G. and Björklund A (1992) Spatial learning impairments in rats with selective immunolesion of the forebrain cholinergic system. *NeuroReport*, 3: 1005–1008.

Nordberg, A. (1992) Neuroreceptor changes in Alzheimer disease. *Cerebrovasc. Brain Met. Rev.*, 4: 303–328.

Potter, P.E., Tedford, C.E., Kindel, G. and Hanin, I. (1989) Inhibition of high affinity choline transport attenuates both cholinergic and non-cholinergic effects of ethylcholine aziridinium (AF64A). *Brain Res.*, 487: 238–244.

Roßner, S., Perez-Polo, J.R., Wiley, R.G., Schliebs, R. and Bigl, V. (1994a) Differential expression of immediate early genes in distinct layers of rat cerebral cortex after selective immunolesion of the forebrain cholinergic system. *J. Neurosci. Res.*, 38: 282–293.

Roßner, S., Schliebs, R. and Bigl, V. (1994b) Ibotenic acid lesion of nucleus basalis magnocellularis differentially affects cholinergic, glutamatergic and GABAergic markers in cortical rat brain regions. *Brain Res.*, 668: 85–99.

Roßner, S., Schliebs, R., Perez-Polo, J.R., Wiley, R.G. and Bigl, V. (1995a) Differential changes in cholinergic markers from selected brain regions after specific immunolesion of rat cholinergic basal forebrain system. *J. Neurosci. Res.*, 40: 31–43.

Roßner, S., Härtig, W., Schliebs, R., Brückner, G., Brauer, K., Perez-Polo, J.R., Wiley, R.G. and Bigl, V. (1995b) 192IgG-saporin immunotoxin-induced loss of cholinergic cells differentially activates microglia in rat basal forebrain nuclei. *J. Neurosci. Res.*, 41: 335–346.

Roßner, S., Schliebs, R., Härtig, W. and Bigl, V. (1995c) 192IgG-saporin-induced selective lesion of cholinergic basal forebrain system: neurochemical effects on cholinergic neurotransmission in rat cerebral cortex and hippocampus. *Brain Res. Bull.*, 38: 371–381.

Roßner, S., Schliebs, R. and Bigl, V. (1995d) 192IgG-saporin-induced immunotoxic lesions of cholinergic basal forebrain system differentially affect glutamatergic and GABAergic markers in cortical rat brain regions. *Brain Res.*, 696: 165–176.

Schliebs, R. and Roßner, S. (1995) Distribution of muscarinic acetylcholine receptors in the CNS. In: T.W. Stone (Ed.), *CNS Neurotransmitters and Neuromodulators. Acetylcholin*, CRC Press, Boca Raton, FL, pp. 67–83.

Schliebs, R., Feist, T., Roßner, S. and Bigl, V. (1994) Receptor function in cortical rat brain regions after lesion of nucleus basalis. *J. Neural Transm.*, 44(Suppl.): 195–208.

Seeburg, P.H. (1993) The molecular biology of mammalian glutamate receptor channels. *Trends Pharmacol. Sci.*, 14: 297–303.

Thomas, L.B., Book, A.A. and Schweitzer, J.B. (1991) Immunohistochemical detection of a monoclonal antibody directed against the NGF receptor in basal forebrain neurons following intraventricular injection. *J. Neurosci. Methods*, 37: 37–45.

Torres, E.M., Perry, T.A., Blokland, A., Wilkinson, L.S., Wiley, R.G., Lappi, D.A. and Dunnett, S.B. (1994) Behavioural, histochemical and biochemical consequences of selective immunolesions in discrete regions of the basal fore-brain cholinergic system. *Neuroscience*, 63: 95–122.

Waite, J.J., Chen, A.D., Wardlow, M.L., Wiley, R.G., Lappi, D.A. and Thal, L.J. (1995) 192 immunoglobulin G-saporin produces graded behavioral and biochemical changes accompanying the loss of cholinergic neurons in the basal forebrain and cerebellar Purkinje cells. *Neuroscience*, 65: 463–476.

Wellman, C.L. and Sengelaub, D.R. (1995) Alterations in dendritic morphology of frontal cortical neurons after basal forebrain lesions in adult and aged rats. *Brain Res.*, 669: 48–58.

Wenk, H., Bigl, V. and Meyer, U. (1980) Cholinergic projections from magnocellular nuclei of the basal forebrain to cortical areas in rats. *Brain Res. Rev.*, 2: 295–316.

Wenk, G.L., Harrington, C.A., Tucker, D.A., Rance, N.E. and Walker, L.C. (1992) Basal forebrain neurons and memory: a biochemical, histological, and behavioral study of differential vulnerability to ibotenate and quisqualate. *Behav. Neurosci.*, 106: 909–923.

Wenk, G.L., Stoehr, J.D., Quintana, G., Mobley, S. and Wiley, R.G. (1994) Behavioral, biochemical, histological, and electrophysiological effects of 192 IgG-saporin injections into the basal forebrain of rats. *J. Neurosci.*, 14: 5986–5995.

Wiley, R.G. (1992) Neural lesioning with ribosome-inactivating proteins: suicide transport and immunolesioning. *Trends Neurosci.*, 15: 285–290.

Wiley, R.G. and Lappi, D.A. (1993) Preparation of anti-neuronal immunotoxins for selective neural immunolesioning. *Neurosci. Prot.*, 93–020–02–01–12.

Wiley, R.G., Oeltmann, T.N. and Lappi, D.A. (1991) Immunolesioning: selective destruction of neurons using immunotoxin to rat NGF receptor. *Brain Res.*, 562: 149–153.

Woolf, N.J., Gould, E. and Butcher, L.L. (1989) Nerve growth factor receptor is associated with cholinergic neurons of the basal forebrain but not the pontomesencephalon. *Neuroscience*, 30: 143–152.

Yan, Q. and Johnson, Jr., E.M. (1989) Immunohistochemical localization and biochemical characterization of nerve growth factor receptor in adult rat brain. *J. Comp. Neurol.*, 290: 585–598.

Zilles, K. (1985) *The Cortex of the Rat. A Stereotaxic Atlas.* Springer, Berlin.

Zilles, K., Wree, A. and Dausch, N.D. (1990) Anatomy of the neocortex: neurochemical organization. In: B. Kolb and R.C. Tees (Eds.), *The Cerebral Cortex of the Rat*, MIT Press, Cambridge, MA, pp. 113–150.

J. Klein and K. Löffelholz (Eds.)
*Progress in Brain Research*, Vol. 109
© 1996 Elsevier Science B.V. All rights reserved.

CHAPTER 26

# Cholinergic drug resistance and impaired spatial learning in transgenic mice overexpressing human brain acetylcholinesterase

Christian Andres[1,*], Rachel Beeri[1], Tamir Huberman[1], Moshe Shani[2] and Hermona Soreq[1]

[1]*Department of Biological Chemistry, The Hebrew University of Jerusalem, 91904 Israel and* [2]*Department of Genetic Engineering, ARO, The Vulcani Center, Bet Dagan, 50250 Israel*

## Introduction

Appropriate functioning of the cholinergic synapse requires a precise balance of its elements. Important contributors towards this balance are the acetylcholine (ACh) synthesizing enzyme choline acetyltransferase, the hydrolyzing enzyme acetylcholinesterase (AChE) and ACh receptors (Taylor, 1990). Balanced cholinergic neurotransmission is disrupted in several pathological situations. For example, organophosphorus (OP) AChE inhibitors like agricultural insecticides or chemical warfare agents lead to increased ACh levels which cause hypothermia, tremor and muscle paralysis (Soreq and Zakut, 1993; Schwarz et al., 1995). Inversely, the level of muscle nicotinic ACh receptor decreases in myasthenia gravis, due to autoimmune antibodies blocking these receptors (Drachman, 1987). Moreover, death of cholinergic neurons and subsequent cholinergic imbalance appears and worsens progressively in the most common cause of human dementia, Alzheimer's disease, as well as in humans with Down's syndrome (Katzman, 1986; Coyle et al., 1988). This explains the efforts invested in creating animal models with cholinergic neurotransmission deficits: surgical lesions (Cuello et al., 1990), chemical lesions (Mantione et al., 1981; Lev-Lehman et al., 1995), selection of

animal strains (Bentivoglio et al., 1994), genetic manipulation of chromosome 16, the equivalent of human trisomy 21 in mouse (Holtzman et al., 1992) or knock-out of specific acetylcholine receptor subtypes (Picciotto et al., 1995). While all of these models display transient or continuous defects in cholinergic neurotransmission, none combines cholinergic imbalance with progressive cognitive impairments, which are the hallmarks of Alzheimer's disease.

## The experimental approach

To examine the in vivo consequences of synaptic cholinergic imbalance we chose to overexpress AChE in transgenic animals. First, we expressed human AChE in transiently transgenic embryos of the South-African frog *Xenopus laevis* (Ben Aziz-Aloya et al., 1993). The recombinant human enzyme accumulated in frog neuromuscular junctions, increased their post-synaptic length and deepened their synaptic cleft (Shapira et al., 1994; Seidman et al., 1995). The structural alterations observed in frog neuromuscular junctions resembled those reported for cholinergic brain synapses in patients at the early stages of Alzheimer's disease (DeKosky et al., 1990). This, in turn, raised the possibility that the latter changes as well could perhaps be caused by imbalanced cholinergic neurotransmission. To examine the consequences of congenital cholinergic imbalance onto mammalian brain functioning, we created transgenic mice ex-

---

[1]Present address: INSERM-Unité 316 "Système nerveux du foetus à l'enfant, développement, circulation, métabolism", 3è étage Bât. Vialle, Faculté de médicine, BP 3223, 2bis, Bd Tonnellé, 37032 Tours Cedex, France.

pressing human AChE in brain neurons and examined pharmacological responses and cognitive functions in these animals. Our findings demonstrate that AChE overexpressing mice acquire resistance to cholinergic drugs and undergo selective progressive decline in their capacity for spatial learning and memory. This, in turn, suggests the use of these mice to search for the regulatory genes whose functioning is modified in conjunction with the cognitive deterioration characteristic of diseases associated with cholinergic imbalance.

## Efficient expression and/or high copy numbers of the ACHE transgene may be lethal

Two DNA constructs encoding the brain and muscle form of human AChE (Ben Aziz-Aloya et al., 1993) were employed for microinjecting mouse eggs. One of these included the potent ubiquitous promoter of cytomegalovirus (CMV) (Schmitt et al., 1990) and the other a 596 nucleotide long fragment from the authentic promoter of the human ACHE gene, followed by its first intron and the same coding sequence (HpACHE). Fig. 1 presents these two DNA sequences and the transcription factor binding sites identified in them. Interestingly, all of the transcription factor binding sites in the CMV promoter-enhancer sequence also exist in the human ACHE promoter, although not necessarily in the same copy numbers and positions within each promoter. Moreover, the ACHE promoter includes at least four additional sequence motifs for binding transcription factors that are not recognized by the CMV promoter. These include nervous system, muscle and embryonically active factors (Fig. 1); however, myogenic factors were recently found to be ineffective with the closely homologous promoter of the mouse ACHE gene (Mutero et al., 1995), suggesting that at least part of the transcription factor binding sites in HpACHE are inactive.

Both HpACHE and the CMV promoter were previously found to direct human AChE production in *Xenopus* oocytes and embryos, however, the CMV promoter was ca. 10-fold more efficient in its capacity to direct production of the AChE protein than the HpACHE human promoter (Ben Aziz-

Aloya et al., 1993; Seidman et al., 1994). When injected into fertilized eggs of FVB/N mice as previously described (Shani, 1985), integration of the transgene into the host genome of founder mice and their progeny was only observed in 4 cases out of 110 microinjections (Table 1). One viable founder mouse carried the CMVACHE transgene, whereas microinjection of HpACHE DNA yielded 3 viable pedigrees with 2, 10 and 15 copies of the transgene (Beeri et al., 1995).

Expression of the transgene was examined in mouse tissues by RNA extraction followed by PCR amplification (Beeri et al., 1994). Species-specific PCR primers were designed to distinguish between human and mouse ACHEmRNA transcripts. Transgenic mRNA was only expressed in that pedigree with the lowest amount (2 copies) of HpACHE DNA. The limited number of viable AChE-transgenic mice and the yet lower success in obtaining ACHEmRNA expression suggest that AChE overexpression may be lethal above certain levels. Moreover, these findings suggest that the additional consensus motifs present in the HpACHE DNA have negative roles, suppressing AChE production. However, the mouse pedigree with ACHEmRNA transcription developed and proliferated normally, pertaining HpACHE DNA with unmodified organization and copy number. All subsequent experiments were carried out with apparently homozygous mice from this pedigree.

## Transgenic HpACHE is not expressed outside the CNS

Reverse transcription and PCR amplification (Beeri et al., 1994) revealed that unlike the host ACHE gene, the transgene is not expressed in muscle, adrenal and bone marrow. In contrast, both human and mouse ACHEmRNA transcripts were observed in dissected brain regions of the transgenic mice. There was no interference with the levels and alternative splicing patterns of host ACHEmRNAs, both remained apparently similar to those observed by others (Rachinsky et al., 1990).

When fixed brain sections from the mice were subjected to in situ hybridization with digoxigenin ACHEcRNA, followed by detection with alkaline

## Human ACHE Promoter

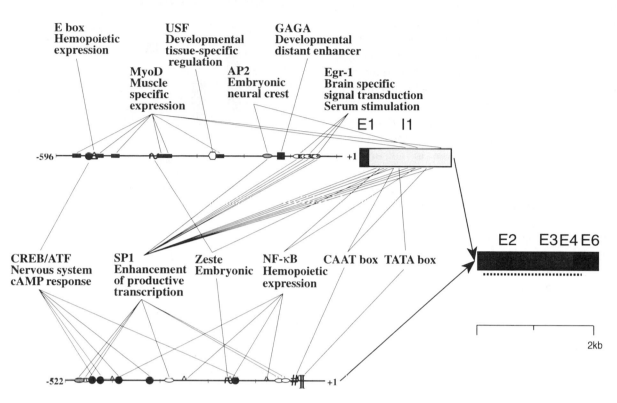

**CMV Enhancer-Promoter**

Fig. 1. Comparison of the human ACHE and the CMV sequences used to direct AChE production in transgenic mice. Sequences of the two promoters are presented with schematic localizations of consensus sequence motifs for binding transcription factors. Sites common to both promoters are presented in the middle, sequences unique to HpACHE are shown on top. For details of each of these motifs, see Ben Aziz-Aloya et al. (1993). The transcribed sequence of the human ACHE gene is presented on the right side and was linked (arrows) to each of these promoters. Exons are noted E1–E6. Note that E1 and intron 1 (I1) were only included in the HpACHE construct with the human ACHE promoter. Open reading frame is noted by a dashed underline. Length of each of the promoter sequences is noted in base pairs. For the transcribed regions of HpACHE, a size bar in kb is included.

phosphatase-conjugated anti-digoxigenin antibody (Boehringer/Mannheim), ACHEcRNA labeling was seen in the same brain neurons in transgenics and controls (Beeri et al., 1995). Particularly intense labeling was observed in cell bodies in the basal forebrain and brainstem nuclei of the transgenic mice and in the cholinoceptive hippocampal neurons, especially in the CA1–CA2 region. Thus, the HpACHE transgene was expressed in the central nervous system neurons but not in the peripheral tissues normally expressing this gene in mammals. This is consistent with findings of others,

who observed that separable promoter elements control neurogenic expression of pan-neural genes (i.e. *Drosophila*, snail) in the central and peripheral nervous system (Ip et al., 1994).

### Multimeric transgenic AChE reaches cholinergic brain synapses

To search for multimeric assembly of the transgenic enzyme, brain region homogenates were subjected to sucrose gradient centrifugation followed by adhesion to immobilized human-selective anti-AChE monoclonal antibodies and measurements of acetylthiocholine hydrolysis levels. AChE from the brain of control and transgenic mice displayed similar sedimentation profiles, demonstrating unmodified assembly into multimeric enzyme forms. Up to 50% of the active enzyme in basal forebrain (Fig. 2) adhered to monoclonal antibodies specific to human AChE. Moreover, catalytic activity measurements of antibody-immobilized AChE from tissue homogenates revealed that the transgenic enzyme was present in higher levels within the basal forebrain, whereas more limited amounts of this protein were detected in cortex, brainstem, cerebellum and spinal cord extracts. There were no age-dependent changes in this pattern. Gel electrophoresis followed by cytochemical staining of enzyme activity (Seidman et al., 1995) revealed similar migration for AChE from the brain of transgenic and control mice, indicating comparable glycosylation patterns (Fig. 3).

Cytochemical staining of AChE activity was observed in $50 \mu$m brain sections from transgenic mice in all of the areas that showed high AChE activity in homogenate assays. Intense staining was detected in the neo-striatum, pallidum and hippocampus. Thus, the extent of excess AChE reflected the brain region distribution characteristic of the primate brain, suggesting that the transgenic mice retained the initial species-specific capacity to regulate human AChE production.

TABLE 1

Effect of promoter selection and transgene copy numbers on the creation of human AChE transgenic mice

|  | Promoters | |
| --- | --- | --- |
|  | CMV | HpACHE |
| Number of microinjections | 70 | 40 |
| Number of pedigrees carrying the transgene | 1 | 3 |
| Number of DNA copies in each pedigree | 3 | 2–10–15 |
| Pedigrees expressing the transgene | 0 | 1–0–0 |

CMV, cytomegalovirus promoter; HpACHE, human AChE promoter. See text for details.

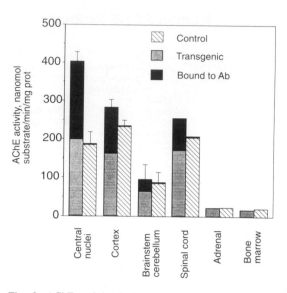

Fig. 2. AChE activity in brain and different organs. AChE activities were measured with acetylthiocholine as substrate (Seidman et al., 1994). AChE of human origin was quantified by binding extracts from the noted tissues and brain regions to specific anti-human AChE monoclonal antibodies.

## Transgenic AChE selectively alters thermoregulatory responses to cholinergic and serotonergic agonists

Among other functions, cholinergic neurotransmission is involved in controlling body temperature in mammals (Simpson et al., 1994). We ascertained that thermoregulation was properly retained in the transgenic mice by checking their cold adaptation, which remained unchanged. We then examined hypothermic responses of these transgenic mice to intraperitoneally-injected hypothermia-inducing cholinergic drugs. Core body temperature was reduced by a limited extent and for shorter duration in the transgenic as compared with control mice. This was first examined with the potent AChE inhibitor diethyl $p$-nitrophenyl phosphate (paraoxon), the toxic metabolite of the agricultural insecticide parathion (Table 2). Most importantly, transgenic mice exposed to 1 mg/kg dose of paraoxon retained apparently normal locomotor activity and behavior, while control mice subjected to this dose presented symptoms characteristic of cholinergic overstimulation.

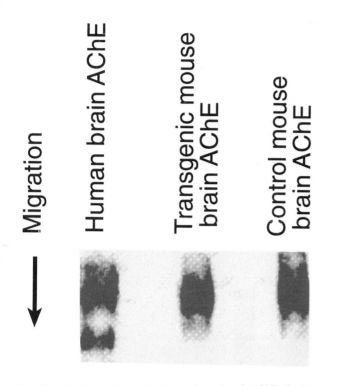

Fig. 3. Non denaturing gel electrophoresis of AChE. Brain homogenates from the noted sources were prepared in the presence of Triton X-100 and were run on a non-denaturing 7% polyacrylamide gel. AChE activity was revealed by an acetylthiocholine precipitating method as described elsewhere (Seidman et al., 1995).

In addition to their improved, yet predictable, capacity for scavenging of the anti-AChE paraoxon, the transgenic mice also displayed resistance to the hypothermic effects of oxotremorine, an effective agonist of muscarinic receptors (Clement, 1991). They were also resistant to the less potent effect of nicotine, and to the serotonergic agonist 8-hydroxy-2-(di-$n$-propylamino) tetralin (8-OH-DPAT), but not to the $\alpha_2$-adrenergic agonist clonidine (Table 2). Serotonergic agonists may act directly on serotonin receptors, involved in thermoregulation (Simpson et al., 1994). However, they may also interact with ACh receptors (Garcia-Colunga and Miledi, 1995). Therefore, the altered drug responses may reflect either transcriptional or post-transcriptional changes in ACh (and perhaps serotonin) receptors within the brain of transgenic mice.

## Transgenic AChE does not affect open-field behavior but induces progressive decline in spatial learning

When compared to matched groups of non-transgenic control mice at the age of 1, 2–3 and 5–7 months, AChE-transgenic mice retained normal behavior in an open field. They covered the same space and distance as their control counterparts

TABLE 2

Effects of different hypothermia-inducing drugs administered intra-peritoneally to control (C) and transgenic (T) 6-month-old mice

| Drugs | Dose (mg/kg) | Minimum temperature in °C in °C × min | | | Area under baseline | | | Number of animals | |
|---|---|---|---|---|---|---|---|---|---|
| | | C | T | Significance | C | T | Significance | C | T |
| Paraoxon | 1 | 30.4 | 32.4 | – | 2868 | 1505 | – | 1 | 1 |
| Oxotremorine | 0.15 | 29.2 ± 0.9 | 32.3 ± 0.3 | 0.02 | 909 ± 211 | 392 ± 91 | 0.04 | 3 | 3 |
| Nicotine | 10 | 33.8 ± 0.6 | 35.5 ± 0.2 | 0.006 | 160 ± 57 | 93 ± 14 | ns | 4 | 4 |
| 8OH-DPAT | 1 | 34.8 ± 0.5 | 35.5 ± 0.3 | ns | 273 ± 97 | 78 ± 25 | 0.02 | 4 | 3 |
| Clonidine | 0.5 | 31.4 ± 1.2 | 31.6 ± 0.8 | ns | 1338 ± 207 | 1696 ± 296 | ns | 3 | 3 |

The baseline for the area above the curve of experimental points is a horizontal line drawn through the temperature of the animals before the injection. Results are expressed in means ± standard deviations, except for the high dose paraoxon experiment, which was performed only once. Statistical significance was tested with Student's $t$-test.

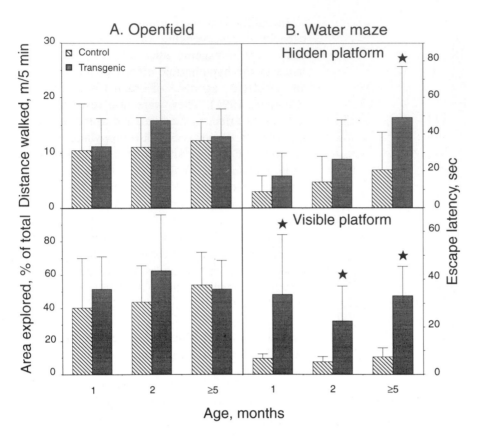

Fig. 4. Behavioural tests. (A) Open field; mice spent 5 min in a void 60 × 60 cm box and their walkpath was traced. The walked distance (top) and the explored fraction of the box floor (bottom) were calculated thereafter. (B) Water maze; mice were tested in a water maze with a hidden platform in a fixed location (top) or with a visible platform in alternate location for each test (bottom). They were introduced into alternating corners of the maze in 4 daily sessions during 4 consecutive days. Means of daily escape latencies are presented for the 4th day of transgenic ($n = 6$ to 10) and age-matched controls. Stars note statistically significant different latencies (ANOVA followed by Neuman–Keuls test, $P < 0.05$).

(Fig. 4). In addition, these mice did not display more anxiety than controls, as evaluated in the frequency of defecation incidents and grooming behavior.

In contrast, memory and learning tests revealed clear differences between transgenic and control mice. We first used the hidden platform version of the Morris water maze (Morris et al., 1981), in which mice are expected to use their spatial orientation to find a platform submerged under opaque water and escape a swimming task. The transgenics' performance in this test was apparently normal at the age of 4 weeks, when they needed a similar time (defined as the escape latency) to reach the platform as age-matched control mice. At the age

of 2–3 months, transgenics already needed more time at the 4th day of training compared to controls. Finally, at the age of 6 months, the escape latency of the transgenic mice was significantly longer than that of controls, showing that they failed to find the platform even after 16 training sessions (Fig. 4). That this deterioration pattern was not caused by locomotion deficiencies was clear from the normal performance of these mice in the open field test (Fig. 4). Moreover, the progressive decline of these transgenic mice in spatial learning and memory was more pronounced than the defects observed in mice with knocked-out glutamate receptor 1 (Conquet et al., 1994), NMDA receptor $\varepsilon 1$ (Sakimura et al., 1995), fyn

(Grant et al., 1992), calcium calmodulin kinase II (Silva et al., 1992) or CRE-binding protein (Bourtchuladze et al., 1994). This, in turn, suggests that a more substantial, yet delayed, and progressive perturbation in learning and memory occurred in our mice than in any of these knock-out strains. Interestingly, the defects observed in our mice resembled those reported in mice treated with blockers of central muscarinic or nicotinic acetylcholine receptors (i.e. atropine, scopolamine or mecamylamine) (McNamara and Skelton, 1993).

An earlier, persistent defect was observed in the visible platform version of the Morris water maze. In this test, mice are trained to escape the swimming task by using short-distance cues. They can then climb on a visible platform decorated by a flag and a paper cone. The transgenics' performance in this test was poor from the age of 4 weeks onward (Fig. 4). This could perhaps be due to early-onset difficulties in short-distance vision or reflect abnormal avoidance behavior. That the defect in this visual memory test occurred in these transgenic mice when they still succeeded in the hidden platform test, demonstrates a dissociation between the visual and the hidden platform test performances.

## Implications for future research

The transgenic mice expressing human AChE can serve as a novel appropriate model system for several lines of research. First and foremost, they can be used as a model to unravel the exact contribution of cholinergic neurotransmission toward learning and memory processes, using electrophysiological tools and behavioral tests. In addition, these mice should be of major assistance for identifying the molecular regulatory mechanisms involved in maintenance of normal cholinergic neurotransmission. These two lines of investigation can therefore lead to development of new physiological concepts involved in control over body temperature as well as in memory impairments in mammals. At the clinical level, such new concepts should be of primary importance toward the development of new strategies for diagnosis and treatment of diseases associated with cholinergic deficits, like Alzheimer's disease.

## Conclusions

Expression of human AChE under control of the authentic human promoter and first intron was observed in central nervous system neurons of transgenic mice. This expression changed responses to hypothermia-inducing drugs acting on cholinergic and probably serotonergic receptors. In addition, it created a progressive spatial learning and memory impairment. In contrast, the open field behavior of these transgenic animals remained normal. These findings suggest that subtle alterations in the cholinergic balance may cause physiologically observable changes and contribute by itself to the memory deterioration in at least part of the patients with cholinergic deficits.

## Acknowledgements

We thank Drs. T. Bartfai (Stockholm), J. Crawley (Washington, DC), A. Ungerer (Strasbourg) and H. Zakut (Tel Aviv) for helpful discussions and Dr. B. Nørgaard-Pedersen (Copenhagen) for antibodies. This work was supported by USARMRDC grant 17-94-C-4031 and the Israel Academy of Sciences and Humanities (to H.S.). C.A. was a recipient of an INSERM, France fellowship, and an INSERM-NCRD exchange fellowship with the Israel Ministry of Science and Arts. The behavioural experiments were performed in the Smith Foundation Institute for Psychobiology at the Life Sciences Institute in the Hebrew University.

## References

Beeri, R., Gnatt, A., Lapidot-Lifson, Y., Ginzberg, D., Shani, M., Soreq, H. and Zakut, H. (1994) Testicular amplification and impaired transmission of human butyrylcholinesterase cDNA in transgenic mice. *Human Reprod.*, 9: 284–292.

Beeri, R., Andres, C., Lev-Lehman, E., Timberg, R., Huberman, T., Shani, M. and Soreq, H. (1995) Transgenic expression of human acetylcholinesterase induces progressive cognitive deterioration in mice. *Curr. Biol.*, 45: 1063–1071.

Ben Aziz-Aloya, R., Seidman, S., Timberg, R., Sternfeld, M., Zakut, H. and Soreq, H. (1993) Expression of a human acetylcholinesterase promoter-reporter construct in developing neuromuscular junctions of *Xenopus* embryos. *Proc. Natl. Acad. Sci. USA*, 90: 2471–2475.

Bentivoglio, A.R., Altavista, M.C., Granata, R. and Albanese,

A. (1994) Genetically determined cholinergic deficiency in the forebrain of C57BL/6 mice. *Brain Res.*, 637: 181–189.

Bourtchuladze, R., Frenguelli, B., Blendy, J., Cioffi, D., Schutz, G. and Silva, A. J. (1994) Deficient long-term memory in mice with a targeted mutation of the cAMP-responsive element-binding protein. *Cell*, 79: 59–68.

Clement, J.G. (1991) Effect of a single dose of an acetylcholinesterase inhibitor on oxotremorine- and nicotine-induced hypothermia in mice. *Pharmacol. Biochem. Behav.*, 39: 929–934.

Conquet, F., Bashir, Z.I., Davies, C.H., Daniel, H., Ferraguti, F., Bordi, F. et al. (1994) Motor deficit and impairment of synaptic plasticity in mice lacking mGluR1. *Nature*, 372: 237–243.

Coyle, J.T., Oster-Granite, M.L., Reeves, R.H. and Gearhart, J.D. (1988) Down syndrome, Alzheimer's disease and the trisomy 16 mouse. *Trends Neurosci.*, 11: 390–394.

Cuello, A.C., Garofalo, L., Maysinger, D., Pioro, E.P. and Da Silva, A.R. (1990) Injury and repair of central cholinergic neurons. *Prog. Brain Res.*, 84: 301–311.

DeKosky, S.T. and Scheff, S.W. (1990) Synapse loss in frontal cortex biopsies in Alzheimer's disease: correlation with cognitive severity. *Ann. Neurol.* 27:457–464.

Drachman, D.B. (1987) Myasthenia gravis: biology and treatment. *Ann. N. Y. Acad. Sci.*, 505.

Garcia-Colunga, J. and Miledi, R. (1995) Effects of serotonergic agents on neuronal nicotinic acetylcholine receptors. *Proc. Natl. Acad. Sci. USA*, 92: 2919–2923.

Grant, S.G.N., O'Dell, T.J., Karl, K.A., Stein, P.L., Soriano, P. and Kandel, E.R. (1992) Impaired long-term potentiation, spatial learning, and hippocampal development in fyn mutant mice. *Science*, 258: 1903–1909.

Holtzman, D.M., Li, Y., DeArmond, S.J., McKinley, M.P., Gage, F.H., Epstein, C.J. and Mobley, W.C. (1992) Mouse model of neurodegeneration: atrophy of basal forebrain cholinergic neurons in trisomy 16 transplants. *Proc Natl Acad Sci USA*, 89: 1383–1387.

Ip, Y.T., Levine, M. and, Bier E. (1994) Neurogenic expression of *snail* is controlled by separable CNS and PNS promoter elements. *Development*, 120: 199–207.

Katzman, R. (1986) Alzheimer's disease. *N. Engl. J. Med.*, 314: 962–973.

Lev-Lehman, E., El-Tamer, A., Yaron, A., Grifman, M., Ginzberg, D., Hanin, I. and Soreq, M. (1994) Cholinotoxic effects on acetylcholinesterase gene expression are associated with brain-region specific alterations in G,C-rich transcripts. *Brain Res.*, 661: 75–82.

Mantione, C.R., Fisher, A. and Hanin, I. (1981) The AF64A-treated mouse: possible model for central cholinergic hypofunction. *Science*, 213: 579–580.

McNamara, R.K. and Skelton, R.W. (1993) The neuropharmacological and neurochemical basis of place learning in the Morris water maze. *Brain Res. Rev.*, 18: 33–49.

Morris, R.G.M., Garrud, P., Rawlins, J.N.P. and O'Keefe, J. (1981) Place navigation impaired in rats with hippocampal lesions. *Nature*, 297: 681–682.

Mutero, A., Camp, S. and Taylor, P. (1995) Promoter elements of the mouse acetylcholinesterase gene. *J. Biol. Chem.*, 270: 1866–1872.

Picciotto, M.R., Zoli, M., Léna, C., Bessis, A., Lallemand, Y., LeNovère, N. et al. (1995) Abnormal avoidance learning in mice lacking functional high-affinity nicotine receptor in the brain. *Nature*, 374: 65–67.

Rachinsky, T.L., Camp, S., Li, Y., Ekström, T.J., Newton, M. and Taylor, P. (1990) Molecular cloning of mouse acetylcholinesterase: tissue distribution of alternatively spliced mRNA species. *Neuron*, 5: 317–327.

Sakimura, K., Kutsuwada, T., Ito, I., Manabe, T., Takayama, C., Kushiya, E., Yagi, T. et al. (1995) Reduced hippocampal LTP and spatial learning in mice lacking NMDA receptor $\varepsilon 1$ subunit. *Nature*, 373: 151–155.

Schmitt, E.V., Christoph, G., Zeller, R. and Leder, P. (1990) The cytomegalovirus enhancer: a pan-active control element in transgenic mice. *Mol. Cell Biol.*, 10: 4406–4411.

Schwarz, M., Glick, D., Loewenstein, Y. and Soreq, H. (1995) Engineering of human cholinesterases explains and predicts diverse consequences of administration of various drugs and poisons. *Pharmacol. Ther.*, 67: 283–322.

Seidman, S., Ben Aziz-Aloya, R., Timberg, R., Loewenstein, Y., Velan, B., Shafferman, A., Liao, J., Nørgaard-Pedersen, B., Brodbeck, U. and Soreq, H. (1994) Overexpressed monomeric human acetylcholinesterase induces subtle ultrastructural modifications in developing neuromuscular junctions of *Xenopus laevis* embryos. *J. Neurochem.*, 62: 1670–1681.

Seidman, S., Sternfeld, M., Ben Aziz-Aloya, R., Timberg, R., Kaufer-Nachum, D. and Soreq, H. (1995) Synaptic and epidermal accumulations of human acetylcholinesterase are encoded by alternative 3'-terminal exons. *Mol. Cell Biol.*, 14: 459–473.

Shani, M. (1985) Tissue-specific expression of rat myosin light-chain 2 gene in transgenic mice. *Nature*, 314: 283–286.

Shapira, M., Seidman, S., Sternfeld, M., Timberg, R., Kaufer, D., Patrick, J. and Soreq, H. (1994) Transgenic engineering of neuromuscular junctions in *Xenopus laevis* embryos transiently overexpressing key cholinergic proteins. *Proc. Natl. Acad. Sci. USA*, 91: 9072–9076.

Silva, A.J., Paylor, R., Wehner, J.M. and Tonegawa, S. (1992) Impaired spatial learning in $\alpha$-calcium-calmodulin kinase II mutant mice. *Science*, 257: 206–211.

Simpson, C.V., Ruwe, W.D. and Myers, R.D. (1994) Prostaglandins and hypothalamic neurotransmitter receptors involved in hypothermia: a critical evaluation. *Neurosci. Behav. Rev.*, 18: 1–20.

Soreq, H. and Zakut, H. (1993) *Human Cholinesterases and Anticholinesterases*. Academic Press, San Diego, CA.

Taylor, P. (1990) Anticholinesterase agents. In: L.S. Goodman, A.G. Gilman, T.W. Rall, A.S. Nies and P. Taylor (Eds.), *Pharmacological Basis of Therapeutics*, Macmillan, New York, 1990, pp. 131–147.

J. Klein and K. Löffelholz (Eds.)
*Progress in Brain Research*, Vol. 109
© 1996 Elsevier Science B.V. All rights reserved.

# CHAPTER 27

# Amyloid β-peptides injection into the cholinergic nuclei: morphological, neurochemical and behavioral effects

G. Pepeu, L. Giovannelli, F. Casamenti, C. Scali and L. Bartolini

*Department of Preclinical and Clinical Pharmacology, University of Florence, Viale Morgagni 65, 50134 Florence, Italy*

## Introduction

Alzheimer's disease (AD) seems to be indissolubly connected with the cholinergic system since the seminal letter to *Lancet* of Davies and Maloney (1976) reporting a decrease in cortical choline acetyltransferase (ChAT) activity in the brain of AD subjects. The connection has been reinforced by the introduction of tacrine as the first therapeutic agent for the treatment of this disease (Farlow et al., 1992).

AD is characterized by a dramatic histopathological picture, whose hallmarks are the senile plaques and the neurofibrillary tangles. Aggregated β-amyloid (Aβ) is the main component of the senile plaque, which consists of a central core of congophylic amyloid fibrils associated with other molecular components, infiltrated by microglial cells and surrounded by an astrocytic reaction and dystrophic neurites (Ulrich et al., 1987). The Aβ protein is a 4 kDa fragment of a much larger transmembrane glycoprotein (APP) partially included in the hydrophobic transmembrane domain (Kang et al., 1987). APP can be processed in at least two alternative proteolytic pathways (Wisniewski et al., 1994), one of which leads to the physiological production and release of soluble Aβ, whose presence has been detected in extracellular fluids of cultured cells during normal metabolism and in the CSF of normal individuals (Haass et al., 1992; Seubert et al., 1992). However, the event leading to the formation of deposits of aggregated Aβ is still unknown. The fact that the presence of an extra copy of the APP gene in patients affected by Down's syndrome results in the development of plaques (Mann, 1989), along with the evidence that at least one of the familiar AD-linked mutated genes can induce overproduction of Aβ when transfected into cultured cells (Cai et al., 1993), has led to the hypothesis that Aβ overproduction may precipitate its aggregation. Recently this view has been further supported by the finding that transgenic mice for either mutated APP or Aβ exhibit histopathological modifications resembling those seen in AD brains (Games et al., 1995; La Ferla et al., 1995). However, the importance of structural alterations of the amyloid protein itself and the role of other plaque core components in promoting the formation of amyloid fibrils cannot be ruled out. In this regard, the importance of apolipoprotein E as an additional component of senile plaques which might be involved in neurodegeneration has been shown (Strittmatter et al., 1993).

The deposition of Aβ might be a crucial event for the initiation of the neuritic and neuronal degeneration which occurs in AD. After the initial demonstration by Yankner et al. (1990) that Aβ is neurotoxic, several reports have shown a correlation between neurotoxic effects and aggregation state of the peptide in vitro (Pike et al., 1990, 1993), and it has been demonstrated that fibril formation is required for maximal neurotoxicity (Lorenzo and Yankner, 1994). In vivo, isolated plaque cores from post-mortem AD brains (Frautschy et al., 1991) as well as synthetic Aβ fragments induce neuronal and neuritic loss and the expression of AD antigens when injected into the rat brain (Emre et al., 1992; Kowall et al., 1992; Rogers et al., 1992). Several mechanisms have been proposed for $A\beta$ toxicity including a loss of $Ca^{2+}$ homeostasis (Mattson et al., 1992), the potentiation of excitatory amino acid neurotoxicity (Koh et al., 1990), the production of reactive peptide free-radicals (Butterfield et al., 1994), the activation of complement (Rogers et al., 1992).

Since the degeneration of forebrain cholinergic neurons is a characteristic neuropathological feature of AD, the aim of the present work was to investigate the toxicity of $\beta$-peptides to the cholinergic neurons by direct injection in the rat forebrain. As it has been reported that the neurotoxic activity of $\beta$-amyloid resides within the 25–35 portion of the peptide (Yankner et al., 1990; Pike et al., 1993), the full-length peptide $\beta$-(1–40) and the $\beta$-(25–35) peptide were studied. In a previous investigation (Abe et al., 1994) we demonstrated that the administration of $\beta$-peptides into the medial septum is followed by a decrease in acetylcholine (ACh) release from the hippocampus. We also studied the histological and functional changes due to the presence in the nucleus basalis (NB) of $\beta$-peptide deposits (Giovannelli et al., 1995). In the present study, we have extended our investigation on the actions of $\beta$-peptides in the NB at longer times post-injection. The effects of $\beta$-peptides on ChAT immunoreactivity (IR), cortical ACh release and behavior were examined.

## Experimental procedures

Male Charles River Wistar rats, 3 months of age at the beginning of the experiment, were used. The rats were individually housed in macrolon cages with food and water ad libitum and maintained on a 12 h light/dark cycle, at 23°C room temperature (RT). All experiments were carried out according to the guidelines of the European Community's Council for Animal Experiments.

$\beta$-Amyloid peptides (25–35) (a gift from Menarini Laboratories, Florence, Italy), (1–40) (Sigma) and the scrambled (25–35) peptide (synthesized by Dr. P. Roveri, CNR, Pisa, Italy) were dissolved in bidistilled water at the concentration of 10 $\mu$g/$\mu$l and the solutions were incubated at 37°C for 1 week before use. A unilateral injection into the right NB was performed under chloral hydrate (400 mg/kg i.p.) anesthesia by means of a stereotaxic apparatus (coordinates: AP = –0.5; L = –2.8 from bregma; H = –7.0 from the dura; Paxinos and Watson, 1982). One microliter of either peptide or saline solution was injected. After surgery, the animals were returned to their home cages, and sacrificed 7, 14, 21, 30, 60 , 120 or 180 days later.

At different times after injection, the rats were tested for object recognition (OR) and passive avoidance conditioned response. OR was carried out according to Ennaceur and Delacour (1988) and Scali et al. (1994). Briefly, the animals were first placed in an arena (70 × 60 × 30 cm) for the time needed to explore for 20 s two identically shaped objects (plastic cubes, pyramids and cylinders) placed in two opposite corners. In a second trial, one of the objects was replaced by a new one and the rats were left in the arena for 5 min. The times spent for exploration of the familiar (F) and new object (N) were recorded and a discrimination index (D) was calculated (N – F/N + F).

For the passive avoidance conditioned response test, a two-compartment step-through apparatus was used (Casamenti et al., 1988). Briefly, in the training trial the rats were placed in the illuminated chamber and the latency to step into the dark one was recorded. Once the rats were in the dark compartment with all 4 paws, a 0.6 mA scrambled shock (5 Hz, 20 ms, for 5 s) was delivered through the grid floor and the rats were returned to their cages. Retesting was performed 24 h later. Again the latency to step into the dark chamber was recorded and the training trial latency subtracted from it.

ACh release from the parietal cortex was measured in vivo by means of the transversal microdialysis technique, as previously reported (Casamenti et al., 1991). After completing the behavioral tests, the rats were anaesthetized with chloral hydrate (400 mg/kg i.p.) and placed in a stereotaxic frame. A microdialysis tube corresponding to the width of the right parietal cortex ipsilateral to the injection was inserted transversally into the brain according to the following coordinates: AP = −0.5, H = −2.5 from bregma. On the following day, the membrane was perfused at a constant flow rate with Ringer solution containing $7\,\mu M$ physostigmine sulfate and the perfusate collected at 20 min intervals for 2 h and directly assayed for ACh and choline. At the end of this period, the normal Ringer was substituted for 20 min with high potassium (100 mM KCl) Ringer and the following sample was collected.

ACh and choline levels of the dialysates were assayed directly by the high-performance liquid chromatographic method with an electrochemical detector as described by Damsma et al. (1987) and Giovannini et al. (1991). Basal ACh release was stable over the entire collection period in all rat groups and was expressed as the mean ± SEM of six collection periods for each rat. K+-stimulated ACh release was expressed as the mean ± SEM of a single 20 min collection period for each rat.

At the end of ACh release experiments, the rats were deeply anaesthetized and perfused through the ascending aorta with 50 ml of saline solution followed by 200–300 ml of ice-cold 4% paraformaldehyde solution in phosphate-buffered saline (PBS). The brains were postfixed in the same fixative, cryoprotected in sucrose and cut in a cryostat from the level of the caudal septum to the level of the caudal caudate-putamen in 20-$\mu$m thick coronal sections. Sections throughout the extension of the area injected with peptides or saline were Nissl-stained in order to check the position of the injection and the general morphology of the area.

For Congo Red staining of the amyloid deposits, the sections were hydrated and the nuclei stained with hematoxylin for 5 min. After a thorough wash in water, they were incubated in a 1% Congo Red solution in water for 1 h at RT. At the end of the incubation the sections were directly passed in a solution of $Li_2CO_3$ (1.5% in water) and incubated at RT for 2 min with gentle shaking. After a further water wash, excess staining was removed by brief immersion in 80% ethanol, the slides were checked under a microscope and then dehydrated and coverslipped.

ChAT IR was visualized by means of a polyclonal rabbit antibody (Chemicon) used at a 1:2000 dilution in PBS. After a blocking step the slides were incubated overnight at RT with the primary antibody in the presence of 2% normal goat serum (NGS) and 0.3% Triton-X-100 (TX). On the following day, the sections were washed in PBS and incubated with the secondary anti-rabbit antibody (Vector) diluted 1:1000 in PBS–2% NGS followed by incubation with the avidin-biotin-peroxidase complex (Vector). The peroxidase reaction was carried out using diaminobenzidine (DAB) as the substrate and $NiCl_2$ as intensifier. At the end the slides were washed, dehydrated and mounted.

For microscopic analysis and photography, a Nikon Labophot-2 microscope was used. Cell counting was performed manually under a 10× objective using a calibrated eyepiece grid. NB ChAT-immunoreactive cells were identified as purple-black bodies with neuronal shape located in the ventral pallidum and the adjoining internal capsule. Only the cells with a well-defined nucleus were included in the counts. Five sections per animal were analyzed, selected for the presence of the peptide deposit or, in the case of the saline and scrambled peptide injection, of the needle track. The number of ChAT-positive cells in the lesioned NB was compared to that in the contralateral unlesioned side of the same section.

Statistical analysis was carried out using the NCSS 5.0 program. For cell counting the paired Student's t-test was used. ANOVA followed by Fisher's post-hoc analysis was applied to the ACh release data. The behavioral studies were analyzed by means of the non-parametric Wilcoxon and Kruskal–Wallis tests.

### Results

Congo-red staining showed that the deposits formed by the $\beta$-(1–40) and $\beta$-(25–35) peptides consisted of fibrillary material as they exhibited the typical birefringency when observed under polarized light (Fig. 1). The deposit formed by the $\beta$-(25–35) peptide was visible under these conditions up to 21 days after the lesion, and was no

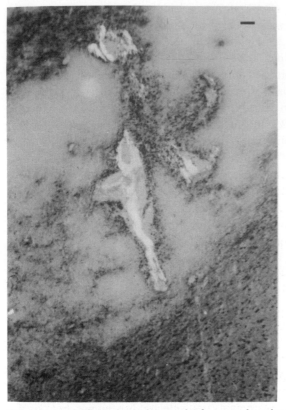

Fig. 1. Polarized light photomicrograph of a coronal section of NB stained with Congo Red. Note the birefringent (white) material present at the $\beta$-(1–40) injection site 7 days after lesion. Scale bar = 50 $\mu$m.

longer present at 1 month while the deposit of $\beta$-(1–40) peptide was seen at the lesion site up to 6 months after lesioning.

The injection of saline and of the scrambled peptide in the NB induced a massive glial reaction which was evident 1 week after the lesion and was reduced but still present at 2 months (Fig. 2, left panel). The injection of $\beta$-peptides (1–40) and (25–35) also induced a strong tissue reaction but, at variance with saline and the scrambled peptide, a lightly stained, ribbon-like deposit was visible at the injection site immediately after surgery. The material inside the lesion appeared infiltrated by glial-like cells, presumably microglia or macrophages. After 4 months, a substantial deposit was still visible at the injection site of the $\beta$-(1–40) peptide, showing a less intense surrounding glial

reaction, mostly localized in close vicinity with the deposit (Fig. 2, middle panel). On the contrary, the deposit formed by the $\beta$-(25–35) peptide was no longer evident after 2 months: at this time the lesion appeared almost entirely filled with reactive cells, while in the surrounding tissue numerous neuronal cell bodies were evident (Fig. 2, right panel).

In the NB ChAT immunoreactivity was localized in intensely labelled neurons located at the border between the internal capsule and the globus pallidum (Fig. 3). In Table 1 the results of ChAT-labelled cell counts are shown. The injection of saline did not affect the number of cholinergic NB neurons 1 week or 2 months after surgery. Instead, the injection of the scrambled (25–35) peptide was followed by a significant reduction (−30%) in the number of ChAT-immunoreactive neurons after 1 week, which was no longer detected 2 months post-lesion. Similarly, a significant decrease in the number of cholinergic neurons was detected 1 (−39%) and 2 (−24%) weeks after the injection of the $\beta$-(25–35) peptide but the number returned to control values 3 weeks post-injection. Examples of analyzed sections are given in Fig. 3. On the contrary, the $\beta$-(1–40) peptide induced a decrease (from −30 to −40%) in the number of cholinergic NB neurons which lasted up to at least 4 months post-lesion.

Table 2 shows the effect of peptide injection on ACh release from the ipsilateral parietal cortex. The saline-injected values of basal release did not significantly differ from those of naive rats (not shown). Instead, a statistically significant decrease in basal ACh release [−29% scrambled, −28% $\beta$-(25–35), −30% $\beta$-(1–40)] was detected 1 week after the injection of the three peptides. However, while the effect of the (25–35) $\beta$-peptide persisted for only 2 weeks (−25%), the decrease in ACh release was still statistically significant 30 days after the $\beta$-(1–40) peptide injection (−38%). At this time, no effect of the scrambled peptide was detected. K⁺-stimulated ACh release was also affected by the different treatments, as shown in Table 2. The $\beta$-(25–35) peptide induced a statistically significant decrease in the stimulated release on day 14 after lesioning (−45%), while a significant

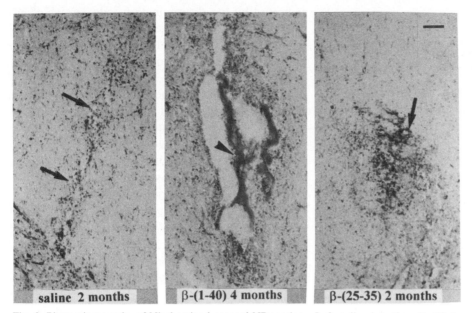

Fig. 2. Photomicrographs of Nissl-stained coronal NB sections. Left: saline injection site 60 days after surgery; middle: deposit of β-(1–40) peptide 4 months after lesion; the ribbon-like deposit is indicated by the arrowhead; right: β-(25–35) peptide injection site 60 days after lesion; note the presence of glial cells infiltrating the scar (thin arrows). Scale bar = 100 μm.

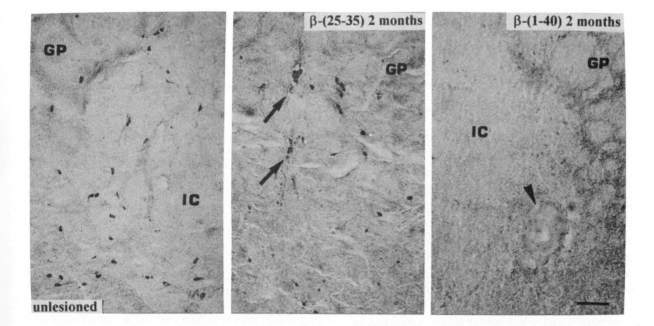

Fig. 3. Photomicrographs of coronal NB sections stained for ChAT IR. Note the presence of numerous immunolabelled neurons at the border between the internal capsule (IC) and the globus pallidus (GP) in the contralateral, unlesioned NB (left). Middle: right NB injected with β-(25–35) peptide 2 months after surgery; right: right NB injected with β-(1–40) peptide 2 months after lesion. The arrowhead indicates the presence of a deposit at the lesion site. Arrows point to the needle track. Scale bar = 100 μm.

TABLE 1

Effect of unilateral nucleus basalis injection of amyloid $\beta$-peptides on ChAT-positive neurons number

| Treatment | Neurons number (%) |
|---|---|
| *Saline* | |
| Day 7–60 | 96 ± 7 |
| *Scrambled* | |
| Day 7 | 70 ± 5** |
| Day 60 | 91 ± 11 |
| *$\beta$-(25–35)* | |
| Day 7 | 61 ± 7** |
| Day 14 | 76 ± 5* |
| Day 21 | 105 ± 5 |
| Day 30 | 86 ± 11 |
| Day 60 | 101 ± 5 |
| *$\beta$-(1–40)* | |
| Day 7 | 61 ± 14* |
| Day 14 | 71 ± 5** |
| Day 21 | 63 ± 11* |
| Day 30 | 69 ± 9* |
| Day 60 | 70 ± 10* |
| Day 120 | 67 ± 9* |

Data are expressed as mean percentage ± SEM of the contralateral unlesioned NB of 3–5 animals (5 sections per each animal). Statistical analysis was performed using the paired Student's $t$ test. *$P < 0.02$; **$P < 0.01$ statistically significant difference from the respective control value.

decrease (−43%) was detected 30 days after the $\beta$-(1–40) peptide injection.

Table 3 shows the effect of peptide injections on object recognition. Naive and saline-injected rats' performances in this test did not significantly differ from each other (not shown). The rats injected with saline and scrambled (25–35) peptide were able to discriminate between a familiar and a new object as demonstrated by the longer time spent in exploring the latter, whereas the rats injected with the $\beta$-(25–35) peptide lost the discrimination ability and spent longer times than controls in exploring the familiar object and an equal time exploring the new object, as shown by their small discrimination index. A tendency towards recovery seemed to begin at day 30 post-injection, as also

indicated by the discrimination indexes, which at this time were no longer significantly different from controls. Conversely, the $\beta$-(1–40)-injected animals showed no OR impairment up to 30 days after the injection. Sixty days post-lesion OR appeared to be impaired since the difference between novel and familiar object was not statistically significant. The impairment of OR was even more pronounced 6 months after injection.

The effects of $\beta$-(25–35) and $\beta$-(1–40) peptides were also tested on passive avoidance conditioned response. The training trial latencies did not vary among the different groups, indicating no difference in exploratory activity and motor behavior induced by treatments. No statistical significant difference in the performance was found among the different groups. However, a tendency towards

TABLE 2

Effect of unilateral nucleus basalis injection of amyloid $\beta$-peptides on acetylcholine release

| Treatment | Basal | K⁺-stimulated |
|---|---|---|
| *Saline* | | |
| Day 7–30 | 3.86 ± 0.26 | 7.03 ± 0.68 |
| *Scrambled* | | |
| Day 7 | 2.74 ± 0.48* | 6.75 ± 1.00 |
| Day 60 | 4.05 ± 0.15 | 7.80 ± 1.12 |
| *$\beta$-(25–35)* | | |
| Day 7 | 2.78 ± 0.29** | 4.99 ± 0.92 |
| Day 14 | 2.89 ± 0.31* | 3.87 ± 0.24** |
| Day 21 | 3.13 ± 0.40 | 6.47 ± 1.03 |
| Day 30 | 3.74 ± 0.31 | 7.38 ± 0.84 |
| Day 60 | 3.28 ± 0.85 | 5.63 ± 0.89 |
| *$\beta$-(1–40)* | | |
| Day 7 | 2.70 ± 0.40* | 5.98 ± 0.42 |
| Day 21 | 3.12 ± 0.48 | 5.70 ± 0.90 |
| Day 30 | 2.39 ± 0.20** | 4.01 ± 0.96* |

Data represent the mean ± SEM basal ACh efflux from the ipsilateral parietal cortex of six collection periods and the mean ± SEM stimulated ACh efflux of a single 20 min collection period for each rat ($n = 4$–5). Statistical analysis was performed using ANOVA ($F_{10-35} = 2.59$; $P < 0.02$ for basal and $F_{10-33} = 2.15$; $P < 0.05$ for stimulated release) followed by Fisher's LSD post-comparison test. *$P < 0.05$; **$P < 0.02$ versus saline-injected rats.

TABLE 3

Effect of unilateral nucleus basalis injection of amyloid $\beta$-peptides on object recognition

| Treatment (no. of animals) | Exploration time (s ± SEM) | | |
|---|---|---|---|
| | F | N | D |
| *Saline* | | | |
| Day 7–30 (11) | 9.7 ± 1.0 | 16.3 ± 1.8** | 0.25 |
| Day 180 (14) | 7.0 ± 1.2 | 14.7 ± 2.4** | 0.35 |
| *Scrambled (25–35)* | | | |
| Day 7 (4) | 9.7 ± 1.2 | 17.8 ± 3.6* | 0.26 |
| *$\beta$-(25–35)* | | | |
| Day 7 (8) | 10.2 ± 0.8 | 10.4 ± 1.6 | 0.08[#] |
| Day 14 (6) | 16.0 ± 2.5 | 17.6 ± 2.4 | 0.09[#] |
| Day 30 (7) | 11.6 ± 1.5 | 15.1 ± 2.1 | 0.16 |
| Day 60 (4) | 12.9 ± 2.1 | 19.4 ± 4.6 | 0.19 |
| *$\beta$-(1–40)* | | | |
| Day 7 (10) | 8.5 ± 1.8 | 13.4 ± 2.4** | 0.26 |
| Day 14 (10) | 6.5 ± 2.1 | 18.1 ± 2.7*** | 0.44 |
| Day 30 (6) | 8.3 ± 1.4 | 12.8 ± 1.3* | 0.24 |
| Day 60 (10) | 8.7 ± 1.1 | 11.6 ± 1.5 | 0.12[#] |
| Day 180 (14) | 4.6 ± 0.8 | 5.1 ± 0.9 | 0.04[#] |

F, exploration time of familiar object; N, exploration time of new object; D, discrimination index (N − F/N + F). Statistically significant differences: ***$P < 0.005$, **$P < 0.01$ and *$P < 0.05$ N versus F values within the same group (Wilcoxon matched pairs non-parametric test); [#]$P < 0.05$ versus saline-injected D value [Kruskal–Wallis non-parametric test ($F_{11} = P < 0.01$) and post-comparison Z value].

a shortening of the retest latency was observed in the first 2 weeks after the injection of the $\beta$-(25–35) peptide, and from 7 up to 180 days following injection of $\beta$(1–40) (data not shown).

## Discussion

In this study we have shown that injections into the NB of equal amounts of amyloid peptides $\beta$-(1–40) and $\beta$-(25–35), but not of the scrambled (25–35) peptide, produce birefringent congophylic deposits surrounded and infiltrated by glial cells as seen in AD senile plaques. Birefringent peptide deposits have also been obtained by Rush et al. (1992) after intrahippocampal injection of the $\beta$-(1–40) peptide.

We also found that all three peptides tested impair the function of NB cholinergic neurons with different time-courses. These differences might depend on the different duration of the deposits formed by the peptides. The scrambled peptide formed no deposit but only induced a non-specific tissue reaction. The $\beta$-(1–40) deposit lasted for up to 6 months with little reduction in its extension. A$\beta$ immunoreactivity in the hippocampus has been detected up to 16 months after $\beta$-(1–42) peptide injection (Winkler et al., 1994). On the other hand, the deposit formed by the $\beta$-(25–35) peptide began to disappear after 21 days, leaving only a glia-enriched scar 1 month after the injection. Burdick et al. (1992) showed that small differences in the hydrophobic domain sequence can influence peptide solubility, and the $\beta$-(25–35) peptide lacks a hydrophobic C-terminal sequence of 5 amino acids as compared to $\beta$-(1–40). Thus, this difference may be the cause of less stable aggregation and shorter deposit duration. A different duration of the deposits formed by the two $\beta$-peptides has been previously demonstrated in the septum (Abe et al., 1994) using acetonitrile/trifluoroacetic acid-solubilized peptides. In the present study, the peptides were dissolved in water and used after 1 week incubation at 37°C. This condition facilitates aggregation (Burdick et al., 1992; Pike et al., 1993), and may increase deposit stability, as indicated by the longer deposit durations detected in this work as compared to our previous results (Abe et al., 1994).

The duration of the deposit and the effects on the cholinergic system seem to be correlated, as the $\beta$-(1–40) peptide induced a long-lasting decrease in cortical ACh release and number of ChAT-immunopositive neurons, while the effect of $\beta$-(25–35) on NB cholinergic neurons faded after 2 weeks. With the scrambled peptide, which did not form a deposit, a reduction in both ACh release and number of ChAT-positive neurons were only detected early after injection. Giordano et al. (1994) also showed that, when injected in the cortex of young rats, the peptide $\beta$-(1–40) was more toxic than $\beta$-(40–1), the first forming fibrillary structures, the second amorphous particles, and Emre et al. (1992) demonstrated that the lesions induced by intracortical injections of $\beta$-(1–40) were larger than those induced by $\beta$-(40–1) peptides. These and our findings suggest that the damage brought about by the peptides results from both non-specific and specific neurotoxicity, the first presumably due to tissue displacement and extracellular environment disorganization. The cholinergic damage caused by the scrambled (25–35) and $\beta$-(25–35) peptides is reversible since no difference in ChAT-IR or ACh release between the saline- and peptide-injected animals is detectable 1–2 months after injection. Apparently, the NB cholinergic neurons lose their ChAT-IR as an immediate consequence of the injection, but regain it later when they recover from the mechanical and toxic insult. This also indicates that the number of cholinergic neurons which are physically destroyed by the injection is actually small. Conversely, the loss of ChAT IR and the hypofunction of cholinergic neurons after the injection of the $\beta$-(1–40) peptide last for as long as the deposit. Further studies at longer times are needed in order to ascertain whether this impairment is still reversible.

The present data show that the decrease in ChAT IR was closely paralleled by a decrease in basal ACh release. However, $K^+$-stimulated release appeared to be less affected than basal by peptide treatment and did not exhibit an equally strong correlation with the loss in ChAT IR. This might indicate that the cholinergic hypofunction induced by peptide injections can be overcome when the system is strongly stimulated, as upon high $K^+$ exposure.

Damage to the cortical acetylcholine-esterase-rich fibers, along with a loss of noradrenergic fibers, have been observed after intracortical injection of $\beta$-(1–40) peptide by Emre et al. (1992), indicating that the cholinergic system is vulnerable to the toxicity of A$\beta$ together with other neuronal systems. The mechanism by which the cholinergic damage is induced by the $\beta$-(1–40) peptide might be the same as that of the $\beta$-(25–35) and scrambled peptides, only longer-lasting, concomitantly with the longer deposit duration. This would argue in favor of a mechanical disruption induced by the amyloid deposits, which could interfere with the

function of the surrounding cells by simply creating a physical barrier to cell-to-cell contacts. However, it is also possible that the $\beta$-peptides exert specific neurotoxic actions, through the mechanisms mentioned in the Introduction. This neurotoxicity might be reversible at first and only later result in neuronal death. The $\beta$-peptides might also act indirectly by means of factors produced by the glial cells at the deposit site (Araujo and Cotman, 1992; Canning et al., 1993) or by activating the complement pathway (Rogers et al., 1992). In our experiments the injection sites of all three peptides, and even of saline, were surrounded by a glial reaction of similar extent, as also reported by Winkler et al. (1994). This finding suggests that the $\beta$-peptides effects observed by us do not depend on the intensity of the inflammatory reaction, but does not rule out the possibility that glial factors produced at the deposit site can mediate the $\beta$-peptides action.

An amnestic effect of synthetic peptides homologous to A$\beta$ has been reported in mice (Flood et al., 1991). In the present work both $\beta$-(25–35) and $\beta$-(1–40) peptides did not significantly affect the working memory assessed by passive avoidance conditioned responses. However, they disrupted the working memory evaluated in the object recognition test. A disruption of both behavioural tests is associated to central cholinergic hypofunction induced by aging (Scali et al., 1994), scopolamine (Ennaceur and Meliani, 1992) and excitotoxic lesions of the NB (Casamenti et al., 1988; Pepeu et al., 1994). In the present experiments the decrease in ChAT-IR and ACh release following $\beta$-peptides injections into the NB is associated with a disruption of object discrimination, although no strict temporal relationship can be drawn between the cholinergic impairment and the behavioral deficit. However, the same tendency towards recovery after the injections of $\beta$-(25–35) peptide is present in behavioral, morphological and functional studies. Moreover, a decrease in ChAT-IR and an impairment in object recognition were concomitantly detected at longer times after the injection of $\beta$-(1–40). Furthermore, the possibility that object recognition deficits may be due at least in part to damage in some other neuronal system cannot be ruled out. The lack of significant effects of β-peptides on passive avoidance response might indicate that the cholinergic damage brought about by NB injection is too limited to effectively impair the cognitive processes involved in the performance of this task.

In conclusion, local injections of amyloid peptides in the rat NB result in local damage and cholinergic hypofunction. However, only the β-(1–40) and β-(25–35) peptides but not the scrambled (25–35) form congophylic fibrillary deposits and induce some behavioral impairment, with the most persistent decrease in the number of cholinergic neurons and ACh release brought about by β-(1–40). Since the deposit formed by the β-(1–40) peptide lasts largely unaltered for a long time, the injection of this peptide may be a useful model for investigating the temporal progression and neurotoxic effects of an amyloid plaque.

## Acknowledgements

We thank Professor Daniele Bani of the Department of Anatomy and Histology of the University of Florence for helpful discussion and Mr. Carlo Lodovico Susini for technical help. This work was supported by grant no. 93 00436 PF 40 from CNR, Target Project on Aging.

## References

Abe, E., Casamenti, F., Giovannelli, L., Scali, C. and Pepeu, G. (1994) Administration of amyloid β-peptides into the medial septum of rats decreases acetylcholine release from hippocampus in vivo. *Brain Res.*, 636: 162–164.

Araujo, D.M. and Cotman, C.W. (1992) Beta-amyloid stimulates glial cells in vitro to produce growth factors that accumulate in senile plaques in Alzheimer's disease. *Brain Res.*, 569: 141–145.

Burdick, D., Soreghan, B., Kwon, M., Kosmoski, J., Knauer, M., Henschen, A., Yates, J., Cotman, C. and Glabe, C. (1992) Assembly and aggregation properties of synthetic Alzheimer's A4/β amyloid peptide analogs. *J. Biol. Chem.*, 267: 546–554.

Butterfield, D.A., Hensley, K., Harris, M., Mattson, M. and Carney, J. (1994) β-amyloid peptide free radical fragments initiate synaptosomal lipoperoxidation in a sequence-specific fashion: implications to Alzheimer's disease. *Biochem. Biophys. Res. Commun.*, 200: 710–715.

Cai, X., Golde, T.E. and Younkin, S.G. (1993) release of ex-

cess amyloid β protein from a mutant amyloid β protein precursor. *Science*, 259: 514–516.

Canning, D.R., McKeon, R.J., DeWitt, D.A., Perry, G., Wujek, J.R., Frederickson, R.C.A. and Silver, J. (1993) β-amyloid of Alzheimer's disease induces reactive gliosis that inhibits axonal growth. *Exp. Neurol.*, 124: 289–298.

Casamenti, F., Di Patre, P.L., Bartolini, L. and Pepeu, G. (1988) Unilateral and bilateral nucleus basalis lesions: differences in neurochemical and behavioural recovery. *Neuroscience*, 24: 209–215.

Casamenti, F., Scali, C. and Pepeu, G. (1991) Phosphatidylserine reverses the age-dependent decrease in cortical acetylcholine release: a microdialysis study. *Eur. J. Pharmacol.*, 194: 11–16.

Damsma, G., Lammerts Van Bueren, D., Westerink, B.H.C. and Horn, A.S. (1987) Determination of acetylcholine in the femtomole range by means of HPLC, a post-column enzyme reactor, and electrochemical detection. *Chromatographia*, 24: 827–831.

Davies, P. and Maloney, A.J.R. (1976) Selective loss of cholinergic neurons in Alzheimer's disease. *Lancet*, ii: 1403.

Emre, M., Geula, C., Ransil, B.J. and Mesulam, M.M. (1992) The acute neurotoxicity and effects upon cholinergic axons of intracerebrally injected β-amyloid in the rat brain. *Neurobiol. Aging*, 13: 553–559.

Ennaceur, A. and Delacour, J. (1988) A new one trial test for neurobiological studies on memory in rats 1: behavioral data. *Behav. Brain Res.*, 31: 47–59.

Ennaceur, A. and Meliani, K. (1992) Effects of physostigmine and scopolamine on rats performances in object-recognition and radial-maze tests. *Psychopharmacology*, 109: 321–330.

Farlow, M., Gracon, S.I., Hershey, L.A., Lewis, K.W., Sadowski, C.H. and Dolan-Ureno, J. (1992) A controlled trial of tacrine in Alzheimer's disease. *J. Am. Med. Assoc.*, 268: 2523–2529.

Flood, J.F., Morley, J.E. and Roberts, E. (1991) Amnestic effects in mice of four synthetic peptides homologous to amyloid β protein from patients with Alzheimer's disease. *Proc. Natl. Acad. Sci. USA*, 88: 3363–3366.

Frautschy, S.A., Baird, A. and Cole, G.M. (1991) Effects of injected Alzheimer β-amyloid cores in rat brain. *Proc. Natl. Acad. Sci. USA*, 88: 8362–8366.

Games, D., Adams, D., Alessandrini, R., Barbour, R., Berthelette, P., Blackwell, C. et al. (1995) Alzheimer-type neuropathology in transgenic mice overexpressing V717F β-amyloid precursor protein. *Nature*, 373: 523–527.

Giordano, T., Bao Pan, J., Monteggia, L.M., Holzman, T.F., Snyder, S.W., Krafft, G., Ghanbari, H. and Kowall, N. (1994) Similarities between β amyloid peptides 1–40 and 40–1: effects on aggregation, toxicity in vitro, and injection in young and aged rats. *Exp. Neurol.*, 125: 175–182.

Giovannelli, L., Casamenti, F., Scali, C., Bartolini, L. and Pepeu, G. (1995) Differential effects of amyloid peptides β-(1–40) and β-(25–35) injections into the rat nucleus basalis. *Neuroscience*, 66: 781–792.

Giovannini, M.G., Casamenti, F., Nistri, A., Paoli, F. and

Pepeu, G. (1991) Effect of thyrotropin releasing hormone (TRH) on acetylcholine release from different brain areas investigated by microdialysis. *Br. J. Pharmacol.,* 102: 363–368.

Haass, C., Schlossmacher, M.G., Hung, A.Y., Vigo-Pelfrey, C., Mellon, A., Ostaszewski, B.L., Lieberburg, I., Koo, E.H., Schenk, D., Teplow, D.B. and Selkoe, D.J. (1992) Amyloid $\beta$-peptide is produced by cultured cells during normal metabolism. *Nature,* 359: 322–325.

Kang, J., Lemaire, H.G., Unterbeck, A., Salbaum, J.M., Masters, C.L., Grzesckik, K.H., Multhaup, G., Beyreuther, K. and Muller-Hill, B. (1987) The precursor of Alzheimer's disease amyloid A4 protein resembles a cell-surface receptor. *Nature,* 325: 733–736.

Koh, J., Yang, L.L. and Cotman, C.W. (1990) $\beta$-amyloid protein increases the vulnerability of cultured cortical neurons to excitotoxic damage. *Brain Res.,* 533: 315–320.

Kowall, N.W., McKee, A.C., Yankner, B.A. and Beal, M.F. (1992) In vivo neurotoxicity of beta-amyloid $\beta(1$–$40)$ and the $\beta(25$–$35)$ fragment. *Neurobiol. Aging,* 13: 537–542.

La Ferla, F.M., Tinkle, B.T., Bieberich, C.J., Haudenschild, C.C. and Jay, G. (1995) The Alzheimer's A$\beta$ peptide induces neurodegeneration and apoptotic cell death in transgenic mice. *Nature Genet.,* 9: 21–30.

Lorenzo, A. and Yankner, B.A. (1994) $\beta$-amyloid neurotoxicity requires fibril formation and is inhibited by Congo Red. *Proc. Natl. Acad. Sci. USA,* 91: 12243–12247.

Mann, D.M.A. (1989) Cerebral amyloidosis, aging and Alzheimer's disease: a contribution from studies on Down's syndrome. *Neurobiol. Aging,* 10: 397–399.

Mattson, M.P., Cheng, B., Davis, D., Bryant, K., Lieberburg, I. and Rydel, R.E. (1992) $\beta$-amyloid peptides destabilize calcium homeostasis and render human cortical neurons vulnerable to excitotoxicity. *J. Neurosci.,* 12: 376–389.

Paxinos, G. and Watson, G. (1982) *The Rat Brain in Stereotaxic Coordinates.* Academic Press, New York.

Pepeu, G., Giovannini, M.G. and Bartolini, L. (1994) Nootropic drugs: effects on brain cholinergic mechanisms and object recognition in the rat. *Abstr. XIX C.I.N.P. Congress,* Washington, DC, S-65-287.

Pike, C.J., Walenecwicz, A.J., Glabe, C.G. and Cotman, C.W. (1990) In vitro aging of $\beta$-amyloid protein causes peptide aggregation and neurotoxicity. *Brain Res.,* 563: 311–314.

Pike, C.J., Burdick, D., Walencewicz, A.J., Glabe, C.G. and Cotman, C.W. (1993) Neurodegeneration induced by $\beta$-amyloid peptides in vitro: the role of peptide assembly state. *J. Neurosci.,* 13: 1676–1687.

Rogers, J., Cooper, N.R., Webster, S., Schultz, J., McGeer, P.L., Styren, S.D., Civin, W.H., Brachova, L., Bradt, B., Ward, P. and Lieberburg, I. (1992) Complement activation by $\beta$-amyloid in Alzheimer's disease. *Proc. Natl. Acad. Sci. USA,* 89: 10016–10020.

Rush, D.K., Aschmies, S. and Merriman, M. (1992) Intracerebral $\beta$-amyloid (25–35) produces tissue damage: is it neurotoxic? *Neurobiol. Aging,* 13: 591–594.

Scali, C., Casamenti, F., Pazzagli, M., Bartolini, L. and Pepeu, G. (1994) Nerve growth factor increases extracellular acetylcholine levels in the parietal cortex and hippocampus of aged rats and restores object recognition. *Neurosci. Lett.,* 170: 117–120.

Seubert, P., Vigo-Pelfrey, C., Esch, F., Lee, M., Dovey, H., Davis, D., Sinha, S., Schlossmacher, M., Whaley, J., Swindlehurst, C., McCormack, R., Wolfert, R., Selkoe, D.J., Lieberburg, I. and Schenk, D. (1992) Isolation and quantification of soluble Alzheimer's $\beta$-peptide from biological fluids. *Nature,* 359: 325–327.

Strittmatter, W.J., Saunders, A.M., Schmechel, D., Pericak-Vance, M., Enghild, J., Salvesen, G.S. and Roses, A.D. (1993) Apolipoprotein E: high avidity binding to $\beta$-amyloid and increased frequency of type 4 allele in late-onset familial Alzheimer disease. *Proc. Natl. Acad. Sci. USA,* 90: 1977–1981.

Ulrich, J., Anderton, B.H. and Probst, A. (1987) Alzheimer's dementia: a study of the senile plaque with antisera and a monoclonal antibody specific for neurofilament proteins. *Acta Histol.,* 34: 115- 121.

Winkler, J., Connor, D.J., Frautschy, S.A., Behl, C., Waite, J.J., Cole, G. and Thal, L. J. (1994) Lack of long-term effects after $\beta$-amyloid protein injections in rat brain. *Neurobiol. Aging,* 15: 601–607.

Wisniewski, T., Ghiso, J. and Frangione, B. (1994) Alzheimer's disease and soluble A$\beta$. *Neurobiol. Aging,* 15: 143–152.

Yankner, B.A., Duffy, L.K. and Kirschner, D.A. (1990) Neurotrophic and neurotoxic effects of amyloid $\beta$-protein: reversal by tachykinin neuropeptides. *Science,* 250: 279–282.

# Section IX

# Cholinergic Dysfunction: 2. Mechanisms in Dementia

J. Klein and K. Löffelholz (Eds.)
*Progress in Brain Research*, Vol. 109

# The systems-level organization of cholinergic innervation in the human cerebral cortex and its alterations in Alzheimer's disease

M.-Marsel Mesulam

*The Cognitive Neurology and Alzheimer's Disease Centre, Departments of Neurology and Psychiatry, Northwestern University Medical School, Chicago, IL 60611, USA*

## Introduction

In the 1920s, Otto Loewi identified acetylcholine (ACh) as the cardioactive substance released by the vagus nerve (Loewi, 1921). It took more than 50 years of additional work to establish that ACh was also a neurotransmitter of the central nervous system. As recently as the mid-1970s, for example, the cholinergic innervation of the cerebral neocortex was poorly understood and even questioned (Silver, 1974). The past 20 years have witnessed dramatic changes. The cholinergic innervation of the cerebral cortex has now become one of the most dynamic areas of research and offers exciting opportunities for exploring the chemical neuroanatomy of cognition, age-related memory impairments, and Alzheimer's disease.

The cholinergic pathway from the basal forebrain is the most massive of all extra-thalamic corticofugal projections of the cerebral cortex. Developments based on electron microscopy, immunocytochemistry, antibodies to recombinant receptor subtypes, in situ hybridization, single unit recordings and selective immunotoxic lesioning have introduced a wealth of new information on the organization and function of this pathway in various animal species, including humans.

## The nucleus basalis of Meynert

The basal forebrain contains four major cholinergic cell groups that project to other telencephalic structures. These cholinergic cell groups do not respect traditional nuclear boundaries and their constituent cells are intermixed with other noncholinergic neurons. We have therefore introduced the Ch1–Ch4 nomenclature in order to designate the cholinergic (i.e. choline acetyltransferase-containing) neurons within these 4 cell groups (Mesulam et al., 1983a,b; Mesulam and Geula, 1988a). According to this nomenclature, Ch1 designates the cholinergic cells associated predominantly with the medial septal nucleus, Ch2 those associated with the vertical nucleus of the diagonal band, Ch3 those associated with the horizontal limb of the diagonal band nucleus, and Ch4 those associated with the nucleus basalis of Meynert. Tracer experiments in a number of animal species have shown that Ch1 and Ch2 provide the major cholinergic innervation for the hippocampal complex, Ch3 for the olfactory bulb, and Ch4 for the rest of the cerebral cortex and the amygdala (Mesulam et al., 1983b).

In the primate brain, the Ch4 group contains a compact component in the nucleus basalis of Meynert and interstitial elements embedded within the internal capsule, medullary laminae of the globus pallidus, ansa peduncularis and ansa lenticularis (Mesulam et al., 1983a; Satoh and Fibiger, 1985; Everitt et al., 1988; Mesulam and Geula, 1988a). Even within the nucleus basalis, cholinergic Ch4 neurons are intermingled with a heterogeneous population of non-cholinergic neurons (Geula et al., 1993a). The terms "Ch4" and "nucleus basalis" are therefore not synonymous. The term "nucleus basalis" can be used to designate all of the components in this nucleus whereas the more restrictive Ch4 designation is reserved for the corresponding contingent of compact and interstitial cholinergic neurons as revealed by ChAT immunohistochemistry.

The human nucleus basalis extends from the level of the olfactory tubercle to that of the posterior amygdala, spanning a distance of 13–14 mm in the antero-posterior axis and attaining a mediolateral width of 18 mm within the substantia innominata (subcommissural gray). Arendt et al. (Arendt et al., 1985) have estimated that the human nucleus basalis contains 200 000 neurons in each hemisphere. On topographical grounds, the human nucleus basalis (and the compact group of Ch4 neurons that it contains) can be subdivided into sectors that occupy its anteromedial (nb-Ch4am), anterolateral (nb-Ch4al), anterointermediate (nb-Ch4ai), intermediate (nb-Ch4i), and posterior (nb-Ch4p) regions (Mesulam and Geula, 1988a).

There are no strict boundaries between the nucleus basalis and adjacent cell groups such as those of the olfactory tubercle, preoptic area, hypothalamic nuclei, nuclei of the diagonal band, amygdaloid nuclei and globus pallidus. In addition to this "open" nuclear structure, the neurons of the nucleus basalis display physiological and morphological heterogeneity. The majority of the Ch4 neurons have an isodendritic morphology with overlapping dendritic fields, many of which extend into fiber tracts traversing the basal forebrain. These characteristics are also present in the nuclei of the brainstem reticular formation and have led to the suggestion that the nb-Ch4 complex could be conceptualized as a telencephalic extension of the brainstem reticular core (Ramon-Moliner and Nauta, 1966). The perikarya of Ch4 neurons display large amounts of cytoplasm with abundant organelles and well developed stacks of short parallel cisternae of RER, indented nuclei, and prominent nucleoli (Walker et al., 1983; Ingham et al., 1985; Palacios, 1990). Synaptic input to the perikaryon and proximal dendrites is sparse but increases distally. The synaptic specializations are mostly asymmetrical in the dendrites and symmetrical in the soma.

In the human brain, in addition to ChAT, Ch4 neurons also express acetylcholinesterase (AChE), calbindin-d28k, the high affinity nerve growth factor receptor trkA, and the low affinity p75 nerve growth factor receptor (NGFr) (Geula et al., 1993a; Kordower et al., 1994). A small minority of Ch4 neurons are NGFr-negative and project preferentially to the amygdala (Mufson et al., 1989; Henderson and Evans, 1991; Heckers et al., 1994). The nucleus basalis also contains a complex mosaic of non-cholinergic neurons that are NADPHd-positive, GABAergic, peptidergic, and tyrosine hydroxylase (TH)-positive (Henderson, 1987; Mesulam et al., 1989; Walker et al., 1989; Gouras et al., 1992; Wisniowski et al., 1992; Gritti et al., 1993). The GABAergic neurons of the nucleus basalis may be as numerous as the cholinergic neurons with which they are intermingled. Some of these GABAergic neurons are interneurons, others project to the cerebral cortex and innervate cortical inhibitory interneurons (Freund and Meskenaite, 1992; Gritti et al., 1993). The TH-positive neurons are of particular interest since they appear to be quite numerous in several non-human species and could provide an alternative source of monoaminergic innervation for the nucleus basalis and cerebral cortex. These neurons have not yet been investigated in the human brain but our preliminary observations indicate that TH-positive neurons are quite rare in the human nucleus basalis.

The neurotransmitter circuitry of the nucleus basalis is complex. Electron microscopic studies show that GABAergic terminals make synaptic contact with Ch4 neurons and that cholinergic terminals make synaptic contact with GABAergic nucleus basalis neurons (Záborzky et al., 1986). Dissociated cell cultures of nucleus basalis neurons show that they are responsive to acetylcholine, neurotensin, substance P, and L-glutamate (Nakajima et al., 1985; Farkas et al., 1994). The monoaminergic innervation of the nucleus basalis has also attracted considerable attention. The nucleus basalis contains receptor sites for 5HT, dopamine, and norepinephrine (Zilles et al., 1991). In the rat, projections from dopaminergic ventral tegmental neurons, serotenergic raphe neurons, and noradrenergic locus coeruleus neurons have been identified but the identity of the nucleus basalis neurons receiving these connections and the details of the associated synaptic morphology remain to be determined (Jones and Cuello, 1989). In the human brain, TH and dopamine beta hydroxylase (DBH) immunoreactivity has been investigated by light microscopy in the septal area but not in the nucleus basalis (Gaspar et al., 1985). Our observations with the light microscope show intense, preterminal-like TH and DBH immunoreactive axonal profiles in the human nucleus basalis (Smiley and Mesulam, 1995). Reliable serotonin immunoreactivity has been difficult to obtain in the human brain. In the monkey brain, we detected a dense plexus of serotonin-immunoreactive axons within the nucleus basalis (Smiley and Mesulam, 1995). The primate nucleus basalis therefore provides a site for extensive cholinergic-mono-aminergic transmitter interactions.

The nucleus basalis also receives cholinergic innervation. Electron microscopic investigations of ChAT-immunolabeled nucleus basalis tissue reveals the existence of cholinergic terminals. The vast majority of these terminals make contact with non-cholinergic neurons (Martinez-Murillo et al., 1990). The precise source of the cholinergic input to the nucleus basalis is unknown but could include collaterals from cholinergic Ch1–Ch4 neurons of the basal forebrain or ascending projections from the Ch5–Ch6 group of pontomesencephalic cholinergic nuclei (Jones and Cuello, 1989). In dissociated nucleus basalis neurons, patch clamp techniques indicate the presence of currents associated with nicotinic as well as m2 muscarinic receptors (Harata et al., 1991). Nucleus basalis neurons in the rat express m2 receptor mRNA and contain m2 and m3 receptors. The m2 receptor appears to be the dominant species of cholinergic receptor in the nucleus basalis (Levey et al., 1991; Vilaró et al., 1992; Levey et al., 1994). This receptor subtype is expressed by approximately a third of Ch4 neurons and by many non-cholinergic neurons of the nucleus basalis. Immunocytochemistry

with an antibody raised against recombinant m2 receptor protein shows that the nucleus basalis contains numerous m2 immunoreactive axonal terminals which contact predominantly unlabeled dendrites and cell bodies, mostly through asymmetrical contacts (Levey et al., 1995). These observations indicate that the m2 receptor subtype may function as a presynaptic and also as a postsynaptic receptor for cholinergic as well as non-cholinergic neurons of the nucleus basalis. It also appears that cholinergic transmission in the nucleus basalis is dominated by the m2 receptor subtype whereas cholinergic transmission in the cerebral cortex is dominated by the m1 subtype.

Research on the nucleus basalis has traditionally emphasized its cholinergic contingent of neurons. The non-cholinergic neurons of the nucleus basalis are likely to play important roles in the synaptic organization of the local circuitry and may provide a substantial component of the corticofugal projections that originate from this nucleus. The importance of this non-cholinergic component has been emphasized by recent evidence which shows that selective immunotoxic lesions of cholinergic neurons in the nucleus basalis yield behavioral and physiological impairments which are distinctly less severe than those obtained by excitotoxic lesions which destroy all cell types (Berger-Sweeney et al., 1994; Wenk et al., 1994).

## Cholinergic innervation of the cerebral cortex

Experimental neuroanatomical methods in the monkey brain have shown that individual cortical areas receive their major cholinergic input from different sectors of the nb-Ch4 complex. Thus, Ch4am provides the major source of cholinergic input to medial cortical areas including the cingulate gyrus; Ch4al to frontoparietal opercular regions, and the amygdaloid nuclei; Ch4i to latero-dorsal frontoparietal, peristriate, and mid-temporal regions; and Ch4p to the superior temporal and temporopolar areas (Mesulam et al., 1983a). The experimental methods that are needed to reveal this topographic arrangement cannot be used in the human brain. However, indirect evidence for the existence of a similar topographical arrangement can be gathered from patients with Alzheimer's disease. We described two patients in whom extensive loss of cholinergic fibers in temporopolar but not frontal opercular cortex was associated with marked cell loss in the posterior (Ch4p) but not the anterior (Ch4am + Ch4al) sectors of Ch4 (Mesulam and Geula, 1988a). This relationship is consistent with the topography of the projections in the monkey brain.

Choline acetyltransferase (ChAT) immunocytochemistry provides one of the most important tools for the morphological investigation of cholinergic pathways since ChAT is a selective and specific marker for cholinergic perikarya and axons. Cholinergic (ChAT-immunoreactive) axons are present in all sectors of the human cerebral cortex, but also display an orderly gradient of density which follows the synaptic hierarchy of information processing systems (Mesulam and Geula, 1992; Mesulam et al., 1992). The density of cholinergic axons is higher in the more superficial layers of the cerebral cortex, suggesting that the axons which enter the cortex from the underlying gray matter undergo branching as they course towards the pial surface, an interpretation which is supported by physiological evidence (Aston-Jones et al., 1985; Mesulam et al., 1992). In the rat and cat, ChAT-immunoreactive fibers are almost exclusively unmyelinated, display numerous varicosities, and make mostly symmetrical synapses on the perikarya and dendrites of pyramidal as well as nonpyramidal neurons (Wainer et al., 1984; Frotscher and Leranth, 1985; DeLima and Singer, 1986).

The ACh which is released by cholinergic axons exerts its influence upon the cerebral cortex by interacting with muscarinic and nicotinic receptors. Antibodies raised against heterogeneous muscarinic and nicotinic receptor preparations have revealed large numbers of immunopositive, presumably cholinoceptive, cortical neurons in the human brain (Schröder et al., 1989; Schröder et al., 1990). In the mammalian brain, the pharmacologically defined M1 (or molecularly identified m1) subtype of muscarinic receptor is the most numerous species of cholinergic receptor in the cerebral cortex. Methods based on receptor autoradiography have detected regional and laminar variations in the distribution of muscarinic receptor subtypes in both the human and monkey brain (Mash et al., 1988; Lidow et al., 1989; Zilles, 1995).

The investigation of cholinergic receptors has entered a new and very productive phase with the molecular identification and cloning of five subtypes (m1–m5) of muscarinic receptors (Kubo et al., 1986; Bonner et al., 1987, 1988; Peralta et al., 1987). All five of these receptors are G protein coupled: m1, m3, and m5 preferentially activate phospholipase C whereas m2 and m4 inhibit adenylyl cyclase activity (Bonner et al., 1988). The availability of oligonucleotide probes and selective antibodies to recombinant receptor subtypes has enabled the definitive investigation of these subtypes with methods of in situ hybridization and immunocytochemistry (Buckley et al., 1988; Levey et al., 1991; Li et al., 1991). Immunocytochemical experiments and in situ hybridization studies show that the m1 receptor subtype is found in the majority of cortical neurons, a distribution which is consistent with its role as the major postsynaptic cholinergic receptor of the cerebral cortex (Buckley et al., 1988; Levey et al., 1991). Perikaryal m1-like immunoreactivity is seen not only in association with symmetrical synapses characteristic of cortical cholinergic pathways but also in association with asymmetrical synapses characteristic of excitatory amino acid pathways, raising the possibility that ACh may modulate excitatory neurotransmission in cortical neurons via an m1 receptor site (Mrzljak et al., 1993). Cortical immunostaining related to the m2 receptor subtype is located mostly in the neuropil but also in some perikarya, suggesting that this receptor subtype may function both as a pre-synaptic autoreceptor and also as a post-synaptic receptor (Levey et al., 1991). The m2 subtype is also associated with non-cholinergic terminals, suggesting that it may additionally act as a presynaptic heteroreceptor through which ACh may modulate the release of other transmitters (Mrzljak et al., 1993). Immunostaining and message for the m3 receptor is detected in several areas of the forebrain, including the pyramidal layer of the hippocampus, entorhinal cortex and the superficial layers of the cerebral cortex (Buckley et al., 1988). In immunocyto-chemical experiments, the staining is located mostly in the neuropil, but faint perikaryal staining can also be discerned, suggesting that the m3 receptor subtype can function as a pre- as well as post-synaptic receptor in the cerebral cortex (Levey et al., 1994). Cortical immunostaining for m4 protein is seen mostly in the neuropil but at a lesser density than that associated with the m1 or m2 subtypes. Immunostaining for the m5 receptor subtype has not yet been detected reliably in the cerebral cortex (Levey et al., 1991). These observations show that the muscarinic receptor subtypes can function not only as post-synaptic receptors at traditional cholinoceptive sites but also as post-synaptic modulators of non-cholinergic transmission and as pre-synaptic autoreceptors and heteroceptors that influence the release of ACh and other transmitters.

Incoming cholinergic axons innervate large numbers of cortical neurons. A full morphological and cytochemical classification of cortical cholinoceptive neurons is not yet available. The one marker that is thought to exist in all cholinoceptive neurons is acetylcholinesterase (AChE), the enzyme which terminates cholinergic neurotransmission through the rapid hydrolysis of ACh into acetate and choline. Although all cholinoceptive neurons probably express some AChE, only a subset of these neurons yields an AChE-rich histochemical staining pattern. In the cerebral cortex of the adult rat, the AChE-rich cytochemical pattern is limited to a few polymorphic neurons (Silver, 1974; Kutscher, 1991; Geula et al., 1993b). The situation is dramatically different in the human cerebral cortex which contains a dense network of AChE-rich cortical neurons, especially in layers III and V of premotor and sensory association cortex (Kostovic et al., 1988; Mesulam and Geula, 1988a,b; Mesulam and Geula, 1991a,b).

The density and staining intensity of these neurons is higher in the human brain than in any other species that we have studied, including the macaque and baboon. These neurons also display a most unusual ontogenetic profile: their AChE-rich staining pattern is not detectable as late as the 10th year of life and becomes fully established during adulthood. These AChE-rich neurons are ChAT-negative and, therefore, non-cholinergic. Paralimbic and limbic areas of the human brain receive a very dense cholinergic input but have very few AChE-rich neurons. Furthermore, AChE-rich intracortical neurons are rare during infancy when the cerebral cortex contains a dense net of cholinergic afferents and, presumably, a correspondingly large number of cholinoceptive neurons. It is therefore reasonable to assume that the AChE-rich cortical neurons constitute a special subset of cholinoceptive neurons and that their high AChE content may reflect affiliations that transcend the necessary requirements of standard cholinergic transmission. In keeping with this possibility, the expression of AChE has been implicated in a number of non-cholinergic phenomena including the promotion of axonal growth, sprouting, synaptogenesis, neural differentiation, neurotrophic influences, plasticity, and cell adhesion (Robertson, 1987; Layer et al., 1993; Small et al., 1995).

The mechanisms for these effects remain enigmatic but are thought to be non-enzymatic since they are not necessarily blocked by inhibitors of AChE enzyme activity. The intense AChE activity of these pyramidal intra-cortical neurons may act as a marker for particularly active events of reorganization, plasticity and maturation, occurring relatively late in development, at a time when the more advanced aspects of behavioral and cognitive skills are being established. The AChE-rich neurons of the cerebral cortex are recognized by monoclonal antibodies to human AChE, they are enriched in non-phosphorylated neurofilament protein, and their density remains stable in advanced senescence, but only in individuals with an intact mental state (Mesulam and Geula, 1988a, 1991a; Mesulam et al., 1991; Rezaki et al., 1994). These neurons are severely depleted in degenerative diseases that affect mental state such as Alzheimer's disease (Heckers et al., 1992). We have raised the possibility that this group of neurons may normally play an important role in the advanced phases of cortical and mental development and that their depletion may provide an anatomical substrate for dementia.

## Physiological and behavioral implications

The nucleus basalis is a phylogenetically progressive nucleus which displays its greatest differentiation in the cetacean and human brains (Gorry, 1963). Even in the human brain where it reaches its largest size, the nucleus basalis contains no more than approximately 200 000 neurons in each hemisphere. Although limited in number, these neurons provide every cytoarchitectonic sector and every cortical layer of the human cerebral cortex with a luxurious innervation of 60–100 cholinergic axons per mm. The high density of cholinergic axons, especially in paralimbic and limbic areas, is truly impressive. Its magnitude alone indicates that this pathway is likely to constitute the single most substantial regulatory afferent system of the cerebral cortex. From the vantage point of comparative neuroanatomy, the nucleus basalis can be considered both as a telencephalic extension of the isodendritic reticular core of the brainstem and also as a constituent of the mediobasal limbic cortex (Mesulam and Geula, 1988a; Mesulam et al., 1989). In keeping with these anatomical affiliations, the two major behavioral specializations that have been attributed to the nucleus basalis are in the realms of attention-arousal and memory-learning.

The major effect of ACh upon cortical neurons is mediated through muscarinic m1 receptors and causes a relatively prolonged reduction of potassium conductance so as to make cortical cholinoceptive neurons more susceptible to other excitatory inputs. Complex inhibitory effects, both direct and also through the mediation of GABAergic interneurons, have also been identified (McCormick, 1990). Stimulation of the nucleus basalis elicits EEG activation via muscarinic receptors, depolarizes cortical neurons and produces a change in subthreshold membrane potential fluctuations from large amplitude slow oscillations to low amplitude fast (20–40 Hz) oscillations (Metherate et al., 1992). In view of the ubiquitous distribution of cortical cholinergic pathways, it is not surprising that all aspects of cortical function can be modulated by cholinergic neurotransmission. In primary visual cortex, for example, cholinergic stimulation increases the likelihood that a neuron will fire in response to its preferred stimulus (Sato et al., 1987). Similar modulatory effects have been described in the somatosensory and auditory cortices (Ma et al., 1989; Metherate and Ashe, 1993). Neurons of the nucleus basalis are sensitive to novel and motivationally relevant sensory events (DeLong, 1971; Wilson and Rolls, 1990), leading to the expectation that cortical cholinergic innervation can modulate the impact of sensory events upon cortical circuitry in a manner that reflects their behavioral relevance and novelty. In keeping with this expectation, the novelty-related P300 potential in the human cerebral cortex is abolished upon the administration of cholinergic blockers (Hammond et al., 1987). Lesions in the nucleus basalis suppress low voltage fast EEG activity, reduce cortical glucose utilization and interfere with attentional tasks (Steward et al., 1984; Kiyosawa et al., 1989; Voytko et al., 1994). These observations provide considerable support for including the corticopetal projections of the nucleus basalis in the ascending reticular activating system and for implicating them in the modulation of cortical arousal states.

Various observations have implicated cortical cholinergic pathways in the neural organization of memory and learning. The systemic administration of the muscarinic blocker scopolamine causes memory impairments in a number of animal species, including humans (Drachman and Leavitt, 1974). During Pavlovian conditioning, approximately half of nucleus basalis neurons show a significantly greater change of activity in response to a tone that predicts the occurrence of a mildly aversive unconditioned stimulus than to a tone that does not (Whalen et al., 1994). Pharmacological experiments have indicated that the cholinergic innervation of the amygdala plays an important role in memory consolidation (McGaugh et al., 1993). Destructive lesions which include the cholinergic as well as the non-cholinergic components of the nucleus basalis lead to memory deficits in some experiments (Ridley et al., 1986) but not in others (Voytko et al., 1994). The recent availability of 192 IgG-saporin (a ribosome inactivating neurotoxin conjugated to an antibody that recognizes NGFr) has made it possible to produce a selective destruction of only the cholinergic neurons in the nucleus basalis (Chapter 25). Such experiments show that the resultant destruction of cortical cholinergic innervation causes severe impairments of learning in spatial navigation but not in delayed alteration or passive avoidance tasks (Berger-Sweeney et al., 1994; Wenk et al., 1994). The behavioral and physiological specializations of the non-cholinergic nucleus basalis neurons remain unexplored.

The mechanisms that link the cholinergic projections of the Ch1–Ch4 cell group to memory function are incompletely understood and have led to several speculations. The role of ACh in hippocampal long-term potentiation (Tanaka et al., 1989; Auerbach and Segal, 1994) may provide a cellular mechanism that underlies the relationship of cholinergic pathways to memory. Brain slice experiments in piriform cortex of the rat have shown that ACh can selectively suppress intrinsic synaptic transmission through a presynaptic mechanism, while leaving extrinsic afferent neurotransmission unaffected. This selective suppression, could prevent interference from previously stored patterns during the learning of new patterns. Hasselmo (1992) has argued that this could provide a novel mechanism through which cortical cholinergic innervation could participate in new learning. Buzsaki (1989) has proposed a model according to which the cholinergic innervation, especially of the hippocampal complex, plays a major role in switching between on-line attentive processing, characterized by the hippocampal theta rhythm, and off-line consolidation, characterized by sharp wave activity.

Another mechanism that links cholinergic axons to memory and learning may be related to the differential regional density of cortical cholinergic innervation. Experimental evidence leads to the conclusion that sensory-limbic pathways play pivotal roles in a wide range of behaviors related to emotion, motivation, and especially memory (Mishkin, 1982; Mesulam, 1985). The process starts within the primary sensory areas of the cerebral cortex which provide a portal for the entry of sensory information into cortical circuitry. These primary areas project predominantly to upstream (parasensory) unimodal sensory association areas which then project to downstream unimodal areas and heteromodal cortex. In turn, heteromodal association areas and downstream sectors of unimodal association areas collectively provide the major sources of sensory information into paralimbic and limbic areas of the brain. Our observations show that the density of cholinergic innervation is lower within unimodal and heteromodal association areas then in paralimbic areas of the brain. In the unimodal areas, moreover, the downstream sectors have a higher density of cholinergic innervation than the upstream sectors. Core limbic areas such as the amygdala and hippocampus contain the highest densities of cholinergic innervation. This gradient of density lead us to suggest that sensory information is likely to come under progressively greater cholinergic influence as it is conveyed along the multisynaptic pathways leading to the limbic system. As a consequence of this arrangement, cortical cholinergic innervation may help to channel (or gate) sensory information into and out of the limbic system in a way that is sensitive to the behavioral relevance of the associated experience (Mesulam et al., 1986). The memory disturbances that arise after damage to the Ch1–Ch4 cell groups or after the systemic administration of cholinergic antagonists may therefore reflect a disruption of sensory-limbic interactions which are crucial for effective memory and learning.

## Changes of cortical cholinergic innervation in aging and Alzheimer's disease

The literature on age-related changes in cortical cholinergic innervation is remarkably inconsistent. Some investigators report severe depletion, whereas others report complete preservation (see Geula and Mesulam, 1994). In our experience, age-related changes in the cholinergic innervation of the human cerebral cortex are regionally selective and relatively modest. We find that most amygdaloid nuclei and the cingulate cortex display virtually no age-related changes whereas the inferotemporal and entorhinal cortices appear to sustain a significant but relatively modest loss of approximately 20% when densities of cholinergic axons in specimens from 22–43-year-old subjects are compared to those from subjects above 68 years of age (Geula and Mesulam, 1989; Emre et al., 1993). Age-related changes in the nucleus basalis are also generally modest and do not become established until advanced senescence (Geula and Mesulam, 1994). An initial perikaryal hypertrophy around the age of 60 followed by shrinkage and cell loss appears to represent a characteristic pattern of age-related change. Although modest, these changes in cortical cholinergic innervation may provide an important anatomical substrate for age-related changes in memory function.

Neurons of the nucleus basalis as well as cortical cholinergic axons are severely depleted in Alzheimer's disease (Pilleri, 1966; Bowen et al., 1976; Davies and Maloney, 1976; Whitehouse et al., 1981; Geula and Mesulam, 1994). The loss of cortical cholinergic innervation in Alzheimer's disease has been reported in more that 30 papers (see Geula and Mesulam, 1994). We find that this depletion is severe (76–85%) in inferotemporal, midtemporal and entorhinal cortex and in parts of the amygdala; modest (40–67%) in prefrontal, posterior parietal, peristriate, orbitofrontal, insular, posterior cingulate, primary auditory and hippocampal cortex; and light (4–28%) in primary visual, primary somatosensory, primary motor, premotor, and anterior cingulate cortex (Emre et al., 1993; Geula and Mesulam, 1994). In general, the cholinergic depletion tends to be the most accentuated within the temporal lobe, including its limbic, paralimbic and association components.

Even in very advanced stages of Alzheimer's disease, and even in areas with a severe depletion of cholinergic innervation, the cerebral cortex still contains some cholinergic axons. The regional densities of these residual cholinergic axons in the temporal lobe appear to reflect differences in premorbid levels of innervation. Thus, regions with a relatively low basal density of cholinergic innervation such as the inferior, middle and superior temporal association areas appear almost completely denuded of cholinergic fibers whereas the entorhinal area which loses an equal proportion of its cholinergic innervation but which has a much higher premorbid density of innervation, retains a considerably higher residual innervation (Geula and Mesulam, 1994). In addition to the depletion of cortical cholinergic axons, muscarinic m2 receptors and nicotinic receptors are also depleted in Alzheimer's disease and remaining cholinoceptive neurons may fail to express nicotinic receptors and AChE enzyme activity (see Geula and Mesulam, 1994).

More than 20 papers in the literature have reported a 30–90% reduction in the number of nucleus basalis neurons in Alzheimer's disease (Geula and Mesulam, 1994). A review of these reports indicates that the depletion is just under 50% in the anterior (nb-Ch4a) and intermediate (nb-Ch4i) sectors whereas it increases to above 60% in the posterior (Ch4p) sector (Geula and Mesulam, 1994). The Ch4p sector is the major source of cholinergic projections to the temporal pole and the superior temporal gyrus (Mesulam et al., 1983), two areas that consistently show a severe depletion of cholinergic axons. The relatively greater magnitude of cell loss in this sector is therefore consistent with the connectivity of Ch4 and the regional distribution of cholinergic axonal loss in Alzheimer's disease. In keeping with the relatively modest loss of cholinergic innervation in the hippocampus and subiculum, the cell loss in Ch1 and Ch2 in Alzheimer's disease is usually reported to be of lesser magnitude than in Ch4. In Alzheimer's disease as well as in aging, neurons of the nucleus basalis seem to undergo an initial perikaryal hypertrophy followed by shrinkage and cell death. Many neurons in the nucleus basalis of patients with Alzheimer's disease display neurofibrillary degeneration.

Alzheimer's disease leads to the degeneration of extrathalamic corticopetal projections emanating not only from the nucleus basalis but also from the hypothalamus (histaminergic and alpha-MSH containing), ventral tegmental area (dopaminergic), raphe nuclei (serotonergic) and the nucleus locus coeruleus (noradrenergic). In general, however, the loss of cortical cholinergic innervation and the associated degeneration of the nucleus basalis occurs earlier in the course of the disease and is generally of greater magnitude (see Geula and Mesulam, 1994).

Some authors have suggested that the nucleus basalis and the depletion of cortical cholinergic innervation play pathogenetic roles in Alzheimer's disease. For example, Arendash et al. (1987) reported that experimentally induced lesions in the nucleus basalis of rats induces the formation of plaque-like deposits in the cerebral cortex and Cohen et al. (1988) suggested that the neurons of the nucleus basalis may overexpress amyloid precursor protein in a way that may promote the formation of plaques in Alzheimer's disease. We have not been able to replicate the Arendash et al. result. We have also not been able to establish a relationship between regional plaque counts and either the severity of cholinergic denervation or the premorbid density of cholinergic innervation (Geula and Mesulam, 1994). Nitsch et al. (1992) reported that muscarinic stimulation of cortical neurons promotes the non-amyloidogenic processing of the amyloid precursor protein, leading to the implication that the cholinergic depletion may promote amyloidogenesis. This hypothesis is under investigation in several laboratories.

The reason for the selective vulnerability of cholinergic innervation in Alzheimer's disease is not fully understood. According to Wurtman et al. (1990), choline is a precursor for both ACh and membrane phosphatidylcholine. An initial cholinergic denervation may lead to the compensatory overactivity of remaining cholinergic neurons in a way that may shunt choline away from membrane synthesis and into the synthesis of ACh. This could lead to the "autocannibalization" of cholinergic neurons and may provide a mechanism for selectively accelerating the degeneration of cholinergic projection pathways.

The vast majority of cholinergic neurons in the nucleus basalis depend for their survival on the retrograde transport of cortically produced nerve growth factor (NGF). In fact, cortical lesions which presumably interfere with the retrograde transport of NGF cause pathological changes in the Ch4 neurons that supply the damaged cortical area (Liberini et al., 1994). Although there is no evidence that the cerebral cortex in Alzheimer's disease is deficient in NGF or that the Ch4 neurons fail to express NGFr, the cortical pathology, including the extracellular deposition of amyloid, could interfere with the appropriate release of NGF, its binding to NGFr, or its retrograde transport.

The projections from the nucleus basalis to the cerebral cortex are much denser than the corticopetal projections from the ventral tegmental area (dopaminergic), raphe nuclei (serotonergic) or nucleus locus coeruleus (noradrenergic). Our unpublished observations also indicate that the nucleus basalis may be one of the brain areas most vulnerable to tau hyperphosphorylation and neurofibrillary tangle formation in the course of aging. Cortical cholinergic projections are therefore in double jeopardy: they have a greater number of axons exposed to the cortical pathology and their cells of origin in the nucleus basalis are located within a zone of selective vulnerability to age-related neuronal pathology. A selective loss of calbindin-d28k immunoreactivity occurs in the Ch4 neurons of non-demented elderly individuals, suggesting that alterations in intracellular calcium regulation may provide a mechanism for the selective vulnerability of these neurons to age-related pathological processes and Alzheimer's disease (Wu et al., 1995).

The cholinergic pathway emanating from the basal forebrain constitutes one of the most important modulatory afferents of the mammalian cortex.The initial expectation that the cholinergic deficiency would provide a unifying pathophysiological basis for Alzheimer's disease and that cholinergic therapies would cure the dementia were clearly too optimistic. Nonetheless, the cortical cholinergic denervation remains one of the earliest, most severe, and most consistent transmitter changes in this disease. The cholinergic depletion may provide an important substrate for the neuropsychological features of Alzheimer's disease, and may eventually yield important clues to its pathogenesis.

## Acknowledgements

Supported in part by a Javits Neuroscience Investigator Award (NS20285) and an Alzheimer's Disease Care Centre Grant (AG13854).

# References

Arendash, G.W., Millard, W.J., Dunn, A.J. and Meyer, E.M. (1987) Long-term neuropathological and neurochemical effects of nucleus basalis lesions in the rat. *Science*, 238: 952–956.

Arendt, T., Bigl, V., Tennstedt, A. and Arendt, A. (1985) Neuronal loss in different parts of the nucleus basalis is related to neuritic plaque formation in cortical target areas in Alzheimer's disease. *Neuroscience*, 14: 1–14.

Aston-Jones, G., Shaver, R. and Dinan, T.G. (1985) Nucleus basalis neurons exhibit axonal branching with decreased impulse conduction velocity in rat cerebrocortex. *Brain Res.*, 325: 271–285.

Auerbach, J.M. and Segal, M. (1994) A novel cholinergic induction of long-term potentiation in rat hippocampus. *J. Neurophysiol.*, 72: 2034–2040.

Berger-Sweeney, J., Heckers, S., Mesulam, M.-M., Wiley, R.G., Lappi, D.A. and Sharma, M. (1994) Differential effects upon spatial navigation of immunotoxin-induced cholinergic lesions of the medial septal area and nucleus basalis magnocellularis. *J. Neurosci.*, 14: 4507–4519.

Bonner, T.I., Buckley, N.J., Young, A.C. and Brann, M.R. (1987) Identification of a family of muscarinic acetylcholine receptor genes. *Science*, 237: 527–532.

Bonner, T.I., Young, A.C., Brann, M.R. and Buckley, N.J. (1988) Cloning and expression of the human and rat m5 muscarinic acetylcholine receptor genes. *Neuron*, 1: 403–410.

Bowen, D.M., Smith, C.B., White, P. and Davison, A.N. (1976) Neurotransmitter-related enzymes and indices of hypoxia in senile dementia and other abiotrophies. *Brain*, 99: 459–496.

Buckley, N.J., Bonner, T.I. and Brann, M.R. (1988) Localization of muscarinic receptor mRNAs in rat brain. *J. Neurosci.*, 8: 4646–4652.

Buszaki, G. (1989) Commentary: two-stage model of memory trace formation: a role for "noisy" brain states. *Neuroscience*, 31: 551–570.

Cohen, M.L., Golde, T.E., Usiak, M.F., Younkin, L.H. and Younkin, S.G. (1988) In situ hybridization of nucleus basalis neurons shows increased β-amyloid mRNA in Alzheimer's disease. *Proc. Natl. Acad. Sci. USA*, 85: 1227–1231.

Davies, p. and Maloney, A.J.F. (1976) Selective loss of central cholinergic neurons in Alzheimer's disease. *Lancet*, ii: 1943.

DeLima, A.D. and Singer, W. (1986) Cholinergic innervation of the cat striate cortex: a choline acetyltransferase immunocytochemical analysis. *J. Comp. Neurol.*, 250: 324–338.

DeLong, M.R. (1971) Activity of pallidal neurons during movement. *J. Neurophysiol.*, 34: 414–427.

Drachman, D.A. and Leavitt, J. (1974) Human memory and the cholinergic system-A relationship to aging? *Arch. Neurol.*, 30: 113–121.

Emre, M., Heckers, S., Mash, D.C., Geula, C. and Mesulam, M.-M. (1993) Cholinergic innervation of the amygdaloid complex in the human brain and its alterations in old age and Alzheimer's disease. *J. Comp. Neurol.*, 336: 117–134.

Everitt, B.J., Sirkiä, T.E., Roberts, A.C., Jones, G.H. and Robbins, T.W. (1988) Distribution and some projections of cholinergic neurons in the brain of the common marmoset, Callithrix jaccus. *J. Comp. Neurol.*, 271: 533–558.

Farkas, R.H., Nakajima, S. and Nakajima, Y. (1994) Neurotensin excites basal forebrain cholinergic neurons: ionic and signal-transduction mechanisms. *Proc. Natl. Acad. Sci. USA*, 91: 2853–2857.

Freund, T.F. and Meskenaite, V. (1992) γ-Aminobutyric acid-containing basal forebrain neurons innervate inhibitory interneurons in the neocortex. *Proc. Natl. Acad. Sci. USA*, 89: 738–742.

Frotscher, M. and Leranth, C. (1985) Cholinergic innervation of the rat hippocampus as revealed by choline acetyltransferase immunocytochemistry: a combined light and electron microscopic study. *J. Comp. Neurol.*, 239: 237–246.

Gaspar, P., Berger, B., Alvarez, C., Vigny, A. and Henry, J.P. (1985) Catecholaminergic innervation of the septal area in man: immunocytochemical study using TH and DBH antibodies. *J. Comp. Neurol.*, 241: 12–33.

Geula, C. and Mesulam, M.M. (1989) Cortical cholinergic fibers in aging and Alzheimer's disease: a morphometric study. *Neuroscience*, 33: 469–481.

Geula, C. and Mesulam, M.-M. (1994) Cholinergic systems and related neuropathological predilection patterns in Alzheimer disease. In: R.D. Terry, R. Katzman and K.L. Bick, (Eds), *Alzheimer Disease*. Raven Press, New York,

Geula, C., Schatz, C.R. and Mesulam, M.M. (1993a) Differential localization of NADPH-diaphorase and calbindin-D28k within the cholinergic neurons of the basal forebrain, striatum and brainstem in the rat, monkey, baboon and human. *Neuroscience*, 54: 461–476.

Geula, G., Mesulam, M.-M., Tokuno, H. and Kuo, C.C. (1993b) Developmentally transient expression of acetylcholinesterase within cortical pyramidal neurons of the rat brain. *Dev. Brain. Res.*, 76: 23–31.

Gorry, J.D. (1963) Studies on the comparative anatomy of the ganglion basale of Meynert. *Acta Anat.*, 55: 51–104.

Gouras, G.K., Rance, N.E., Young III, W.S. and Koliatsos, V.E. (1992) Tyrosine-hydroxylase containing neurons in the primate basal forebrain magnocellular complex. *Brain. Res.*, 584: 287–293.

Gritti, I., Mainville, L. and Jones, B.E. (1993) Codistribution of GABA- with acetylcholine-synthesizing neurons in the basal forebrain of the rat. *J. Comp. Neurol.*, 329: 438–457.

Hammond, E.J., Meador, K.J., Aunq-Din, R. and Wilder, B.J. (1987) Cholinergic modulation of human P3 event-related potentials. *Neurology*, 37: 346–350.

Harata, N., Tateishi, N. and Akaike, N. (1991) Acetylcholine receptors in dissociated nucleus basalis of Meynert neurons of the rat. *Neurosci. Lett.*, 130: 153–156.

Hasselmo, M.E. (1992) Cholinergic modulation of cortical associative memory function. *J. Neurophysiol.*, 67: 1230–1246.

Heckers, S., Geula, C. and Mesulam, M.M. (1992) Acetylcholinesterase-rich pyramidal neurons in Alzheimer's disease. *Neurobiol. Aging*, 13: 455–460.

Heckers, S., Ohtake, T., Wiley, R.G., Lappi, D.A., Geula, C. and Mesulam, M.-M. (1994) Complete and selective cholinergic denervation of rat neocortex and hippocampus but not amygdala by an immunotoxin against the p75 NGF receptor. *J. Neurosci.*, 14: 1271–1289.

Henderson, Z. (1987) A small proportion of cholinergic neurones in the nucleus basalis magnocellularis of ferret appear to stain positively for tyrosine hydroxylase. *Brain. Res.*, 412: 363–369.

Henderson, Z. and Evans, S. (1991) Presence of a cholinergic projection from ventral striatum to amygdala that is not immunoreactive for NGF receptor. *Neurosci. Lett.*, 127: 73–76.

Ingham, C.A., Bolam, J.P., Wainer, B.H. and Smith, A.D. (1985) A correlated light and electron microscopic study of identified cholinergic basal forebrain neurons that project to the cortex in the rat. *J. Comp. Neurol.*, 239: 176–192.

Jones, B. and Cuello, A.C. (1989) Afferents to the basal forebrain cholinergic cell area from pontomesencephalic-catecholamine, serotonin and acetylcholine-neurons. *Neuroscience*, 31: 37–61.

Kiyosawa, M., Baron, J.-C., Hamel, E., Pappata, S., Duverger, D., Riche, D., Mazoyer, B., Naquet, R. and MacKenzie, E.T. (1989) Time course of effects of unilateral lesions of the nucleus basalis of Meynert on glucose utilization by the cerebral cortex. *Brain*, 112: 435–455.

Kordower, J.H., Chen, E.-Y., Sladek Jr., J.R. and Mufson, E.J. (1994) TRK-immunoreactivity in the monkey central nervous system: forebrain. *J. Comp. Neurol.*, 349: 20–35.

Kostovic, I., Skavic, J. and Strinovic, D. (1988) Acetylcholinesterase in the human frontal associative cortex during the period of cognitive development: early laminar shifts and late innervation of pyramidal neurons. *Neurosci. Lett.*, 90: 107–112.

Kubo, T., Fukuda, K., Mikami, A., Maeda, A., Takahashi, H., Mishina, M., Haga, T., Haga, K., Ichiyama, A., Kangawa, k., Kojima, M., Matsuo, H., Hirose, T. and Numa, S. (1986) Cloning, sequencing and expression of complementary DNA encoding the muscarinic acetylcholine receptor. *Nature*, 323: 411–416.

Kutscher, C.L. (1991) Development of transient acetylcholinesterase staining in cells and permanent staining in fibers in cortex of rat brain. *Brain. Res. Bull.*, 27: 641–649.

Layer, P.G., Weikert, T. and Alber, R. (1993) Cholinesterases regulate neurite growth of chick nerve cells in vitro by means of a non-enzymatic mechanism. *Cell Tissue Res.* 273: 219–226.

Levey, A.I., Kitt, C.A., Simonds, W.F., Price, D.L. and Brann, M.R. (1991) Identification and localization of muscarinic acetylcholine receptor protein in brain with subtype-specific antibodies. *J. Neurosci.*, 11: 3218–3226.

Levey, A.I., Edmunds, S.M., Heilman, C.J., Desmond, T.J. and Frey, K.A. (1994) Localization of muscarinic m3 receptor protein and M3 receptor binding in the rat brain. *Neuroscience*, 63: 207–221.

Levey, A.I., Edmunds, S.M., Hersch, S.M., Wiley, R.G. and Heilman, C.J. (1995) Light and electron microscopic study of m2 muscarinic acetylcholine receptor in the basal forebrain of the rat. *J. Comp. Neurol.*, 351: 339–356.

Li, M., Yasuda, R.P., Wall, S.J., Wellstein, A. and Wolfe, B.B. (1991) Distribution of m2 muscarinic receptors in rat brain using antisera selective for m2 receptors. *Mol. Pharmacol.*, 40: 28–35.

Liberini, P., Pioro, E.K., Maysinger, D. and Cuello, A.C. (1994) Neocortical infarction in subhuman primates leads to restricted morphological damage of the cholinergic neurons in the nucleus basalis of Meynert. *Brain. Res.*, 648: 1–8.

Lidow, M.S., Gallager, D.W., Rakic, P. and Goldman-Rakic, P.S. (1989) Regional differences in the distribution of muscarinic cholinergic receptors in the macaque cerebral cortex. *J. Comp. Neurol.*, 289: 247–259.

Loewi, O. (1921) Über humorale Übertragbarkeit der Herznervenwirkung. *Pflügers Arch. Ges. Physiol.*, 189: 239–242.

Ma, W., Höhmann, C.F., Coyle, J.T. and Juliano, S.L. (1989) Lesions of the basal forebrain alter stimulus-evoked metabolic activity in mouse somatosensory cortex. *J. Comp. Neurol.*, 288: 414–427.

Martinez-Murillo, R., Villalba, R.M. and Rodrigo, J. (1990) Immunocytochemical localization of cholinergic terminals in the region of the nucleus basal magnocellularis of the rat: a correlated light and electron microscopic study. *Neuroscience*, 36: 361–376.

Mash, D.C., White, W.F. and Mesulam, M.M. (1988) Distribution of muscarinic receptor subtypes within architectonic subregions of the primate cerebral cortex. *J. Comp. Neurol.*, 278: 265–274.

McCormick, D.A. (1990) Cellular mechanisms of cholinergic control of neocortical and thalamic neuronal excitability. In: M. Steriade and D. Biesold (Eds.), *Brain Cholinergic Systems*. Oxford University Press, Oxford, UK, pp. 236–264.

McGaugh, J.L., Introini-Collison, I.B., Cahill, L.F., Castellano, C., Dalmaz, C., Parent, M.B. and Williams, C.L. (1993) Neuromodulatory systems and memory storage; role of the amygdala. *Behav. Brain. Res.*, 58: 81–90.

Mesulam, M.-M. (1985) Patterns in behavioral neuroanatomy; Association areas, the limbic system, and hemispheric specialization. In: M.-M. Mesulam (Eds.), *Principles of Behavioral Neurology*. F.A. Davis, Philadelphia, PA, pp. 1–70.

Mesulam, M.M. and Geula, C. (1988a) Nucleus basalis (Ch4) and cortical cholinergic innervation in the human brain: observations based on the distribution of acetylcholinesterase and choline acetyltransferase. *J. Comp. Neurol.*, 275: 216–240.

Mesulam, M.-M. and Geula, C. (1988b) Acetylcholinesterase-rich pyramidal neurons in the human neocortex and hippocampus: absence at birth, development during the life span, and dissolution in Alzheimer's disease. *Ann. Neurol.*, 24: 765–773.

Mesulam, M. and Geula, C. (1991a) Differential distribution of a neurofilament protein epitope in acetylcholinesterase-rich neurons of human cerebral neocortex. *Brain Res.*, 544: 169–173.

Mesulam, M.M. and Geula, C. (1991b) Acetylcholinesterase-rich neurons of the human cerebral cortex: cytoarchitectonic and ontogenetic patterns of distribution. *J. Comp. Neurol.*, 306: 193–220.

Mesulam, M.M. and Geula, C. (1992) Overlap between acetylcholinesterase-rich and choline acetyltransferase-positive (cholinergic) axons in human cerebral cortex. *Brain Res.*, 577: 112–120.

Mesulam, M.-M., Mufson, E.J., Levey, A.I. and Wainer, B.H. (1983a) Cholinergic innervation of cortex by the basal forebrain: cytochemistry and cortical connections of the septal area, diagonal band nuclei, nucleus basalis (substantia innominata), and hypothalamus in the rhesus monkey. *J. Comp. Neurol.*, 214: 170–197.

Mesulam, M.M., Mufson, E.J., Wainer, B.H. and Levey, A.I. (1983b) Central cholinergic pathways in the rat: an overview based on an alternative nomenclature (Ch1-Ch6) *Neuroscience*, 10: 1185–1201.

Mesulam, M.M., Volicer, L., Marquis, J.K., Mufson, E.J. and Green, R.C. (1986) Systematic regional differences in the cholinergic innervation of the primate cerebral cortex: distribution of enzyme activities and some behavioral implications. *Ann. Neurol.*, 19: 144–151.

Mesulam, M.-M., Geula, C., Bothwell, M.A. and Hersh, L.B. (1989) Human reticular formation: cholinergic neurons of the pedunculopontine and laterodorsal tegmental nuclei and some cytochemical comparisons to forebrain cholinergic neurons. *J. Comp. Neurol.*, 283: 611–633.

Mesulam, M.M., Geula, C., Cosgrove, R., Mash, D. and Brimijoin, S. (1991) Immunocytochemical demonstration of axonal and perikaryal acetylcholinesterase in human cerebral cortex. *Brain Res.*, 539: 233–238.

Mesulam, M.M., Hersh, L.B., Mash, D.C. and Geula, C. (1992) Differential cholinergic innervation within functional subdivisions of the human cerebral cortex: a choline acetyltransferase study. *J. Comp. Neurol.*, 318: 316–328.

Metherate, R. and Ashe, J.H. (1993) Nucleus basalis stimulation facilitates thalamocortical synaptic transmission in the rat auditory cortex. *Synapse*, 14: 132–143.

Metherate, R., Cox, C.L. and Ashe, J.H. (1992) Cellular bases of neocortical activation: modulation of neuronal oscillations by the nucleus basalis and endogenous acetylcholine. *J. Neurosci.*, 12: 4701–4711.

Mishkin, M. (1982) A memory system in the monkey. *Phil. Trans. R. Soc. London B*, 298: 85–92.

Mrzljak, L., Levey, A.I. and Goldman-Rakic, P.S. (1993) Association of m1 and m2 muscarinic receptor proteins with asymmetric synapses in the primate cerebral cortex: morphological evidence for cholinergic modulation of excitatory neurotransmission. *Proc. Natl. Acad. Sci. USA*, 90: 5194–5198.

Mufson, E.J., Bothwell, M., Hersh, L.B. and Kordower, J.H. (1989) Nerve growth factor receptor immunoreactive profiles in the normal, aged human basal forebrain: colocalization with cholinergic neurons. *J. Comp. Neurol.*, 285: 196–217.

Nakajima, Y., Nakajima, S., Obata, K., Carlson, C.G. and Yamaguchi, K. (1985) Dissociated cell culture of cholinergic neurons from nucleus basalis of Meynert and other basal forebrain nuclei. *Proc. Natl. Acad. Sci. USA*, 82: 6325–6329.

Nitsch, R.M., Slack, B.E., Wurtman, R.J. and Growdon, J.H. (1992) Release of Alzheimer amyloid precursor derivatives stimulated by activation of muscarinic acetylcholine receptors. *Science*, 258: 304–307.

Palacios, G. (1990) The endomembrane system of cholinergic and non-cholinergic neurons in the medial septal nucleus and vertical limb of the diagonal band of Broca: a cytochemical and immunocytochemical study. *J. Histochem. Cytochem.*, 38: 563–571.

Peralta, E.G., Ashkenazi, A., Winslow, J.W., Smith, D.H., Ramachandran, J. and Capon, D.J. (1987) Distinct primary structures, ligand-binding properties and tissue-specific expression of four human muscarinic acetylcholine receptors. *EMBO J.*, 6: 3923–3929.

Pilleri, G. (1966) The Klüver-Bucy syndrome in man- a clinicoanatomical contribution to the function of the medial temporal lobe structures. *Psychiatr. Neurol. (Basel)*, 152: 65.

Ramon-Moliner, E. and Nauta, W.J.H. (1966) The isodendritic core of the brain. *J. Comp. Neurol.*, 126: 311–336.

Rezaki, M., Geula, C. and Mesulam, M.-M. (1994) Acetylcholinesterase-reactive pyramidal neurons of the human cerebral cortex; a quantitative study in young and aged brains. *Soc. Neurosci. Abstr.*, 20: 50.

Ridley, R.M., Murray, T.K., Johnson, J.A. and Baker, H.F. (1986) Learning impairment following lesion of the basal nucleus of Meynert in the marmoset: modification by cholinergic drugs. *Brain. Res.*, 376: 108–116.

Robertson, R.T. (1987) A morphogenetic role for transiently expressed acetylcholinesterase in developing thalamocortical systems? *Neurosci. Lett.*, 75: 259–264.

Sato, H., Hata, V., Hagihara, K. and Tsumoto, T. (1987) Effects of cholinergic depletion on neuron activities in the cat visual cortex. *J. Neurophysiol.*, 58: 781–794.

Satoh, K. and Fibiger, H.C. (1985) Distribution of central cholinergic neurons in the baboon (*Papio papio*) I. General morphology. *J. Comp. Neurol.*, 236: 197–214.

Schröder, H., Zilles, K., Maelicke, A. and Hajós, F. (1989) Immunohisto- and cytochemical localization of cortical cholinoceptors in rat and man. *Brain. Res.*, 502: 287–295.

Schröder, H., Zilles, K., Luiten, P.G.M. and Strosberg, A.D. (1990) Immunocytochemical visualization of muscarinic cholinoceptors in the human cerebral cortex. *Brain. Res.*, 514: 249–258.

Silver, A. (1974) *The Biology of Cholinesterases*, Elsevier, New York.

Small, D.H., Reed, G., Whitefield, B. and Nurcombe, V. (1995) Cholinergic regulation of neurite outgrowth from isolated chick sympathetic neurons in culture. *J. Neurosci.*, 15: 144–151.

Smiley, J.F. and Mesulam, M.-M. (1995) Dopamine, norepinephrine, and serotonin axons in the human and monkey nucleus basalis, and in Alzheimer's disease. *Soc. Neurosci. Abstr.*, 21.

Steward, D.F., Macfabe, D.F. and Vanderwolf, C.H. (1984) Cholinergic activation of the EEG: role of the substantia innominata and effects of atropine and quinuclidinyl benzilate. *Brain. Res.*, 322: 219–232.

Tanaka, Y., Sakurai, M. and Hayashi, S. (1989) Effect of scopolamine and HP029, a cholinesterase inhibitor, on long-term potentiation in hippocampal slices of guinea pig. *Neurosci. Lett.*, 98: 179–183.

Vilaró, M.T., Wiederhold, K.-H., Palacios, J.M. and Mengod, G. (1992) Muscarinic M2 receptor mRNA expression and receptor binding in cholinergic and non-cholinergic cells in the rat brain: a correlative study using in situ hybridization histochemistry and receptor autoradiography. *Neuroscience*, 47: 367–393.

Voytko, M.L., Olton, D.S., Richardson, R.T., Gorman, L.K., Tobin, J.R. and Price, D.L. (1994) Basal forebrain lesions in monkeys disrupt attention but not learning and memory. *J. Neurosci.*, 14: 167–186.

Wainer, B.H., Bolam, J.P., Freund, T.F. and Henderson, Z. (1984) Cholinergic synapses in the rat brain: a correlated light and electron microscopic immunohistochemical study employing a monoclonal antibody against choline acetyltransferase. *Brain. Res.*, 308: 69–76.

Walker, L.C., Tigges, M. and Tigges, J. (1983) Ultrastructure of neurons in the nucleus basalis of Meynert in squirrel monkey. *J. Comp. Neurol.*, 217: 158–166.

Walker, L.C., Koliatsos, V.E., Kitt, C.A., Richardson, R.T., Rökaeus, Å. and Price, D.L. (1989) Peptidergic neurons in the basal forebrain magnocellular complex of the rhesus monkey. *J. Comp. Neurol.*, 280: 272–282.

Wenk, G.L., Stoehr, J.D., Quintana, G., Mobley, S. and Wiley, R.G. (1994) Behavioral, biochemical, histological, and electrophysiological effects of 192 IgG-saporin injections into the basal forebrain of rats. *J. Neurosci.*, 14: 5986–5995.

Whalen, P.J., Knapp, B.S. and Pascoe, J.P. (1994) Neuronal activity within the nucleus basalis and conditioned neocortical electroencephalographic activation. *J. Neurosci.*, 14: 1623–1633.

Whitehouse, P.J., Price, D.L., Clark, A.W., Coyle, J.T. and DeLong, M.R. (1981) Alzheimer disease: evidence for selective loss of cholinergic neurons in the nucleus basalis. *Ann. Neurol.*, 10: 122–126.

Wilson, F.A.W. and Rolls, E.T. (1990) Neuronal responses related to novelty and familiarity of visual stimuli in the substantia innominata, diagonal band of Broca and periventricular region of the primate basal forebrain. *Exp. Brain. Res.*, 80: 104–120.

Wisniowski, l., Ridley, R.M., Baker, H.F. and Fine, A. (1992) Tyrosine hydroylase-immunoreactive neurons in the nucleus basalis of the common marmoset (Callithrix jaccus) *J. Comp. Neurol.*, 325: 379–387.

Wu, C.-K., Mesulam, M.-M. and Geula, C. (1995) Aging causes selective loss of calbindin-d28k from the cholinergic neurons of the human basal forebrain. *Soc. Neurosci. Abstr.*, 21.

Wurtman, R.J., Blusztajn, J.K. and Ulus, I.H. (1990) Choline metabolism in cholinergic neurons: implications for the pathogenesis of neurodegenerative diseases. In: R.J. Wurtman, S. Corkin, J.H. Growdon and E. Ritter-Walker (Eds.), *Advances in Neurology. Alzheimer's Disease*. Raven Press, New York, pp. 117–125.

Záborzky, L., Heimer, L., Eckenstein, F. and Leranth, C. (1986) GABAergic input to cholinergic forebrain neurons: an ultrastructural study using retrograde tracing of HRP and double immunolabeling. *J. Comp. Neurol.*, 250: 282–295.

Zilles, K. (1995) Codistribution of receptors in the human cerebral cortex. In: F.A.O. Mendelsohn and G. Paxinos (Eds.), *Receptors in the Human Nervous System*. Academic Press, New York.

Zilles, K., Werner, L., Qü, M., Schleicher, A. and Gross, G. (1991) Quantitative autoradiography of 11 different transmitter binding sites in the basal forebrain region of the rat-evidence of heterogeneity in distribution patterns. *Neuroscience*, 42: 473–481.

J. Klein and K. Löffelholz (Eds.)
*Progress in Brain Research*, Vol. 109
© 1996 Elsevier Science B.V. All rights reserved.

# CHAPTER 29

# Cholinomimetic treatment of Alzheimer's disease

Leon J. Thal

*Department of Neurosciences, University of California at San Diego School of Medicine, 9500 Gilman Drive, La Jolla, CA 92093-0624,
USA and Department of Neurology, San Diego Veterans Administration Medical Center, San Diego, CA, USA*

## Introduction

In the United States and other developed nations, diseases of the aged are a major public health concern because of increased life expectancy. In 1990, 12.5% of the US population was over the age of 65. These population figures will change dramatically so that by the year 2040, approximately 22.5% of the US population will exceed 65 years of age. Since 15% of individuals over the age of 65 suffer from acquired cognitive loss and two-thirds of these have dementia secondary to Alzheimer's disease (AD), by 2040, approximately 6 million Americans will be afflicted with this disorder. The economic consequences are considerable with estimated costs of $67 billion in 1991 (Ernst and Hay, 1994) and the expectation is that this figure will double within the next 50 years. These demographic and economic factors, coupled with our expanding biological knowledge of the disease, have promoted a major focus on developing effective treatments for AD.

A number of other factors have contributed to drug research in this area during the past two decades. The prevalence and malignancy of AD as a major killer was recognized in the mid-1970s (Katzman, 1976). The concept that AD and senile dementia represented a single entity was suggested. The formation of the Alzheimer's Disease and Related Disorders Association (ADRDA) and the commitment of the National Institute on Aging (NIA) supporting basic biomedical research in aging and dementia was also initiated in the early 1980s. By 1984, the NIA recognized a need to

expand beyond basic laboratory studies to define the clinical pathological characteristics of this disorder. The NIA therefore established the first five AD centers and by 1991 this number had increased to 28. In 1991, the NIA also funded a series of basic science groups aimed at new drug discovery and the Alzheimer Disease Cooperative Study, a consortium of academic medical centers working collectively to design, implement and analyze clinical trials for promising new agents in AD. Most recently, the NIA has put forth a 5–5, 10–10 plan, a proposal to slow the progression or delay the onset of disease by 5 years within the next 5 years and by 10 years within the next decade (Khachaturian, 1992).

## Biological considerations and design difficulties inherent in the conduct of clinical trials in AD

### Diagnostic accuracy

There is no biological marker for AD during life. Diagnosis is generally achieved by the use of standardized criteria, such as DSM-IV for dementia and the NINCDS-ADRDA criteria for AD. Using these criteria, diagnostic accuracy for AD is approximately 85% in large clinical series with autopsy verification (Molsa et al., 1985; Wade et al., 1987; Jellinger et al., 1990; Galasko et al., 1994). When difficult patients are excluded, as is commonly done for clinical drug trials, diagnostic accuracy probably exceeds 90% for individuals enrolled in clinical drug trials. Nevertheless, approximately 10% of individuals enrolled in clinical drug trials are likely to have other etiological explanations for their dementia. In addition, diagnosis is less certain for early cases. Thus, the development of meaningful biological markers would be most useful for identifying early or pre-clinical subjects.

### Variations in presentation and course

The course of AD can be extremely variable. This is not surprising considering the varying cognitive skills including memory, language, intelligence, abstract thinking and judgment present in each individual. During the course of AD, difficulties with memory, language, visuospatial relations, mood and behavioral changes occur in different combinations and progress at different rates. While on average individuals with AD survive for 8–10 years after diagnosis, some individuals decline quite precipitously and expire in as little as 3–4 years, while others decline quite slowly and survive for 20 or more years following diagnosis.

Decline can be assessed in many ways. The simplest and most widely used involve rates of change on cognitive scales. Four global scales have been employed including the Blessed Information Memory Concentration Test (BIMC) (Blessed et al., 1968), the Dementia Rating Scale (DRS) (Mattis, 1976), the Mini Mental State Examination (MMSE) (Folstein et al., 1975), and the Alzheimer Disease Assessment Scale (ADAS) (Rosen et al., 1984), particularly its cognitive subcomponent (ADAS-Cog). These instruments have floor and ceiling effects, are best suitable for patients with mild to moderate dementia, and are relatively insensitive for individuals in the very early or very late stages of the disorder. For all of these instruments, the average 1 year rate of change of AD patients is approximately equal to the standard deviation of the 1 year rate of change (Table 1). More precise estimates can be obtained by increasing the period of observation beyond one year (Morris et al., 1993). Further inspection of rates of change clearly indicates that the most rapid and predictable change occurs during the middle stages of the disease. Although the standard deviation is large, the rate of change is quite

TABLE 1

Annual rate of change in AD

| Scale | Study | No. | Rate of change |
|-------|-------|-----|----------------|
| BIMC | Katzman et al., 1988 | 161 | 4.4 ± 3.6 |
| | Thal et al., 1988 | 40 | 4.5 ± 3.1 |
| | Ortof and Crystal, 1989 | 54 | 4.1 ± 3.0 |
| | Salmon et al., 1990 | 50 | 3.2 ± 3.0 |
| DRS | Salmon et al., 1990 | 55 | 11.4 ± 11.1 |
| MMSE | Uhlman et al., 1986 | 120 | 2.2 ± 5.0 |
| | Salmon et al., 1990 | 55 | 2.8 ± 4.3 |
| | Teri et al., 1990 | 106 | 2.8 ± 4.6 |
| ADAS | Kramer-Ginsberg et al., 1988 | 60 | 9.3 ± 9.8 |
| | Stern et al., 1994 | 93 | 9.1 ± 8.4 |

predictable for groups allowing for accurate computation of sample size for AD clinical trials. In addition to measuring cognitive scales, some work has been carried out measuring decline on global staging systems (Berg et al., 1988) and on activities of daily living (ADL) (Green et al., 1993).

Although all of the factors controlling the rate of cognitive decline is not fully understood, some have been identified. These include age at onset with younger patients progressing somewhat more rapidly than older subjects (Burns et al., 1991). Genetic factors clearly control the rate of progression and in some cases of familial AD, progression is rapid in all afflicted members. Recently, the gene dose of the cholesterol carrying protein, apolipoprotein $E_4$ was demonstrated to have a major effect on age of onset of AD. Individuals carrying two alleles for apolipoprotein $E_4$ develop the disease approximately 10 years earlier than individuals who carry no alleles for this form of the cholesterol carrying protein (Corder et al., 1993). Whether or not apolipoprotein $E_4$ status affects the rate of decline is as yet unknown. Other factors suspected of affecting the rate of decline include clinical features such as psychotic behavior, and extrapyramidal features (Stern et al., 1987a). Recently, approximately 20–25% of AD patients have been found to have cortical Lewy bodies at autopsy, the pathological hallmark of Parkinson's disease. The rate of decline for this cohort is approximately 50% more rapid than for individuals with AD pathology without accompanying Lewy bodies (Klauber et al., 1992).

## Treatment strategies for individuals with cognitive decline

At least five approaches can be envisioned for the treatment of individuals with cognitive decline. The first would be the treatment of behavioral symptoms associated with the disorder. These include depression, anxiety, insomnia, agitation and psychotic symptomatology. The second level of treatment would be the treatment of cognitive symptoms. This strategy has primarily employed neurotransmitter replacement therapies in an attempt to bolster cognition. While worthwhile, neurotransmitter replacement therapy is likely to produce only small and temporary improvements in patients with AD. At present, there is no available evidence suggesting the use of such agents will alter the inevitable downhill course. Use of symptomatic therapy should therefore result in a "shift effect" where the slope representing the rate of progression does not change but cognitive test scores are temporarily improved (Fig. 1a). A third strategy would be to prevent decline in AD. This should result in a change of slope in the downward progression of the disease (Fig. 1b). A fourth approach would be to delay the onset of appearance of the disease. Since prevalence figures indicate that the number of AD cases doubles with every 5 year epoch, a treatment that delayed appearance of the disease by 5 years would half the prevalence in one generation. If disease appearance could be delayed by 10 years, its prevalence would diminish by 75% in a generation. Development of such a treatment would have far reaching medical, economical and ethical consequences. A fifth approach would be actual disease prevention. Clinical trials for the latter two categories have not yet been attempted.

## Treatment with cholinergic agents

Cholinergic abnormalities in AD were first described in 1976 and are well documented in previous chapters. Additionally, a large body of litera-

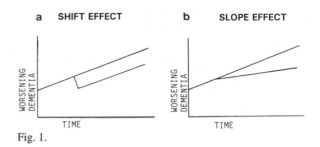

Fig. 1.

ture supports the concept of cholinergic involvement in learning and memory. Observations with scopolamine, a cholinergic antagonist, revealed that this agent could lead to memory impairment in humans (Drachman and Leavitt, 1974). This amnestic property of scopolamine was recognized and utilized in obstetrical anesthesia to produce a state of "twilight sleep." Women given scopolamine during labor and delivery were often amnesic for the details of the delivery. Both young (Drachman and Leavitt, 1974) and elderly individuals (Beatty et al., 1986) demonstrate neuropsychological impairment following scopolamine administration. However, the profile of this impairment is not the same as that seen in AD patients (Beatty et al., 1986).

Animal studies using anticholinergic agents also support the involvement of cholinergic neurotransmission in learning and memory. Beginning with the sentinel studies of Deutsch (1971), who demonstrated that memory consolidation in rats was interrupted by scopolamine administration, numerous investigators have demonstrated that both lesions of known cholinergic pathways (Olton and Feustle, 1981; and reviewed by Dekker et al., 1991) and cholinergic antagonists impair acquisition of new memory (Whishaw, 1989) in animals.

In AD, neurons of the nucleus basalis, the major cholinergic input to the cortex, undergo degeneration (Whitehouse et al., 1981). In animals, lesions of the nucleus basalis and/or septum clearly impair learning and memory and result in reduced levels of cortical choline acetyltransferase. These impairments can be partially ameliorated by the administration of the cholinomimetic agents physostigmine (Mandel and Thal, 1988) and tetrahydroaminoacridine (THA; tacrine) (Kwo-On-Yuen et al., 1990). On the basis of these clinical and experimental observations, many investigators hypothesize that memory impairment in AD may at least in part be caused by alterations of cholinergic neurotransmission.

Initial attempts at improving cholinergic neurotransmission focused on the use of choline and lecithin based on preclinical studies indicating that administration of choline to rats increased rat brain acetylcholine levels. However, neither choline nor lecithin improved performance on a wide variety of learning, memory, construction, and self-care tasks in AD (Table 2). Failure to improve cognition with choline and lecithin is probably secondary to saturation of high affinity choline uptake and the inability to further increase acetylcholine production by administration of exogenous choline.

Cholinesterase inhibitors prolong the action of acetylcholine at the postsynaptic cholinergic receptor by preventing its hydrolysis. Physostigmine is a tertiary amine which is poorly absorbed after oral administration but crosses the blood–brain barrier. Because of its safety, physostigmine has been widely studied in AD.

Early studies used intravenous physostigmine to overcome the problem of poor gastrointestinal absorption. The majority of double-blind studies of parental physostigmine demonstrated improvement on verbal and nonverbal memory tasks (Table 3). Most negative studies included patients with severe dementia or the use of a low fixed dose of physostigmine without dose titration.

Studies of oral physostigmine in AD have been carried out (Table 3) and cognitive improvement has been reported in seven studies in verbal memory on the ADAS (Beller et al., 1985; Mohs et al., 1985; Harrell et al., 1986; Mitchell et al., 1986; Sano et al., 1988, 1993; Thal et al., 1993). Five studies reported negative results, including one in which patients were too demented to be tested on a verbal learning task (Wettstein, 1983) and a second short-term study (Stern et al., 1987b) where the same patients subsequently responded after a longer duration of treatment (Sano et al., 1988). The efficacy of physostigmine clearly depends upon intact presynaptic terminals; consequently the amelioration of memory in AD is likely to be restricted to patients with at least a partially functioning cholinergic system.

Immediate-release preparations of physostigmine have a very short half-life of less than 30 min. The pharmacokinetics of physostigmine have been markedly improved by the development of controlled-release preparations that produce sustained blood levels for up to 7 h after ingesting

TABLE 2

Acetylcholine precursors in AD

| Study | Daily dose (g) | Duration (weeks) | No. of subjects | Results |
|---|---|---|---|---|
| *Choline* | | | | |
| Smith et al., 1978 | 9 | 2 | 10 | − |
| Christie et al., 1979 | 2–5 | 1 | 12 | − |
| Renvoize and Jerram, 1979 | 15 | 8 | 18 | − |
| Fovall et al., 1980 | 8–16 | 2 | 5 | ±[a] |
| Thal et al., 1981 | 4–16 | 2 | 7 | − |
| | | | | |
| *Lecithin* | | | | |
| Brinkman et al., 1982 | 35 | 2 | 10 | − |
| Dysken et al., 1982 | 15–30 | 2 | 10 | − |
| Weintraub et al., 1983 | 10–20 | 9 | 13 | − |
| Little et al., 1985 | 20–25 | 26 | 51 | − |

[a]Minimal improvement on word recognition. All studies are double-blind, placebo-controlled.

TABLE 3

Physostigmine in AD

| Study | Daily dose (mg) | Duration (days) | No. of subjects | Result |
|---|---|---|---|---|
| *Intravenous administration* | | | | |
| Peters and Levin, 1979 | 0.015 | Acute[a] | 5 | + |
| Ashford et al., 1981 | 0.5 | Acute | 6 | − |
| Christie et al., 1981 | 0.25–1 | Acute | 11 | + |
| Davis and Mohs, 1982 | 0.125–0.5 | Acute | 10 | + |
| Muramoto et al., 1984 | 0.3–0.8 | Acute | 6 | + |
| Schwartz and Kohlstaedt, 1986[b] | 0.004–0.013[c] | Acute | 6 | + |
| | | | | |
| *Oral administration* | | | | |
| Thal et al., 1983 | 3–16 | 6–8 | 8 | + |
| Wettstein, 1983 | 3–10 | 14–42 | 8 | − |
| Schmechel et al., 1984 | 10–12.5 | 4 | 12 | − |
| Beller et al., 1985 | 7–14 | 2 | 8 | + |
| Mohs et al., 1985 | 4–16 | 14–19 | 10 | + |
| Harrell et al., 1986 | Max tolerated | 14 | 15 | + |
| Mitchell et al., 1986 | Max tolerated | 4 | 16 | − |
| Stern et al., 1987b | 12–16 | 3 | 22 | − |
| Sano et al., 1988 | 12–16 | 42 | 17 | + |
| Thal et al., 1989 | 8–16 | 42 | 10 | + |
| Jenike et al., 1990 | 6–16 | 8–20 | 12 | − |
| Sano et al., 1993 | 10–16 | 42 | 29 | + |

[a]Single administration of drug.
[b]Intramuscular administration.
[c]mg/kg.
All studies are double-blind, placebo-controlled.

a single dose (Thal et al., 1989). A large multicenter trial examining the efficacy of controlled-release physostigmine in AD was initiated in the United States in 1989 and completed in 1992. Results of this trial, which are reported as being positive, are not yet published.

Recently, a second cholinesterase inhibitor, THA, marketed as tacrine, has attracted considerable attention. Several small studies suggested improvement in verbal memory (Table 4). In one of three studies of 89 subjects, a 2.6 point improvement on the MMSE but no ADL change was noted while on tacrine (Eagger et al., 1991). Subsequently, three large multicenter studies were conducted (Table 5). The first of these involved 632 patients with probable AD using an enriched population design. Of these, 215 met putative beneficial response criteria during a 6-week, double-blind, dose-titration phase and were randomized to receive placebo or their best dose of tacrine (40 or 80 mg daily) in a subsequent 6-week, double-blind parallel group trial. On average, treated patients showed a 2.4 point lesser decline on the ADAS-Cog, a 70 point scale of cognition, when compared to placebo. However, tacrine had no effect on the Clinical Global Impression of Change (CGIC). The treatment effect size was quite small and almost one-half of patients developed side effects, including asymptomatic elevation of liver enzymes during treatment (Davis et al., 1992). Based on the results of this study, the US Food and Drug Administration granted a treatment IND for the use of tacrine in AD. These results were corroborated in a second 12-week parallel group design study involving 468 patients treated with 20, 40 and 80 mg of tacrine daily. Significant im-

TABLE 4

Tetrahydroaminoacridine (THA; tacrine) in AD

| Study | Daily dose (mg) | Duration (days) | No. of subjects | Result |
|---|---|---|---|---|
| Kaye et al., 1982 | 30 + L | 3 doses | 10 | + |
| Summers et al., 1986 | Up to 200 + L | 3–26 months | 17 | + |
| Gauthier et al., 1990 | 100 + L | 18 weeks | 52 | − |
| Fitten et al., 1990 | Up to 250 | 7 | 10 | − |
| Chatellier and Lacombez, 1990 | Up to 125 | 28 | 67 | − |
| Davies et al., 1990 | Up to 200 + L | 4 | 10 | − |
| Molloy et al., 1991 | Up to 100 + L | 63 | 34 | − |
| Eagger et al., 1991 | Up to 150 + L | 30 weeks | 89 | + |
| Davis et al., 1992 | 40 or 80 | 6 weeks | 632 | +/− |
| Farlow et al., 1992 | 40 or 80 | 12 weeks | 468 | + |
| Maltby et al., 1994 | Up to 10 + L | 36 weeks | 53 | − |
| Knapp et al., 1994 | 80, 120, 160 | 30 weeks | 663 | + |

L, lecithin. All studies are double-blind, placebo-controlled.

provement was found on both the ADAS-Cog and the CGIC (Farlow et al., 1992). A third study explored higher daily doses of tacrine up to 160 mg in a 30-week study of 663 patients. Patients who could tolerate the 160 mg dose ended the study with ADAS-Cog scores about 5 points higher than placebo-treated patients and the global instrument utilized, a Clinician Interview Based Impression, was also improved. Unfortunately, of 238 patients in the highest dose group, only 64 (27%) were

evaluable at the end of the 30-week trial. The majority of dropouts were secondary to nausea, vomiting, and elevations in serum transaminase (Knapp et al., 1994).

Cholinergic agonists act directly at the postsynaptic receptor and several have been tried in AD (Table 6). Their use has often been impeded by peripheral side effects and lack of selectivity for brain muscarinic receptors. Using arecoline, slight improvement was noted on a picture recognition

TABLE 5

Recent large tacrine trials in AD

| Study | Design | Duration (weeks) | No. of subjects | ADAS-Cog[a] | Global scale[b] |
|---|---|---|---|---|---|
| Davis et al., 1992 | Enrichment | 6 | 632 entered 215 DB | 2.4[c] (0.001) 40 mg–1.2 80 mg–3.4 | 0.1 (0.30) |
| Farlow et al., 1992 | Parallel | 12 | 468 entered 273 week 12 evaluable | 40 mg–1.4 (0.36) 80 mg–3.8 (0.015) | 0.1 (0.70) 0.5 (0.015) |
| Knapp et al., 1994 | Parallel | 30 | 663 entered 263 completed | 80 mg–2.3 (0.11) 120 mg–1.8 (0.11) 160 mg–5.3 (0.001) | 0.3 (0.20) 0.3 (0.13) 0.5 (0.002) |

[a]ADAS-Cog difference between treated patients and controls.
[b]Difference between treated patients and controls on a global scale.
[c]Statistical analysis reported on combination of both doses. DB, double-blind.

TABLE 6

Cholinergic agonists in AD

| Study | Daily dose (mg) | Duration (days) | No. of subjects | Result |
|---|---|---|---|---|
| *Arecoline* | | | | |
| Christie et al., 1981 | 4 | Acute | 11 | + |
| Tariot et al., 1988 | 1–4 mg/h | Acute | 12 | – |
| Raffaele et al., 1991 | 0.5–40 | 14 | 8 | + |
| Soncrant et al., 1993 | 0.008–1.7 mg/h | 14 | 9 | + |
| | | | | |
| *Oxotremorine* | | | | |
| Davis et al., 1987 | 0.5–1 | Acute | 7 | – |
| | | | | |
| *RS-86* | | | | |
| Wettstein, 1984 | 2–25 | 14 | 6 | – |
| Bruno et al., 1986 | 4–5 | 8 | 8 | – |
| Hollander et al., 1987 | 2.25–4.5 | 7 | 12 | – |
| Mouradian et al., 1988 | 10 | 2 | 7 | – |
| | | | | |
| *Pilocarpine* | | | | |
| Caine, 1980 | 40 | 14 | 2 | – |
| | | | | |
| *Intraventricular bethanechol* | | | | |
| Harbaugh et al., 1984 | 0.05–0.07 | 21 | 4 | +* |
| Penn et al., 1988 | 0.35–1.75 | 56 | 10 | – |

*Subjective improvement. All of these studies were double-blind except Harbaugh et al., 1984.

task (Christie et al., 1981) and on verbal learning (Raffaele et al., 1991). Verbal memory was reported to improve with an inverted U-shaped dose response curve (Soncrant et al., 1993). A fourth study demonstrated no improvement (Tariot et al., 1988). Trials with other cholinergic agonists, including oxotremorine (Davis et al., 1987), RS-86 (Wettstein, 1984; Bruno et al., 1986; Hollander et al., 1987; Mouradian et al., 1988) and pilocarpine (Caine, 1980) have also been negative. A single blind encouraging trial of intraventricular bethanediol in AD (Harbaugh et al., 1984) was followed by a larger double-blind study (Penn et al., 1988) which failed to demonstrate improvement in cognition. Trials with more selective agonists are currently underway.

Why has cholinergic therapy not been more successful in AD? In addition to the changes in the cholinergic system, there are alterations in multiple additional neurotransmitter systems including loss of neurons containing noradrenaline, serotonin, somatostatin, substance P, neuropeptide Y, and corticotropin-releasing factor. It is therefore unlikely that correction of a single neurotransmitter abnormality in isolation would markedly enhance function. Given the marked and widespread pathology in AD, it seems likely that, at best, small improvements in learning and memory can occur with cholinergic augmentation. Whether these small changes will translate into meaningful functional improvement in ADL in patients with widespread neuropathology remains to be empirically demonstrated.

## The future of cholinomimetic therapy

At present, at least ten cholinesterase inhibitors and seven cholinergic agonists are under development. First generation cholinesterase inhibitors such as tacrine may be characterized by short duration of action and lack of specificity. Second generation agents will correct one of these issues

while third generation agents will have high specificity for acetylcholinesterase and a long duration
of activity. Nevertheless, it seems clear that only
small improvements in cognition will be obtained
using these agents in view of the widespread pathology and involvement of multiple neurotransmitter systems in the brains of patients with AD. In
addition, true clinical efficacy of these agents has
not yet been firmly established. While the large
tacrine trials demonstrated improvements on
cognitive scales and a very small improvement in
global functioning, improvement in ADL have not
been adequately demonstrated. Attempts to measure the cost effectiveness of these agents and their
potential role in health care utilization have not
been carried out. Undoubtedly safer cholinesterase
inhibitors and cholinergic agonists will be developed, but attention to overall clinical efficacy including ADL and cost should be addressed in future trials of these agents.

## Acknowledgements

Supported by NIH grants #AG05131 and 10483
and a grant from the Veterans Administration.

## References

Ashford, J.W., Soldinger, S., Schaeffer, J., Cochran, L. and
Jarvik, L.F. (1981) Physostigmine and its effects on six pa-
tients with dementia. *Am. J. Psychiatry*, 138: 829–830.
Beatty, W.W., Butters, N. and Janowsky, D.S. (1986) Patterns
of memory failure after scopolamine treatment: implications for cholinergic hypothesis of dementia. *Behav. Neural
Biol.*, 45: 196–211.
Beller, S.A., Overall, J.E. and Swann, A.C. (1985) Efficacy of
oral physostigmine in primary degenerative dementia. *Psychopharmacology*, 87: 147–151.
Berg, L., Miller, J.P., Storandt, M., Duchek, J., Morris, J.C.,
Rubin, E.H., Burke, W.J. and Coben, L.A. (1988) Mild senile dementia of the Alzheimer type 2. Longitudinal assessment. *Ann. Neurol.*, 23: 477–484.
Blessed, G., Tomlinson, B.E. and Roth, M. (1968) The asso-
ciation between quantitative measures of dementia and of
senile change in the cerebral grey matter of elderly subjects.
*Br. J. Psychiatry*, 114: 797–822.
Brinkman, S.D., Smith, R.C., Meyer, J.S., Vroulis, G., Shaw,
T., Gordon, J.R. and Allen, R.H. (1982) Lecithin and memory training in suspected Alzheimer's disease. *J. Gerontol.*,
37: 4–9.
Bruno, G., Mohr, E., Gillespie, M., Fedio, P. and Chase, T.N.

(1986) Muscarinic agonist therapy of Alzheimer's disease.
*Arch. Neurol.*, 43: 659–661.
Burns, A., Jacoby, R. and Levy, R. (1991) Progression of
cognitive improvement in Alzheimer's disease. *J. Am.
Geriatr. Soc.*, 39: 39–45.
Caine, E. (1980) Cholinomimetic treatment fails to improve
memory disorders. *N. Engl. J. Med.*, 303: 585–586.
Chatellier, G. and Lacomblez, L. (1990) Tacrine (tetrahydro-
aminoacridine; THA) and lecithin in senile dementia of the
Alzheimer type: a multicentre trial. *Br. Med. J.*, 300: 395–
399.
Christie, J.E., Blackburn, I.M., Glenn, A.I.M., Zeiglel, S.,
Shering, A. and Yates, C.M. (1979) Effects of choline and
lecithin on CSF choline levels and on cognitive function in
patients with presenile dementia of the Alzheimer type. In:
A. Barbeau, J.H. Growdon and R.J. Wurtman (Eds.), *Nutrition and the Brain*, Vol. 5, Raven Press, New York, pp.
377–387.
Christie, J.E., Shering, A., Ferguson, J. and Glen, A.I. (1981)
Physostigmine and arecoline: effects of intravenous infusions in Alzheimer presenile dementia. *Br. J. Psychiatry*,
138: 46–50.
Corder, E.H., Saunders, A.M., Strittmatter, W.J., Schmechel,
D.E., Gaskell, P.C., Small, G.W., Roses, A.D., Haines, J.L.
and Pericak-Vance, M.A. (1993) Gene dose of Apolipoprotein E Type 4 Allele and the risk of Alzheimer's disease in
late onset families. *Science*, 261: 921–923.
Davies, B., Andrewes, S., Stargatt, R., Ames, D., Tuckwell, V.
and Davis, S. (1990) Tetrahydro-aminoacridine in Alzheimer's disease. *Int. J. Geriatr. Psychiatry*, 5: 317–321.
Davis, K.L. and Mohs, R. (1982) Enhancement of memory
processes in Alzheimer's disease with multiple-dose intravenous physostigmine. *Am. J. Psychiatry*, 139: 1421–
1424.
Davis, K.L., Hollander, E., Davidson, M., Davis, B.M., Mohs,
R.C. and Horvath, T.B. (1987) Induction of depression with
oxotremorine in patients with Alzheimer's disease. *Am. J.
Psychiatry*, 144: 468–471.
Davis, K.L., Thal, L., Gamzu, E., Davis, C.S., Woolson, R.F.
and Gracon, S.I. (1992) A double-blind, placebo-controlled
multicenter study of tacrine for Alzheimer's disease. *N.
Engl. J. Med.*, 327: 1253–1259.
Dekker, A.J.A.M., Connor, D.M. and Thal, L.J. (1991) The
role of cholinergic projections from the nucleus basalis in
memory. *Neurosci. Biobehav. Rev.*, 15: 299–317.
Deutsch, D.A. (1971) The cholinergic synapse and the site of
memory. *Science*, 174: 788–790.
Drachman, D.A. and Leavitt, J. (1974) Human memory and
the cholinergic system: relationship to aging? *Arch. Neurol.*, 30: 113–121.
Dysken, M.W., Fovall, P., Harris, C.M., Davis, J.M. and No-
ronha, A. (1982) Lecithin administration in Alzheimer's
disease. *Neurology*, 32: 1203–1204.
Eagger, S.A., Levy, R. and Sahakian, B.J. (1991) Tacrine in
Alzheimer's disease. *Lancet*, 337: 989–992.
Ernst, R.L. and Hay, J.W. (1994) The US economic and social

costs of Alzheimer's disease revisited. *Am. J. Publ. Health*, 84: 1261.

Farlow, M., Gracon, S., Hershey, L., Lewis, K., Sadowsky, C. and Dolan-Ureno, J. (1992) A controlled trial of tacrine in Alzheimer's disease. *J. Am. Med. Assoc.*, 268: 2523–2529.

Fitten, L.J., Perryman, K.M., Gross, P., Fine, H., Cummins, J. and Marshall, E. (1990) Treatment of Alzheimer's disease with short- and long-term oral THA and lecithin: a double-blind study. *Am. J. Psychiatry*, 147: 239–242.

Folstein, M.F., Folstein, S.E. and McHugh, P.R. (1975) Mini-mental state: a practical method for grading the cognitive status of patients for the clinician. *J. Psychiatr. Res.*, 12: 189–198.

Fovall, P., Dysken, M.W., Lazarus, L.W., Davis, J.M., Kahn, R.L., Jope, R. and Finekl, S. (1980) Choline bitartrate treatment of Alzheimer-type dementias. *Commun. Psycho pharmacol.*, 4: 141–145.

Galasko, D., Hansen, L.A., Katzman, R., Wiederholt, W.C., Masliah, E., Terry, R., Hill, R., Lessin, P. and Thal, L.J. (1994) Clinical-neuropathological correlations in Alzheimer's disease and related dementias. *Arch. Neurol.*, 51: 888–895.

Gauthier, S., Bouchard, R., Lamontagne, A., Bailey, P., Bergman, H., Ratner, J., Tesfaye, Y., Saint-Martin, M., Bacher, Y and Carrier, L. (1990) Tetrahydroaminoacridine-lecithin combination treatment in patients with intermediate-stage Alzheimer's disease. *N. Engl. J. Med.*, 322: 1272–1276.

Green, C.R., Mohs, R.C., Schmeidler, J., Aryan, M. and Davis, K.L. (1993) Functional decline in Alzheimer's disease: a longitudinal study. *J. Am. Geriatr. Soc.*, 41: 654–661.

Harbaugh, R.E., Roberts, D.W., Coombs, D.W., Saunders, R.L. and Reeder, T.M. (1984) Preliminary report: intracranial cholinergic drug infusion in patients with Alzheimer's disease. *Neurosurgery*, 15: 514–518.

Harrell, L., Falgout, J., Leli, D., Jope, R., McLain, C., Spiers, M., Callaway, R. and Halsey, J. (1986) Behavioral effects of oral physostigmine in Alzheimer's disease patients. *Neurology*, 36 (Suppl. 1): 269.

Hollander, E., Davidson, M., Mohs, R.C., Horvath, T.B., Davis, B.M., Zemishlany, Z and Davis, K.L. (1987) RS 86 in the treatment of Alzheimer's disease: cognitive and biological effects. *Biol. Psychiatry*, 22: 1067–1078.

Jellinger, K., Danielczyk, W., Fischer, P. and Gabriel, E. (1990) Clinicopathological analysis of dementia disorders in the elderly. *J. Neurol. Sci.*, 95: 239–258.

Jenike, M.A., Albert, M.S., Heller, H., Gunther, J. and Goff, D. (1990) Oral physostigmine treatment for patients with presenile and senile dementia of the Alzheimer's type: a double-blind placebo-controlled trial. *J. Clin. Psychiatry*, 51: 3–7.

Katzman, R. (1976) The prevalence and malignancy of Alzheimer disease: a major killer. *Arch. Neurol.*, 33: 217–218.

Katzman, R., Brown, T., Thal, L.J., Fuld, P.A., Aronson, M., Butters, N., Klauber, M.R., Wiederholt, W.C., Pay, M. and

Xiong, R.B. (1988) Comparison of rate of annual change of mental status score in four independent studies of patients with Alzheimer's disease. *Ann. Neurol.*, 34: 384–389.

Kaye, W.H., Sitaram, N., Weingartner, H., Ebert, M.H., Smallberg, G. and Gillin, J.C. (1982) Modest facilitation of memory in dementia with combined lecithin and anticholinesterase treatment. *Biol. Psychiatry*, 17: 275–280.

Khachaturian, Z. (1992) The five-five, ten-ten plan for Alzheimer's disease (editorial). *Neurobiol. Aging*, 13: 197–198.

Klauber, M.R., Hofstetter, C.R., Hill, L.R. and Thal, L.J. (1992) Patterns of decline in the Lewy body variant of Alzheimer's disease. *Shanghai Arch. Psychiatry*, 4: 50–53.

Knapp, M.J., Knopman, D.S., Solomon, P.R., Pendlebury, W.W., Davis, C.S. and Gracon, S.I. (1994) A 30-week randomized controlled trial of high-dose tacrine in patients with Alzheimer's disease. *J. Am. Med. Assoc.*, 271: 985–991.

Kramer-Ginsberg, E., Mohs, R.C., Aryan, M., Lobel, D., Silverman, J., Davidson, M. and Davis, K.L. (1988) Clinical predictors of course for Alzheimer patients in a longitudinal study: a preliminary report. *Psychopharmacol. Bull.*, 24: 458–462.

Kwo-On-Yuen, P.F., Mandel, R., Chen, A.D. and Thal, L.J. (1990) Tetrahydroaminoacridine improves the spatial acquisition deficit produced by nucleus basalis lesions in rats. *Exp. Neurol.*, 108: 221–228.

Little, A., Levy, R., Chuaqui-Kidd, P. and Hand, D. (1985) A double-blind, placebo-controlled trial of high-dose lecithin in Alzheimer's disease. *J. Neurol. Neurosurg. Psychiatry*, 48: 736–742.

Maltby, N., Broe, G.A., Creasey, H., Jorm, A.F., Christensen, H. and Brooks, W.S. (1994) Efficacy of tacrine and lecithin in mild to moderate Alzheimer's disease: double blind trial. *Br. J. Med.*, 308: 879–83.

Mandel, R.J. and Thal, L.J. (1988) Physostigmine improves water maze performance following nucleus basalis magnocellularis lesions in rats. *Psychopharmacology*, 96: 421–425.

Mattis, S. (1976) Mental status examination for organic mental syndrome in the elderly patient. In: L. Bellack, T.B. Karasu (Eds.), *Geriatrics Psychiatry: A Handbook for Psychiatrists and Primary Care Physicians*, Grune and Stratton, New York, 1976, pp. 77–121.

Mitchell, A., Drachman, D., O'Donnell, B. and Glosser, G. (1986) Oral physostigmine in Alzheimer's disease. *Neurology*, 36: 295.

Mohs, R.C., Davis, B.M., Johns, C.A., Mathe, A.A., Greenwald, B.S., Horvath, T.B. and Davis, K.L. (1985) Oral physostigmine in treatment of patients with Alzheimer's disease. *Am. J. Psychiatry*, 142: 28–33.

Molloy, D.W., Guyatt, G.H., Wilson, D.B., Duke, R., Rees, L. and Singer, J. (1991) Effect of tetrahydroaminoacridine on cognition, function and behaviour in Alzheimer's disease. *Can. Med. Assoc. J.*, 144: 29–34.

Molsa, P.K., Paljarvi, L., Rinne, J.O., Rinne, U.K. and Sako, E. (1985) Validity of clinical diagnosis in dementia: a prospective clinico-pathological study. *J. Neurol. Neurosurg. Psychiatry,* 48: 1085–1090.

Morris, J.C., Edland, S., Clark, C., Galasko, D., Koss, E., Mohs, R., van Belle, G., Fillenbaum, G. and Heyman, A. (1993) The consortium to establish a registry for Alzheimer's disease (CERAD). Part IV. Rates of cognitive change in the longitudinal assessment of probable Alzheimer's disease. *Neurology,* 43: 2457–2465.

Mouradian, M.M., Mohr, E., Williams, J.A. and Chase, T.N. (1988) No response to high-dose muscarinic agonist therapy in Alzheimer's disease. *Neurology,* 38: 606–608.

Muramoto, O., Sugishita, M. and Ando, K. (1984) Cholinergic system and constructional praxis: a further study of physostigmine in Alzheimer's disease. *J. Neurol. Neurosurg. Psychiatry,* 47: 485–491.

Olton, D.S. and Feustle, W.A. (1981) Hippocampal function required for nonspatial working memory. *Exp. Brain Res.,* 41: 380–389.

Ortof, E. and Crystal, H.A. (1989) Rate of progression of Alzheimer's disease. *J. Am. Geriatr. Soc.,* 37: 511–514.

Penn, R.D., Martin, E.M., Wilson, R.S., Fox, J.H. and Savoy, S.M. (1988) Intraventricular bethanechol infusion for Alzheimer's disease: results of double-blind and escalating-dose trials. *Neurology,* 38: 219–222.

Peters, B. and Levin, H.S. (1979) Effects of physostigmine and lecithin on memory in Alzheimer's disease. *Ann. Neurol.,* 6: 219–221.

Raffaele, K.C., Berardi, A., Asthana, S., Morris, P., Haxby, J.V. and Soncrant, T.T. (1991) Effects of long-term continuous infusion of the muscarinic cholinergic agonist arecoline on verbal memory in dementia of the Alzheimer type. *Psychopharmacol. Bull,* 27: 315–319.

Renvoize, E.B. and Jerram, T. (1979) Choline in Alzheimer's disease. *N. Engl. J. Med.,* 301: 330.

Rosen, W.G., Mohs, R.C. and Davis, K.L. (1984) A new rating scale for Alzheimer's disease. *Am. J. Psychiatry,* 141: 1356–1364.

Salmon, D.P., Thal, L.J., Butters, N. and Heindel, W.C. (1990) Longitudinal evaluation of dementia of the Alzheimer type: a comparison of three standardized mental status examinations. *Neurology,* 40: 1225–1230.

Sano, M., Stern, Y., Stricks, L., Marder, K. and Mayeux, R. (1988) Physostigmine response in probable Alzheimer's disease is related to duration of exposure. *Neurology,* 38(Suppl. 1): 373.

Sano, M., Bell, K., Marder, K., Stricks, L., Stern, Y. and Mayeux, R. (1993) Safety and efficacy of oral physostigmine in the treatment of Alzheimer's Disease. *Clin. Neuropharmacol.,* 16: 61–69.

Schmechel, D.E., Schmitt, F., Horner, J., Wilkinson, W.E., Hurwitz, B.J. and Heyman, A. (1984) Lack of effect of oral physostigmine and lecithin in patients with probably Alzheimer's disease. *Neurology,* 34: 280.

Schwartz, A.S. and Kohlstaedt, E.V. (1986) Physostigmine effects in Alzheimer's disease: relationship to dementia severity. *Life Sci.,* 38: 1021–1028.

Smith, C.M., Swash, M., Exton-Smith, A.N., Phillips, M.J., Overslall, P.W., Piper, M.E. and Bailey, M.E. (1978) Choline therapy in Alzheimer's disease. *Lancet,* ii: 318.

Soncrant, T.T., Raffaele, K.C., Asthana, S., Berardi, A., Morris, P.P. and Haxby, J.V. (1993) Memory improvement without toxicity during chronic, low dose intravenous arecoline in Alzheimer's disease. *Psychopharmacology,* 112: 421–427.

Stern, Y., Mayeux, R., Sano, M., Hauser, W.A. and Bush, T. (1987a) Predictors of disease course in patients with probable Alzheimer's disease. *Neurology,* 37: 1649–1653.

Stern, Y., Sano, M. and Mayeux, R. (1987b) Effects of oral physostigmine in Alzheimer's disease. *Ann. Neurol.,* 22: 306–310.

Stern, R.G., Mohs, R.C., Davidson, M., Schmeidler, J., Silverman, J., Kramer-Ginsberg, E., Searcey, T., Bierer, L. and Davis, K.L. (1994) A longitudinal study of Alzheimer's disease: measurement, rate, and predictors of cognitive deterioration. *Am. J. Psychiatry,* 151: 390–396.

Summers, W.K., Majorski, L.V., Marsh, G.M., Tachiki, K. and Kling, A. (1986) Oral tetrahydroaminoacridine in long-term treatment of senile dementia, Alzheimer type. *N. Engl. J. Med.,* 315: 1241–1245.

Tariot, P., Cohen, R., Welkowitz, J., Sunderland, T., Newhouse, P.A., Murphy, D.L. and Weingartner, H. (1988) Multiple-dose arecoline infusions in Alzheimer's disease. *Arch. Gen. Psychiatry,* 45: 901–905.

Teri, L., Hughes, J.P. and Larson, E.B. (1990) Cognitive deterioration in Alzheimer's disease: behavioral and health factors. *J. Gerontol.,* 45: P58–63.

Thal, L.J., Rosen, W., Sharpless, S. and Crystal, H. (1981) Choline chloride fails to improve cognition in Alzheimer's disease. *Neurobiol. Aging,* 44: 24–29.

Thal, L.J., Fuld, P.A., Masur, D.M. and Sharpless, N.S. (1983) Oral physostigmine and lecithin improve memory in Alzheimer's disease. *Ann. Neurol.,* 13: 491–496.

Thal, L.J., Grundman, M. and Klauber, M.R. (1988) Dementia: characteristics of a referral population and factors associated with progression. *Neurology,* 38: 1083–1090.

Thal, L.J., Lasker, B., Sharpless, N.S., Bobotas, G., Schor, J.M. and Nigalye, A. (1989) Plasma physostigmine concentrations after controlled-release oral administration. *Arch. Neurol.,* 46: 13.

Thal, L.J., Masur, D.M., Blau, A.D., Fuld, P.A. and Klauber, M.R. (1989) Chronic oral physostigmine without lecithin improves memory in Alzheimer's disease. *J. Am. Geriatr. Soc.,* 37: 42–48.

Uhlman, R.F., Larson, E.B. and Koepsell, T.D. (1986) Hearing impairment and cognitive decline in senile dementia of the Alzheimer's type. *J. Am. Geriatr. Soc.,* 34: 207–210.

Wade, J.P.H., Mirsen, T.R., Hachinski, V.C., Fisman, M., Lau, C. and Merskey, H. (1987) The clinical diagnosis of Alzheimer's disease. *Arch. Neurol.,* 44: 24–29.

Weintraub, S., Mesulam, M.M., Auty, R., Baratz, R., Cholakos, B.N., Kapust, L., Ransil, B., Tellers, J.G., Albert, M.S., LoCastro, S. and Moss, M. (1983) Lecithin in the treatment of Alzheimer's disease. *Arch. Neurol.*, 40: 527–528.

Wettstein, A. (1983) No effect from double-blind trial of physostigmine and lecithin in Alzheimer's disease. *Ann. Neurol.*, 13: 210–212.

Wettstein, A. and Spiegel, R. (1984) Clinical trials with the cholinergic drug RS 86 in Alzheimer's disease (AD) and senile dementia of the Alzheimer type (SDAT). *Psychopharmacology*, 84: 572 573.

Whishaw, I.Q. (1989) Dissociating performance and learning deficits on spatial navigation tasks in rats subjected to cholinergic muscarinic blockade. *Brain Res. Bull.*, 23: 347–358.

Whitehouse, P.J., Price, D.L., Clark, A.W., Coyle, J.T. and DeLong, M.R. (1981) Alzheimer's disease: evidence for selective loss of cholinergic neurons in the nucleus basalis. *Ann. Neurol.*, 10: 122–126.

J. Klein and K. Löffelholz (Eds.)
*Progress in Brain Research*, Vol. 109
© 1996 Elsevier Science B.V. All rights reserved.

# CHAPTER 30

# New trends in cholinergic therapy for Alzheimer disease: nicotinic agonists or cholinesterase inhibitors?

Ezio Giacobini[1]

*Department of Pharmacology, Southern Illinois University School of Medicine, PO Box 19230, Springfield, IL 62794-1222, USA*

## Introduction

Cholinesterase inhibitors (ChEI) are, so far, the only drugs demonstrating clinical efficacy in the treatment of Alzheimer's disease (AD) (cf. Giacobini, 1994). The principle used in indirect cholinomimetic therapy is to reduce acetylcholine (ACh) hydrolysis in central nervous system (CNS) nerve terminals by means of acetylcholinesterase (AChE) inhibition (Chapter 29; Becker et al., 1991). The resulting increase in extracellular ACh concentration could restore the central cholinergic hypofunction and improve memory and cognition (Becker and Giacobini, 1988). The use of a ChEI (THA, tacrine, tetrahydroaminoacridine) has resulted in a dose-dependent clinical efficacy in 20–30% of AD patients (Knapp et al., 1994; Chapter 29). In addition to tacrine, 12–15 new ChEI are at different phases of clinical development (cf. Giacobini and Cuadra, 1994). A novel and interesting alternative to ChEI are new nicotinic agonists which are discussed here. The study of the cholinergic system has been made possible through the development of sensitive micromethods during the last 40 years (Table 1) with a sensitivity for ACh in the low femtomole (fmol) range (Fig. 1).

## Cholinesterase inhibitors effect on extracellular concentrations of cortical neurotransmitters

Clinical and experimental evidence indicates in-
volvement and interactions between the cholinergic system and the biogenic amine systems in the cognitive impairments observed in AD (Hardy et al., 1985; Decker and McGaugh, 1991). A brain region of particular interest is the frontal cortex because in both humans and rodents it represents the major cholinergic projection of the nucleus basalis magnocellularis of the basal forebrain (Mesulam and Geula, 1988). Of the nucleus basalis magnocellularis neurons that project to the cerebral cortex, 80–90% are cholinergic in the rat (Rye et al., 1984). Similarly, the major, if not sole, noradrenergic projection to the cortex is the locus coeruleus (LC) (Parnavelas, 1990). Pharmacological alleviation of combined cholinergic (nucleus basalis magnocellularis)/noradrenergic (locus coeruleus) lesion-induced memory deficits in rats has been reported (Santucci et al., 1991).

Table 2 compares the effects on ACh levels and AChE inhibition of five ChEI studies in our laboratory [physostigmine (PHY); heptylphysostigmine (HEP); MF-268 (2,6-dimethylmorfolin-octyl carbamoyl eseroline); E2020 [(R,S)-1-benzyl-4-(5,6 dimethoxy-1-idanon)-2-yl-methylpiperidine]; and metrifonate [0,0-dimethyl-(1-hydroxy-2,2,2 trichloroethyl-phosphate)] administered subcutaneously (s.c.).

Our results show a significant increase in endogenous ACh levels measured with in vivo microdialysis in rat frontal cortex for all five ChEIs investigated (Table 2). This is in agreement with the results of previous studies demonstrating an increase of extracellular ACh and other neurotransmitters (NE, DA) after systemic administra-

---

[1]Present address: Department of Geriatrics, University of Geneva, School of Medicine, Route de Mon Idée, CH-1226 Thonex-Geneva, Switzerland.

TABLE 1

Cholinergic system: 40 years of micromethod development (1955–1995)

|  | Method | Sample size | Sensitivity (mol) | Reference |
|---|---|---|---|---|
| AChE activity | Microdiver Gasometric | One cell |  $10^{-12}$  | Giacobini and Zajicek, 1956 Giacobini, 1957 |
|  | Radiometric | One cell | $10^{-12}$ | Koslow and Giacobini, 1969 |
| ChAT activity | Radiometric | One cell | $10^{-12}$ | Buckley et al., 1967 |
| Acetylcholine | Radiometric | One cell | $10^{-12}$ | McCaman and Hunt, 1965 |
|  |  |  | $10^{-12}$ | McCaman and Dewhurst, 1970 Goldberg and McCaman, 1973 |
|  | HPLC-ECD | $10 \mu l$ of dialysate | $10^{-14}$ $10^{-15}$ | Cuadra et al., 1994 Giacobini (this publication) |

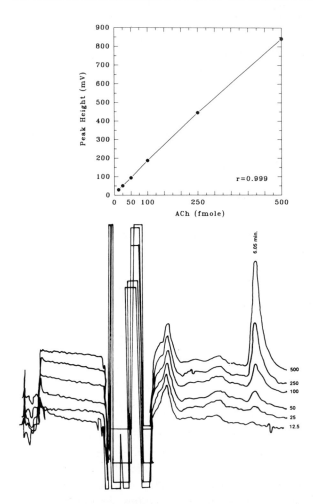

Fig. 1. Representative chromatogram (lower panel) showing standard dilution curves for ACh (500, 250, 100, 50, 25 and 12.5 fmol) and linearity of standards ($n = 6$) (insert) even at low fmol ACh levels.

tion of PHY and HEP (Giacobini and Cuadra, 1994). With the exception of MF-268 (not tested), they have all shown clinical efficacy. The difference in chemical structure among these compounds is due to a striking characteristic of new ChEI.

We observed differences between PHY and its analogue HEP in peak values of ACh, in time to pea and in return to baseline (Table 2). PHY ($300 \mu g/kg$) produced its maximal ACh elevation between 30 and 60 min, whereas HEP (2 mg/kg s.c.) induced maximal effect peaked at 60 min. Among the five ChEI, the longest-lasting effect on ACh levels after single s.c. administration was seen with HEP. At 360 min after the injection, ACh concentration was still significantly increased above baseline (Cuadra et al., 1994).

Inhibition of AChE after PHY and HEP was dose-dependent; however, it differed markedly between the two ChEIs (Cuadra et al., 1994). Differences were observed in the level of inhibition as well as in duration of AChE inhibition (Table 2). The results reported in Table 2 suggest that extracellular ACh levels are not directly related to the degree of AChE inhibition, supporting results of previous microdialysis studies showing comparable elevations of ACh levels in spite of different magnitudes of AChE inhibition (Messamore et al., 1993). As a consequence, CNS AChE inhibition cannot be considered as a reliable predictor of its effect on concentrations of extracellular ACh in cerebral cortex.

TABLE 2

ChEI effects on ACh levels and AChE activity in rat brain cortex (s.c. administration)

| Compound | Dose (mg/kg) | ACh max. increase (%) | ACh increase duration (h) | ChE max. inhibition % (h) | ChE inhibition duration (h) |
|---|---|---|---|---|---|
| Physostigmine | 0.3 | 4100 | 2 | 57 (0.5) | 2 |
| Heptyl-physostigmine | 2 | 2600 | 6 | 74 (1.0) | 6 |
| MF-268[a] | 2 | 2500 | 5.5 | 41 (1.0) | 6 |
| E2020[b] | 2 | 2100 | 5 | 35 (1.0) | 6 |
| Metrifonate[c] | 20 | 170 | 3 | 20 (1.0) | 2 |

[a]MF-268, 2,6-dimethylmorfolin-octyl carbamoyl eseroline.
[b]E2020, ($R,S$)-1-benzyl-4-(5,6 dimethoxy-1-idanon)-2-yl-methylpiperidine.
[c]Metrifonate, $O,O$-dimethyl-(1-hydroxy-2,2,2 trichloroethyl-phosphate).

A new aspect of ChEI pharmacology is their effect on neurotransmitters other than ACh (Cuadra et al., 1994). This effect depends not only on dose but also on the type of compound.

After systemic administration of PHY and HEP, we observed that the increase in ACh levels was followed in time by a significant elevation of extracellular NE levels. However, only PHY at both doses examined (0.03 mg/kg and 0.3 mg/kg) elicited a significant NE increase. This effect was weaker after HEP (2 mg/kg) and not measurable at a higher dose (5 mg/kg). Physostigmine and HEP administered locally in the cortex did not significantly modify NE levels (Cuadra et al., 1994). This suggests a subcortical effect of the drugs.

After systemic administration of both PHY and HEP, DA levels were also increased over control values (Cuadra et al., 1994). MF-268 (2 mg/kg) produced an increase in cortical ACh of the same order of magnitude of that of an equal dose of HEP however, maximal AChE inhibition was only 41% as compared to 74% for HEP. MF-268 (s.c.) did not produce any effect on either NE or DA, however, oral administration (2 mg/kg) of MF-268 produced an increase of both biogenic amines. The effect caused by a single dose of E2020 on ACh levels was almost as strong and as sustained as after an equal dose of HEP. Maximal AChE inhibition, however, was only 50% of the one following HEP.

Metrifonate, a long-term acting organophosphate-type ChEI, did increase ACh to a much smaller extent at a higher dose (20 mg/kg) than other ChEI (Mori et al., 1994) (Table 2). It is noteworthy that metrifonate is not per se a ChEI but needs to be metabolized to DDVP (dichlorvos, 2,2-dichlorovinyldimethyl phosphate) in order to become active. AChE inhibition was also lower after metrifonate than after the other inhibitors. Only ACh and DA levels were increased after this dose. These studies demonstrate that all the ChEI studied so far interact with ACh as well as with other cortical neurotransmitters, mainly NE and DA. This secondary effect may be of therapeutic interest.

## Co-administration of ChEIs with adrenergic agonists and antagonists demonstrates the interaction between cholinergic and adrenergic systems

Several studies have indicated close interactions between cholinergic and noradrenergic systems (Decker and McGaugh, 1991). Norepinephrine decreases the release of ACh from cholinergic terminals in cortex (Vizi, 1980; Moroni et al., 1983) (Fig. 2). This effect is mediated both directly via $\alpha$-adrenergic receptors on cholinergic terminals and indirectly via NE modulation of gamma aminobutyric acid (GABA) release (Beani et al., 1986) (Fig. 2). There is also evidence that NE and ACh interact with each other, influencing learning and memory (Santucci et al., 1991). The interaction between ACh and NE appears to be reciprocal as ACh is also able to modulate NE function (Roth et al., 1982; Egan and North, 1985;

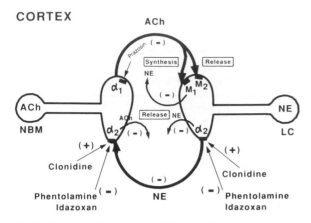

CORTEX

Fig. 2. Diagram of cholinergic (NBM, nucleus basalis magno-cellularis)-noradrenergic (LC, locus coeruleus) synaptic inter-actions in the rat cortex. The mutual effect of release and synthesis of ACh and NE on cholinergic muscarinic ($M_1$ and $M_2$) receptors as well on adrenergic ($\alpha_1$ and $\alpha_2$) receptors is depicted together with the action of adrenergic agonists (clonidine) and antagonists (phentolamine, idazoxan).

Hörtnagl et al., 1987). In a previous study (Cuadra et al., 1994), we have shown that systemic ad-ministration of low doses of PHY and HEP elicit a significant and simultaneous increase in ACh and NE levels. It is possible that the NE elevation seen in our studies could down-regulate ACh levels and decrease the therapeutic effect of these drugs.

## Effect of co-administration of adrenergic antagonist

To investigate this putative cholinergic-adrenergic interaction, we studied the effect of PHY and its analog HEP in animals pretreated with idazoxan, a selective $\alpha_2$-antagonist (Fig. 2), on the extracellu-lar levels of ACh, NE, DA and 5-hydroxytryp-tamine (serotonin, 5-HT) in cerebral cortex using microdialysis (Cuadra and Giacobini, 1995a).

In this study, we found that idazoxan adminis-tered either systemically or locally into the brain has no effect on the extracellular levels of ACh (Table 3). This suggests that NE may not be in-volved in tonic regulation of cortical cholinergic activity. The increase of cortical NE release that we saw after local or systemic idazoxan admini-stration is in agreement with the results of

L'Heureux et al. (1986) and Dennis et al. (1987). This suggests that the effects of idazoxan on NE release are mediated primarily by $\alpha_2$-adrenoceptors located presynaptically on noradrenergic nerve terminals (Fig. 2).

On the other hand, NE and DA were also in-creased after both types of administration. Phy-sostigmine administered to rats pretreated with idazoxan (systemically or locally) induced a simi-lar increase in the ACh levels as demonstrated for PHY alone. Conversely, co-administration of HEP with idazoxan intraperitoneally (i.p.) produced a further enhancement of the ACh cortical levels than HEP alone (Fig. 3) (Table 3).

Differences between HEP and PHY on ACh elevation may reflect the complex pharmacological profile of these drugs, such as different rates of metabolism and membrane permeability, effects on release of active compounds and rate of dissocia-tion of the inhibitor from the enzyme (Brufani et al., 1986; Cuadra et al., 1994).

These results are consistent with our previous data (Cuadra et al., 1994) in demonstrating a more favorable profile for HEP than PHY as a potential drug for treatment of cognitive dysfunctions asso-ciated with impaired cortical cholinergic transmis-sion. HEP produces a long-lasting effect on ACh in cerebral cortex (Fig. 3). These pharmacological data are also supported by preliminary findings (Imbimbo and Lucchelli, 1994). The possibility of further prolonging this effect with co-administra-

TABLE 3

Multiple effects of ChEI and idazoxan and clonidine co-administration on ACh and NE in rat cortex in vivo[a]

| Compound | Adrenergic $\alpha_2$-receptors | ACh | NE |
|---|---|---|---|
| ChEI (PHY or HEP) | 0 | ↑↑ | ↑ |
| Clonidine | + | 0 | ↓↓↓ |
| Idazoxan | − | 0 | ↑↑↑ |
| Clonidine + ChEI | + | ↑(↑) | ↓↓↓ or 0 |
| Idazoxan + ChEI | − | ↑↑↑ | ↑↑↑ |

0, no effect; +, agonistic effect; −, antagonistic effect; ↑, increased level; ↓, decreased level.
[a]Based on data from Cuadra and Giacobini (1995a,b).

IDAZOXAN          HEPTYLPHYSOSTIGMINE     IDAZOXAN  +  HEPTYLPHYSOSTIGMINE

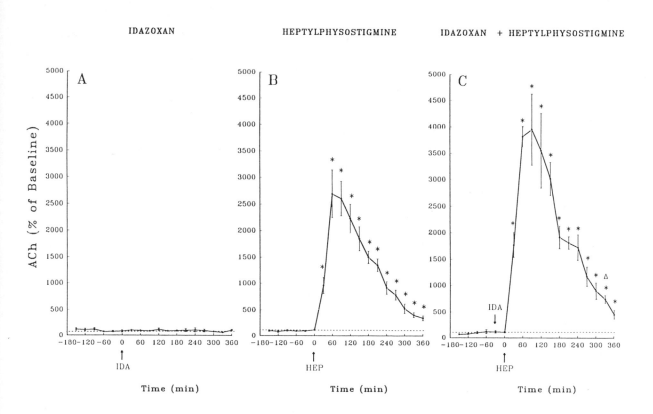

Fig. 3. Effect of local administration of idazoxan (IDA) (A), heptylphysostigmine (HEP) 2 mg/kg s.c. (B) and co-administration of IDA and HEP (C) on extracellular levels of ACh, in microdialysis samples from cerebral cortex of conscious, freely moving rats. IDA was perfused via the dialysis probe into the cortex at a concentration of $10^{-4}$ M, starting at time zero ($\uparrow$) or 30 min before PHY ($\downarrow$) and throughout the experiment. HEP was injected subcutaneously at time zero ($\uparrow$). Data are expressed as percent of control levels (average of six samples prior to injection = 100%); mean ± SEM; $n = 6$. *$P < 0.05$ compared to baseline by paired $t$-test. $^{\Delta}P < 0.05$ compared to HEP alone (Fig. 3B) (Split-Plot ANOVA followed by post hoc independent $t$-test).

tion of a selective $\alpha_2$-antagonist and additive DA-ACh interaction may also be of therapeutic interest.

Specifically, these data suggest that a combination of cholinergic and adrenergic drugs may improve the pharmacological effects of ChEI on several cortical neurotransmitter functions which may represent a significant advantage in AD treatment because of the multiple transmitter deficits seen in the disease.

### Effect of co-administration of adrenergic agonist

In order to obtain further information on cortical neurotransmitter interaction, we evaluated the ef-

fect of PHY and its analogue HEP on the extracellular levels of ACh, NE, DA and 5-HT in animals pre-treated with idazoxan, a selective $\alpha_2$-agonist (Fig. 2) (Cuadra and Giacobini, 1995b).

In agreement with our previous observations (Cuadra et al., 1994; Cuadra and Giacobini, 1995a), which suggested that NE may not be involved in the tonic regulation of cortical cholinergic activity, we detected no effect on extracellular levels of ACh after either systemic or local administration of clonidine; in contrast NE (Table 3), DA and 5-HT levels were all decreased. Clonidine co-administration reduced the effect of PHY on ACh levels (Fig. 4), however, HEP administered to animals pre-treated with clonidine produced a stronger effect than HEP alone. A possible expla-

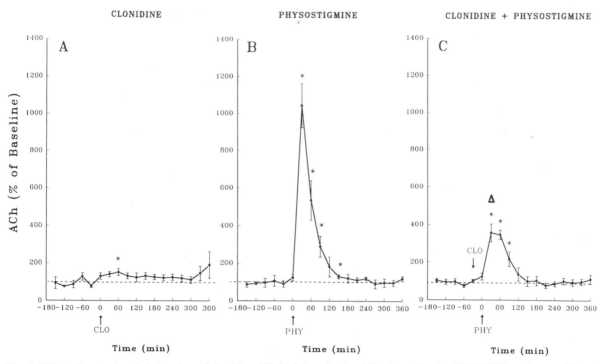

CLONIDINE          PHYSOSTIGMINE         CLONIDINE + PHYSOSTIGMINE

Fig. 4. Effect of systemic administration of clonidine (CLO) 0.1 mg/kg i.p. (A), physostigmine (PHY) 0.03 mg/kg s.c. (B) and co-administration of clonidine and PHY (C) on extracellular levels of ACh, in microdialysis samples from cerebral cortex of conscious, freely moving rats. CLO was injected i.p. at time zero (↑) or 30 min before PHY (↓). PHY was injected subcutaneously at time zero (↑). Data are expressed as percent of control levels (average of six samples prior to injection = 100%); mean ± SEM; baseline values were: (A) 2.40 ± 0.26 nM ($n = 6$); (B) 2.86 ± 0.04 nM ($n = 6$) and (C) 5.21 ± 0.37 nM ($n = 6$), respectively. *$P < 0.05$ compared to baseline by paired $t$-test. $^\Delta P < 0.05$ compared to PHY alone (B) (Split-Plot ANOVA followed by independent $t$-test).

nation for this difference is the variation in duration on ACh elevation and receptor desensitization for the two drugs.

The fact that there were no differences on ACh levels between two routes of co-administration of clonidine and PHY (s.c. and local through the probe) may suggest a cortical localization of clonidine effect and stimulation of $\alpha_2$-hetero-receptors localized on cortical cholinergic terminals (Fig. 2). Activation of these receptors by clonidine may down-regulate ACh levels.

The reduction in cortical NE release observed after local or systemic clonidine (54% and 57%, respectively) is in agreement with the results previously reported by L'Heureux et al. (1986) and Van Veldhuizen et al. (1993). The clonidine data, together with our previous results (Cuadra and Giacobini, 1995b) obtained in rats pre-treated with

idazoxan, suggest that ChEI effects on cortical NE release might be mainly mediated by $\alpha_2$-autoreceptors located on noradrenergic nerve terminals (Fig. 2) (Ong et al., 1991; Coull, 1994).

In analogy, both routes of clonidine administration (s.c. and local through the probe) also decreased extracellular levels of DA. This effect of clonidine on cortical release of DA might indicate an activation of $\alpha_2$-heteroreceptors localized presynaptically on terminals of dopaminergic neurons which have been demonstrated to modulate its release (Ueda et al., 1983; Dubocovich, 1984). It is well established that DA participates in the control of cognitive function (Brozoski et al., 1979) and plays a role in attention and reward mechanisms (Wise, 1978; Beninger, 1983).

Systemic and intracortical administration of clonidine elicited a similar decrease on cortical 5-

317

HT levels. The $\alpha_2$-heteroreceptors located on serotoninergic nerve terminals may play a major role in these effects. This is in agreement with previous findings which suggested that presynaptic $\alpha_2$-adrenoceptors mediated inhibition of 5-HT release in rat cerebral cortex (Maura et al., 1992).

In conclusion, our data suggest that co-administration of a selective $\alpha_2$-agonist such as clonidine with ChEI, does not represent a favorable pharmacological and therapeutic alternative. Furthermore, the decrease of extracellular DA may represent a negative effect in the treatment of cognitively impaired AD patients. Considering our previous results with idazoxan (Cuadra and Giacobini, 1995a), we suggest that a combination of an $\alpha_2$-antagonist with HEP may represent a more favorable approach to improve the clinical efficacy of ChEIs in AD treatment.

### ChEIs and APP secretion: a possible effect on patient deterioration?

The $\beta$-amyloid peptide ($\beta$A4), one of the major constituent proteins of neuritic plaques in the brain of AD patients, originates from a larger polypeptide denominated Alzheimer amyloid precursor protein (APP) (Kang et al., 1987). APP is widely distributed throughout the mammalian brain including rat brain with a prevalent neuronal localization (Beeson et al., 1994). APP can be processed by several alternative pathways but the mechanisms responsible for this processing are not completely understood. A secretory pathway is believed to generate non-amyloidogenic soluble derivatives (APPs) following cleavage within the $\beta$A4 segment (Sisodia et al., 1990; Esch et al., 1990). Cholinergic agonists regulating processing and secretion of APPs by increasing, as demonstrated in vitro, protein kinase C activity of target cells (Nitsch et al., 1992; Buxbaum et al., 1992; Nitsch and Growdon, 1994) could decrease potentially amyloidogenic derivatives. We suggested that long-term inhibition of AChE having the effect of increasing the level of synaptic ACh may result in the activation of normal APP processing in AD brain (Giacobini, 1994) (Fig. 5). This phe-

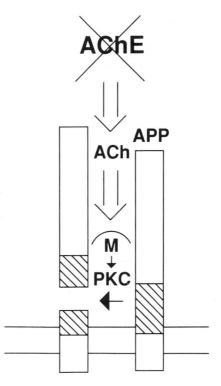

Fig. 5. Mechanism of action of AChE on normal APP processing through muscarinic (M) receptor and protein kinase C activation.

nomenon could slow down the formation of amyloidogenic APP fragments. To determine whether ChEI could alter the release of APP we used superfused brain cortical slices of the rat following the method described by Nitsch et al. (1993).

Three short- and long-lasting ChEI were tested for their ability to enhance the release of non-amyloidogenic soluble derivatives (APPs). These included: PHY, HEP and DDVP at concentrations producing AChE inhibitions ranging from 5% to 95%. All three ChEI elevated APPs release significantly above control levels (Table 4). Electrical field stimulation significantly increased the release of APPs within 50 min. Similar increase was observed after muscarinic receptor stimulation with bethanecholnechol (Table 4). Tetrodotoxin completely blocked the effect of electrical stimulation.

Level of total APP RNAs in rat cortical slices did not change after incubation with bethanechol,

TABLE 4

Drug-stimulated changes of basal APPs release and APP-KPI mRNA from rat brain[a]

| Drug | Concentration ($\mu$M) | Increase (% of basal) | ChE activity (% inhibition) | APP-KPI mRNA variation (% of basal) |
|------|------|------|------|------|
| Bethanechol | 1 | 48 | 0 | – |
|  | 100 | 53 | 0 | – |
| Physostigmine | 0.1 | 48 | 25 | – |
| Heptyl-physostigmine | 0.1 | 41 | 61 | –35[b] |
| DDVP | 0.02 | 33 | 95 | – |
| Phorbol myristate | 0.1 | – | – | +50 |

[a]From Mori et al., 1995.

[b]From Giacobini et al., 1995 (5 mg/kg s.c. 48 h).

DDVP and PHY, but activation of protein kinase C with phorbol 12-myristate 13-acetate (100 nM) increased the level of total APP mRNA by 50% (Table 4). PHY and metrifonate administration (0.3 mg/kg and 80 mg/kg s.c., respectively) for 3–48 h did not significantly change the levels of APP 695 and APP-KPI (containing the Kunitz-type protease inhibitor) mRNAs (Table 4). HEP administration (5 mg/kg s.c., 3–48 h) decreased by 35% the level of APP-KPI mRNA in rat cerebral cortex. AD pathology has been associated with an increase of the KPI-containing forms of APP and the propensity across species to develop neuritic plaques in the cortical regions (Anderson et al., 1989). Our findings suggest that administration of ChEI to AD patients by increasing secretion of APP and inhibiting formation of specific APP mRNAs may exert a neuroprotective effect by activating normal APP processing through a muscarinic mechanism and decreasing amyloid deposition in brain cells (Fig. 5).

The observed relationship among AChE inhibition, extracellular ACh increase and APPs processing in brain cortex raises two important questions. First, loss of synapses in human cortex, particularly cholinergic ones, represents one of the major correlates of cognitive impairment in AD (Terry et al., 1991; DeKosky et al., 1994). As a result of the cortical cholinergic deafferentation seen in AD, the normal neurotransmitter signal (ACh) which stimulates the release of APPs through muscarinic receptor activation could be attenuated. A conse-

quence of cholinergic denervation could be an accelerated amyloid formation in the brain as observed in AD. Secondly, pharmacological activation of cholinergic muscarinic receptors subsequent to ChEI-induced increase of ACh levels may increase secretion of non-amyloidogenic APPs. This effect would not only enhance the trophic action of APP but also inhibit amyloid deposition and cellular damage. The result would be a slower cognitive deterioration of the AD patient treated with ChEI. The recent finding of a relationship between choline acetyltransferase (ChAT) activity in frontal cortex and apolipoprotein E (apoE) epsilon 4 allele (APOE$_4$ genotype) in AD patients adds a new, interesting dimension to the relationship between cholinergic system, ChEI and $\beta$-amyloid formation (Soininen et al., 1995).

## An alternative to Alzheimer disease therapy: new nicotinic agonists

Several investigations have reported that systemically administered nicotine improves cognitive function in patients with AD (Newhouse et al., 1986; Jones and Sahakian, 1989; Newhouse et al., 1990; Warburton, 1990). Nicotine has many potential adverse effects including cardiovascular and gastrointestinal disturbances that limit its clinical use. Consequently, attempts are being made to investigate novel nicotinic receptor agonists devoid of these side effects that may prove to be therapeutically useful. Development of new nico-

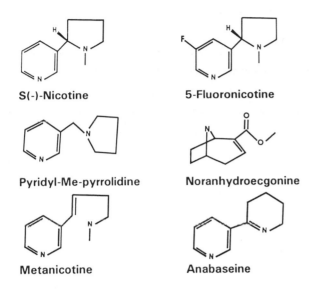

**S(-)-Nicotine**

**5-Fluoronicotine**

**Pyridyl-Me-pyrrolidine**

**Noranhydroecgonine**

**Metanicotine**

**Anabaseine**

Fig. 6. Chemical structure of S(−)-nicotine and nicotine-like compounds.

tinic agonists will also help to characterize the neuronal nicotinic receptor pharmacology (Summers and Giacobini, 1995). In a recent study, Perry et al. (1995) showed that, compared to other neurotransmitter receptors, the nicotinic receptor in cortex is significantly decreased in number of binding sites declining 50% with aging. This change occurs earlier and is greater than the decrease in ChAT activity. In addition, amyloid deposition in entorhinal cortex is inversely related to nicotinic binding and to smoking (Perry et al., 1995).

We studied six novel nicotine-like compounds which were selected based on structural diversity and binding affinities (Fig. 6). 5-Fluoronicotine is a direct nicotine derivative. Pyridyl-methylpyrrolidine represents an N-linked pyridine/pyrrolidine structure and noranhydroecgonine is structurally similar to anatoxin-a which is a natural nicotinic agonist produced by freshwater cyanobacterium (Fig. 6). Metanicotine was chosen because it has been reported to produce less cardiovascular toxicity than nicotine (Dominiak et al., 1985) while maintaining the ability to increase DA release in striatal synaptosomes and $Rb^+$ efflux in oocytes (Lippiello et al., 1994) (Fig. 6). All compounds tested have affinity for both major subtypes of nicotinic receptors, the high affinity

[3H]nicotine binding site and the high affinity [125I]$\alpha$-bungarotoxin binding site (Lippiello et al., 1994). In our study, a single dose level (3.6 $\mu$mol/kg) was selected at a concentration equal to that of nicotine for producing the maximal release of ACh in vivo (Summers and Giacobini, 1995). This dose produces plasma levels of nicotine in the range of those found in smokers (Pratt et al., 1983).

Using transcortical in vivo microdialysis in the rat, we have determined that different pharmacological profiles exist for these agonists with the same administered dose.

As shown in Table 5, of the six compounds investigated metanicotine and 5-fluoronicotine appeared to produce the greatest release of ACh (90% and 76%, respectively). An equivalent dose of (−)nicotine (3.6 $\mu$mol/kg) produced a 106% increase of extracellular ACh levels over basal values (Summers et al., 1994). Pyridyl-methylpyrrolidine, GTS-21 and anabaseine administration also resulted in a 40–50% increase in extracellular ACh levels, but noranhydroecgonine had no effect on the release of this neurotransmitter (Table 5) (Lippiello et al., 1995; Summers et al., 1995).

Pyridyl-methylpyrrolidine and anabaseine increased the release of NE but not of DA while 5-fluoronicotine increased the levels of DA but not NE (Table 5). Noranhydroecgonine significantly elevated both of these neurotransmitters with the strongest effect apparent on the dopaminergic system. With the exception of metanicotine, none of these compounds elevated 5-HT levels although, in an earlier study of the cortical response to subcutaneously administered (−)nicotine (1.2 $\mu$mol/kg), a slight increase of release was observed (Summers and Giacobini, 1995). Unlike (−)nicotine administration (Summers and Giacobini, 1995), no seizures were observed following metanicotine administration. Binding studies in rat brain have revealed that metanicotine has a lower affinity for the [125I]$\alpha$-bungarotoxin site than (−)nicotine (35–0.5 $\mu$M). The absence of convulsant activity of metanicotine corroborates earlier reports of a lower toxicity of metanicotine as compared to (−)nicotine (Dominiak et al., 1985; Lippiello et al., 1994) and makes this compound particularly attractive.

TABLE 5

Effects of nicotine and nicotinic receptor agonists on acetylcholine and biogenic amine levels in rat cerebral cortex[a]

| Compound[b] | % Increase and peak effect in min ( )[a] | | | |
|---|---|---|---|---|
| | ACh | NE | DA | 5-HT |
| S(−)-Nicotine | 106 (90) | 115 (60) | 103 (90) | c |
| D(+)-Nicotine | 48 (30) | 43 (60) | c | c |
| Noranhydroecgonine | c | 64 (150) | 147 (120) | c |
| Metanicotine | 90 (90) | 50 (90) | 50 (90) | 70 (90) |
| 5-Fluoronicotine | 76 (60) | c | 69 (90) | c |
| Anabaseine | 50 (60) | 62 (120) | c | c |
| GTS-21[3-(2,4-dimethoxybenzylidene) anabaseine] | 41 (30) | 83 (270) | 96 (210) | c |
| Pyridyl-methylpyrrolidine | 40 (30) | 63 (60) | c | c |

[a]Reported data are statistically significant ($P < 0.05$–$0.01$) (Lippiello et al., 1995; Summers et al., 1995).
[b]3.6 $\mu$mol/kg = 0.58 mg/kg s.c. (all compounds were administered at the same dose).
[c]No significant effect.

The strong variation of effects of the four neurotransmitters examined may reflect the structural diversity of the compounds tested. 5-Fluoronicotine is a direct nicotine derivative. Pyridyl-methylpyrrolidine represents an N-linked pyridine/pyrrolidine structure, and noranhydroecgonine is structurally similar to anatoxin-a. The results may also reflect a diversity of functional nicotinic receptors in vivo and their localization.

All responses were blocked by prior administration of mecamylamine, a nicotinic receptor antagonist, indicating that the effects were mediated by nicotinic receptors. Abundant evidence exists for the presence of nicotinic receptors not only in the cortex but also in major areas projecting to the cortex such as nucleus basalis, locus coeruleus and the ventral tegmental area (Shimohama et al., 1983; Egan and North, 1986; Deutch et al., 1987; Schröder et al., 1994). These areas supply ACh, NE and DA to the cortex, respectively, supporting our evidence of multi-neurotransmitter effects.

We conclude that several new nicotinic receptor agonists increase the release of ACh, NE and DA in rat neocortex. There appears to be distinctive pharmacological profiles for these agonists suggestive of effects on specific nicotinic receptor subtypes, an effect that may be related to the diverse structures of the tested compounds. These data support the hypothesis that selective nicotinic receptor agonists may be developed that are therapeutically beneficial for some neurodegenerative disorders such as AD. This is a new field of AD therapy that awaits exploration. Neuropsychological testing in humans suggests that the cognitive effect of ChEI and nicotinic agonists may be different (Newhouse et al., 1986; Jones and Sahakian, 1989; Newhouse et al., 1990; Warburton, 1990).

## Acknowledgements

The author wishes to thank Elizabeth Williams for technical assistance, and Diana Smith for typing and editing the manuscript. This work was supported in part by National Institutes of Aging grant P30 AG08014; Mediolanum Farmaceutici Inc. (Milan, Italy); Bayer (West Haven, CT); and R.J. Reynolds Tobacco Company.

## References

Anderson, J.P., Refolo, L.M., Wallace, W., Mehta, P., Krishnamurthi, M., Gotlib, J., Bierer, L., Haroutunian, V., Perl, D. and Robakis, N.K. (1989) Differential brain expression of the Alzheimer's amyloid precursor protein. *EMBO J.*, 8: 3627–3632.

Beani, L., Tanganelli, S., Antonelli, T. and Bianchi, C. (1986) Noradrenergic modulation of cortical acetylcholine release is both direct and gamma-aminobutyric acid-mediated. *J. Pharmacol. Exp. Ther.*, 236: 230–236.

Becker, R.E. and Giacobini, E. (1988) Mechanisms of cholinesterase inhibition in senile dementia of the Alzheimer type: clinical, pharmacological and therapeutic aspects. *Drug Dev. Res.*, 12: 163–195.

Becker, R.E., Moriearty, P. and Unni, L. (1991) The second generation of cholinesterase inhibitors: clinical and pharmacological effects. In: E. Giacobini and R. Becker (Eds.), *Cholinergic Basis for Alzheimer Therapy*, Birkhauser Boston, Cambridge, MA, pp. 263–296.

Beeson, J.G., Shelton, E.R., Chan, H.W. and Gage, F.H. (1994) Differential distribution of amyloid protein precursor immunoreactivity in the rat brain studied by using five different antibodies. *J. Comp. Neurol.*, 342: 78–96.

Beninger, R.J. (1983) The role of dopamine in locomotor activity and learning. *Brain Res. Rev.*, 6: 173–196.

Brozoski, T.J., Brown, R.M., Rosvold, H.E. and Goldman, P.S. (1979) Cognitive deficit caused by regional depletion of dopamine in the prefrontal cortex of rhesus monkey. *Science*, 205: 929–932.

Brufani, M., Marta, M. and Pomponi, M. (1986) Anticholinesterase activity of a new carbamate, heptylphysostigmine, in view of its use in patients with Alzheimer-type dementia. *Eur. J. Biochem.*, 157: 115–120.

Buckley, G., Consolo, S., Giacobini, E. and McCaman, R. (1967) A micromethod for the determination of choline acetylase in individual cells. *Acta Physiol. Scand.*, 71: 341–347.

Buxbaum, J.D., Oishi, M., Chen, H.I., Pinkas-Kramarski, R., Jaffe, E.A., Gandy, S.E. and Greengard, P. (1992) Cholinergic agonists and interleukin 1 regulate processing and secretion of the Alzheimer $\beta$A4 amyloid protein precursor. *Proc. Natl. Acad. Sci. USA*, 89: 10075–10078.

Coull, J.T. (1994) Pharmacological manipulation of the $\alpha_2$-noradrenergic system - effects on cognition. *Drugs Aging*, 5: 116–126.

Cuadra, G. and Giacobini, E. (1995a) Co-administration of cholinesterase inhibitors and idazoxan: effects of neurotransmitters in rat cortex *in vivo. J. Pharmacol. Exp. Ther.*, 273: 230–240.

Cuadra, G. and Giacoibni, E. (1995b) Effects of cholinesterase inhibitors and clonidine co-administration on rat cortex neurotransmitters *in vivo. J. Pharmacol. Exp. Ther.*, 275: 228–236.

Cuadra, G., Summers, K. and Giacobini, E. (1994) Cholinesterase inhibitor effects on neurotransmitters in rat cortex *in vivo. J. Pharmacol. Exp. Ther.*, 270: 277–284.

Decker, M.W. and McGaugh, J.L. (1991) The role of interactions between cholinergic system and other neuromodulatory systems in learning and memory. *Synapse*, 7: 151–168.

DeKosky, S.T., Styren, S.D. and O'Malley, M.E. (1994) Cholinergic changes and synaptic alterations in Alzheimer's disease. In E. Giacobini and R. Becker (Eds.), *Alzheimer Disease: Therapeutic Strategies*, Birkhauser Boston, Cambridge, MA, pp. 93–98.

Dennis, T., L'Heureux, R., Carter, C. and Scatton, B. (1987) Presynaptic $\alpha_2$- adrenoceptors play a major role in the effects of idazoxan on cortical noradrenaline release (as measured by *in vivo* dialysis) in the rat. *J. Pharmacol. Exp. Ther.*, 241: 642–649.

Deutch, A.Y., Holliday, J., Roth, R.H., Chun, L.L.Y. and Hawrot, E. (1987) Immunohistochemical localization of a neuronal nicotinic acetylcholine receptor in mammalian brain. *Proc. Natl. Acad. Sci. USA*, 84: 8697–8701.

Dominiak, G., Fuchs, G., von Toth, S. and Grobecker, H. (1985) Effects of nicotine and its major metabolites on blood pressure in anesthetized rats. *Klin. Wochensch.*, 63: 90–92.

Dubocovich, M.L. (1984) Presynaptic $\alpha$-adrenoceptors in the central nervous system. *Ann. N. Y. Acad. Sci.*, 430: 7–25.

Egan, T.M. and North, R.A. (1985) Acetylcholine acts on $M_2$-muscarinic receptors to excite rat locus coeruleus neurones. *Br. J. Pharmacol.*, 85: 733–735.

Egan, T.M. and North, R.A. (1986) Actions of acetylcholine and nicotine on rat locus coeruleus neurons *in vitro. Neuroscience*, 19: 565–571.

Esch, F.S., Keim, P.S., Beattie, E.C., Blacher, R.W., Culwell, A.R., Oltersdorf, T., McClure, D. and Ward, P.J. (1990) Cleavage of amyloid beta peptide during constitutive processing of its precursor. *Science*, 248: 1122–1124.

Giacobini, E. (1957) Quantitative determination of cholinesterase in individual sympathetic cells. *J. Neurochem.*, 1: 234–244.

Giacobini, E. (1994) Cholinomimetic therapy of Alzheimer disease: does it slow down deterioration? In: G. Racagni, N. Brunello and S.Z. Langer (Eds.), *Recent Advances in the Treatment of Neurodegenerative Disorders and Cognitive Dysfunction. Int. Acad. Biomed. Drug Res.*, Vol. 7, Karger, New York, pp. 51–57.

Giacobini, E. and Cuadra, G. (1994) Second and third generation cholinesterase inhibitors: from preclinical studies to clinical efficacy. In: E. Giacobini and R. Becker (Eds.), *Alzheimer Disease: Therapeutic Strategies*, Birkhauser Boston, Cambridge, MA, pp. 155–171.

Giacobini, E. and Zajicek, J. (1956) Quantitative determination of acetylcholinesterase activity in individual nerve cells. *Nature*, 177: 185–186.

Giacobini, E., Mori, F., Buznikov, A. and Becker, R. (1995) Cholinesterase inhibitors alter APP secretion and APP mRNA in rat cerebral cortex. *Soc. Neuroscience Abstr.*, in press.

Goldberg, A.M. and McCaman, R.E. (1973) The determination of picomole amounts of acetylcholine in mammalian brain. *J. Neurochem.*, 20: 1–8.

Hardy, J., Adolfsson, R., Alafuzoff, I., Bucht, G., Marcusson, J., Nyberg, P., Perdahl, E., Wester, P. and Winblad, B. (1985) Transmitter deficits in Alzheimer's disease. *Neurochem. Int.*, 7: 545–563.

Hörtnagl, H., Potter, P.E. and Hanin, I. (1987) Effect of cholinergic deficit induced by ethylcholine aziridinium (AF64A) on noradrenergic and dopaminergic parameters in rat brain. *Brain Res.*, 421: 75–84.

Imbimbo, B.P. and Lucchelli, P.E. (1994) A pharmacodynamic strategy to optimize the clinical response to eptastigmine (MF-201). In: E. Giacobini and R. Becker (Eds.), *Alzheimer Disease: Therapeutic Strategies*, Birkhauser Boston, Cambridge, MA, pp. 103–107.

Jones, G.M.M. and Sahakian, B.J. (1989) The effects of nicotine in patients with dementia of the Alzheimer type. In: T. Kewitz, T. Thomsen and E. Bickel (Eds.), *Pharmacological Interventions on Central Cholinergic Mechanisms in Senile Dementia*, Zuckschwerdt, San Francisco, CA, pp. 89–96.

Kang, J., Lemaire, H-G., Unterbeck, A., Salbaum, J. M., Master, C. L., Grzeschil, K-H., Multaup, G., Beyreuther, K. and Muller-Hill, B. (1987) The precursor of Alzheimer disease amyloid A4 protein resembles a cell-surface receptor. *Nature*, 325: 733–736.

Knapp, M.J., Knopman, D.S., Solomon, P.R., Pendlebury, W.W., Davis, C.S. and Gracon, S.I. (1994) A 30-week randomized controlled trial of high-dose Tacrine in patients with Alzheimer's disease. *J. Am. Med. Assoc.*, 271: 985–991.

Koslow, S.H. and Giacobini, E. (1969) An isotopic micromethod for the measurement of cholinesterase activity in individual cells. *J. Neurochem.*, 16: 1523–1528.

L'Heureux, R., Dennis, T., Curet, O. and Scatton, B. (1986) Measurement of endogenous noradrenaline release in the rat cerebral cortex in vivo by transcortical dialysis: Effects of drugs affecting noradrenergic transmission. *J. Neurochem.*, 46: 1794–1801.

Lippiello, P.M., Caldwell, W.S., Marks, M.J. and Collins, A.C. (1994) Development of nicotinic agonists for the treatment of Alzheimer's disease. In: E. Giacobini and R. Becker (Eds.), *Alzheimer Disease: Therapeutic Strategies*, Birkhauser Boston, Cambridge, MA, pp. 186–190.

Lippiello, P.M., Gray, J.A., Peters, S., Grigoryan, G., Hodges, H., Summers, K., Giacobini, E. and Collins, A.C. (1995) Metanicotine: a nicotine analog with CNS selectivity - *in vivo* characterization. *Soc. Neuroscience Abstr.*, in press.

Maura, G., Bonanno, G. and Raiteri, M. (1992) Presynaptic $\alpha_2$-adrenoceptors mediating inhibition of noradrenaline and 5-hydroxytryptamine release in rat cerebral cortex: further characterization as different $\alpha_2$-adrenoceptors subtypes. *Naunyn-Schmiedeberg's Arch. Pharmacol.*, 345: 410–416.

McCaman, R.E. and Dewhurst, S.A. (1970) Choline acetyltransferase in individual neurons of *Aplysia californica*. *J. Neurochem.*, 17: 1421–1426.

McCaman, R.E. and Hunt, J.M. (1965) Microdetermination of choline acetylase in nervous tissue. *J. Neurochem.*, 12: 253–259.

Messamore, E., Warpman, U., Ogane, N. and Giacobini, E. (1993) Cholinesterase inhibitor effects on extracellular acetylcholine in rat cortex. *Neuropharmacology*, 32: 745–750.

Mesulam, M.-M. and Geula, C. (1988) Nucleus basalis (Ch4) and cortical cholinergic innervation in the human brain: observations based on the distribution of acetylcholinesterase and choline acetyltransferase. *J. Comp. Neurol.*, 275: 216–240.

Mori, F., Cuadra, G., Williams, E., Giacobini, E. and Becker, R. (1994) Effects of metrifonate on acetylcholine and monoamine levels in rat cortex. *Soc. Neuroscience Abstr.*, 20: 83 (No. 40.16).

Mori, F., Lai, C-C., Fusi, F. and Giacobini, E. (1995) Cholinesterase inhibitors increase secretion of APPs in rat brain cortex. *NeuroReport*, 6: 633–636.

Moroni, F., Tanganelli, S., Antonelli, T., Carlá, V., Bianchi, C. and Beani, L. (1983) Modulation of cortical acetylcholine and gamma-aminobutyric acid release in freely moving guinea pigs: effects of clonidine and other adrenergic drugs. *J. Pharmacol. Exp. Ther.*, 236: 230–236.

Newhouse, P.A., Sunderland, T., Thompson, K., Tariot, P.N., Weingartner, H., Mueller, E.R., Cohen, R.M. and Murphy, D.L. (1986) Intravenous nicotine in a patient with Alzheimer's disease. *Am. J. Psychiatr.*, 143: 1494–1495.

Newhouse, P., Sunderland, T., Narang, P., Mellow, A.M., Fertig, J.B., Lawlor, B.A. and Murphy, D.L. (1990) Neuroendocrine, physiologic and behavioral responses following intravenous nicotine in non-smoking healthy volunteers and patients with Alzheimer's disease. *Psychoneuroendocrinology*, 15: 471–484.

Nitsch, R.M. and Growdon, J.H. (1994) Role of neurotransmission in the regulation of amyloid beta-protein precursor processing. *Biochem. Pharmacol.*, 47: 1275–1284.

Nitsch, R.M., Slack, B.E., Wurtman, R.J. and Growdon, J.H. (1992) Release of Alzheimer precursor derivatives stimulated by activation of muscarinic acetylcholine receptors. *Science*, 258: 304–307.

Nitsch, R.M., Farber, S.A., Growdon, J.H. and Wurtman, R.J. (1993) Release of amyloid beta-protein precursor derivatives by electrical depolarization of rat hippocampal slices. *Proc. Natl. Acad. Sci. USA*, 90:191–193.

Ong, M.L., Ball, S.G. and Vaughan, P.F.T. (1991) Regulation of noradrenaline release from rat occipital cortex tissue chops by $\alpha_2$-adrenergic agents. *J. Neurochem.*, 56: 1387–1393.

Parnavelas, J.G. (1990) Neurotransmitters in the cerebral cortex. In: H.B.M. Uylings, C.G. Van Eden, J.P.C. De Bruin, M.A. Corner and M.G.P. Feenstra (Eds.), *Progress in Brain Research*, Vol. 85, Elsevier, Amsterdam, pp. 13–29.

Perry, E.K., Court, J.A., Lloyd, S., Johnson, M., Griffiths, M.H., Spurden, D., Piggott, M.A., Turner, J. and Perry, R.H. (1995) Beta-amyloidosis in normal aging and transmitter signalling in human temporal lobe. In: J.H. Growdon, R.M. Nitsch, S. Corkin and R.J. Wurtman (Eds.), *The Neurobiology of Alzheimer Disease*, Center for Brain Sciences, Zurich, pp. 451–455.

Pratt, J.A., Stolerman, I.P., Garcha, H.S., Giardini, V. and Feyerabend, C. (1983) Discriminative stimulus properties of nicotine: further evidence for mediation at a cholinergic receptor. *Psychopharmacology*, 32: 291–296.

Roth, K.A., McIntire, S.L. and Barchas, J.D. (1982) Nicotinic-catecholaminergic interactions in rat brain: evidence for cholinergic nicotinic and muscarinic interactions with hypothalamic epinephrine. *J. Pharmacol. Exp. Ther.*, 221: 416–420.

Rye, D.B., Wainer, B.H., Mesulam, M.-M., Mufson, E.J. and Saper, C.B. (1984) Cortical projections arising from the basal forebrain: a study of cholinergic and noncholinergic components employing combined retrograde tracing and immunohistochemical localization of choline acetyltransferase. *Neuroscience*, 13: 627–643.

Santucci, A.C., Haroutunian, V. and Davis, K. L. (1991) Pharmacological alleviation of combined cholinergic - noradrenergic lesion-induced memory deficits in rats. *Clin. Neuropharmacol.*, 14: 1–8.

Schröder, H., Wevers, A. and Birtsch, C. (1994) Nicotinic receptors in human brain. In: E. Giacobini and R. Becker (Eds.), *Alzheimer Disease: Therapeutic Strategies*, Birkhauser Boston, Cambridge, MA, pp. 181–185.

Shimohama, S., Taniguchi, T., Fujiwara, M. and Kameyama, M. (1985) Biochemical characterization of the nicotinic cholinergic receptor in human brain: binding of [$^3$H]-nicotine. *J. Neurochem.*, 45: 604–610.

Sisodia, S.S., Koo, E.H., Beyreuther, K., Unterbeck, A. and Price, D.L. (1990) Evidence that beta amyloid protein in Alzheimer disease is not derived by normal processing. *Science*, 248: 492–495.

Soininen, H., Kosunen, O., Helisalmi, S., Mannermaa, A., Paljarvi, L., Talasniemi, S., Ryynanen, M. and Riekkinen, Sr., P. (1995) A severe loss of choline acetyltransferase in frontal cortex of Alzheimer patients carrying apolipoprotein e4 allele. *Neurosci. Lett.*, 187: 79–82.

Summers, K.L. and Giacobini, E. (1995). Effects of local and repeated systemic administration of (-)nicotine on release of acetylcholine, norepinephrine, dopamine and serotonin in rat cortex. *J. Neurosci. Res.*, 20: 683–689.

Summers, K.L., Cuadra, G., Naritoku, D. and Giacobini, E. (1994) Effects of nicotine on levels of acetylcholine and biogenic amines in rat cortex. *Drug Dev. Res.*, 31: 108–119.

Summers, K.L., Lippiello, P.M., Verhulst, S. and Giacobini, E. (1995) 5-Fluoronicotine, noranhydroeccgonine and pyridylmehtylpyrrolidine release acetylcholine and biogenic amines in rat cortex *in vivo*. *Neurochem. Res.*, 20: 1089–1094.

Terry, R.D., Masliah, E., Salmon, D.P., Butters, N., DeTeresa, R., Hill, R., Hansen, L.A. and Katzman, R. (1991) Physical basis of cognitive alterations in Alzheimer disease: synapse loss is the major correlate of cognitive impairment. *Ann. Neurol.*, 30: 572–580.

Ueda, H., Goshima, Y. and Misu, Y. (1983) Presynaptic mediation by alpha$_2$-, beta$_1$- and beta$_2$-adrenoceptors of endogenous noradrenaline and dopamine release from slices of rat hypothalamus. *Life Sci.*, 33: 371–376.

Van Veldhuizen, M.J., Feenestra, M.G., Heinsbroek, R.P. and Boer, G.J. (1993) In vivo microdialysis of noradrenaline overflow: effects of alpha-adrenoceptor agonists and antagonists measured by cumulative concentration-response curves. *Br. J. Pharmacol.*, 109: 655–660.

Vizi, E.S. (1980) Modulation of cortical release of acetylcholine by noradrenaline released from nerves arising from the rat locus coeruleus. *Neuroscience*, 5: 2139–2144.

Warburton, D.M. (1990) Nicotine as a cognitive enhancer. In: I.I. Yamashita, M. Toru and A.J. Copen (Eds.), *Clinical Neuropharmacology*, Vol. 13, Raven Press, New York, pp. 579–580.

Wise, R.A. (1978) Catecholamine theories of reward: a critical review. *Brain Res.*, 152: 215–247.

# Section X


# Cholinergic Dysfunction: 3.
# Trophic Factors

J. Klein and K. Löffelholz (Eds.)
*Progress in Brain Research*, Vol. 109
© 1996 Elsevier Science B.V. All rights reserved.

# CHAPTER 31

# Neurotrophic factors

## Leon J. Thal

*Department of Neurosciences, University of California at San Diego School of Medicine, 9500 Gilman Drive, La Jolla, CA 92093-0624,*
*USA and Department of Neurology, San Diego Veterans Administration Medical Center, San Diego, CA, USA*

## Introduction

Growth or neurotrophic factors promote the differentiation, growth and survival of many populations of peripheral and central nervous system neurons during development and adulthood. Significant advances have been made in understanding the biological role of these compounds during the last several decades. Their molecular characterization, regulation, and signaling mechanisms have been partially defined. The responsiveness of neuronal subpopulations to these molecules has also been identified in a variety of animal model systems. These agents may be potentially useful as therapeutic agents in a wide variety of neurologic diseases including Alzheimer's disease (AD), Parkinson's disease, (PD) amyotrophic lateral sclerosis (ALS), and peripheral neuropathies.

## Neurotrophic factors

During development, cell death occurs in about one-half of all neurons. Developing neurons generally require contact with their proper target cells to survive (Hamburger and Levi-Montalcini, 1949). A number of neurotrophic factors have been identified which stimulate survival, growth and plasticity of different neurons. The first identified polypeptide growth factor was nerve growth factor (NGF) (Levi-Montalcini and Hamburger, 1951). NGF supports neuronal maintenance and survival through target-derived retrograde transport (Levi-Montalcini, 1987; Thoenen et al., 1987). In addition, a host of other growth factors including brain-derived neurotrophic factor (BDNF) (Barde et al., 1982) and ciliary neurotrophic factor (CNTF) (Barbin et al., 1984), as well as acidic and basic fibroblast growth factor (aFGF, bFGF) (Lemmon and Bradshaw, 1983; Bohlen et al., 1984), and insulin-like growth factor-I (IGF-I) (Rinderknecht and Humbel, 1976) were also found to exert trophic effects on neurons. Growth factors elicit their biological action by binding to specific cell surface receptors (Ullrich et al., 1985; Coughlin et al., 1988; Kaplan et al., 1991; Ip and Yancopoulos, 1993). Activation of these receptors results in the transduction of the signal across the membrane.

## Common neurodegenerative disorders

In modern society, diseases of the aged have assumed major importance. Among these are degenerative neurologic disorders such as AD, PD, and ALS. These neurodegenerative diseases are characterized by a defined clinical syndrome, unknown etiology, increasing risk with age, and a progressive and chronic course (Calne, 1994). The vulnerability of certain groups of neurons is the common pathological hallmark of these disorders.

## Role of growth factors in the treatment of neurological disorders

### Alzheimer's disease (AD)

AD is characterized by neuronal and synaptic loss as well as the presence of amyloid containing plaques and neurofibrillary tangles (for review see Katzman, 1986). Among the earliest changes in this disorder are loss of basal forebrain cholinergic neurons with their primary projections to the hippocampus and cortex (Whitehouse et al., 1982). Cognitive decline, especially involving memory, correlates with degeneration of the cholinergic system (Wilcock et al., 1982). The cholinergic cells of the basal forebrain complex composed of the medial septum and nucleus basalis possess NGF receptors (Batchelor et al., 1989). The interest in NGF in the treatment of AD (Phelps et al., 1989) is centered on the therapeutic potential to enhance the function of residual cholinergic neurons.

Over the last decade, several animal models (Smith, 1988; Shiosaka, 1992; Wenk, 1993) have been used to address NGF responsiveness in cholinergic basal forebrain neurons. These models include: (1) cholinergic deafferentation of the hippocampus using fimbria-fornix transection in rodents and non-human primates; (2) cholinergic deafferentation of the cortex by lesioning the nucleus basalis magnocellularis (NBM) in rodents and non-human primates; (3) aging models in rodents and primates (Table 1).

Intraventricular administration of NGF can partially reverse lesion and age-related behavioral, biochemical and histological deficits in all of these animal models. NGF treatment ameliorates memory and learning deficits in the Morris water maze task (Fischer et al., 1987; Dekker et al., 1992), increases the activity of choline acetyltransferase (Dekker et al., 1992) and increases cortical acetylcholine synthesis (Dekker et al., 1991). Histologi-

cal reversal of cellular atrophy due to the lesioning (Gage et al., 1988) or aging (Fischer et al., 1987) of cholinergic basal forebrain neurons has also been demonstrated after intraventricular NGF administration.

Nevertheless, a number of caveats exist regarding the potential use of NGF for the treatment of humans. NGF must be administered intraventricularly, a complex invasive procedure. Adverse events may occur including: an increase in the expression of amyloid precursor protein (Mobley et al., 1988), increased sympathetic innervation of cerebral blood vessels (Isaacson et al., 1990), the stimulation of neurite outgrowth from sensory neurons and dorsal root ganglia (Lewin et al., 1993), and stimulation of other central nervous system NGF responsive neurons (Williams, 1991).

### Parkinson's disease (PD)

In PD, there is a relatively selective loss of mesocephalic dopaminergic neurons of the substantia nigra (Jellinger, 1987) with an accompanying reduction in the release of dopamine. Several growth factors including FGF, BDNF, and glial-derived neurotrophic factor (GDNF) have been investigated in different animal models of PD. However, contradictory data currently exists regarding the effect of these compounds on striatal dopaminergic function (Altar et al., 1992; Lapchak et al., 1993).

### Amyotrophic lateral sclerosis (ALS)

ALS is a degenerative disease in which motor

TABLE 1

NGF in cholinergic animal models

| Model | Species | Anatomy | Structure of change | Behavioral change |
|---|---|---|---|---|
| Fimbria-fornix section | Rat, primate | Medial septum → hippocampus | Rescues MS cholinergic neurons | |
| NBM lesion | Rat | NBM → CTX | $\uparrow$ size NBM neurons | $\uparrow$ H$_2$O maze acquisition |
| Aging | Rat, primate | Cholinergic basal forebrain | $\uparrow$ Chol. cell size | $\uparrow$ H$_2$O maze retention (rat) |

NGF, nerve growth factor; MS, medial septum; NBM, nucleus basalis magnocellularis; CTX, cortex.

neurons of the central nervous system degenerate and die (Swash and Schwartz, 1992). CNTF supports the survival of developing cholinergic and parasympathetic motor neurons and promotes the survival of cultured cranial and spinal motor neurons (Arakawa et al., 1990; Magal et al., 1991). It also rescues motor neurons after axotomy (Sendtner et al., 1990). These preliminary results have led to clinical trials with CNTF in patients with ALS.

*Peripheral nervous system disorders*

In models of peripheral nerve damage, neurotrophic factors produce a variety of beneficial effects (Fitzgerald et al., 1985). Clinical trials of NGF, CNTF, and IGF-I for diabetic and vincristine-induced peripheral neuropathy in humans are currently underway.

## Conclusions

The etiology and pathogenesis of most neurodegenerative disorders remains unknown. Nevertheless, major advances have been made in the understanding of the physiological role and therapeutic potential for neurotrophic factors. However, it is only through the conduct of controlled clinical trials that definitive answers regarding the utility of these compounds for the treatment of neurodegenerative disorders can be gained.

## Acknowledgements

Supported by NIH grant nos. AG05131 and 10483.

## References

Altar, C.A., Boylan, C.B., Jackson, C., Hershenson, S., Miller, J., Wiegand, S.J., Lindsay, R.M. and Hyman, C. (1992) Brain-derived neurotrophic factor augments rotational behavior and nigrostriatal dopamine turnover in vivo. *Proc. Natl. Acad. Sci. USA,* 89: 11347–11351.

Arakawa, Y., Sendtner, M. and Thoenen, H. (1990) Survival effect of ciliary neurotrophic factor (CNTF) on chick embryonic motoneurons in culture: comparison with other neurotrophic factors and cytokines. *J. Neurosci.,* 10: 3507–3715.

Barbin, G., Manthorpe, M. and Varon, S. (1984) Purification of the chick eye ciliary neuronotrophic factor (CNTF). *J. Neurochem.,* 43: 1468–1478.

Barde, Y.-A., Edgar, D. and Thoenen, H. (1982) Purification of a new neurotrophic factor from mammalian brain. *EMBO J.,* 1: 549–53.

Batchelor, P.E., Armstrong, D.M., Blaker, S.N. and Gage, F.H. (1989) Nerve growth factor receptor and choline acetyltransferase co-localization in neurons within the rat forebrain: Response to fimbria-fornix transection. *J. Comp. Neurol.,* 284: 187–204.

Bohlen, P., Baird, A., Esch, F., Ling, N. and Gospodarowicz, D. (1984) Isolation and partial molecular characterization of pituitary fibroblast growth factor. *Proc. Natl. Acad. Sci. USA,* 81: 5364–5368.

Calne, S. (Ed.) (1994) *Neurodegenerative Disease.* Saunders, Philadelphia, PA.

Coughlin, S.R., Barr, P.J., Cousens, L.S., Fretto, L.J. and Williams, L.T. (1988) Acidic and basic fibroblast growth factors stimulate tyrosine kinase activity in vivo. *J. Biol. Chem.,* 263: 988–993.

Dekker, A.J., Langdon, D.J., Gage, F.H. and Thal, L.J. (1991) NGF increases cortical acetylcholine *release* in rats with lesions of the nucleus basalis. *NeuroReport,* 2: 577–580.

Dekker, A.J., Gage, F.H. and Thal, L.J. (1992) Delayed treatment with nerve growth factor improves acquisition of a spatial task in rats with lesions of the nucleus basalis magnocellularis: evaluation of the involvement of different neurotransmitter systems. *Neuroscience,* 48: 111–119.

Fischer, W., Wictorin, K., Bjorklund, A., Williams, L.R., Varon, S. and Gage, F.H. (1987) Amelioration of cholinergic neuron atrophy and spatial memory impairment in aged rats by nerve growth factor. *Nature,* 329: 65–68.

Fitzgerald, M., Wall, P.D., Goedert, M. and Emson, P.C. (1985) Nerve growth factor counteracts the neurophysiological and neurochemical effects of chronic sciatic nerve section. *Brain Res.,* 332: 131–141.

Gage, F.H., Armstrong, D.M., Williams, L.R. and Varon, S. (1988) Morphological response of axotomized septal neurons to nerve growth factor. *J. Comp. Neurol.,* 269: 147–155.

Hamburger, V. and Levi-Montalcini, R. (1949) Proliferation, differentiation, and degeneration in the spinal ganglia of the chick embryo under normal and experimental conditions. *J. Exp. Zool.,* 11: 457–502.

Ip, N.Y. and Yancopoulos, G.D. (1993) Receptors and signaling pathways of ciliary neurotrophic factor and the neurotrophins. *Semin. Neurosci.,* 5: 249–257.

Isaacson, L.G., Saffran, B.N. and Crutcher, K.A. (1990) Intracerebral NGF infusion induces hyperinnervation of cerebral blood vessels. *Neurobiol. Aging,* 11: 51–55.

Jellinger, K. (1987) The pathology of parkinsonism. In: C.D. Marsden and S. Fahn (Eds.), *Movement Disorders,* Vol. 2, Butterworths, London, pp. 124–165.

Kaplan, D.R., Hempstead, B.L., Martin-Zanca, D., Chao, M.V. and Prada, L.F. (1991) The trk proto-oncogene prod-

330

uct: a signal transducing receptor for nerve growth factor. *Science,* 525: 554–558.

Katzman, R. (1986) Alzheimer's disease. *N. Engl. J. Med.* 314: 964–973.

Lapchak, P.A., Beck, K.D., Aroujo, D.M., Irwin, I., Langston, J.W. and Hefti, F. (1993) Chronic intranigral administration of brain-derived neurotrophic factor produces striatal dopaminergic hypofunction in unlesioned rats and fails to attenuate the decline of striatal dopaminergic function following medial forebrain bundle transection. *Neuroscience,* 53: 639–650.

Lemmon, S.K. and Bradshaw, R.A. (1983) Purification and partial characterization of bovine pituitary fibroblast growth factor. *J. Cell. Biochem.,* 21: 195–208.

Levi-Montalcini, R. (1987) The nerve growth factor 35 years later. *Science,* 237: 1154–1162.

Levi-Montalcini, R. and Hamburger, V. (1951) Selective growth stimulating effects of mouse sarcoma on the sensory and sympathetic nervous system of the chick embryo. *J. Exp. Zool.,* 116: 321–363.

Lewin, G.R., Ritter, A.M. and Mendell, L.M. (1993) Nerve growth factor-induced hyperalgesia in the neonatal and adult rat. *J. Neurosci.,* 13: 2136–2148.

Magal, E., Burnham, P. and Varon, S. (1991) Effects of ciliary neurotrophic factor on rat spinal cord neurons in vitro: survival and expression of choline acetyltransferase and low-affinity nerve growth factor receptors. *Dev. Brain Res.,* 63: 141–150.

Mobley, W.C., Neve, R.L., Prusiner, S.B. and McKinley, M.P. (1988) Nerve growth factor increased mRNA levels for the prion protein and the $\beta$-amyloid protein precursor in developing hamster brain. *Proc. Natl. Acad. Sci. USA,* 85: 9811–9815.

Phelps, C.H., Gage, F.H., Growdon, J.H., Hefti, F., Harbaugh, R., Johnston, M.V., Khachaturian, Z.S., Mobley, W.C., Price, D.L. and Raskind, M. (1989) Potential use of nerve growth factor to treat Alzheimer's disease. *Neurobiol Aging,* 10: 205–207.

Rinderknecht, E. and Humbel, R.E. (1976) Polypeptides with nonsuppressible insulin-like and cell-growth promoting activities in human serum: isolation, chemical characterization, and some biological properties of forms I and 11. *Proc. Natl. Acad. Sci. USA,* 73: 2365–2369.

Sendtner, M., Kreutzberg, G.W. and Thoenen, H. (1990) Ciliary neurotrophic factor (CNTF) prevents the degeneration of motoneurons after axotomy. *Nature,* 345: 440–441.

Shiosaka, S. (1992) Attempts to make models for Alzheimer's disease. *Neurosci. Res.,* 13: 237–255.

Smith, G. (1988) Animal models of Alzheimer's disease: experimental cholinergic denervation. *Brain Res. Rev.,* 13: 103–118.

Swash, M. and Schwartz, M.S. (1992) What do we really know about amytrophic lateral sclerosis? *J. Neurol. Sci.,* 113: 4–16.

Thoenen, H., Bandtlow, C. and Heuman, R. (1987) The physiological function of nerve growth factor in the central nervous system: comparison with the periphery. *Rev. Physiol. Biochem. Pharmacol.,* 109: 145–178.

Ullrich, A., Bell, J.R., Chen, E.Y., Herrera, R., Petruzzelli, L.M., Dull, T.J., Gray, A., Coussens, L., Liao, Y.C. and Tsubokawa, M. (1985) Human insulin receptor and its relationship to the tyrosine kinase family of oncogenes. *Nature,* 313: 756–761.

Wenk, G.L. (1993) A primate model of Alzheimer's disease. *Behav. Brain Res.,* 57: 117–122.

Whitehouse, P.J., Price, D.L., Struble, R.G., Clark, A.W., Coyle, J.T. and Delon, M.R. (1982) Alzheimer's disease and senile dementia: loss of neurons in the basal forebrain. *Science,* 215: 1237–1239.

Wilcock, G.K., Esiri, M.M., Bowen, D.M. and Smith, C.C. (1982) Alzheimer's disease: correlation of cortical choline acetyltransferase activity with severity of dementia and histological abnormalities. *J. Neurol. Sci.,* 57: 407–417.

Williams, L. (1991) Hypophagia is induced by intracerebroventricular administration of nerve growth factor. *Exp. Neurol.,* 113: 31–37.

J. Klein and K. Löffelholz (Eds.)
*Progress in Brain Research*, Vol. 109
© 1996 Elsevier Science B.V. All rights reserved.

CHAPTER 32

# Development, survival and regeneration of rat cholinergic septohippocampal neurons: in vivo and in vitro studies

M. Frotscher, B. Heimrich, M. Plaschke, R. Linke and T. Naumann

*Institute of Anatomy, University of Freiburg, PO Box 111, D-79001 Freiburg, Germany*

## Introduction

It has been known for almost 150 years that axotomy invariably leads to the degeneration of the distal axonal stump (Waller, 1850). It has also been observed that the parent cell bodies of the axotomized neurons are affected by the transection of their axons. This retrograde neuronal degeneration may eventually result in cell death and may even lead to subsequent secondary degeneration (retrograde transneuronal degeneration) of afferent neurons (Cowan, 1970). Originally it was assumed that this degeneration of axotomized neurons was a direct effect of damage to the cell. More recently, evidence has been accumulated that not axotomy itself, but the resulting disconnection from target-derived neurotrophins, may play a major role. This neurotrophic hypothesis has first been formulated for the peripheral autonomic nervous system (for review see Levi-Montalcini and Angeletti, 1968; Thoenen and Barde, 1980; Levi-Montalcini, 1987), but has more recently also been applied to some central neurons (Cowan et al., 1984). Thus, it has been shown that cholinergic septohippocampal projection neurons express the low-affinity nerve growth factor receptor (LNGFR, p75 receptor; review: Chao, 1994) as well as the trk receptor (Barbacid, 1994; Sobreviela et al., 1994), take up NGF synthesized by their target neurons and transport it back to their parent cell bodies in the septal complex (Schwab et al., 1979; Seiler and Schwab, 1984; review: Gage et al., 1989). A role of target-derived neurotrophins for the survival of

central cholinergic neurons was further supported by the observation of a dramatic loss of choline acetyltransferase (ChAT)-immunoreactive medial septal neurons following disconnection from the target region by fimbria-fornix transection (Gage et al., 1986, 1988, 1989; Hefti, 1986; Williams et al., 1986; Armstrong et al., 1987; Kromer, 1987; Frotscher, 1988; Naumann et al., 1994a).

However, loss of immunostaining does not necessarily indicate cell death (Lams et al., 1988; Peterson et al., 1990). In fact, in experiments in which the septohippocampal projection neurons were prelabeled by retrograde tracing prior to axotomy, a much larger number of surviving neurons was found (Naumann et al., 1992b; Peterson et al., 1992). This discrepancy might have two explanations: (1) the fluorescent retrograde tracer may stay for extended periods of time in the degenerated cells, or (2) the axotomized neurons do not degenerate, but lose their capability to express the transmitter-synthesizing enzyme ChAT. In a series of experiments to be reported here, we have tested these two possibilities. Septohippocampal projection neurons were prelabeled prior to axotomy by retrograde tracing with Fluoro-Gold which stays in the labeled cells for extended periods of time (Naumann et al., 1992a). Following axotomy, the animals were allowed to survive for varying periods of time, before they were sacrificed by transcardial perfusion fixation. Next, in vibratome sections through the septal complex the prelabeled and then axotomized cells were intracellularly injected with a second tracer (Lucifer

Yellow) which allowed us to study not only the dendritic arbor of these neurons, but also their fine structure, since Lucifer Yellow-injected cells may be made electron-dense by photoconversion (Maranto, 1982). With this approach we were able to demonstrate that many more septohippocampal neurons survive axotomy and disconnection from target-derived neurotrophins than was concluded from ChAT-immunocytochemical studies (Naumann et al., 1992b; Frotscher and Naumann, 1992).

All studies mentioned before were carried out in adult rats. It remained an open question to what extent target-derived neurotrophins would be essential for differentiating cholinergic septohippocampal neurons. In order to attack this question, two alternative approaches were chosen. First, in young postnatal rats, i.e. at the time when the majority of septohippocampal fibers invades the hippocampal formation (Linke and Frotscher, 1993), the target of the septal cells was removed by an excitotoxic lesion of the hippocampus. Second, in newborn rats septohippocampal neurons were retrogradely labeled and, after a survival time of 2 days to allow for the retrograde transport of the tracer, slice cultures of the septal complex of these animals were prepared. In line with our studies in adult rats, both approaches clearly demonstrated that young septohippocampal neurons may survive and differentiate following target removal by excitotoxic lesion and axotomy by culture preparation, respectively (Plaschke et al., 1994; Linke et al., 1995).

Septohippocampal cholinergic cells surviving axotomy form a population of neurons that may regenerate an axonal process eventually reinnervating the target region. In order to test this capability of axotomized septohippocampal neurons, we have recently prepared slice cultures of septum containing retrogradely prelabeled cells (thereby axotomizing these projection neurons) and have then cocultivated the septal slices with slices of hippocampus from other littermates. Having shown before that both adult and young septohippocampal projection neurons may survive axotomy and disconnection from target-derived neurotrophins, we will demonstrate with these experiments that the axotomized cells have the capacity of regenerating an axonal process and reinnervating their normal target tissue.

## Methodological approaches

The methods described in this report have been published elsewhere (Naumann et al., 1992a,b, 1994a,b; Frotscher and Naumann, 1992; Peterson et al., 1992; Frotscher and Heimrich, 1993; Heimrich and Frotscher, 1993, 1994; Plaschke et al., 1994; Linke et al., 1995). Therefore, the reader is referred to these papers for any details on the methods.

### Retrograde prelabeling of septohippocampal neurons

Following anesthesia (cocktail of ketamine, xylazine, and acepromazine, i.m.), the animals received up to 5 injections of 2.5% Fluoro-Gold (adult animals) or Fast-Blue (newborn rats) into the hippocampus. Following a survival time of at least 5 days (adult animals) or 2 days (young postnatal rats) to allow for the retrograde labeling of the cell bodies in the septal region, the animals were either subjected to bilateral fimbria-fornix transection or used for the preparation of slice cultures of the septal complex.

### Fimbria-fornix transections in adult rats

Following anesthesia, the fimbria-fornix was aspirated bilaterally which resulted in the transection of the supracallosal striae, the dorsal fornix, and the fimbria proper. The animals were then allowed to survive for 8, 11 days, 3, 6, 10 weeks and 6 and 11 months.

### Treatment with NGF via osmotic mini-pumps

Following anesthesia and bilateral fimbria-fornix transection, the animals received mini-osmotic pumps (Alzet, Model 2002) immediately after the operation for the continuous infusion of NGF or vehicle fluid into the lateral ventricle. The flow rate was 0.47 ml/h and maintained in both experimental groups for 20 days. The first group of animals received mini-pumps containing NGF (0.55 mg/ml, 1:55 diluted in 0.9% NaCl and 1 mg/ml rat serum albumin). A second group received mini-pumps with vehicle fluid (0.9% NaCl containing 1 mg/ml rat serum albumin). A third group was lesioned, but not treated with NGF or vehicle. Part of the animals was perfusion-fixed after 20 days following the lesion. In the remaining rats, the pumps were removed on postlesional day 20, and the animals were allowed to survive for 6 months following the bilateral fimbria transection. Vibratome sections of the septal complex from these animals were either stained for ChAT or parvalbumin (PARV), which is known to be a marker for septohippocampal GABAergic neurons (Freund, 1989).

### Excitotoxic lesions in young postnatal rats

Following anesthesia, lesions in 5, 10, and 20 day-old rats were placed by stereotaxic injections of NMDA (0.3–0.4 $\mu$l each, 5 mg/ml NMDA in 0.9% NaCl) into the hippocampus. Operated animals and age-matched unoperated controls were allowed to survive up to postnatal day 70 and were then reanesthesized and transcardially perfusion-fixed. Vibratome sections of the septal region were immunostained for ChAT and PARV.

### Lucifer Yellow injection and photoconversion

The animals were fixed in anesthesia by transcardial perfusion with 4% paraformaldehyde, 0.1% glutaraldehyde, and 15% saturated picric acid in 0.1 M phosphate buffer. Vibratome sections containing retrogradely Fluoro-Gold-labeled septal neurons were placed in a Petri dish which was transferred to a fluorescence microscope. Glass capillaries filled with Lucifer Yellow (5%) were attached to a micromanipulator and advanced towards Fluoro-Gold-labeled neurons under visual guidance. The cells were impaled, and Lucifer Yellow was iontophoretically injected until the cell and all dendrites appeared brightly fluorescent. Next, the cells were photoconverted under the fluorescence microscope (Maranto, 1982) by incu-

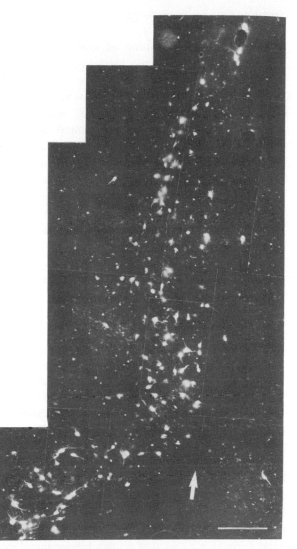

Fig. 1. Retrogradely Fluoro-Gold-labeled and then axotomized neurons in the medial septum of rats that survived the lesion (bilateral fimbria-fornix transection) for 11 months. Note numerous normally appearing neurons and many shrunken cells. This finding contrasts with previous reports of a dramatic loss of ChAT-immunoreactive neurons after this kind of lesion. However, the viability of the fluorescent cells cannot be judged by light microscopic analysis. Arrow marks midline. Scale bar: 200 $\mu$m.

bating the sections in phosphate buffer containing 1.5 mg/ml diaminobenzidine (DAB). Thereafter, the sections were osmicated and flat-embedded in Epon for subsequent correlated light and electron microscopic analysis.

*Immunostaining of septohippocampal neurons*

For the immunostaining of cholinergic septo-hippocampal neurons, we have applied a mono-clonal antibody against ChAT (Type I, Boehringer, Mannheim, Germany; see Eckenstein and Thoe-nen, 1982; Frotscher and Leranth, 1985, 1986), and for GABAergic septohippocampal neurons a monoclonal antibody against PARV (Celio and Heizmann, 1981). For visualization of the immu-nolabeling, the ABC technique (Hsu et al., 1981) was employed. Following osmication, the sections were flat-embedded in Epon, allowing for a corre-lated light and electron microscopic study of the immunostained cells.

*Preparation of septohippocampal co-cultures*

Newborn rats were prelabeled with Fast Blue (see above) and allowed to survive for 2 days. Then the animals were decapitated, and static slice cultures of the septal complex were prepared as described in detail elsewhere (Heimrich and Frot-scher, 1994). Briefly, slices (400 $\mu$m) were cut by means of a tissue chopper and placed onto mois-tened, uncoated Millipore membranes. Slices of septum and hippocampus were placed next to each other and incubated in humidified, $CO_2$-enriched atmosphere at 37°C for 2 weeks. Then, a second tracer, Latex beads coupled to Lucifer Yellow, was injected into the hippocampal culture for the retrograde labeling of septal neurons that may have formed a projection to the hippocampal co-culture in vitro. Septal cultures were then studied for solely Fast Blue-labeled neurons, cells solely labeled with Latex beads, and double-labeled cells.

## Survival of cholinergic septohippocampal neurons following axotomy in adult rats

In accordance with other investigators, we found a dramatic loss of ChAT-immunopositive neurons in the medial septum (MS) after fimbria-fornix tran-section (Frotscher, 1988; Naumann et al., 1994a). In our recent series of experiments, the loss of ChAT-positive neurons was larger than that ob-served by most other investigators, since we per-formed bilateral fimbria transections in order to include crossed projections (Peterson, 1989). In each animal the completeness of the lesion was confirmed by the loss of histochemical staining for acetylcholinesterase (AChE) of cholinergic fibers in the hippocampal formation (cf. Naumann et al., 1994a). Animals with incomplete fimbria-fornix transections were excluded from the study.

Despite this dramatic loss of ChAT-immuno-positive neurons, a surprisingly large number of retrogradely prelabeled cells were still found in the medial septum after various survival times follow-ing fimbria-fornix transection (Fig. 1). However, prelabeling with a retrograde fluorescent tracer prior to axotomy did not allow us to draw far-reaching conclusions as to the viability of the stained neurons. Moreover, due to the shrinkage of many cells, it was often difficult to decide whether a certain cell was a neuron or a glial cell. There-fore, we have performed an electron microscopic study of these neurons. In the beginning we did not know to what extent the lysosome-associated

Fig. 2. Photoconversion and ChAT immunostaining of Fluoro-Gold-prelabeled cells 11 months following bilateral fimbria-fornix transection. (a) Sections containing retrogradely labeled medial septal neurons (cf. Fig. 1) are first immersed in phosphate buffer containing 1.5 mg/ml DAB and then illuminated under the fluorescence microscope for about 20 min in order to enhance the stain-ing intensity of lysosomes associated with Fluoro-Gold in the retrogradely labeled cells. As a result of the photoconversion, a dark halo corresponding to the illuminated field has formed. Then, the sections are immunostained for ChAT. The framed area contain-ing many ChAT-positive neurons is shown at higher magnification in b. Scale bar: 300 $\mu$m. (b) Higher magnification of part of the illuminated section shown in a (framed area). As can be seen, numerous medial septal neurons have regained their immunoreactiv-ity for ChAT after this long survival time. Note normally appearing ChAT-positive neurons among shrunken cells. Scale bar: 150 $\mu$m. (c) Electron micrograph of a ChAT-positive neuron of the section shown in (a,b). The cell is shrunken, but otherwise displays normal fine-structural characteristics of a septal cholinergic neuron. The cell is identified as a former septohippocampal projection neuron by the presence of Fluoro-Gold particles associated with lysosomes (arrows), which are heavily electron-dense resulting from the photoconversion. Scale bar: 2 $\mu$m.

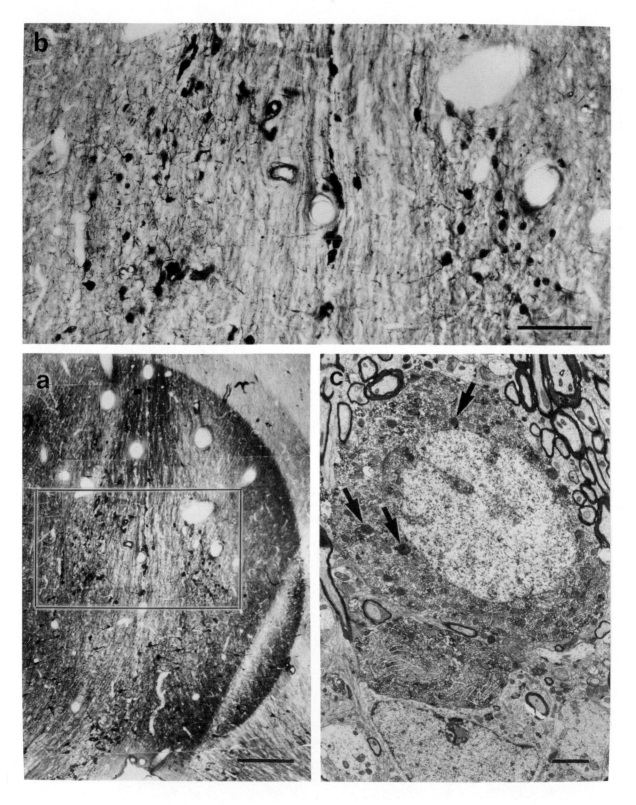

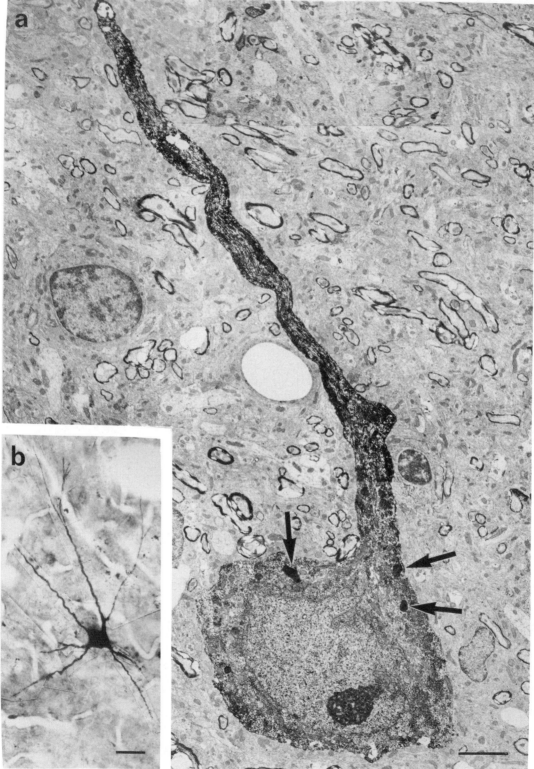

Fluoro-Gold particles would allow us to identify prelabeled somata and dendrites at the electron microscopic level. We thus looked for an additional electron-dense staining technique of the prelabeled and axotomized neurons. Such a method was found with the intracellular injection of Lucifer Yellow into the prelabeled cells under the fluorescence microscope following fixation by transcardial perfusion of the animal and sectioning of the septal region with a vibratome (Naumann et al., 1992a). Lucifer Yellow diffuses in the fixed neurons and thus stains their dendritic processes. Moreover, when illuminated in the presence of DAB, the Lucifer Yellow fluorescence is converted to an electron-dense, stable reaction product (Maranto, 1982). We found it advantageous that not only the Lucifer Yellow, but also the Fluoro-Gold particles associated with lysosomes became electron-dense after photoconversion. This allowed us to combine retrograde Fluoro-Gold tracing and photoconversion with ChAT immunolabeling (Fig. 2). The obvious advantage of this procedure was that many Fluoro-Gold-labeled cells (and not only single Lucifer Yellow-stained neurons) could be photoconverted simultaneously and subsequently immunostained to determine the transmitter identity of these septohippocampal neurons.

With these combined approaches (Figs. 2 and 3) we were able to show that the majority (approximately 70%, see Naumann et al., 1992b; Peterson et al., 1993) of prelabeled neurons survived axotomy for extended periods of time (exceeding 1 year), and many of them even regained the capability to synthesize ChAT (Naumann et al., 1994a). We also observed neuronal alterations such as shrinkage of cell somata, the electron-dense degeneration of some neurons (Frotscher and Naumann, 1992; Naumann et al., 1992b), and a variety of tissue reactions in the sur-

rounding neuropil including astrocytic and microglial hypertrophy. As far as the axotomized neurons are concerned, we have to conclude that many more cells survived the lesion than one would have expected with regard to the results of previous ChAT-immunocytochemical studies after short survival times (Gage et al., 1989).

Most pronounced neuronal alterations following fimbria-fornix transection were observed within the first 3 postlesional weeks (Naumann et al., 1992b, 1994a). By electron microscopic analysis of identified septohippocampal neurons, we have shown that the cells were transiently altered but recovered, the majority of them displaying fine-structural characteristics of normal medial septal neurons in control rats (Naumann et al., 1992a). In line with these observations, little further changes were noticed after the 3 week-stage, i.e. a similar co-existence of normally appearing and shrunken cells was observed after 3 weeks and 6 months postlesion, respectively. These findings prompted us to study a possible rescue effect of NGF, exogenously applied via mini-pumps during the first 3 critical postlesional weeks (Naumann et al., 1994b). As shown in Fig. 4a, NGF treatment had a dramatic rescue effect on the number of ChAT-positive neurons after 3 weeks post-lesion when compared with vehicle-treated animals. However, when the mini-pumps were removed and the animals allowed to survive for 6 months, there was no difference between NGF-treated rats and lesioned controls (Naumann et al., 1994b; Fig. 4a). Moreover, these studies revealed a substantial, highly significant increase in the number of ChAT-immunoreactive neurons in lesioned, but not NGF-treated animals with increasing survival time (Naumann et al., 1994a). This indicates that many axotomized cholinergic neurons have regained the capability of synthesizing ChAT after long sur-

Fig. 3. (a) Intracellularly Lucifer Yellow-stained and photoconverted medial septal neuron. The cell was prelabeled with Fluoro-Gold prior to axotomy by bilateral fimbria-fornix transection which the animal survived for 6 weeks. Note long, normally appearing aspiny dendrites. Scale bar: 25 $\mu$m. (b) Electron micrograph of the cell shown in a. The cell appears slightly shrunken, but otherwise displays a rather intact fine structure. Note nuclear infoldings and distinct nucleolus. Fluoro-Gold particles associated with lysosomes (arrows) identify the cell as a septohippocampal projection neuron. Scale bar: 3 $\mu$m (adapted from Frotscher and Naumann, 1992).

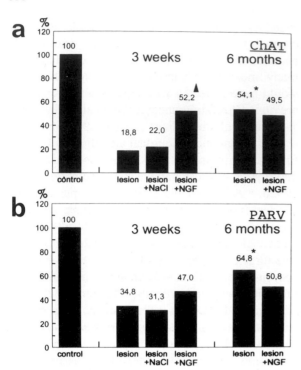

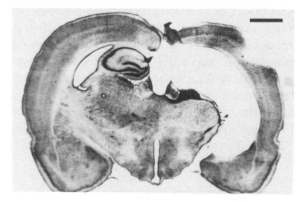

Fig. 4. Short-term and long-term effects of intraventricular NGF application for the first 3 postlesional weeks on the number of ChAT-immunoreactive neurons (a) and PARV-immunoreactive neurons (b) in the medial septum. Cell counts in lesioned rats were compared with those in unoperated animals (% of control). Lesion = bilateral fimbria-fornix transection and no further treatment; lesion + NaCl = bilateral fimbria-fornix transection + treatment with vehicle via intraventricular mini-pumps; lesion + NGF = bilateral fimbria-fornix transection + treatment with NGF for the first 3 postlesional weeks. A first group of animals was perfused and immunostained 3 weeks after the lesion, whereas a second group was allowed to survive the lesion for 6 months. Like in the first group, the NGF treatment was stopped after the third postlesional week. Since no difference was found between lesioned rats and lesioned rats treated with vehicle after short survival time, no experiments with vehicle treatment were performed for the long survival period. ▲Statistically significant difference between lesioned and lesioned + NGF-treated rats 3 weeks postlesion ($P < 0.01$). *Statistically significant difference between lesioned rats 3 weeks and 6 months after fimbria-fornix transection ($P < 0.01$) (from Naumann et al., 1994b).

vival periods. Since only animals were used in the present studies that were devoid of AChE-positive fibers in the hippocampus (cf. Naumann et al., 1994a), it can be excluded that the axotomized

neurons have got access to hippocampus-derived neurotrophic factors. It remains to be investigated whether the axotomized cells require neurotrophins from other sources than the hippocampus, for instance from local neurotrophin-producing cells, for their ChAT re-expression. Interestingly enough, the number of PARV-immunoreactive, supposedly GABAergic septohippocampal neurons (Freund, 1989) was significantly decreased after short survival time, and there was a similar, significant increase in immunolabeled cells after 6 months survival as found in the number of ChAT-positive neurons (Fig. 4a,b). However, treatment with NGF, which had an effect on the number of ChAT-immunoreactive neurons after short survival time (Fig. 4a), did not act on PARV-positive cells (Fig. 4b).

## Development of septal cholinergic cells despite target removal in the early postnatal period

In order to monitor hippocampal degeneration after NMDA injection, some animals were perfused after short survival times (1, 2, 7 and 14 days following NMDA injection at P 5 and P 10, respectively). Most hippocampal neurons underwent degeneration within the first 24 h after injection of the excitotoxin. Thus, at the time of perfusion, i.e., after a

Fig. 5. Cresyl-Violet-stained frontal vibratome section through the brain of a 70 day-old rat which had received unilateral NMDA injections at postnatal day 5. The hippocampus of the right hemisphere is ablated completely at all rostrocaudal levels. Scale bar: 2 mm.

survival period of 65 days, there was an almost complete lack of hippocampal tissue on the side of the NMDA injection (Fig. 5), with the exception of a fimbrial remnant (Plaschke et al., 1994).

Despite this virtually complete removal of the target region in early development, a large number of ChAT-positive neurons was found in the septal complex of rats perfused on postnatal day 70 (approximately 60% of unlesioned control rats; Fig. 6, Table 1). This value may be an underestimation, since many neurons were heavily shrunken (Fig. 6), and only immunoreactive profiles larger than 10 $\mu$m were counted. The most interesting observation was that GABAergic neurons in the septal complex, immunostained for PARV in adjacent sections through the septal region of the same animals, were reduced to a similar extent and also showed shrinkage of their cell bodies (Plaschke et al., 1996). However, these GABAergic septohippocampal projection cells are unlikely to express LNGFR and trk (Heckers et al., 1994; Sobreviela et al., 1994). In line with this, exogenous NGF application via mini-pumps increased the number of ChAT-positive neurons after short survival times, but did not have an effect on the number of PARV-immunoreactive cells (Naumann et al., 1994b; Fig. 4a,b). We accordingly tend to conclude that the very similar neuronal loss in both cell populations is not closely related to the absence of hippocampus-derived neurotrophins. At present it cannot be excluded that the development of cholinergic septal cells is affected by the lack of hippocampus-derived NGF and that of GABAergic septohippocampal neurons by the lack of another, yet unknown target-derived neurotrophic factor. It may be added here that mice lacking the coding sequence of the NGF gene develop normally appearing basal forebrain cholinergic neurons (Crowley et al., 1994).

## Regeneration of axotomized septohippocampal projection neurons in vitro

In the preceding paragraphs we have reported studies which have shown that (1) septohippocampal projection neurons in adult animals survive axotomy by bilateral fimbria-fornix transection,

and (2) target removal in the early postnatal period has only a limited effect on the survival of cholinergic septal cells. It remained to be investigated to what extent axotomy in young postnatal rats would induce retrograde degeneration of the parent cell bodies in the septal complex.

In recent studies on the effects of axotomy in young postnatal rats, we have followed a similar approach as described before for adult animals. Thus, in newborn rats septohippocampal projection neurons were prelabeled by injection of a retrograde tracer into the hippocampus (Fig. 7). In these young animals we have used Fast Blue, since we found out that this tracer is better transported in young animals than Fluoro-Gold. After a survival time of 2 days, the animals were killed by decapitation, and slice cultures of the septal region were prepared (Stoppini et al., 1991; Frotscher and Heimrich, 1993; Heimrich and Frotscher, 1993, 1994) which were co-cultivated with slices of the hippocampus obtained from other, non-injected littermates. Following co-cultivation for 2 weeks, a second retrogradely transported tracer, Latex beads coupled to Lucifer Yellow, was injected into the hippocampal cultures in order to label those cells in the septal cultures retrogradely that had formed a septohippocampal projection in vitro (Fig. 7). Previous studies had in fact shown that a septohippocampal cholinergic projection develops under these in vitro conditions (Gähwiler and Brown, 1985; Gähwiler et al., 1989; Heimrich and Frotscher, 1993; Fig. 8a,b) and that normal cholinergic (ChAT-immunopositive) synapses are established with neurons in the hippocampal target culture (Heimrich and Frotscher, 1993; Fig. 8c). However, these earlier studies did not clarify whether the cholinergic projection in vitro was formed by cholinergic neurons whose axons had been transected by culture preparation or from only those cells that were not affected, since they had not yet projected to the hippocampus by the time of culture preparation. This issue was addressed by the present double-labeling experiments.

Following co-cultivation we found all three possible types of retrogradely labeled cells (Linke et al., 1995; Fig. 7): (1) solely Fast Blue-labeled cells which already had projected to the hippo-

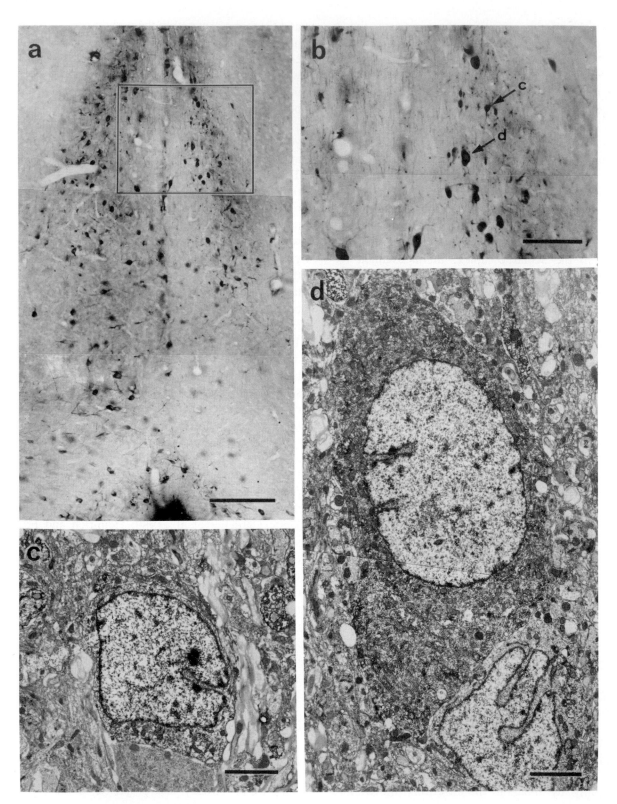

campus at the time of Fast Blue injection in vivo. These cells had to be axotomized for culture preparation. Their presence in the co-cultures indicated survival of young postnatal septohippocampal neurons following axotomy; (2) cells in the septal culture solely labeled with Latex beads. Since Latex beads coupled to Lucifer Yellow were injected into the hippocampal culture, the presence of these neurons demonstrated the in vitro formation of a septohippocampal projection; (3) septal neurons double-labeled with Fast Blue and Latex beads. The presence of these neurons provided convincing evidence that septohippocampal neurons in these young animals did not only survive axotomy by culture preparation, but were capable of regenerating an axonal process and reinnervating their normal target tissue.

## Conclusions and perspectives

The various experiments presented here have demonstrated that septohippocampal projection neurons have a much greater capacity to survive in the absence of hippocampus-derived neurotrophins than was assumed before on the basis of ChAT-immunocytochemical studies. By using retrograde tracing and intracellular staining in combination, we were able to show that the majority of axotomized neurons in adult animals display fine-structural characteristics of vital septal cells even after prolonged survival periods. Experiments with excitotoxic lesions of the hippocampus as well as experiments with slice cultures have proven that young postnatal cholinergic septal cells may also differentiate normally in the absence of target-derived neurotrophins. Although these studies do not support an essential role of hippocampus-

TABLE 1

Number of ChAT-positive neurons in the medial septum after excitotoxic lesion of the hippocampus

| | Lesioned animals ($n = 13$) (%) | Controls ($n = 10$) (%) |
|---|---|---|
| Ipsilateral septum | 60.7 ± 15.3* (60) | 100.6 ± 19.3 (100) |
| Contralateral septum | 84.5 ± 10.6** (83) | 102.0 ± 17.0 (100) |

Lesioned animals had received unilateral injections of NMDA at P 5, P 10 and P 20 and were perfused and immunostained at P 70. Controls are untreated littermates. Values are given as mean ± SD as well as in % of controls (from Plaschke et al., 1994).
*$P < 0.01$
**Not significant.

derived neurotrophins for the survival and differentiation of cholinergic septal cells, they do not disprove a possible role of local neurotrophic factors synthesized by cells in the septal complex. In fact, there are reports that NGF protein is increased in the septal complex following transection of the fimbria-fornix (Weskamp et al., 1986). Moreover, Saporito and Carswell (1995) have recently shown relatively high levels of NGF mRNA synthesis in the septal region by applying a sensitive RNase protection method. Clearly, further studies on the role of neurotrophins for differentiation and survival of septal cholinergic cells have to focus on these local factors and the molecules governing their regulation. Further studies in this direction may include non-neuronal cells, astrocytes and microglial cells, that show a substantial hypertrophy after fimbria-fornix lesion.

Fig. 6. ChAT-immunostained frontal section through the medial septum of a 70 day-old rat which had received unilateral intrahippocampal NMDA injections at P 5 (same animal as shown in Fig. 5). (a) Immunolabeling revealed numerous cholinergic neurons in the medial septum, both ipsilateral (right-hand side) and contralateral to the hippocampal lesion. Scale bar: 200 μm. (b) Higher magnification of framed area in (a). Besides normally appearing immunostained neurons (d), many of the ChAT-positive cells are shrunken (c) and show a reduced dendritic arborization when compared with controls. Scale bar: 100 μm. (c) Electron micrograph of the small immunoreactive profile marked by arrow in (b). Electron microscopy revealed that numerous of these small profiles are heavily shrunken neurons. Scale bar: 2 μm. (d) Electron micrograph of the ChAT-positive neuron marked by arrow in (b). This neuron displays normal ultrastructural characteristics of medial septal neurons. Scale bar: 2 μm.

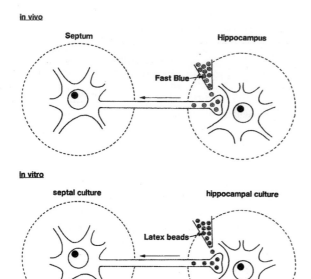

in vivo

Septum           Hippocampus

Fast Blue

in vitro

septal culture       hippocampal culture

Latex beads

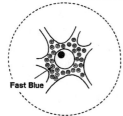

**1. Survival of axotomized septohippocampal neurons**

Fast Blue

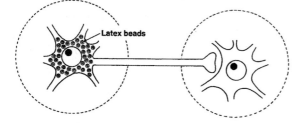

**2. Formation of the septohippocampal projection in vitro**

Latex beads

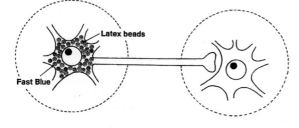

**3. Regeneration of the septohippocampal projection in vitro**

Latex beads

Fast Blue

We regard it as a main result of the present series of experiments that septohippocampal projection neurons not only survive axotomy, but restore their ChAT synthesis and have the capacity for axonal regeneration and reinnervation of the proper target cells, i.e. the neurons in the hippocampal formation. So far, this regenerative capacity has only been shown for prelabeled septohippocampal cells in culture, and it remains to be found out to what extent the axotomized septohippocampal neurons of adult rats may regenerate. Further studies employing intracellular staining techniques will have to show the axonal plexus of septohippocampal projection neurons after various periods of time following axotomy. It remains to be discovered along which structures the regenerating axon is able to grow and which structures are non-permissive for the growth cone. Another question concerns the changes in gene expression underlying the structural changes in the axotomized neuron and in the surrounding non-neuronal cells. Undoubtedly, there is a long way to go before we will be able to understand the complex mechanisms governing the reaction of septohippocampal neurons to axotomy.

Fig. 7. Schematic diagram illustrating the survival and regeneration of septohippocampal projection neurons in co-cultures of septum and hippocampus. First, injections of the retrogradely transported tracer Fast Blue were made into the hippocampal region of young rats in order to label septohippocampal projection neurons retrogradely. Following a survival time of 2 days, slice cultures of the septal complex of these animals were prepared which were cultivated together with hippocampal slices from other, non-injected littermates. Following incubation in vitro for 2 weeks, a second retrogradely transported tracer, Latex beads coupled to Lucifer Yellow, was injected into the hippocampal culture. As a result of this experiment, 3 different types of labeled cells were found: (1) Solely Fast Blue-labeled cells in the septal culture, indicating the survival of septohippocampal projection neurons. These neurons had to be axotomized for culture preparation, since they had already projected to the hippocampus and were thus retrogradely labeled by the Fast Blue injection in vivo. (2) Septal neurons solely labeled with Latex beads. The presence of these cells proves the formation of a septohippocampal projection in vitro. (3) Double-labeled cells indicating the survival and regeneration of axotomized septohippocampal projection neurons.

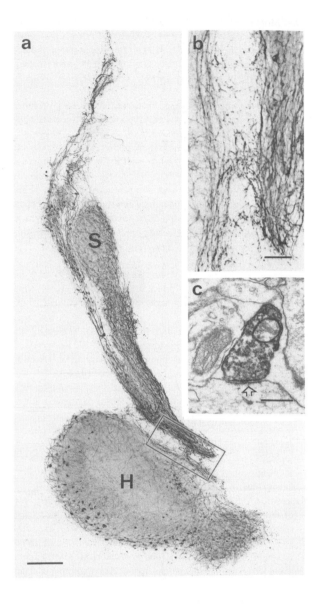

Fig. 8. (a) Co-culture of a septal (S) and a hippocampal (H) slice incubated in vitro for 2 weeks followed by histochemical staining for AChE. Note that AChE-stained fibers originating from the septal culture are directed towards the hippocampal co-culture. The framed area demonstrating fibers invading the hippocampal co-culture is shown at higher magnification in (b). Scale bar: 500 $\mu$m. (b) Higher magnification of framed area in a. Scale bar: 100 $\mu$m. (c) ChAT-immunopositive presynaptic terminal establishing symmetric synaptic contact (open arrow) on a dendritic shaft in a hippocampal culture co-cultivated with septum. Scale bar: 0.25 $\mu$m (adapted from Heimrich and Frotscher, 1993).

## Acknowledgements

The authors wish to thank H. Hildebrandt, C. Hofmann and M. Winter for technical assistance. The studies reported in this article have been supported by grants from the Deutsche Forschungsgemeinschaft (Fr 620/4-2, SFB 505, and Leibniz Program).

## References

Armstrong, D.M., Terry, R.D., Deteresa, R.M., Bruce, G., Hersh, L.B. and Gage, F.H. (1987) Response of septal cholinergic neurons to axotomy. *J. Comp. Neurol.*, 264: 421–436.

Barbacid, M. (1994) The trk family of neurotrophin receptors. *J. Neurobiol.*, 25: 1386–1403.

Celio, M.R. and Heizmann, C.W. (1981) Calcium-binding protein parvalbumin as a neuronal marker. *Nature*, 293: 300–302.

Chao, M.V. (1994) The p75 neurotrophin receptor. *J. Neurobiol.*, 25: 1373–1385.

Cowan, W.M. (1970) Anterograde and retrograde transneuronal degeneration in the central and peripheral nervous system. In: W.J.H. Nauta and S.O.E. Ebbesson (Eds.), *Contemporary Research Methods in Neuroanatomy*, Springer, Berlin, pp. 217–251.

Cowan, W.M., Fawcett, J.W., O'Leary, D.D.M. and Stanfield, B.B. (1984) Regressive events in neurogenesis. *Science*, 225: 1258–1265.

Crowley, C., Spencer, S.D., Nishimura, M.C., Chen, K.S., Pitts-Meek, S., Armanini, M.P., Ling, L.H., McMahon, S.B., Shelton, D.L., Levinson, A.D. and Phillips, H.S. (1994) Mice lacking nerve growth factor display perinatal loss of sensory and sympathetic neurons yet develop basal forebrain cholinergic neurons. *Cell*, 76: 1001 1011.

Eckenstein, F. and Thoenen, H. (1982) Production of specific antisera and monoclonal antibodies to choline acetyltransferase: characterization and use for identification of cholinergic neurons. *EMBO J.*, 1: 363–368.

Freund, T.F. (1989) GABAergic septohippocampal neurons contain parvalbumin. *Brain Res.*, 478: 375–381.

Frotscher, M. (1988) Cholinergic neurons in the rat hippocampus do not compensate for the loss of septohippocampal cholinergic fibers. *Neurosci. Lett.*, 87: 18–22.

Frotscher, M. and Heimrich, B. (1993) Formation of layer-specific fiber projections to the hippocampus in vitro. *Proc. Natl. Acad. Sci. USA*, 90: 10400–10403.

Frotscher, M. and Leranth, C. (1985) Cholinergic innervation of the rat hippocampus as revealed by choline acetyltransferase immunocytochemistry: a combined light and electron microscopic study. *J. Comp. Neurol.*, 239: 237–246.

Frotscher, M. and Leranth, C. (1986) The cholinergic innervation of the rat fascia dentata: identification of target structures on granule cells by combining choline acetyl-transferase immunocytochemistry and Golgi impregnation. *J. Comp. Neurol.,* 243: 58–70.

Frotscher, M. and Naumann, T. (1992) Septohippocampal cholinergic neurons: synaptic connections and survival following axotomy. *Rev. Neurosci.,* 3: 233–248.

Gage, F.H., Wictorin, K., Fischer, W., Williams, L.R., Varon, S. and Björklund, A. (1986) Retrograde cell changes in medial septum and diagonal band following fimbria-fornix transection: quantitative temporal analysis. *Neuroscience,* 19: 241–255.

Gage, F.H., Armstrong, D.M., Williams, L.R. and Varon, S. (1988) Morphological response of axotomized septal neurons to nerve growth factor. *J. Comp. Neurol.,* 269: 147–155.

Gage, F.H., Tuszynski, M.H., Chen, K.S., Armstrong, D. and Buzsaki, G. (1989) Survival, growth and function of damaged cholinergic neurons. In: M. Frotscher and U. Misgeld (Eds.), *Central Cholinergic Synaptic Transmission,* Birkhäuser, Basel, pp. 259–274.

Gähwiler, B.H. and Brown, D.A. (1985) Functional innervation of cultured hippocampal neurons by cholinergic afferents from co-cultured septal explants. *Nature,* 313: 577–579.

Gähwiler, B.H., Brown, D.A., Enz, A. and Knöpfel, T. (1989) Development of the septohippocampal projection in vitro. In: M. Frotscher and U. Misgeld (Eds.), *Central Cholinergic Synaptic Transmission,* Birkhäuser, Basel, pp. 236–250.

Heckers, S., Ohtake, T., Wiley, R.G., Lappi, D.A., Geula, C. and Mesulam, M.-M. (1994) Complete and selective cholinergic denervation of rat neocortex and hippocampus but not amygdala by an immunotoxin against the p75 NGF receptor. *J. Neurosci.,* 14: 1271–1289.

Hefti, F. (1986) Nerve growth factor promotes survival of septal cholinergic neurons after fimbrial transections. *J. Neurosci.,* 6: 2155–2162.

Heimrich, B. and Frotscher, M. (1993) Formation of the septohippocampal projection in vitro: an electron microscopic immunocytochemical study of cholinergic synapses. *Neuroscience,* 52: 815–827.

Heimrich, B. and Frotscher, M. (1994) Slice cultures as a tool to study neuronal development and the formation of specific connections. *Neurosci. Protocols,* 94–030–05–01–09.

Hsu, S.-M., Raine, L. and Fanger, H. (1981) Use of avidin-biotin-peroxidase complex (ABC) in immunoperoxidase techniques: a comparison between ABC and unlabeled antibody (PAP) procedures. *J. Histochem. Cytochem.,* 29: 577–580.

Kromer, L.F. (1987) Nerve growth factor treatment after brain injury prevents neuronal death. *Science,* 235: 214–216.

Lams, B.E., Isacson, O. and Sofroniew, M.V. (1988) Loss of transmitter-associated enzyme staining following axotomy does not indicate death of brainstem cholinergic neurons. *Brain Res.,* 475: 401–406.

Levi-Montalcini, R. (1987) The nerve growth factor: thirty-five years later. *EMBO J.,* 6: 1145–1154.

Levi-Montalcini, R. and Angeletti, P.U. (1968) Nerve growth factor. *Physiol. Rev.,* 48: 534–569.

Linke, R. and Frotscher, M. (1993) Development of the rat septohippocampal projection: Tracing with DiI and electron microscopy of identified growth cones. *J. Comp. Neurol.,* 332: 69–88.

Linke, R., Heimrich, B. and Frotscher, M. (1995) Axonal regeneration of identified septohippocampal projection neurons in vitro. *Neuroscience,* 68: 1–4.

Maranto, A.R. (1982) Neuronal mapping: a photooxidation reaction makes Lucifer Yellow useful for electron microscopy. *Science,* 217: 953–955.

Naumann, T., Linke, R. and Frotscher, M. (1992a) Fine structure of rat septohippocampal neurons: I. Identification of septohippocampal projection neurons by retrograde tracing combined with electron microscopic immunocytochemistry and intracellular staining. *J. Comp. Neurol.,* 325: 207–218.

Naumann, T., Peterson, G.M. and Frotscher, M. (1992b) Fine structure of rat septohippocampal neurons: II. A time course analysis following axotomy. *J. Comp. Neurol.,* 325: 219–242.

Naumann, T., Kermer, P. and Frotscher, M. (1994a) Fine structure of rat septohippocampal neurons: III. Recovery of choline acetyltransferase immunoreactivity after fimbria-fornix transection. *J. Comp. Neurol.,* 350: 161–170.

Naumann, T., Kermer, P., Seydewitz, V., Ortmann, R., D'Amato, F. and Frotscher, M. (1994b) Is there a long-lasting effect of a short-term nerve growth factor application on axotomized rat septohippocampal neurons? *Neurosci. Lett.,* 173: 213–215.

Peterson, G.M. (1989) A quantitative analysis of the crossed septohippocampal projection in the rat brain. *Anat. Embryol.,* 180: 421–425.

Peterson, G.M., Lanford, G.W. and Powell, E.W. (1990) Fate of septohippocampal neurons following fimbria-fornix transection: a time course analysis. *Brain Res. Bull.,* 25: 129–137.

Peterson, G.M., Naumann, T. and Frotscher, M. (1992) Identified septohippocampal neurons survive axotomy: a fine-structural analysis in the rat. *Neurosci. Lett.,* 138: 81–85.

Peterson, G.M., Naumann, T. and Frotscher, M. (1993) Light- and electron microscopic studies of identified septohippocampal neurons surviving axotomy. *Ann. N. Y. Acad. Sci.,* 679: 291–298.

Plaschke, M., Kasper, E.M., Naumann, T. and Frotscher, M. (1994) Survival and transmitter expression of rat cholinergic medial septal neurons despite removal of hippocampus in the early postnatal period. *Neurosci. Lett.,* 176: 243–246.

Plaschke, M., Naumann, T., Kasper, E., Bender, R. and Frotscher, M. (1996) Development of cholinergic and GABAergic neurons in the rat medial septum: effect of target removal in early postnatal development. *J. Comp. Neurol.,* in press.

Saporito, M.S. and Carswell, S. (1995) High levels of synthesis and local effects of nerve growth factor in the septal region of the adult rat brain. *J. Neurosci.,* 15: 2280–2286.

Schwab, M.E., Otten, U., Agid, Y. and Thoenen, H. (1979) Nerve growth factor (NGF) in the rat CNS: absence of specific retrograde axonal transport and tyrosine hydroxylase induction in locus coeruleus and substantia nigra. *Brain Res.,* 168: 473–483.

Seiler, M. and Schwab, M.E. (1984) Specific retrograde transport of nerve growth factor (NGF) from neocortex to nucleus basalis in the rat. *Brain Res.,* 300: 33–39.

Sobreviela, T., Clary, D.O., Reichardt, L.F., Brandabur, M.M., Kordower, J.H. and Mufson, E.J. (1994) TrkA-immunoreactive profiles in the central nervous system: colocalization with neurons containing p75 nerve growth factor receptor, choline acetyltransferase, and serotonin. *J. Comp. Neurol.,* 350: 587–611.

Stoppini, L., Buchs, P.-A. and Muller, D. (1991) A simple method for organotypic cultures of nervous tissue. *J. Neurosci. Methods,* 37: 173–182.

Thoenen, H. and Barde, Y.-A. (1980) Physiology of nerve growth factor. *Physiol. Rev.,* 60: 1284–1335.

Waller, A.V. (1850) Experiments on the section of the glossopharyngeal and hypoglossal nerves of the frog, and observations of the alterations produced thereby in the structure of their primitive fibers. *Philos. Trans.,* 140: 423–469.

Weskamp, G., Gasser, U.E., Dravid, A.R. and Otten, U. (1986) Fimbria-fornix lesion increases nerve growth factor content in adult rat septum and hippocampus. *Neurosci. Lett.,* 70: 121–126.

Williams, L.R., Varon, S., Peterson, G.M., Wictorin, K., Fischer, W., Björklund, A. and Gage, F.H. (1986) Continuous infusion of nerve growth factor prevents basal forebrain neuronal death after fimbria fornix transection. *Proc. Natl. Acad. Sci. USA,* 83: 9231–9235.

J. Klein and K. Löffelholz (Eds.)
*Progress in Brain Research*, Vol. 109
© 1996 Elsevier Science B.V. All rights reserved.

CHAPTER 33

# Effects of trophic factors on the CNS cholinergic phenotype

A. Claudio Cuello

*Department of Pharmacology and Therapeutics, McGill University, 3655 Drummond, suite 1325, Montreal, QC, Canada, H3G 1Y6*

## Introduction

In recent years it has become evident that the CNS cholinergic phenotype in mature animals can be significantly modulated by neurotrophic factors, notably by NGF (nerve growth factor). This idea was facilitated by the discovery of NGF within the CNS, particularly in areas rich in cholinergic innervation (Korsching et al., 1985). Furthermore, basal forebrain cholinergic neurons have been shown to retrogradely transport NGF (Seiler and Schwab, 1984) and to respond to this peptide, following axotomy, by increasing choline acetyltransferase (ChAT) activity (Hefti et al., 1984). These early studies alluded to the possible neurotrophic role of NGF for adult CNS cholinergic neurons, a notion reinforced by further work which identified binding sites for NGF in the adult rat brain (Taniuchi et al., 1986) and which showed that NGF prevents the loss of phenotype and preserves the morphology of cholinergic neurons following injury (Hefti, 1986; Williams et al., 1986; Kromer, 1987; Hagg et al., 1988; Cuello et al., 1989). Interestingly, in addition to NGF, other factors such as basic and acidic fibroblast growth factor (Andersson et al., 1988; Figueiredo et al., 1993) and ciliary neurotrophic factor also ameliorate cholinergic deficits which occur following brain injury (Hagg et al., 1992) or aging (Fischer et al., 1987). Furthermore, cytokines such as interleukin 3 (Kamegai et al., 1990) and other substances not classically considered as growth factors have shown some efficacy in protecting cholinergic neurons. Among these agents, the gangliosides are particularly interesting as they do not act on

a specific "ganglioside receptor," but have been shown to possess putative trophic actions both in vitro and in vivo, including effects on cholinergic phenotype (for reviews see: Ledeen, 1989; Skaper et al., 1989; Cuello, 1990; Cuello et al., 1994).

The well established cholinergic deficits observed in Alzheimer's disease, including the loss of cholinergic neurons of the nucleus basalis of Meynert (for review see Rossor et al., 1982; Whitehouse et al., 1982) along with evidence for a role of cholinergic neurotransmission in learning and memory (Bartus et al., 1982), have provoked particular interest in trophic factor research in this system. The septum to hippocampus pathway has been used extensively for these investigations, mostly by producing lesions of the fimbria-fornix pathway leading to apparent cellular loss in the medial septum and anterograde axonal degeneration in the hippocampus (for review see Hefti, 1986; Williams et al., 1986; Hagg et al., 1988; Kromer, 1987). By contrast, studies involving forebrain cholinergic neurons (NBM), which provide the bulk of the cholinergic innervation to the neocortex, have focused mainly on an anterograde injury paradigm (DiPatre et al., 1989; Haroutunian et al., 1989; Mandel et al., 1989; Dekker et al., 1992). Lesions of the rat NBM produce deficits in cholinergic markers in the cortex and affect performance in passive avoidance and spatial memory based tasks (for review see: Dekker et al., 1991; Fibiger, 1991). Very recently it has been shown that the grafting of cells transfected with the *Drosophila* ChAT gene in animals bearing lesions of the NBM correct these deficits, illustrating the cholinergic participation in higher cortical functions (Winkler et al., 1995).

## Cortical lesions and the cholinergic phenotype

We have investigated the effects of partial, unilateral experimental infarctions of the rodent cerebral cortex on the cholinergic phenotype of the NBM and the consequences of trophic factor therapy in preventing or reverting the retrograde and anterograde degeneration suffered by these neurons after the application of cortical lesions.

## Specificity and dose dependent effects of NGF on ChAT activity in nucleus basalis cell bodies and terminals

We have demonstrated that 30 days post-lesion, animals which receive unilateral cortical lesions develop deficits in ChAT activity in the ipsilateral microdissected NBM (Stephens et al., 1985; Cuello et al., 1989). The activity of this enzyme in lesioned rats is apparently not affected in the contralateral NBM nor in other ipsilateral or contralateral brain areas examined (striatum, septum, hippocampus, cortex) (Stephens et al., 1985; Garofalo and Cuello, 1995). However, ChAT activity in the ipsilateral NBM of lesioned rats is usually decreased by 40–50%. We have observed that treatment with NGF beginning immediately post-lesion and continuing for 7 days prevents, in a dose-dependent manner, this lesion-induced deficit (Fig. 1B). In addition, NGF applied directly in the lateral ventricle caused an increase above control levels in ipsilateral cortical ChAT activity, an area which contains the nerve terminals of the NBM neurons (Fig. 1A). The $ED_{50}$ for NGF to produce these effects was 0.1 $\mu$g/day for the NBM and 1 $\mu$g/day for stimulation of cortical ChAT activity. These short-term (7 days) treatments with NGF did not affect ChAT activity in the contralateral NBM or cortex, nor in the septum or hippocampus. NGF has also been shown to revert the atrophy of NBM cholinergic neurons in a dramatic fashion as shown with the assistance of computer assisted image analysis (Cuello et al., 1989). However, NBM neurons also responded positively to acidic fibroblast

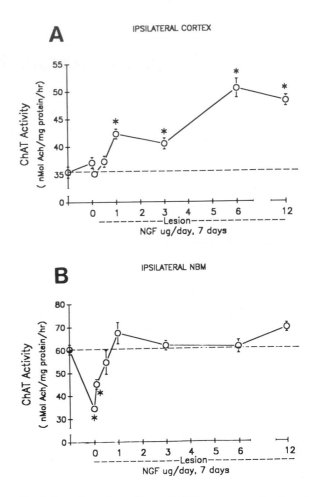

Fig. 1. Dose dependent effects of NGF treatment on ipsilateral (A) Cortical and (B) NBM ChAT activity of unilaterally decorticated rats. Animals were cortically lesioned and received, i.c.v. via minipump, immediately post-lesion either vehicle [0], consisting of artificial CSF + 0.1 % BSA, or NGF at dosages of 0.1–12 $\mu$g/day for 7 days. Rats were sacrificed 30 days post-lesion (i.e. 23 days after the end of drug treatment) and ChAT activity was determined in microdissected NBM and cortices according to the method of Fonnum (1975). Protein content was assessed using the method of Bradford (1976). $n = 6$ animals/group. Error bars represent standard error of the mean (SEM). *$P < 0.05$ from control, ANOVA, post-hoc Newman–Keuls test. Reprinted with permission from Garofalo and Cuello (1995).

growth factor (a-FGF), for which these neurons apparently do not possess receptors, and they responded negatively to related neurotrophins such as brain derived neurotrophic factor (BDNF) or

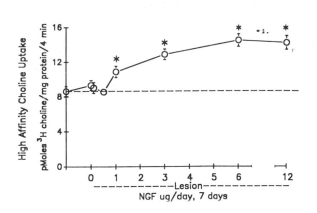

Fig. 2. Dose dependent effects of NGF on high affinity choline uptake (HACU) in the ipsilateral cortex of decorticated rats. Animals were cortically lesioned and received, i.c.v. via minipump, immediately post-lesion either vehicle [0] (artificial CSF + 0.1 % BSA), or NGF at dosages of 0.1–12 $\mu$g/day for 7 days. Rats were sacrificed 30 days post-lesion (i.e. 23 days after the end of drug treatment) and HACU of cortical synaptosomes was determined. Protein content was assessed using the method of Bradford (1976). $n = 6$ animals/group. Error bars represent SEM. *$P < 0.05$ from control, ANOVA, post-hoc Newman–Keuls test. Reprinted with permission from Garofalo and Cuello (1995).

Neurotrophin 3 (NT3) (Skup et al., 1994). These findings stress, on the one hand, the exquisite specificity of these neurons to trophic stimulation and, on the other, the possibility of indirect trophic effects, in the case of a-FGF. This possibility has been reinforced by the in vitro observations of FGF stimulating NGF synthesis by astrocytes (Yoshida and Gage, 1991), and by the increment change in both the NGF mRNA and the peptide in the remaining cortex in our lesion model, when FGF is exogenously administered (Figueiredo et al., 1995b).

## NGF application alters cortical choline uptake in lesioned animals in a dose-dependent manner

Although ChAT is a good cholinergic marker, the high affinity choline uptake (HACU) is considered the rate limiting step for the synthesis of ACh and therefore it should faithfully reflect the functional capacity of the cholinergic terminal. We have analyzed choline uptake into synaptosomes prepared from the ipsilateral or contralateral cortices, hippocampi or striata of control non-operated and cortically lesioned rats which received, i.c.v. via minipump for 7 days, either vehicle or various concentrations of NGF (Garofalo and Cuello, 1991, 1995). As shown by Fig. 2, NGF treatment caused a dose-dependent stimulation of cortical HACU. $ED_{50}$ value was 1.5 $\mu$g/day which is close to what was noted for the stimulation of cortical ChAT activity. These effects require the integrity of the cholinergic neurons as NGF did not directly affect HACU in vitro (Fig. 3). These findings, along with the enhanced release of endogenous cortical acetyl-choline in lesioned NGF-treated rats (Maysinger et al., 1992), would favor the idea that some growth factors can profoundly affect cortical cholinergic presynaptic function.

## Time dependence of NGF effects over the NBM cholinergic phenotype after lesions

The effect of NGF on injured cholinergic neurons on biochemical parameters is time dependent. In

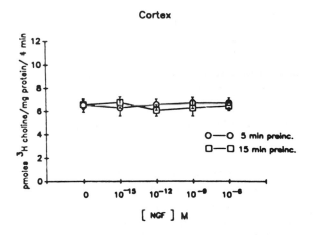

Fig. 3. In vitro effects of NGF on HACU in cortical synaptosomes. Synaptosomes from lesioned animals were preincubated (preinc) for 5 or 15 min with vehicle (0) or NGF ($10^{-6}$ to $10^{-15}$M) before the addition of 0.5 $\mu$M [$^3$H]choline. Values represent differences between mean uptake at 37°C and 0°C of five to six experiments. Error bars indicate SEM, ANOVA showed no differences between groups. Reprinted with permission from Garofalo and Cuello (1995).

our studies (Garofalo and Cuello, 1995), this has been observed in adult rats which were unilaterally lesioned (partial cortical infarctions) and received either vehicle or maximal doses of NGF (12 $\mu$g/day) for a period of 7 days. These animals were sacrificed at either 1, 5, 15 or 30 days post-lesion. Interestingly, an increase in ChAT activity was noted in the ipsilateral NBM at day 1 post-lesion, which coincides with the temporary up-regulation of the RNA message for the low affinity NGF receptor (Figueiredo et al., 1995a). Thereafter a time dependent decrease in ChAT activity in the NBM takes place. NGF treatment altered both the early up-regulation and the late down-regulation of ChAT activity in cholinergic cell bodies. Thus, NGF administration attenuated the 1-day post-lesion increase in NBM ChAT activity as well as the decreases observed at later times.

A time dependent stimulation of cortical ChAT activity and cortical HACU was also noted. As was observed for NBM ChAT activity, the activity of this enzyme in the cortex was found to be significantly increased above control values on post-lesion day 1. By contrast, cortical HACU in day 1 post-lesion rats was not significantly different from control values. At subsequent post-lesion times, cortical ChAT activity and HACU in vehicle-treated animals did not vary from control values.

### Late administration of NGF rescues NBM cholinergic phenotype; biochemical evidence

We have examined the effects of late or early onset of trophic therapy as well as the duration of treatment in this lesion model. The effects of short-term (7 days) versus chronic (28 days) administration of maximal doses of NGF on NBM and cortical ChAT activity and cortical HACU were examined. As shown in Fig. 4, a 7- or 28-day treatment with NGF, when initiated immediately post-lesion, maintained NBM ChAT activity and augmented cortical ChAT activity to an equal degree. Differences were noted, however, in the treatment time onset necessary to induce these effects. For example, if the ganglioside GM1 was employed in a short-term treatment paradigm, the onset of GM1 administration could be delayed no more than 24–48 h for a full preservation of NBM cell morphology or ChAT activity (Stephens et al., 1987; Garofalo and Cuello, 1995). On the other hand, NGF treatment could be delayed up to 6 (short-term treatment) or 14 (chronic treatment) days post-lesion and still prevent the decrease in NBM ChAT activity. However, a 14-day delay in the initiation of NGF treatment failed to stimulate ChAT activity in the remaining cortex (Fig. 4). Initiating NGF treatment at 30 days post-lesion and continuing treatment for 2 weeks failed to recover or stimulate, respectively, ChAT activity in the NBM and cortex. However, if treatment was continued for 4 weeks a partial recovery in NBM ChAT activity occurred. Alterations in cortical HACU following these various treatment paradigms followed what was noted for cortical ChAT activity (Table 1).

### Cortical lesions and trophic factor therapy differentially affect the steady state levels of RNA message for the low and high affinity NGF receptor in NBM cholinergic neurons

Cholinergic neurons of the NBM are well endowed with both the low (p75$^{NGFR}$) and high affinity (p140$^{trk}$) NGF receptors as numerous immunohistochemical and in situ hybridization studies have revealed (for review see Cuello, 1994). Changes in the expression of both low and high affinity NGF receptors presumably involve a number of factors induced by neuronal injury. We have observed that the expression of p75$^{NGFR}$ and p140$^{trk}$ mRNAs in the NBM are both affected by cortical infarctions and the application of NGF (Figueiredo et al., 1995a). However, these manipulations affected the corresponding mRNAs differentially. Cortical devascularization elicited increases in p75$^{NGFR}$ mRNA at 3 days post-lesion, declining 7 days after the lesion. This transient increase might be the consequence of the transitory increase of NGF which has been reported to occur in the NBM following neocortical suction ablation (Lorez et al., 1988) and also in the remaining cerebral cortex following cortical infarction (Figueiredo et al., 1995b). Furthermore, similar increments in NGF expression have been observed

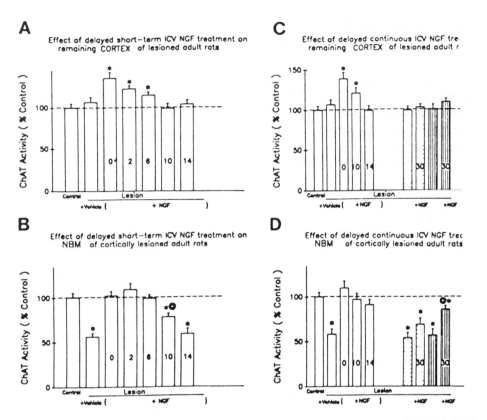

Fig. 4. (A,B) ChAT activity in the NBM of lesioned animals after late application of NGF. Animals were decorticated and received short-term (7 days) treatment with maximal doses of NGF (1 μg/day), i.c.v. via minipump, beginning immediately (0) post-lesion or after a 2, 6, 10 or 14-day delay as indicated within bar graphs. (C,D) Animals were decorticated and received chronic (28 days) treatment with maximal doses of NGF (12 μg/day), i.c.v. via minipump, beginning immediately (0) post-lesion or after a 10 or 14-day delay as indicated within bar graphs. Dotted bar graphs represent decorticated animals treated with vehicle or NGF for 2 weeks after a 30 day delay in treatment onset. Bar graphs with vertical lines indicate decorticated animals which received vehicle or NGF treatment for 28 days following a 30 day delay in treatment time onset. *$P < 0.05$ from control, @$P < 0.05$ from respective lesion vehicle treated group. ANOVA post-hoc Newman–Keuls test. Reprinted with permission from Garofalo and Cuello (1995).

in the septum following fimbria-fornix transection (Gasser et al., 1986).

However, the increase in endogenous NGF as well as the early increase of its low affinity receptor mRNA without exogenous application of the neurotrophic factor is insufficient to protect these NBM neurons from atrophy. In our basolo-cortical lesion model, a differential cellular response to the devascularizing lesion was exhibited by p140$^{trk}$ whose mRNA levels in the NBM was not increased after cortical lesion but decreased showing a late spontaneous recovery (Figueiredo et al., 1995). Also, a divergence between levels of p75$^{NGFR}$ and p140$^{trk}$ was noted in response to NGF administered at doses which ensured a trophic re-

sponse. Exogenous NGF increased p75$^{NGFR}$ mRNA levels in the NBM over control levels during administration, only to decline to levels relative to sham-operated animals on cessation (Table 2), confirming previous studies conducted in other neuronal cell types (Cavicchioli et al., 1989; Gage et al., 1989; Higgins et al., 1989; Holtzman et al., 1992; Verge et al., 1992). At 3 and 7 post-lesion days, p140$^{trk}$ mRNA levels were maintained relative to sham-operated animals, however we observed a significant increase of p140$^{trk}$ mRNA on post-lesion day 15 (8 days after ceasing the NGF treatment) (Table 3 and Fig. 5). The spontaneous up-regulation of mRNA for p140$^{trk}$ in degenerating, atrophic neurons offers an excellent explanation for the successful therapy with NGF we and others (Hagg et al., 1988) have observed when the neurotrophin is offered to cholinergic neurons at late stages, when the degeneration processes are well underway. It would be interesting to determine if the up-regulation of high affinity receptors in degenerating neurons is an universal phenomenon. If so, the therapeutic opportunities for neurotrophic factors are probably better than so far anticipated.

## The trophic factor therapy of injured CNS cholinergic neurons extends to the formation of new cortical synapses

As discussed above, extensive light and electron

TABLE 1

Effect of delayed short-term or chronic NGF treatment on high affinity [$^3$H]choline uptake in remaining cortex of lesioned rats

|  | Uptake of [$^3$H]choline (% of control) | |
| --- | --- | --- |
|  | 7 days of NGF treatment | 28 days of NGF treatment |
| Control | 100 ± 10 | 100 ± 8 |
| Lesioned + vehicle | 105 ± 8 | 98 ± 5 |
| Lesioned + 0 day delay | 144 ± 6* | 149 ± 8* |
| Lesioned + 2 day delay | 131 ± 4* | – |
| Lesioned + 6 day delay | 115 ± 4 | – |
| Lesioned + 10 day delay | 106 ± 7 | 129 ± 5* |
| Lesioned + 14 day delay | 99 ± 5 | 103 ± 4 |
| Lesioned + vehicle | 99 ± 5 | 101 ± 6 |
| Lesioned + 30 day delay | 107 ± 6[a] | 113 ± 5 |

Animals were decorticated and received either vehicle or NGF (1 μg/day) for 7 or 28 days, i.c.v. via minipump, beginning immediately post-lesion or after the delays indicated. Values represent high affinity uptake of [$^3$H]choline, expressed as percent control, of synaptosomes isolated from the remaining ipsilateral cortex. Values are expressed as percentage of control. Reprinted with permission from Garofalo and Cuello (1995).
*$P < 0.05$ from control, ANOVA post-hoc Newman–Keuls test, $n + 6$ animals/group.
[a]14 days of NGF treatment.

TABLE 2

p75$^{NGFR}$ mRNA levels in nucleus basalis magnocellularis

| Post-lesion day | Mean number of silver grains (pixels)/ cross-sectional cell area (μm$^2$) | | |
| --- | --- | --- | --- |
|  | Control | Lesion + vehicle | Lesion + NGF |
| 1 | 8.70 ± 0.75 | 10.91 ± 0.50 | 12.11 ± 0.62* |
| 3 | 7.13 ± 0.48 | 11.42 ± 1.43* | 13.29 ± 1.29* |
| 7 | 8.62 ± 0.39 | 7.26 ± 0.78** | 14.25 ± 0.76*,[†] |
| 15 | 8.53 ± 0.55 | 6.24 ± 0.42** | 9.98 ± 0.59[†] |

Values represent mean ± SEM, $n = 4$–5 animals/group. Reproduced with permission from Figueiredo et al. (1995a).
*$P < 0.05$ from control.
**$P < 0.05$ from day 1 values within the same group.
[†]$P < 0.05$ from lesion + vehicle (ANOVA followed by post-hoc Newman-Keuls' test).

microscopy investigations on ChAT immunoreactive (IR) fibers and their presynaptic elements in the cerebral cortex has yielded evidence for the profound effects of growth factors on the morphology and biochemistry of cell bodies of the NBM cholinergic neurons and the up-regulation of the biochemical markers provoked by NGF in terminal areas of these neurons in the cerebral cortex. Thus, our studies (Garofalo et al., 1992)

TABLE 3

p140$^{trk}$ mRNA levels in nucleus basalis magnocellularis

| Post-lesion day | Mean number of silver grains (pixels)/ cross-sectional cell area (μm$^2$) | | |
| --- | --- | --- | --- |
|  | Control | Lesion + vehicle | Lesion + NGF |
| 1 | 11.53 ± 0.33 | 9.32 ± 0.62 | 12.06 ± 0.61 |
| 3 | 10.22 ± 0.54 | 6.94 ± 1.14* | 11.48 ± 1.09[†] |
| 7 | 9.21 ± 0.78 | 5.19 ± 0.40*,** | 10.17 ± 0.88[†] |
| 15 | 9.40 ± 0.1.10 | 9.67 ± 1.03 | 14.09 ± 1.76*,[†] |

Values represent mean ± SEM, $n = 4$–5 animals/group. Reprinted with permission from Figueiredo et al. (1995a).
*$P < 0.05$ from control.
**$P < 0.05$ from post-lesion day 1 values within the same group.
[†]$P < 0.05$ from lesion + vehicle (ANOVA followed by post-hoc Newman–Keuls test).

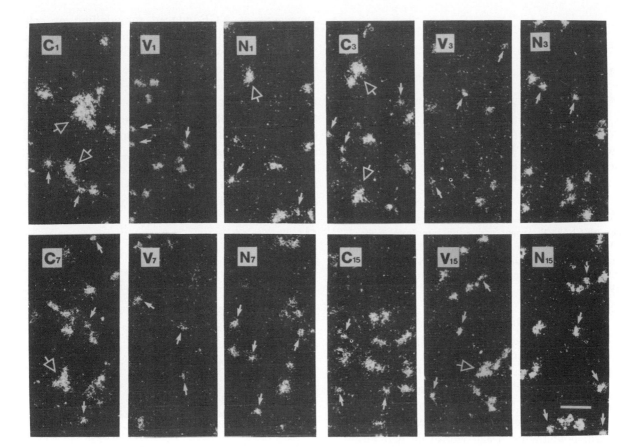

Fig. 5. Dark-field photomicrographs showing silver grains detected in the mid-portion of the NBM. Sections from sham-operated control (C), lesion + vehicle (V) and lesion + NGF (N) treated animals shown at 1, 3, 7 and 15 days post-lesion as indicated by the subscripted number. Clusters of 3 or more cells are commonly found in the NBM, thus with the exception of control photomicrographs (C1, C3, C7 and C15), most of the other panels show a higher proportion of isolated cells, allowing more accurate comparisons. Note the decrease in p140$^{trk}$ mRNA expression in V7. Labelling of isolated cells in N15 is higher than in V15 and C15. Small solid arrows point to single cells while open arrows point to clusters of 4 or more cells. Scale bar in N15 = 90 $\mu$m, applies to all panels. Reproduced with permission from Figueiredo et al. (1995a).

have revealed that the ChAT-IR fiber network is expanded in the cerebral cortex of animals bearing cortical lesions which have been treated with NGF. Furthermore, the number of varicosities (i.e. the light microscopy presynaptic sites) increases considerably in these circumstances (Fig. 6). It has been shown that some CNS sprouting might occur in the fimbria-fornix lesion model in NGF-treated rats by the demonstrable accumulation of ChAT-IR material in the septum from the flow interrupted from the hippocampus (Gage et al., 1989). Our studies extended this work by demonstrating that, when observed under the electron microscopy, the newly formed varicosities are indeed presynaptic cholinergic elements (Garofalo et al., 1992), and that they are enlarged (hypertrophic) in the NGF-treated rat, a finding consistent with the biochemical observations of enhanced ChAT activity and HACU (see above). Furthermore, the number of synaptic differentiations in cholinergic boutons are greater in lesioned and NGF-treated cortices than in control (Garofalo et al., 1992) (Fig. 7). These findings have been more recently corroborated by the painstaking 3-D reconstruction of cholinergic terminals in these trophic factor-treated situations (Garofalo et al, 1993).

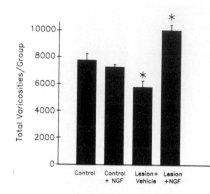

Fig. 6. Number of varicosities per group. Grey plus detected images of ChAT-IR axonal varicosities are from lesioned vehicle-treated and NGF-treated animals. *$P < 0.01$ from control (analysis of variance post-hoc Tukey test; operator was blinded to treatment groups). Reprinted from Garofalo et al. (1992).

It has long been known that considerable plasticity occurs as a consequence of septal nucleus deafferentation (Raisman, 1969) and also that in the dentate gyrus new synapses are formed after entorhinal lesions (Cotman and Nadler, 1978). Plasticity within the fully differentiated CNS has also been documented in the hippocampus after long-term potentiation, (Van Herreveld and Fifkova, 1975; Applegate et al., 1987). Since Ramón y Cajal's time (Ramón y Cajal, 1928), learning and memory have been postulated to involve changes in the strength or efficiency of synaptic transmission. The possibility that growth factors are involved in these phenomena has long been speculated upon. In the peripheral nervous system, exogenous NGF has been shown to modify synaptic numbers after axotomy of sympathetic ganglia (Purves and Nja, 1976). Our studies have demonstrated that a drug (a growth factor in this instance) is capable of inducing alterations in the number and morphology of CNS axonal varicosities and synapses in the fully differentiated cortex of adult lesioned animals and that this has profound behavioral consequences (see below).

## Cortical cholinergic synapses and behavioral correlates

In order to assess and compare the functional con-

sequences of NGF-induced remodelling of cell bodies and terminals, the behavior of treated rats was studied in two tasks – passive avoidance and the Morris water maze – extensively used in NBM-cortex lesion paradigms. An intact cortex is thought important for proper spatial navigation in the rat, since Kolb et al. (1983) demonstrated that damage to the medial frontal, orbital frontal and cingulate cortices produces significant impairments in spatial memory while injury to parietal areas result in only minor deficits. However, when a larger extent of the rat parietal cortex is lesioned

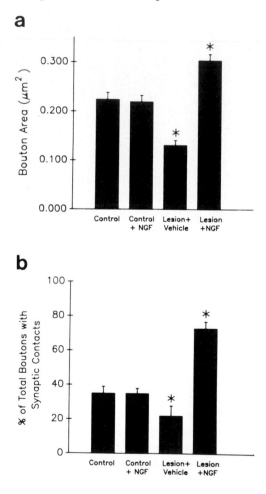

Fig. 7. Cross-sectional area of ChAT-IR boutons in cortical layer V (a) and percentage of varicosity profiles quantified with visible synaptic contacts (b). *$P < 0.01$ from control (analysis of variance post-hoc Tukey test; operator was blinded to treatment groups). Reprinted from Garofalo et al. (1992).

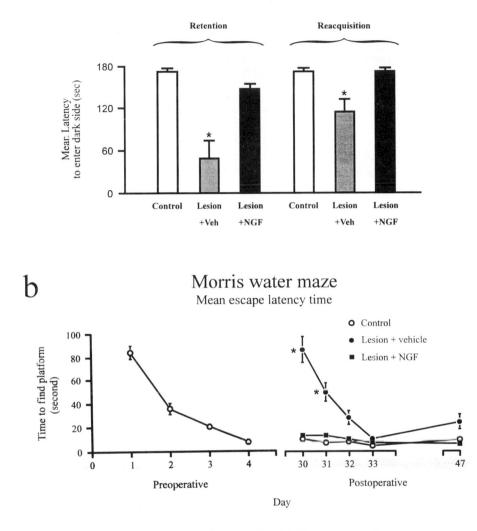

a    Passive Avoidance

b    Morris water maze
Mean escape latency time

Fig. 8. Passive avoidance; retention and reacquisition (a). Preoperative and postoperative mean escape latency times in the Morris water maze (b). *P < 0.05 (analysis of variance post-hoc Newman–Keuls test, n = 9–11 animals per group). Reprinted with permission from Garofalo and Cuello (1994).

a somewhat greater degree of disruption of spatial memory has also been observed (DiMattia and Kesner, 1988). We previously reported that extensive unilateral cortical lesions also cause deficits, albeit mild, in spatial navigation and passive avoidance retention and reacquisition (Elliott et al., 1989) which can be partially corrected with cholinergic agents. The cholinergic participation in these tasks have been confirmed by our investigations, and further emphasized by the recent report

of Winkler et al. (1995) in which behavioral deficits provoked by lesions of the NBM are corrected by grafting ACh synthesizing cells in the cerebral cortex.

In our hands, cortical infarctions involving primarily parietal and frontal cortices affected performance of lesioned animals in the retention of passive avoidance and Morris water maze tasks. In these animals, NGF treatment clearly abolished the retention deficits (Garofalo and Cuello, 1994). The extent of the retention of these learnt tasks by the application of NGF to cortically lesioned rats, in which a profound cholinergic synaptic remodelling occurs, is illustrated in Fig. 8. Cortical synaptic remodelling might very well be the biological substrate which explains the amelioration of spatial memory impairment in aged rats (Fischer et al., 1987).

## Conclusions

Our studies on the maintenance of phenotypic characteristics of the forebrain cholinergic neurons of the NBM subjected to degeneration induced by cortical lesions and treated with experimental neurotrophic factor therapy indicates that the plasticity of these neurons in adulthood is very pronounced. NTFs can act on these neurons with great potency in a dose dependent manner. Their administration alters biochemical markers of cholinergic function in cell body rich areas and induces the expression of the mRNAs for low and high affinity NGF receptors. Delayed treatment with NGF rescues the affected cholinergic phenotype. These changes are extensive to synaptic cholinergic remodelling at the cerebral cortex level, expressed as hypertrophy of presynaptic sites and a higher number of cholinergic synaptic elements. The synaptic remodelling of the cerebral cortex induced by growth factors could explain the improved behavioral performance observed in NGF-treated rats bearing cortical infarcts.

## Acknowledgements

I would like to acknowledge my many collaborators and former students who made this research possible. Their work is referred to in this review. For assistance in preparing this particular paper I would like to thank Sylvain Coté for technical and graphic arts assistance, Alan Foster for photographic expertise, and Sid Parkinson for editorial assistance. This work has been largely supported by the Canadian MRC and more recently by the NIH (USA).

## References

Andersson, K.J., Dam, D., Lee, S. and Cotman, C.W. (1988) Basic fibroblast growth factor prevents neuronal death of lesioned cholinergic neurons in vivo. *Nature*, 332: 360–361.

Applegate, M.D., Kerr, D.S. and Landfield, P.W. (1987) Redistribution of synaptic vesicles during long-term potentiation in the hippocampus. *Brain Res.*, 401: 401–406.

Bartus, R.T., Dean, R.L., Beer, B. and Lippa, A.S. (1982) The cholinergic hypothesis of geriatric memory dysfunction. *Science*, 217: 408–412.

Bradford, M.M. (1976) A rapid and sensitive method for the quantitation of microgram quantities of protein using the principal of dye binding. *Ann. Biochem.*, 72: 248–254.

Cavicchioli, L., Flanigan, T.P., Vantini, G., Fusco, M., Polato, P., Toffano, G., Walsh, F.S. and Leon, A. (1989) NGF amplifies expression of NGF receptor messenger RNA in forebrain cholinergic neurons of rats. *Eur. J. Neurosci.*, 1: 258–262.

Cotman, C.W. and Nadler, J.V. (1978) Reactive synaptogenesis in hippocampus. In: C.W. Cotman (Ed.), *Neuronal Plasticity*, Plenum Press, New York, pp. 227–271.

Cuello, A.C. (1990) Glycosphingolipids that can regulate nerve growth and repair. *Adv. Pharmacol.*, 21: 1–50.

Cuello, A.C. (1994) Trophic factor therapy in the adult CNS: remodelling of injured basalo-cortical neurons. In: F. Bloom (Ed.), *Progress in Brain Research*, Vol. 100, Elsevier, Amsterdam, pp. 213–221.

Cuello, A.C., Garofalo, L., Kenigsberg, R.L. and Maysinger, D. (1989) Gangliosides potentiate *in vivo* and *in vitro* effects of nerve growth factor on central cholinergic neurons. *Proc. Natl. Acad. Sci. USA*, 86: 2056–2060.

Cuello, A.C., Garofalo, L., Liberini, P. and Maysinger, D. (1994) Cooperative effects of gangliosides on trophic factor-induced neuronal cell recovery and synaptogenesis: studies in rodents and subhuman primates. In: L. Svennerholm, A.K. Ashbury, R.A. Reisfeld, K. Sandhoff, K. Suzuki, G. Tettamanti and G. Toffano (Eds.), *Progress in Brain Research*, Vol. 101, Elsevier, Amsterdam, pp. 337–355.

Dekker, A.J, Connor, D.J. and Thal, L.J. (1991) The role of cholinergic projections from the nucleus basalis in memory. *Neurosci. Biobehav. Rev.*, 15: 299–317.

Dekker, A.J., Gage, F.H. and Thal, L.J. (1992) Delayed treatment with nerve growth factor improves acquisition of a spatial task in rats with lesions of the nucleus basalis magnocellularis: involvement of different neurotransmitter systems. *Neuroscience*, 48: 111–119.

DiMattia, B.D. and Kesner, R.P. (1988) Spatial and cognitive maps: differential role of parietal cortex and hippocampal formation. *Behav. Neurosci.*, 102: 471–480.

DiPatre, P.L., Casamenti, F., Cenni, A. and Pepeu, G. (1989) Interaction between nerve growth factor and GM1 ganglioside in preventing cortical choline acetyltransferase and high affinity choline uptake decrease after lesion of the nucleus basalis. *Brain Res.*, 480: 219–224.

Elliott, P.J., Garofalo, L. and Cuello, A.C. (1989) Limited neocortical devascularizing lesions causing deficits in memory retention and choline acetyltransferase activity: effects of the monosialoganglioside GM1. *Neuroscience*, 31: 63–76.

Fibiger, H.C. (1991) Cholinergic mechanisms in learning, memory and dementia: a review of recent evidence. *Trends Neurosci.*, 14: 220–223.

Figueiredo, B.C., Piccardo, P., Maysinger, D., Clarke, P.B.S. and Cuello, A.C. (1993) Effects of acidic fibroblast growth factor on cholinergic neurons of nucleus basalis magnocellularis and in a spatial memory task following cortical devascularization. *Neuroscience*, 56: 955–963.

Figueiredo, B.C., Skup, M., Bedard, A.M., Tetzlaff, W. and Cuello, A.C. (1995a) Differential expression of p140$^{trk}$, p75$^{NGFR}$ and GAP-43 genes in nucleus basalis magnocellularis, thalamus and adjacent cortex following neocortical infarction and NGF treatment. *Neuroscience,* 68: 29–45.

Figueiredo, B.C., Pluss, K., Skup, M., Otten, U. and Cuello, A.C. (1995b) Acidic FGF induces NGF and its mRNA in the neocortex of adult animals. *Mol. Brain Res.*, 33: 1–6.

Fischer, W., Wictorin, K., Björklund, A., Williams, L.R., Varon, S. and Gage, F.H. (1987) Amelioration of cholinergic neuron atrophy and spatial memory impairment in aged rats by nerve growth factor. *Nature*, 329: 65–68.

Fonnum, F. (1975) A rapid radiochemical method for the determination of choline acetyltransferase. *J. Neurochem.*, 24: 407–409.

Gage, F.H., Batchelor, P., Chen, K.S., Chin, D., Higgins, G.A., Koh, S., Deputy, S., Rosenberg, M.B., Fischer, W. and Björklund, A. (1989) NGF receptor reexpression and NGF-mediated cholinergic neuronal hypertrophy in the damaged adult neostriatum. *Neuron*, 2: 1177–1184.

Garofalo, L. and Cuello, A.C. (1991) Nerve growth factor and the monosialoganglioside GM1 modulate cholinergic markers and affect behavior of decorticated rats. *Eur. J. Pharmacol.*, 183: 934–935.

Garofalo, L. and Cuello, A.C. (1994) Nerve growth factor and the monosialoganglioside GM1 analogous and different *in vivo* effects on biochemical, morphological and behavioural parameters of adult cortically lesioned rats. *Exp. Neurol.*, 125: 195–217.

Garofalo, L. and Cuello, A.C. (1995) Pharmacological characterization of nerve growth factor and/or monosialoganglioside GM1 effects on cholinergic markers in the adult lesioned brain. *J. Pharm. Exp. Ther.*, 272: 527–545.

Garofalo, L., Ribeiro-da-Silva, A. and Cuello, A.C. (1992) Nerve growth factor induced synaptogenesis and hypertrophy of cortical cholinergic terminals. *Proc. Natl. Acad. Sci. USA*, 89: 2639–2643.

Garofalo, L., Ribeiro-da-Silva, A. and Cuello, A.C. (1993) Potentiation of nerve growth factor-induced alterations in cholinergic fibre length and presynaptic terminal size in cortex of lesioned rats by the monoganglioside GM. *Neuroscience,* 57: 21–40.

Gasser, U.E., Weskamp, G., Otten, U. and Dravid, A.R. (1986) Time course of the elevation of nerve growth factor (NGF) content in the hippocampus and septum following

lesions of the septohippocampal pathways in rats. *Brain Res.*, 376: 351–356.

Hagg, T.F., Manthorpe, M., Vahlsing, H.L. and Varon, S. (1988) Delayed treatment with nerve growth factor reverses the apparent loss of cholinergic neurons after acute brain damage. *Exp. Neurol.*, 101: 303–312.

Hagg, T.F., Quon, D., Higaki, J. and Varon, S. (1992) Ciliary neurotrophic factor prevents neuronal degeneration and promotes low affinity NGF receptor expression in the adult rat CNS. *Neuron*, 8: 145–158.

Haroutunian, V., Kanof, P. and Davis, K.L. (1989) Attenuation of nucleus basalis of Meynert lesion induced cholinergic deficits by nerve growth factor. *Brain Res.*, 487: 200–203.

Hefti, F. (1986) Nerve growth factor promotes survival of septal cholinergic neurons after fimbrial transections. *J. Neurosci.*, 6: 2155–2162.

Hefti, F., Dravid, A. and Hartikka, J. (1984) Chronic intraventricular injections of nerve growth factor elevate hippocampal choline acetyltransferase activity in adult rats with partial septo-hippocampal lesions. *Brain Res.*, 293: 305–311.

Higgins, G.A., Koh, S., Chen, K.S. and Gage, F.H. (1989) NGF induction of NGF receptor gene expression and cholinergic neuronal hypertrophy within the basal forebrain of the adult rat. *Neuron*, 3: 247–256.

Holtzman, D.M., Li, Y., Parada, L.F., Kinsman, S., Chen, C.-K., Valletta, J.S., Zhou, J., Long, J.B. and Mobley, W.C. (1992) p140$^{trk}$ mRNA marks NGF-responsive forebrain neurons: evidence that *trk* gene expression is induced by NGF. *Neuron*, 9: 465–478.

Kamegai, M., Niijima, K., Kunishita, T., Nishizawa, M., Ogawa, M., Araki, M., Ueki, A., Konishi, Y. and Tabira, T. (1990) Interleukin-3 as a trophic factor for central cholinergic neurons in vitro and in vivo. *Neuron*, 2: 429–436.

Kolb, B., Sutherland, R.J. and Whishaw, I.Q. (1983) A comparison of the contributions of the frontal and parietal association cortex to spatial localization in rats. *Behav. Neurosci.*, 97: 13–27.

Korsching, S., Auburger, G., Heumann, R., Scott, J. and Thoenen, H. (1985) Levels of nerve growth factor mRNA in the central nervous system of the rat correlate with cholinergic innervation. *EMBO J.*, 4: 1389–1393.

Kromer, L.F. (1987) Nerve growth factor treatment after brain injury prevents neuronal death. *Science*, 235: 352–355.

Ledeen, R.W. (1989) Biosynthesis, metabolism and biological effects of gangliosides. In: R.U. Margolis and R.K. Margolis (Eds.) *Neurobiology of Glycoconjugates*, Plenum Press, New York, pp. 43–82.

Lorez, H., vonFrankenber, M., Weskamp, G. and Otten, U. (1988) Effect of bilateral decortication on nerve growth factor content in nucleus basalis and neostriatum of adult rat brain. *Brain Res.*, 454: 355–360.

Mandel, R.J., Gage F.H. and Thal L.J. (1989) Enhanced detection of nucleus basalis magnocellularis lesioned-induced spatial learning deficit in rats by modification of training regimen. *Behav. Brain Res.*, 31: 221–229.

Maysinger, D., Herrera-Marschitz, M., Goiny, M., Ungerstedt, U. and Cuello, A.C. (1992) Effects of nerve growth factor on cortical and striatal acetylcholine and dopamine release in rats with cortical devascularizing lesions. *Brain Res.*, 577: 300–305.

Purves, D. and Nja, Å. (1976) Effect of nerve growth factor on synaptic depression after axotomy. *Nature*, 260: 535–536.

Raisman, G. (1969) Neuronal plasticity in the septal nuclei of the adult rat. *Brain Res.*, 14: 25–48.

Ramón y Cajal, S. (1928) *Degeneration and Regeneration of the Nervous System.* R.M May (translator), Oxford University Press, London.

Rossor, M.N., Svendsen, C., Hunt, S.F., Mountjoy, C.G., Roth, M. and Iversen, L.L. (1982) The substantia innominata in Alzheimer's disease: a histochemical and biochemical study of cholinergic marker enzymes. *Neurosci. Lett,* 28: 217–222.

Seiler, M. and Schwab, M.E. (1984) Specific retrograde transport of nerve growth factor (NGF) from neocortex to nucleus basalis in the rat. *Brain Res.*, 300: 34–39.

Skaper, S.D., Leon, A. and Toffano, G. (1989) Gangliosides function in the development and repair of the nervous system. *Mol. Neurobiol.*, 3: 173–198.

Skup, M.H., Figueiredo, B.C. and Cuello, A.C. (1994) Intraventricular application of BDNF and NT-3 Failed to Protect NBM Cholinergic Neurons. *NeuroReport*, 5: 1105–1109.

Stephens, P.H., Cuello, A.C., Sofroniew, M.V., Pearson, R.C. and Tagari, P. (1985) Effect of unilateral decortication on choline acetyltransferase activity in the nucleus basalis and other areas of the rat brain. *J. Neurochem.*, 45: 1021–1026.

Stephens, P.H., Tagari, P.C., Garofalo, L., Maysinger, D.,

Piotte, M. and Cuello, A.C. (1987) Neural plasticity of basal forebrain cholinergic neurons: effects of gangliosides. *Neurosci. Lett.*, 80: 80–84.

Taniuchi, M., Schweitzer, J.B. and Johnson, Jr., E.M. (1986) Nerve growth factor receptor molecules in the rat brain *Prod. Natl. Acad. Sci. USA*, 83: 1950–1954.

Van Herreveld, A. and Fifkova, E. (1975) Swelling of dendritic spines in the fascia dentata after stimulation of the perforant fibers as a mechanism of post-tetanic potentiation. *Exp. Neurol.*, 49: 736–749.

Verge, V.M.K., Merlio, J.-P., Grondin, J., Ernfors, P., Persson, H., Riopelle, R.J., Hökfelt, T. and Richardson, P.M. (1992) Co-localization of NGF binding sites, trk mRNA and low-affinity NGF receptor mRNA in primary sensory neurons: Responses to injury and infusion of NGF. *J. Neurosci.*, 12: 4011–4022.

Whitehouse, P.J., Price, D.L., Struble, R.G., Clark, A.W., Coyle, J.T. and Delong, M.R. (1982) Alzheimer's disease and senile dementia: loss of neurons in basal forebrain. *Science*, 215: 1237–1239.

Williams, L.R., Varon, S., Peterson, G.M., Wictorin, K., Fischer, W., Björklund, A. and Gage, F.H. (1986) Continuous infusion of nerve growth factor prevents basal forebrain neuronal death after fimbria-fornix transection. *Proc. Natl. Acad. Sci. USA*, 83: 9231–9235.

Winkler, J., Suhr, S.T., Gage, F.H., Thal, L.J. and Fisher, L.J. (1995) Essential role of neocortical acetylcholine in spatial memory. *Science*, 375: 484–487.

Yoshida, K. and Gage, F.H. (1991) Fibroblast growth factors stimulate nerve growth factor synthesis and secretion by astrocytes. *Brain Res.*, 538: 118–126.

J. Klein and K. Höffelholz (Eds.)
*Progress in Brain Research*, Vol. 109
© 1996 Elsevier Science B.V. All rights reserved.

CHAPTER 34

# Synaptic modulation by neurotrophic factors

Ron Stoop and Mu-ming Poo

*Department of Biological Sciences, Columbia University, New York, NY 10027, USA*

## Introduction

Studies on neurotrophic factors in the past have been primarily concerned with their long-term effects in promoting neurite outgrowth and neuronal survival. However, it has been known for some time that nerve growth factor, the prototypic neurotrophic factor, can exert acute actions on the motility of nerve growth cones (Connolly et al., 1981). Since neurotrophic factors are known to induce cytosolic kinase activity soon after their binding to the cell (Heumann, 1994), it is reasonable to expect that elevated kinase activity may affect functions of proteins involved in synaptic transmission. Several recent studies have indeed shown rapid synaptic modulation by a number of neurotrophic factors, including neurotrophins (Thoenen, 1991), a family of factors related to nerve growth factors (NGF), as well as ciliary neurotrophic factor (CNTF), a factor in the cytokine family known to promote the survival of a variety of neuronal populations (Ip and Yancopoulos, 1992). In this review, we summarize these recent results on synaptic modulation by neurotrophic factors and discuss their physiological implications.

## Actions of neurotrophins on developing neuromuscular synapses

The best known neurotrophic factor is nerve growth factor (NGF), which turns out to be one member of a family of factors now collectively known as neurotrophins. Three other members of the neurotrophin family have so far been identi-

fied: brain-derived neurotrophic factor (BDNF), neurotrophin-3 (NT-3) and neurotrophin-4/5. Each of the neurotrophins exert specific growth and survival promoting actions on distinct but overlapping populations of developing and/or mature neurons, via specific interactions with their cognate receptors (Ip et al., 1993). The effects of these neurotrophins on the synaptic functions of developing neuromuscular junctions in *Xenopus* nerve-muscle cultures have been examined by recording synaptic currents before and after application of the neurotrophins (Lohof et al., 1993). Within minutes after bath application of NT-3 and BDNF (at 50 ng/ml) to the cells, the frequency of spontaneous synaptic currents (SSCs) was markedly increased, while the mean amplitude or amplitude distribution of these currents showed no significant change. Impulse-evoked synaptic currents (ESCs) also increased in amplitude, although the effect was less dramatic than that on SSC frequency. Nerve growth factor, which was known to be specific for its trophic effect on sensory and sympathetic neurons, produced no effect on either spontaneous or evoked synaptic currents. This specificity in the synaptic actions of neurotrophins was consistent with the findings that BDNF and NT-3, but not NGF, were also effective in promoting neurite growth and survival of these *Xenopus* neurons (Lohof, 1994).

The effects of neurotrophins on synaptic currents can be accounted for by presynaptic actions of the factors, namely, a potentiation of transmitter secretion, either in the availability of transmitter-containing vesicles or in the probability of vesicular exocytosis. The quantal size, or the transmitter packet, is unchanged. The absence of postsynaptic action on the ACh response was further confirmed by the findings that the amplitude, rise and decay times of ACh-induced currents were the same before and after exposure to BDNF and NT-3 (Lohof, 1994).

One obvious implication of the presynaptic effect of neurotrophins is that endogenous release of these factors from either presynaptic neurons or postsynaptic target cells may serve to regulate the efficacy of transmitter secretion. Adult muscle fiber is known to express mRNA for BDNF and NT-3 (Hohn et al., 1990; Maisonpierre et al., 1990a,b; Henderson et al., 1993). Whether these neurotrophins are indeed translated, packaged and secreted from the muscle cell is unknown. It is known that electrical activity may upregulate the transcription of neurotrophin genes (Thoenen et al., 1991; Patterson et al., 1992), but whether the activity could also regulate the secretion of neurotrophins is also unclear. Relevant to the potential neurotrophin function in vivo, Lohof et al. (1993) found that the synaptic potentiation effect of NT-3 requires constant extracellular presence of the factor. Removal of the factor 30–60 min after exposure of the cells to NT-3, at a time maximal synaptic potentiation had occurred, led to a gradual recovery of original low level of synaptic activity. In contrast, synaptic potentiation induced by neurotrophins at hippocampal synapses appears to be more long-lasting (Kang and Schumann, 1995; see below).

The potential role of neurotrophins in regulating the maturation of synaptic functions has also been addressed in cell culture. Wang et al. (1995) treated *Xenopus* nerve-muscle cultures with neurotrophins for prolonged periods (1–3 days) and observed an accelerated maturation of synaptic functions as determined by the amplitude of spontaneous and evoked synaptic currents and by the immunocytochemical staining of synaptic vesicles proteins. While acute treatment of these cultures with NT-3 and BDNF induced an immediate rise in SSC frequency without any changes in the MEPC amplitude, prolonged treatment with these neurotrophins was found to increase the mean amplitude of the SSCs and the appearance of a population of larger, more "mature" quanta, as compared to parallel untreated cultures. The neurotrophin-treated cultures also showed an enhanced expression of synaptic vesicle proteins synapsin 1 and synaptophysin. These long-term effects on the maturation of *Xenopus* neuromuscular synapses were not observed after NGF treatment, in agreement with the specificity of acute potentiation effects.

## Actions of CNTF on developing neuromuscular synapses

Ciliary neurotrophic factor (CNTF) is a protein factor related to cytokines. It promote survival of motoneuron survival of in vitro (Arakawa et al., 1990) and in vivo (Oppenheim et al., 1991; Forger et al., 1993). Mice carrying a null mutation in the CNTF gene show progressive motor neuron atrophy and postnatal neuron loss (Masu et al., 1993). CNTF also promotes expression of choline acetyltransferase in rat spinal cord neurons in vitro (Magal et al., 1991) and motor neuron sprouting in vivo (Gurney et al., 1992). It has been suggested CNTF derived from glial cells may help the repair of injuries to the nervous system (Thoenen, 1991; Friedman et al., 1992; Rende, et al., 1992; Sendtner et al., 1992a,b).

The effects of CNTF on neuromuscular synapses in *Xenopus* nerve-muscle cultures have been examined (Stoop and Poo, 1995). Extracellular application of CNTF (100 ng/ml) to the culture resulted in a marked increase in the frequency of spontaneous synaptic currents (SSCs) and the amplitude of evoked synaptic currents (Fig. 1), similar to that produced by BDNF and NT-3. The potentiation effects of CNTF appeared to be due to an enhancement of presynaptic transmitter release, rather than postsynaptic increase of receptor sensitivity, since there was no change in the mean amplitude nor amplitude distribution of the spontaneous events.

Neurotrophins and CNTF are known to exert their actions through different surface receptors and signal transduction mechanisms (Ip and Yancopoulos, 1994). The potentiation of transmitter secretion is thus likely to reflect their actions at

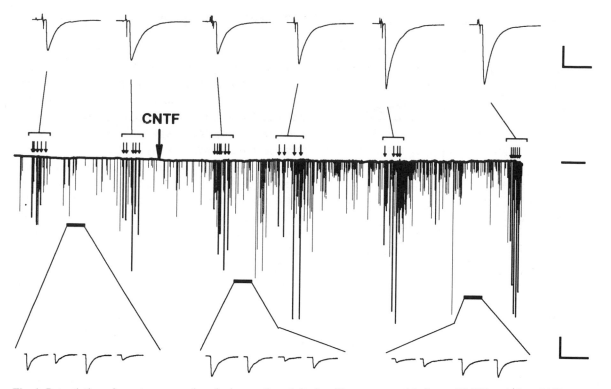

Fig. 1. Potentiation of spontaneous and evoked synaptic activity by ciliary neurotrophic factor (CNTF) at 1-day-old *Xenopus* nerve-muscle synapses. The continuous trace depicts membrane current of an innervated myocyte recorded by voltage-clamp whole-cell recording method ($V_h = -70$ mV). Downward events are inward spontaneous synaptic currents (SSCs) resulting from spontaneous secretion of ACh from the presynaptic neuron and evoked synaptic currents (ESCs) elicited at a low frequency at the time marked by the arrows. The large arrow marks the onset of bath application of CNTF (final bath conc. 100 ng/ml). Bar = 2.5 min. Samples of SSCs are shown below at a higher time resolution, the insets above depict the computer-average ESCs at a higher time resolution for each group of ESC's indicated. Scales: 1 nA and 20 ms.

different steps of the transmitter secretion pathway. Several lines of evidence suggest that the sites of intracellular actions of BDNF and CNTF are distinctly different. Firstly, synaptic potentiation of CNTF appears to require signaling with the cell body, while that of BDNF does not. Focal perfusion of CNTF at the cell body is effective in potentiating its distant synapse, with a delay of onset of potentiation shorter than that induced by direct focal perfusion of CNTF at the synapse, suggesting that additional time is required for signaling from the synapse to the cell body (Fig. 2A). Focal perfusion of BDNF at the cell body, on the other hand, was without effect (Fig. 2B). Transection of the neurite that connects the synapse to the soma abolished the potentiation effect of CNTF at the synapse, while BDNF remained effective (Fig.

2C,D). Second, maximal synaptic potentiation induced by a high level of BDNF (100 ng/ml) does not occlude further potentiation by CNTF, and vice versa (Fig. 3A). Third, pair-pulse facilitation, a short term plasticity in evoked transmitter secretion (Zucker, 1991), was reduced after synaptic potentiation by BDNF, but was unchanged after CNTF-induced potentiation.

While the sites of actions of BDNF and CNTF appear to be distinctly different, they are likely to be different steps of the same secretion pathway. This was suggested by the finding of significant synergistic effects between the two factors when they were applied together to the synapse at low concentrations. As shown in Fig. 3B, treatment of the synapse with 1 ng/ml of either BDNF or CNTF alone was ineffective, but combined treatment

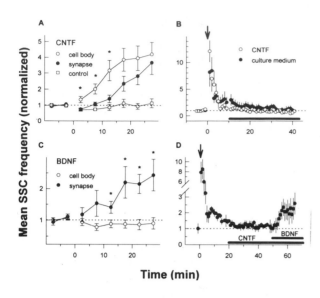

Fig. 2. Local perfusion of the soma and the synapse with CNTF and BDNF and effects of both factors on "cut-loose" synapses (concentration used for both in each experiment 100 ng/ml). (A) Changes in the mean SSC frequency with time after the onset of CNTF application (at time 0, 100 ng/ml) to either the soma ($n = 22$), the synapse ($n = 23$), or a cell-free region at a distance 150 $\mu$m away from either the soma and the synapse ($n = 11$). The values for each synapse were normalized by the mean value prior to the CNTF application before averaging. Data marked by * were significantly different from the corresponding values for synapse perfusion ($P < 0.05$, $t$-test). (B) Changes in the mean SSC frequency following local perfusion of BDNF to either the soma ($n = 21$) or the synapse ($n = 29$). Data marked by * were significantly different from the corresponding values for soma perfusion ($P < 0.05$, $t$-test). (C) Change in the mean SSC frequency with time after application of CNTF or culture medium to synapses that had been "cut-loose" from their soma at $t = 0$ (marked by arrow). The frequency values at each synapse were normalized to the mean value prior to the application of CNTF before averaging. The duration of CNTF or culture medium application is indicated by the bar below. Data represent mean ± SEM ($n = 5$). (D) Change in the mean SSC frequency with time after application of CNTF and later BDNF to synapses that had previously (at $t = 0$ min, marked by arrow) been "cut-loose" from their soma (mean ± SEM, $n = 5$). The duration of CNTF and BDNF application is indicated by the bars below (modified from Stoop and Poo, 1995).

of both factors at the same concentration markedly potentiated the spontaneous transmitter secretion.

## Actions of neurotrophins on central synapses

The effect of neurotrophins on the synaptic activity at central synapses has recently been examined. Using hippocampal slices Kang and Schumann (1995) found that application of BDNF and NT-3 enhanced the strength of Schaffer collateral-CA1 synapses, while NGF was without effect. The effect of the neurotrophins appeared to be long-lasting, at least for 2–3 h. The enhancement was accompanied by a concomitant decrease in paired-pulse facilitation, suggesting the site of neurotrophin action may be presynaptic. Interestingly, tetanus-induced long-term potentiation (LTP) could still be elicited effectively in slices previously potentiated by the neurotrophins, suggesting that neurotrophin-induced potentiation and LTP may be caused by different cellular mechanism. Since stimulus paradigms inducing LTP in the CA1 region of the hippocampus have been shown to evoke significant increases in BDNF and NT-3 mRNAs within 4 h after induction of potentiation (Patterson et al., 1992), it is possible that enhanced expression of BDNF in target cells may function in the long-term maintenance of LTP.

In dissociated cultures of rat cortical neurons, Kim et al. (1994) found that NT-3 can potentiate synaptic activity. This potentiation was attributed to a reduction of the strength of inhibitory GABA-ergic synapses in the culture. No effect of NT-3 on excitatory glutamatergic synapses was found. Acute actions of neurotrophins on synaptic efficacy may be related to neuromodulatory effects of other neuropeptides. Epidermal growth factor and fibroblast growth factor have been shown to enhance the induction of LTP in hippocampus CA1 region (Abe et al., 1991; Seifert et al., 1990). Cytokines, e.g. interleukin-1$\beta$ and TNF-$\alpha$, interferon and interleukin-2 have been shown to exert an inhibitory effect on LTP in the CA1 region (Bindoni et al., 1988; D'Arcangelo et al., 1991; Tancredi et al., 1992; Zeise et al., 1992; Bellinger et al., 1993). While different cytoplasmic transduction mechanisms are triggered by various factors, their actions are likely to converge on common effector proteins involved in the presynaptic transmitter secretion or postsynaptic responses. Our understanding of the

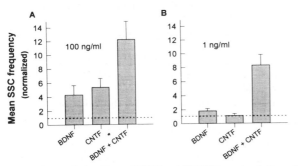

Fig. 3. Synergistic effects CNTF and BDNF in potentiation of spontaneous ACh secretion. (A) Changes in the frequency of spontaneous synaptic currents (SSCs) after 30 min application of either BDNF (100 ng/ml) or CNTF (100 ng/ml), or both. (B) The same as (A), except the concentration of the factors was lowered to 1 ng/ml. All SSC frequencies were normalized on the SSC frequency during 10 min preceding application, all error bars indicate SEM ($n = 6–9$).

mechanisms underlying the synaptic actions of neurotrophic factors thus will depend ultimately on the elucidation of the complex molecular machinery involved in the transmitter secretion pathway and of the mechanisms regulating the number and sensitivity of postsynaptic transmitter receptors.

## Summary and outlook

The study of acute and long-term synaptic actions of neurotrophic factors has only a brief history. The molecular machinery responsible for neurotransmitter release in the presynaptic nerve terminal is now beginning to be understood (Catsicas et al., 1994; Jahn and Sudhof, 1994). Putative proteins involved in the mobilization, docking and exocytosis of synaptic vesicles are likely targets of downstream protein kinases activated by neurotrophic factors. Study of synaptic proteins affected by these factors may in fact help to determine functional roles of these proteins at the synapse.

An important issue that remains to be addressed is the biological relevance of the synaptic modulation by neurotrophic factors observed in various in vitro preparations. What are the endogenous source of these factors within the nervous system? How are these factors packaged and secreted to the extracellular space? Can the secretion of endoge-

nous factors be regulated by synaptic activity? What is the consequence of blocking the secretion of these factors or their receptors on the development and maintenance of synaptic functions? With the availability of molecular probes associated with neurotrophic factors and their receptors, many of these questions are now ready to be addressed experimentally.

## References

Abe, K., Xie, F.J. and Saito, H. (1991) Epidermal growth factor enhances short-term potentiation and facilitates induction of long-term potentiation in rat hippocampal slices. *Brain Res.*, 547: 171–174.

Arakawa, Y., Sendtner, M. and Thoenen, H. (1990) Survival effect of ciliary neurotrophic factor (CNTF) on chick embryonic motoneurons in culture: comparison with other neurotrophic factors and cytokines. *J. Neurosci.*, 10: 3507–3515.

Bellinger, F.P., Madamba, S. and Siggins, G.R. (1993) Interleukin 1 beta inhibits synaptic strength and long-term potentiation in the rat CA1 hippocampus. *Brain Res.*, 628: 227–234.

Bindoni, M., Perciavalle, V., Berretta, S., Belluardo, N. and Diamantstein, T. (1988) Interleukin 2 modifies the bioelectric activity of some neurosecretory nuclei in the rat hypothalamus. *Brain Res.*, 462: 10–14.

Catsicas, S., Grenningloh, G. and Pich, E.M. (1994) Nerve-terminal proteins: to fuse to learn. *Trends Neurosci.*, 17: 368–373.

Connolly, J.L., Green, S. and Greene, L.A. (1981) Pit formation and rapid changes in surface morphology of sympathetic neurons in response to nerve growth factor. *J. Cell Biol.*, 90: 176–180.

D'Arcangelo, G., Grassi, F., Ragozzino, D., Santoni, A., Tancredi, V. and Eusebi, F. (1991) Interferon inhibits synaptic potentiation in rat hippocampus. *Brain Res.*, 564: 245–248.

Forger, N.G., Roberts, S.L., Wong, V. and Breedlove, S.M. (1993) Ciliary neurotrophic factor maintains motoneurons and their target muscles in developing rats. *J. Neurosci.*, 13: 4720–4726.

Friedman, B., Scherer, S.S., Rudge, J.S., Helgren, M., Morrisey, D., McClain, J., Wang, D.Y., Wiegand, S.J., Furth, M.E., Lindsay, R.M. et al. (1992) Regulation of ciliary neurotrophic factor expression in myelin-related Schwann cells in vivo. *Neuron*, 9: 295–305.

Gurney, M.E., Yamamoto, H. and Kwon, Y. (1992) Induction of motor neuron sprouting in vivo by ciliary neurotrophic factor and basic fibroblast growth factor. *J. Neurosci.*, 12: 3241–3247.

Henderson, C.E., Camu, W., Mettling, C. et al. (1993) Neurotrophins promote motor neuron survival and are present in embryonic limb bud. *Nature*, 363: 266–270.

364

Heumann, R. (1994) Neurotrophin signalling. *Curr. Opin. Neurobiol.*, 4: 668–679.

Hohn, A., Leibrock, J., Bailey, K. and Barde, Y.A. (1990) Identification and characterization of a novel member of the nerve growth factor/brain-derived neurotrophic factor family. *Nature*, 344: 339–341.

Ip, N.Y. and Yancopoulos, G.D. (1992) Ciliary neurotrophic factor and its receptor complex. *Prog. Growth Factor Res.*, 4: 139–155.

Ip, N.Y. and Yancopoulos, G.D. (1994) Neurotrophic factors and their receptors. *Ann. Neurol.*, 35 (Suppl.): S13–S16.

Ip, N.Y., Stitt, T.N., Tapley, P., Klein, R., Glass, D.J., Fandl, J., Greene, L.A., Barbacid, M. and Yancopoulos, G.D. (1993) Similarities and differences in the way neurotrophins interact with the Trk receptors in neuronal and non-neuronal cells. *Neuron*, 10: 137–149.

Jahn, R. and Sudhof, T.C. (1994) Synaptic vesicles and exocytosis. *Annu. Rev. Neurosci.*, 17: 219–246.

Kang, H. and Schumann, E.M. (1995) Long-lasting neurotrophin-induced enhancement of synaptic transmission in the adult hippocampus. *Science*, 267: 1658–1662.

Kim, H.G., Wang, T., Olafsson, P. and Lu, B. (1994) Neurotrophin 3 potentiates neuronal activity and inhibits gamma-aminobutyratergic synaptic transmission in cortical neurons. *Proc. Natl. Acad. Sci. USA*, 91: 12341–12345.

Lohof, A.M. (1994) Modulation of neurite outgrowth and synaptic activity of embryonic *Xenopus* spinal neurons. Ph.D. Dissertation, Columbia University, New York.

Lohof, A.M., Ip, N.Y. and Poo, M-m. (1993) Potentiation of developing neuromuscular synapses by the neurotrophins NT-3 and BDNF. *Nature*, 363: 350–353.

Magal, E., Burnham, P. and Varon, S. (1991) Effects of ciliary neurotrophic factor on rat spinal cord neurons in vitro: survival and expression of choline acetyltransferase and low-affinity nerve growth factor receptors. *Dev. Brain Res.*, 63: 141–150.

Maisonpierre, P.C., Belluscio, L., Friedman, B. et al. (1990a) NT-3, BDNF, and NGF in the developing rat nervous system: parallel as well as reciprocal patterns of expression. *Neuron*, 5: 501–509.

Maisonpierre, P.C., Belluscio, L., Squinto, S. et al. (1990b) Neurotrophin-3: a neurotrophic factor related to NGF and BDNF. *Science*, 247: 1446–1451.

Masu, Y., Wolf, E., Holtmann, B., Sendtner, M., Brem, G. and Thoenen, H. (1993) Disruption of the CNTF gene results in motor neuron degeneration. *Nature*, 365: 27–32.

Oppenheim, R.W., Prevette, D., Yin, Q.W., Collins, F. and MacDonald, J. (1991) Control of embryonic motoneuron survival in vivo by ciliary neurotrophic factor. *Science*, 251: 1616–1618.

Patterson, S.L., Grover, L.M., Schwartzkroin, P.A. and Bothwell, M. (1992) Neurotrophin expression in rat hippocampal slices: a stimulus paradigm inducing LTP in CA1 evokes increases in BDNF and NT-3 mRNAs. *Neuron*, 9: 1081–1088.

Rende, M., Muir, D., Ruoslahti, E., Hagg, T., Varon, S. and Manthorpe, M. (1992) Immunolocalization of ciliary neuronotrophic factor in adult rat sciatic nerve. *Glia*, 5: 25–32.

Seifert, W., Forster, F., Flott, B. and Terlau, H. (1990) Effects of a neurotrophic factor (FGF) on development, regeneration and synaptic plasticity of central neurons. *Adv. Exp. Med. Biol.*, 268: 395–399.

Sendtner, M., Kreutzberg, G.W. and Thoenen, H. (1990) Ciliary neurotrophic factor prevents the degeneration of motor neurons after axotomy. *Nature*, 345: 440–441.

Sendtner, M., Schmalbruch, H., Stöckli, K.A., Carroll, P., Kreutzberg, G.W. and Thoenen, H. (1992a) Ciliary neurotrophic factor prevents degeneration of motor neurons in mouse mutant progressive motor neuronopathy. *Nature*, 358: 502–504.

Sendtner, M., Stöckli, K.A. and Thoenen, H. (1992b) Synthesis and localization of ciliary neurotrophic factor in the sciatic nerve of the adult rat after lesion and during regeneration. *J. Cell Biol.*, 118: 139–148.

Stoop, R. and Poo, M.-m. Potentiation of transmitter release by ciliary neurotrophic factor requires somatic signaling. *Science*, 267: 695–699.

Tancredi, V., D'Arcangelo, G., Grassi, F., et al. (1992) Tumor necrosis factor alters synaptic transmission in rat hippocampal slices. *Neurosci. Lett.*, 146: 176–178.

Thoenen, H. (1991) The changing scene of neurotrophic factors. *Trends Neurosci.*, 14: 165–170.

Thoenen, H., Zafra, F., Hengerer, B. and Lindholm, D. (1991) The synthesis of nerve growth factor and brain-derived neurotrophic factor in hippocampal and cortical neurons is regulated by specific transmitter systems. *Ann. N. Y. Acad. Sci.*, 640: 86–90.

Wang, T., Xie, K.W. and Lu, B. (1995) Neurotrophins promote maturation of developing neuromuscular synapses. *J. Neurosci.*, in press.

Zeise, M.L., Madamba, S. and Siggins, G.R. (1992) Interleukin-1$\beta$ increases synaptic inhibition in rat hippocampal pyramidal neurons in vitro. *Regul. Peptides*, 39: 1–7.

Zucker, R.S., Delaney, K.R., Mulkey, R. and Tank, D.W. (1991) Presynaptic calcium in transmitter release and post-tetanic potentiation. *Ann. N. Y. Acad. Sci.*, 635: 191–207.

J. Klein and K. Löffelholz (Eds.)
*Progress in Brain Research*, Vol. 109
© 1996 Elsevier Science B.V. All rights reserved.

CHAPTER 35

# Neurotrophic factors for experimental treatment of motoneuron disease

## M. Sendtner

*Department of Neurology, University of Würzburg, Josef-Schneider-Strasse 11, D-97080 Würzburg, Germany*

## Introduction

The cholinergic motoneurons located in the ventral horn of the spinal cord and motor nuclei of the brain stem were among the first populations of neurons identified to undergo cell death during development at a significant proportion. In chick embryos, about half of the motoneurons generated in the lateral motor column of the lumbar spinal cord until day 6 are lost during the following 5 days of development (Oppenheim, 1991). The target tissue, the innervated skeletal muscle, plays a crucial role during this process. Ablation of the target tissue by limb amputation at early embryonic stages results in significant increase of cell death in the population of innervating motoneurons, and transplantation of an additional limb thus leading to augmentation of the innervated muscle mass leads to reduced loss of motoneurons, suggesting that factors from skeletal muscle prevent the cell death of innervating motoneurons. Cell death rates are also affected by experimental blockade of neuromuscular transmission during this critical time of development. Treatment with curare or $\alpha$-bungarotoxin reduces cell death rates of motoneurons, and chronic electrical stimulation of motoneurons of the hindlimbs of developing chicks leads to enhanced rates of cell death in lumbar motoneurons (Oppenheim and Nunez, 1982).

The underlying molecular mechanisms determining survival or cell death of motoneurons are still not fully understood. The establishment of improved cell culture conditions for embryonic motoneurons has helped to identify a variety of neurotrophic molecules including members of different gene families of neurotrophic factors and cytokines. The molecular cloning of ciliary neurotrophic factor (CNTF) and the identification of receptor components mediating its cellular effects has shown that pluripotent cytokines can support motoneurons both in vitro and in vivo, acting through gp130 and leukemia inhibitory factor receptor $\beta$ (LIFR$\beta$). The purification and molecular cloning of brain-derived neurotrophic factor (BDNF) led to the identification of the still growing gene family of neurotrophins, involving now at least 5 members (for review see Götz et al., 1994; Snider, 1994). Moreover, the identification of the trk oncogene as the high-affinity receptor for nerve-growth factor (NGF) and the subsequent identification of the trk tyrosine kinase gene family of receptor molecules for the neurotrophins provided the basis for the molecular analysis of effects of the neurotrophin gene family on motoneurons. During the last years, the use of homologous recombination in embryonic stem cells of the mouse and the generation of mice lacking gene expression of these factors and their receptors was a major step towards the understanding of the physiological function of such molecules on motoneurons.

## Ciliary neurotrophic factor

Ciliary neurotrophic factor (CNTF) was originally identified as an active compound present in chick eye extracts which supports the survival of embryonic chick ciliary neurons in culture (Manthorpe et al., 1986). Subsequent studies with semi-purified CNTF showed that this neurotrophic factor supports not only ciliary neurons, but also sympathetic, sensory, nodose, trigeminal, and particularly motoneurons. The final purification and subsequent cloning of the CNTF cDNA has shown that this factor lacks a hydrophobic leader sequence typical of secretory proteins such as the neurotrophins and Insulin-like growth factor. Thus it is not yet known how CNTF is made available to responsive neurons from synthesizing cells. The temporal and spatial expression of CNTF mRNA and protein (for review see Sendtner et al., 1994) suggest that this factor does not play any major role during embryonic development. In the developing rat, CNTF expression is absent during the time period when neurons differentiate or when a significant proportion of postmitotic peripheral neurons undergo cell death. The time period of natural motoneuron cell death, which takes place between embryonic day 15 and postnatal day 1 in rats, is followed by an augmented sensitivity to axonal damage. Transsection of motor nerves such as the facial nerve leads to cell death of more than 80% of lesioned motoneurons. The local administration of CNTF after facial nerve transsection in newborn rats prevents the degeneration of most motoneuron cell bodies. In the adult peripheral nervous system, very high amounts of CNTF are localized in the cytoplasm of myelinating Schwann cells. It seems plausible that CNTF could be released from the cytoplasm of these cells after nerve lesion and thus act on the maintenance of lesioned motor and sensory neurons, at least during the early stages following peripheral nerve lesion.

Mice lacking endogenous CNTF develop normally until 3 weeks after birth (Masu et al., 1993). No behavioral differences can be observed in comparison to control litter mates. Histological analysis of motoneurons reveals a gradually increasing atrophy and degeneration of motoneurons (Fig. 1) accompanied by reactive gliosis, starting at the eighth postnatal week. At 6 months of age, a 22% reduction of the number of motoneurons in the facial nucleus is detectable. Measurement of grip strength as a parameter of motoneuron function revealed a small but statistically significant reduction at that time. The lack of major functional deficits both in CNTF deficient mice and in patients with a gene defect of the CNTF gene leading to a comparable lack of CNTF expression (Takahashi et al., 1994) suggests that compensatory mechanisms such as enhanced production of other neurotrophic factors come into play. The predicted three-dimensional structures of CNTF, LIF, Interleukin-6, Oncostatin-M, and Cardiotrophin-1 (Pennica et al., 1995) show common structural features which are reflected by the observation that CNTF shares common receptor subunits with these molecules. Thus it is likely that these factors could, at least in part, compensate for the deficiency of CNTF in mice and man.

## Neurotrophins

This family of molecules includes at the moment five different molecules: the prototypic nerve growth factor (NGF), brain-derived neurotrophic factor (for review see Snider, 1994), neurotrophin-3, neurotrophin-4/5 and neurotrophin-6 (Götz et al., 1994). Brain-derived neurotrophic factor and NT-4/5 have been shown to act on responsible cells through trkB tyrosine kinase receptors, whereas NT-3 exerts its cellular effects through trkC. Both trkB and trkC, but not trkA, are expressed by brain stem and spinal motoneurons innervating skeletal muscle. This correlates well with the observation that motoneurons in cell culture and in vivo are responsive to BDNF, NT-4 and NT-3, but not to NGF.

Among these neurotrophins, NT-3 is most abundantly expressed in skeletal muscle, both during development and in the adult. Levels of BDNF expression in muscle are very low in particular after birth (Griesbeck et al., 1995). The upregulation of BDNF mRNA expression found in denervated skeletal muscle is apparently due to increased BDNF expression in Schwann cells ensheathing nerve fibers within the muscle, but not to increased expression in skeletal muscle (Koliatsos et al., 1993), as observed by in situ hybridization (Griesbeck et al., 1995).

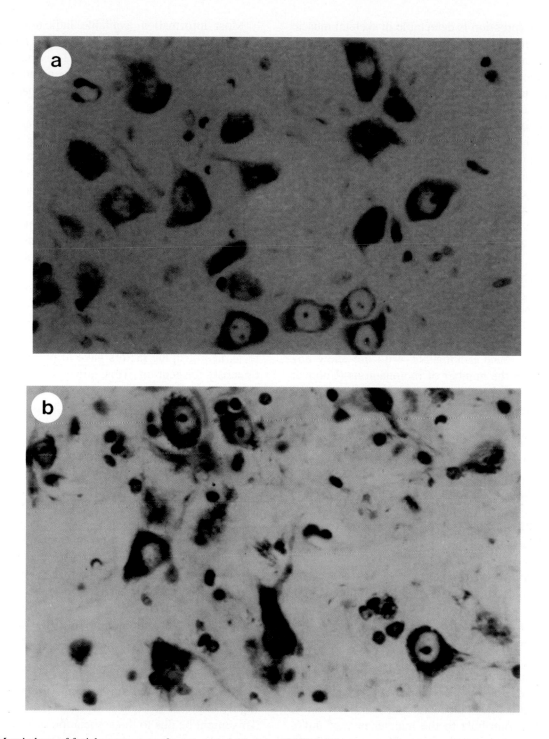

Fig. 1. Morphology of facial motoneurons from a control (a) and a CNTF deficient mouse (b). At an age of 6 months, facial motoneurons appear shrunken and show signs of degeneration such as chromatolysis and reduced size in mice lacking endogenous
CNTF (Masu et al., 1993).

NT-4 expression is detectable in skeletal muscle both during development and in the adult. After muscle denervation either by transsection of innervating motor nerves or by transient blockade of neuronal transmission by injection of tetrodotoxin into the sciatic nerve, NT-4 expression is rapidly down-regulated in adult rats (Griesbeck et al., 1995). Electrical stimulation of motor nerves leads to the opposite effect, a significant up-regulation of NT-4 expression (Funakoshi et al., 1995). These data suggest that neuronal activity at the neuromuscular synapse has a significant influence on the regulation of NT-4 expression in skeletal muscle.

Mice lacking BDNF expression survive for a few days, some mice up to 4 weeks after birth. Histologically, there is a marked loss of neurons in vestibular and nodose ganglia, and a loss of specific subpopulations of myelinated sensory nerve fibers which most probably includes the mechanoreceptors, but not BDNF-responsive motoneurons. Also, mice lacking NT-4 do not show any reduction in the number of motoneurons (Conover et al., 1995; Liu et al., 1995). The possibility that BDNF and NT-4 could compensate for each other in such mice, thus explaining the apparent lack of motoneuron loss, was excluded by results obtained through the analysis of mice deficient for both BDNF and NT-4 (Conover et al., 1995; Liu et al., 1995). Motoneuron numbers were not reduced in such double deficient mice.

In contrast, mice lacking trkB expression show a significant reduction of motoneurons (Klein et al., 1993). The reason for this discrepancy between mice lacking either BDNF, NT-4 or the corresponding receptor, trkB, is not clear so far.

## The use of neurotrophic factors for treatment of degenerative motoneuron disease

The availability of mouse mutants lacking functional gene expression for neurotrophic factors and their receptors has helped to understand how these molecules contribute to the maintenance of motoneuron survival and function. However, with the exception of CNTF (Takahashi et al., 1994), similar gene defects have not been identified in humans so far.

Most information available indicates that degenerative diseases of motoneurons are not caused by such specific gene defects. Thus any use of such factors for treatment of degenerative motoneuron diseases represents a symptomatic rather than causal therapy, and it is therefore essential to collect information in animal models that could be of relevance for the human degenerative disorders. It was shown that motoneurons can be protected against lesion effects and degenerative changes in animal mutants such as the *pmn* mouse (Fig. 2), the wobbler mouse (Ikeda et al., 1995), and the *mnd* mouse. Interestingly, the combination of CNTF and BDNF in wobbler mice seems to have a dramatic influence on the motor performance in such mice (Mitsumoto et al., 1994), exceeding the effect of each factor tested in the absence of the other.

In vitro and in vivo experiments have demonstrated that, in order to exhibit their protective effects, neurotrophic factors have to be present continuously for neurons. This, however, is difficult to achieve by intermittent systemic injections. Moreover, very little is known about the quantities of subcutaneously or intravenously injected CNTF, BDNF, or IGF-I reaching the nerve and becoming available to responsive neurons. To be of therapeutic value in the treatment of degenerative motoneuron disorders, such factors should be accessible to motoneurons, for example via their neuromuscular endplates, where proteins can be taken up from the circulation. After intravenous injection in adult rats, CNTF shows a biphasic clearance with a fast initial phase (half-life of 2.9 min). Analysis of tissue distribution of labeled CNTF after intravenous injection revealed a correlation between the fast elimination from the blood and accumulation in the liver. No accumulation of injected CNTF was detected in skeletal muscle or nerve tissue, the sites where CNTFR$\alpha$, the low affinity specific binding subunit for CNTF, is highly expressed (Stahl and Yancopoulos, 1994). This might be due to the association of CNTF with a soluble receptor component in the circulation. This complex can bind to many cell types expressing gp130 and LIFR$\beta$, including liver cells, macrophages and other cell types (Dittrich et al.,

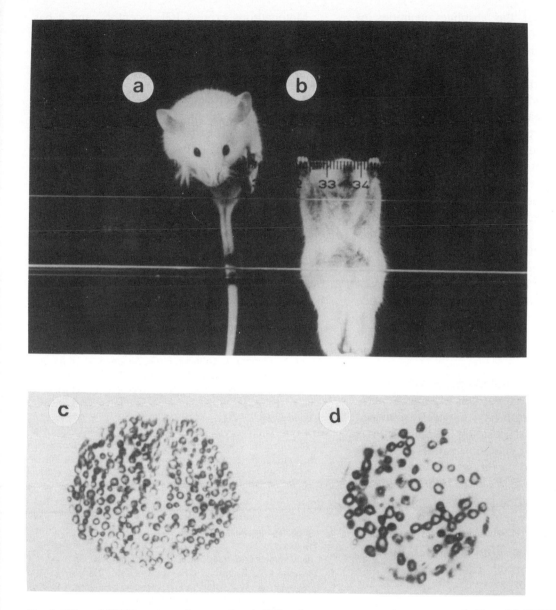

Fig. 2. Effect of CNTF treatment in pmn mice. (a,b) Two homozygous pmn mice were treated either with CNTF secreting D3 cells (a) or with untransfected control cells (b) (Sendtner et al., 1992). The cells were injected intraperitoneally at postnatal day 20. (a) and (b) show the mice at an age of 36 days. CNTF treated mice showed much better motor performance. (c,d) Morphology of the phrenic nerve of a CNTF-treated (c) and untreated (d) pmn mouse. The nerve of the CNTF treated pmn mouse contains more myelinated axons than that of the untreated animal.

1994). It is likely that side-effects observed in CNTF-treated amyotrophic lateral sclerosis (ALS) patients like fever and cachexia are caused by the induction of acute phase responses in liver cells (Dittrich et al., 1994), and that respiratory side-effects in CNTF-treated ALS patients (Brooks et al., 1993) such as cough are caused by action of the CNTF/CNTFR$\alpha$ complex on macrophages in the lung, which express gp130 and LIFR$\beta$.

BDNF protein and bioactivity have been detected in platelets obtained from human blood samples (Yamamoto and Gurney, 1990). It is not known so far what the absolute quantities of BDNF protein in platelets are and whether BDNF can be released from platelets constitutively or in a regulated manner. Subcutaneous injection of BDNF to animal models of motoneuron disease has been shown to be highly effective in slowing the disease, thus leading to clinical studies in patients with amyotrophic lateral sclerosis. These first studies, although still not conclusive, look very promising. Patients receiving BDNF showed only half of the deterioration of breathing capacity as measured by forced vital capacity in comparison with patients treated with placebo. Future studies are planned to confirm these results and to overcome pharmacological problems associated with the subcutaneous injection of BDNF. One of the future strategies will be to apply recombinant BDNF through pumps in the subarachnoidal space of the lumbar spinal cord, from where it can diffuse in the cerebrospinal fluid and reach motoneurons by retrograde transport after being taken up from axons.

Similar efforts are currently made with Insulin-like growth factor I (IGF-I). This pluripotent mitogen and survival factor is physiologically produced by a variety of tissues, including liver and nervous system, and relatively high concentrations are detectable in serum and cerebrospinal fluid. The actions of IGF-I and its homologue IGF-II are modified by at least 6 binding proteins. It is speculated that these binding proteins inhibit the actions of IGF-I and II in body fluids, but on the other hand they could prevent the diffusion of IGF from tissues and thus increase the availability to IGF receptors present on responsive cells such as the motoneurons.

IGF-I has been shown to be a potent survival factor for embryonic chick (Arakawa et al., 1990) and rat motoneurons (Hughes et al., 1993). At maximal concentrations, its potency in serum-free cultures of rat motoneurons is similar to that of BDNF. Interestingly, although IGF-I is also active on lesioned facial motoneurons in newborn rats (Hughes et al., 1993), the effects are significantly lower than those observed with CNTF or BDNF. It could be that this is due to the presence of inhibitory IGF binding proteins in nerve tissue.

Motoneurons express IGF-receptors, and it has been observed that the expression of IGFR is increased in motoneurons of patients suffering from amyotrophic lateral sclerosis (Adem et al., 1994). These data suggest that indeed IGF could play a role in the maintenance of motoneurons during postnatal life. Experimental treatment of patients with motoneuron disease has shown that subcutaneous injection of IGF has beneficial effects. However, the role of IGF-I for motoneurons is still far from being understood.

Experiments designed to elucidate the pharmacokinetics, binding proteins, and availability of these factors for motoneurons are necessary in order to improve strategies for planning clinical trials with such factors and thus to provide a basis for evaluating the potential usefulness of these trophic factors for the treatment of human disease.

# References

Adem, A., Ekblom, J., Gillberg, P.-G., Jossan, S.S., Höög, A., Winblad, B., Aquilonius, S.-M., Wang, L.-H. and Sara, V. (1994) Insulin-like growth factor-1 receptors in human spinal cord: changes in amyotrophic lateral sclerosis. *J. Neural Transm.*, 97: 73–84.

Arakawa, Y., Sendtner, M. and Thoenen, H. (1990) Survival effect of ciliary neurotrophic factor (CNTF) on chick embryonic motoneurons in culture: comparison with other neurotrophic factors and cytokines. *J. Neurosci.*, 10: 3507–3515.

Brooks, B.R., Sanjak, M., Mitsumoto, H. et al. (1993) Recombinant human ciliary neurotrophic factor (rhCNTF) in amyotrophic lateral sclerosis (ALS) patients: dose selection strategy in phase I-II safety, tolerability and pharmacokinetic studies (Abstract). *Can. J. Neurol. Sci.*, 20: 83.

Conover, J.C., Erickson, J.T., Katz, D.M., Bianchi, L.M., Poueymirou, W.T., McClain, J., Pan, L., Helgren, M., Ip, N.Y., Boland, P., Friedman, B., Wiegand, S., Vejsada, R., Kato, A.C., DeChiara, T.M. and Yancopoulos, G.D. (1995) Neuronal deficits, not involving motor neurons, in mice lacking BDNF and/or NT4. *Nature*, 375: 235–238.

Dittrich, F., Thoenen, H. and Sendtner, M. (1994) Ciliary neurotrophic factor: pharmacokinetics and acute-phase response in rat. *Ann. Neurol.*, 35: 151–163.

Funakoshi, H., Belluasdo, N., Arenas, E., Yamamoto, Y., Casabona, A., Persson, H. and Ibanez, C.F. (1995) Muscle-derived Neurotrophin-4 as an activity-dependent trophic signal for adult motor neurons. *Science*, 268: 1495–1499.

Götz, R., Köster, R., Winkler, C., Raulf, F., Lottspeich, F., Schartl, M. and Thoenen, H. (1994) Neurotrophin-6 is a new member of the nerve growth factor family. *Nature*, 372: 266–269.

Griesbeck, O., Parsadanian, A., Sendtner, M. and Thoenen, H. (1995) Expression of neurotrophins in skeletal muscle: quantitative comparison and significance for motoneuron survival and maintenance of function. *J. Neurosci. Res.*, 42: 21–33.

Hughes, R.A., Sendtner, M. and Thoenen, H. (1993) Members of several gene families influence survival of rat motoneurons in vitro and in vivo. *J. Neurosci. Res.*, 36: 663–671.

Ikeda, K., Wong, V., Holmlund, T.H., Greene, T., Cedarbaum, J.M., Lindsay, R.M. and Mitsumoto, H. (1995) Histometric effects of ciliary neurotrophic factor in wobbler mouse motor neuron disease. *Ann. Neurol.*, 37: 47–54.

Klein, R., Smeyne, R.J., Wurst, W., Long, L.K., Auerbach, B.A., Joyner, A.L. and Barbacid, M. (1993) Targeted disruption of the trkB neurotrophin receptor results in nervous system lesions and neonatal death. *Cell*, 75: 113–122.

Koliatsos, V.E., Clatterbuck, R.E., Winslow, J.W., Cayouette, M.H. and Price, D.L. (1993) Evidence that brain-derived neurotrophic factor is a trophic factor for motor neurons in vivo. *Neuron*, 10: 359–367.

Liu, X., Ernfors, P., Wu, H. and Jaenisch, R. (1995) Sensory but not motor neuron deficits in mice lacking NT4 and BDNF. *Nature*, 375: 238–240.

Manthorpe, M., Skaper, S.D., Williams, L.R. and Varon, S. (1986) Purification of adult rat sciatic nerve ciliary neuronotrophic factor. *Brain Res.*, 367: 282–286.

Masu, Y., Wolf, E., Holtmann, B., Sendtner, M., Brem, G. and Thoenen, H. (1993) Disruption of the CNTF gene results in motor neuron degeneration. *Nature*, 365: 27–32.

Mitsumoto, H., Ikeda, K., Klinkosz, B., Cedarbaum, J.M., Wong, V. and Lindsay, R.M. (1994) Arrest of motor neuron disease in *wobbler* mice cotreated with CNTF and BDNF. *Science*, 265: 1107–1110.

Oppenheim, R.W. (1991) Cell death during development of the nervous system. *Annu. Rev. Neurosci.*, 14: 453–501.

Oppenheim, R.W. and Nunez, R. (1982) Electrical stimulation of hindlimb increases neuronal death in chick embryo. *Nature (London)*, 295: 57–59.

Pennica, D., King, K.L., Shaw, K.J., Luis, E., Rullamas, J., Luoh, S.M., Darbonne, W.C., Knutzon, D.S., Yen, R., Chien, K.R. et al. (1995) Expression cloning of cardiotrophin 1, a cytokine that induces cardiac myocyte hypertrophy. *Proc. Natl. Acad. Sci. USA*, 92: 1142–1146.

Sendtner, M., Schmalbruch, H., Stöckli, K.A., Carroll, P., Kreutzberg, G.W. and Thoenen, H. (1992) Ciliary neurotrophic factor prevents degeneration of motor neurons in mouse mutant progressive motor neuronopathy. *Nature*, 358: 502–504.

Sendtner, M., Carroll, P., Holtmann, B., Hughes, R.A. and Thoenen, H. (1994) Ciliary neurotrophic factor. *J. Neurobiol.*, 25: 1436–1453.

Snider, W.D. (1994) Functions of the neurotrophins during nervous system development: what the knockouts are teaching us. *Cell*, 77: 627–638.

Stahl, N. and Yancopoulos, G.D. (1994) The tripartite CNTF receptor complex: activation and signaling involves components shared with other cytokines. *J. Neurobiol.*, 25: 1454–1466.

Takahashi, R., Yokoji, H., Misawa, H., Hayashi, M., Hu, J. and Deguchi, T. (1994) A null mutation in the human CNTF gene is not causally related to neurological diseases. *Nature Genet.*, 7: 79–84.

Yamamoto, H. and Gurney, M.E. (1990) Human platelets contain brain-derived neurotrophic factor. *J. Neurosci.*, 10: 3469–3478.

# Subject index
Page numbers in *italics* refer to figures

374